T0351079

# PRINCIPLES OF DATA ASSIMILATION

Data assimilation is theoretically founded on probability, statistics, control theory, information theory, linear algebra, and functional analysis. At the same time, data assimilation is a very practical subject, given its goal of estimating the posterior probability density function in realistic high-dimensional applications. This puts data assimilation at the intersection between the contrasting requirements of theory and practice. Based on over 20 years of teaching courses in data assimilation, *Principles of Data Assimilation* introduces a unique perspective that is firmly based on mathematical theories, but also acknowledges practical limitations of the theory. With the inclusion of numerous examples and practical case studies throughout, this new perspective will help students and researchers to competently interpret data assimilation results and to identify critical challenges of developing data assimilation algorithms. The benefit of information theory also introduces new pathways for further development, understanding, and improvement of data assimilation methods.

SEON KI PARK is Professor of Meteorology at Ewha Womans University, Seoul, Republic of Korea. His research focuses on storm-scale to mesoscale analysis, parameter estimation, and data assimilation to improve numerical weather and climate prediction. He coedited a series of four volumes titled *Data Assimilation for Atmospheric, Oceanic and Hydrologic Applications* (2009, 2013, 2017, 2022).

MILIJA ŽUPANSKI is Senior Research Scientist at Colorado State University, Fort Collins. He is a principal developer of two four-dimensional variational data assimilation systems and the maximum likelihood ensemble filter. His research focuses on data assimilation development and applications, including the atmosphere, land surface, aerosols, and combustion.

# PRINCIPLES OF DATA ASSIMILATION

SEON KI PARK
*Ewha Womans University*

MILIJA ŽUPANSKI
*Colorado State University*

Shaftesbury Road, Cambridge CB2 8EA, United Kingdom

One Liberty Plaza, 20th Floor, New York, NY 10006, USA

477 Williamstown Road, Port Melbourne, VIC 3207, Australia

314–321, 3rd Floor, Plot 3, Splendor Forum, Jasola District Centre,
New Delhi – 110025, India

103 Penang Road, #05–06/07, Visioncrest Commercial, Singapore 238467

Cambridge University Press is part of Cambridge University Press & Assessment,
a department of the University of Cambridge.

We share the University's mission to contribute to society through the pursuit of
education, learning and research at the highest international levels of excellence.

www.cambridge.org
Information on this title: www.cambridge.org/9781108831765

DOI: 10.1017/9781108924238



First published 2022

*A catalogue record for this publication is available from the British Library.*

*Library of Congress Cataloging-in-Publication data*
Names: Park, Seon Ki, 1961– author. | Zupanski, Milija, 1955– author.
Title: Principles of data assimilation / Seon Ki Park, Ewha Womans
University, Milija Zupanski, Colorado State University.
Description: New York : Cambridge University Press, 2022. |
Includes bibliographical references and index.
Identifiers: LCCN 2022009537 (print) | LCCN 2022009538 (ebook) |
ISBN 9781108831765 (hardback) | ISBN 9781108924238 (epub)
Subjects: LCSH: Meteorology–Data processing. | Meteorology–Observations.
Classification: LCC QC874.3 .P37 2022 (print) | LCC QC874.3 (ebook) |
DDC 551.630285/63–dc23/eng20220517
LC record available at https://lccn.loc.gov/2022009537
LC ebook record available at https://lccn.loc.gov/2022009538

ISBN 978-1-108-83176-5 Hardback

*To the memory of our mentor,*

Yoshi K. Sasaki,

and

*to our families,*

Hyunmi, Jun Bum, Shinwon and Kyung Keun;

Dusanka and Sandra

# Contents

# Preface

Data assimilation has become a major component of today's numerical weather prediction (NWP). It followed the development of numerical modeling in 1950s–1970s, mostly in terms of estimating the initial conditions. This is not surprising given that NWP models are a system of partial differential equations (PDEs). However, in later applications, in particular with climate models, data assimilation has included parameter estimation, either standalone or in addition to initial conditions. From the 1990s the role of data assimilation in NWP has been steadily increasing, closely related to the advancement of numerical modeling with ever increasing complexity. In addition to meteorology, data assimilation has also been getting more attention within other disciplines. One may follow the number of published books addressing data assimilation and notice an almost exponential increase starting with only a couple of books in the 1990s, to probably 10–15 books in the 2010s. This indirectly suggests that data assimilation has transformed from obscure to mainstream over the last 30 years.

The relevance of data assimilation to the big data paradigm also increases, since realistic data assimilation is a method of processing large data sets and modeling of physical phenomena requires powerful high-performance computing. The interest in estimating the uncertainty of these data is also common to big data science and data assimilation. Our approach in this book is to relate data assimilation to big data and high-dimensional problems. While there is a substantial and critical data assimilation development suitable for low-dimensional applications, we are mostly interested in practical data assimilation methods that have a proven utility and/or potential for realistic high-dimensional applications. While in some interpretations "practical" may imply favoring the solutions that work in practice but only loosely follow mathematical reasoning and proofs, we are trying to rather strictly follow the mathematical methods that can work in practice. Although this may appear as a play on words, we believe that closely following mathematical formalism is fundamental for further progress of data assimilation. Adopting such an approach does not necessarily imply that this is always possible, but that all attempts should be made to satisfy mathematical requirements and to never be satisfied with an ad hoc solution that may work in practice.

In addition to big data, one should not dismiss the relevance of artificial intelligence (AI) for data assimilation. This is quite an important new area of research and development that is also based on big data and high-performance computing, and is currently seen as a major new step in data assimilation development. However, we decided not to include the applications of AI in data assimilation because it is quite an intense area of development with fast-changing applications that may be practically impossible to present fairly without missing something important. Current AI applications are very broad and include the following: observation operators, derived observation products, data thinning, bias and model error corrections, as well as model dynamics. New areas may be added as we write this. In an eventual further edition of this book, we will likely consider adding a chapter on AI in data assimilation since by that time it will hopefully become clear what main areas of data assimilation can benefit from AI.

As will be seen in the book, data assimilation is multidisciplinary and it is closely related to applications that may be known under different names in different disciplines. For example, data assimilation can be interpreted as an inverse problem with applicability in optics, acoustics, communication and signal processing, medical imaging, geophysics, astronomy, and other disciplines.

This book originated from the tutorials, presentations, and exercises that the authors and their colleagues conducted during the five years of the *Ewha International School of Data Assimilation* (*EISDA 2012–2016*), which gathered students and young researchers mostly from Asia and Africa, as well as the graduate-level courses taught by the authors during last 20 years or so at Ewha Womans University and Colorado State University, respectively. Therefore, this book is also written for use by students and scientists interested in applying data assimilation in realistic high-dimensional problems. Some mathematical aspects may be too advanced for undergraduate students and may require some teacher assistance. Given the focus on methods for high-dimensional applications, this book can be of help to users and developers of data assimilation. Although meteorological applications dominate, we believe that the topics presented can also benefit researchers from various other disciplines.

We begin by presenting a general background of data assimilation with a brief historical overview and basic estimation problem, and by introducing common terminologies and notation in Chapter 1. Chapter 2 introduces the basic formalism of data assimilation through probability and the Bayesian paradigm. Fundamental data assimilation algorithms, filters, and smoothers, are presented in Chapter 3. Mathematical and practical details of tangent linear and adjoint models, which are fundamental for understanding variational and hybrid ensemble-variational data assimilation methods, are given in Chapter 4. Their practical implementation and development is closely related to automatic differentiation described in Chapter 5. Numerical minimization algorithms that are fundamental for applying data assimilation in realistic nonlinear problems are described in Chapter 6. In Chapter 7, variational methods, still the workhorse in NWP operational data assimilation, are introduced in various forms along with the optimal control theory. Recent methods of ensemble and hybrid data assimilation are presented in Chapter 8. Coupled data assimilation that is applied to coupled modeling systems is described in Chapter 9. The relation between dynamics and data assimilation, quite important but sometimes neglected given the dominant role of observations, is presented in Chapter 10. Some special topics, such as sensitivity analysis, and adaptive observations in Chapter 11 and satellite data assimilation in Chapter 12, are also discussed given their relevance in numerous applications. The Appendices are composed in order to introduce relevant tools and methods that are inherent in data assimilation, such as linear algebra and functional analysis in Appendix A, discretization of PDEs in Appendix B, and the lab practices aimed at understanding the intricate details of a practical data assimilation algorithm in Appendices C and D.

Our gratitude is extended to all our students and colleagues that contributed, directly or indirectly, in developing the material for this book. In particular, we would like to thank our colleagues, Liang Xu and Adrian Sandu, who delivered excellent lectures in EISDA. We also appreciate the Ewha graduate students who had attended EISDA and have been the main resources of our long-time and ongoing collaborative research: Sujeong Lim, Ebony Lee, Ji Won Yoon, Seung Yeon Lee, Yoonjin Lee, Sojung Park, Seungwon Chung, and Mariam Hussain.

# Part I

## General Background

# 1
# Data Assimilation: General Background

## 1.1 Introduction

Data assimilation includes two main components: simulation model and data. The simulation model is defined as a mathematical/numerical system that can simulate an event or a process. In most typical settings the simulation model is a prediction model based on partial differential equations (PDEs) that often includes empirical parameters. Data are generally associated with observations made by a measuring instrument, although data could also imply a product obtained by processing observations. Using an example from meteorology, data include observations such as atmospheric temperature and satellite radiances. The goal of data assimilation is to combine the information from a simulation model and data in order to improve the knowledge of the system, described by the simulation model. Apparently, the formulation of data assimilation will depend on interpretation of the *knowledge of the system*. Before we attempt to clarify a possible interpretation, it is useful to further understand the simulation model and data.

In agreement with common applications in geosciences and engineering, we narrow our discussion to a dynamic-stochastic PDE-based prediction model. Prediction models are developed with the general idea of improving the prediction of various phenomena of interest. From the theory of PDEs it is known that various parameters can impact the result of PDE integration, such as initial conditions (ICs), model errors (MEs), and empirical model parameters (EMPs). It is widely recognized that our knowledge of these parameters is never perfect, implying uncertainty of these parameters and uncertainty of the prediction calculated using such uncertain parameters.

Since the ultimate goal of using prediction models is to produce an improved prediction, it is natural to prefer a prediction that is in some way optimal. Such a prediction should be reliable, implying a desire to have a very small uncertainty associated with prediction. Then, the question is: How can the prediction be improved? First, it is anticipated that by improving the mentioned parameters (ICs, MEs, EMPs) and reducing their uncertainty would result in a desirable prediction. One could also try to improve model equations by including missing physical processes, coupling relevant components, and/or improving spatiotemporal resolution (if the prediction model is discretized). However, the only way to improve prediction is to introduce new information about the model parameters or model equations. The new information could come from another model with superior performance, but the most common source of new information about the real world comes from observations. An additional source of information could be introduced from past model performances if it is believed that the prediction model has some skill. If the prediction model has no skill, then observations are the only source of information, and one has to rely on using purely statistical methods. If the prediction model has some skill, however, then it is possible to combine the information from observations and from past model performances and then rely on using data assimilation.

Note that all sources of information, from observations and from prediction models, are uncertain. We already suggested that imperfect knowledge of model parameters (ICs, MEs, and/or EMPs), as well as model equations, implies an imperfect prediction. Information from observation is also not perfect. There are instrument errors, transmission errors, local errors, as well as the so-called representativeness errors. The instrument error is associated with every measuring instrument and can vary depending on the accuracy of the instrument. The errors created during a transmission from observation site to central location may not be detected in some instances and will contribute to observation error. Local errors refer to unforeseen errors of the local observation site, such as artificial heat sources and the impact of local vegetation. The representativeness error is the error caused by model prediction that is not representative of the actual observation. This can refer to inadequate model resolution, volume-averaged model variable versus point observation, etc. Therefore, observations also have errors, i.e., uncertainties.

Given that the two main components of data assimilation, prediction model and data, are inherently uncertain, then the output of data assimilation, the knowledge of the system, is expected to be uncertain as well. Uncertainty can be measured in many different ways. One can think of uncertainty as a measure of the difference between an estimate and the truth, if the truth is known. Unfortunately, the true value of the field is rarely known, except in a controlled experiments such as an observation system simulation experiment (OSSE). The theory of probability offers a mathematically consistent, formal way of dealing with uncertainties, and is used in our approach to data assimilation. A comprehensive object that describes the probabilistic system is the probability density function (PDF). Therefore, one can think of the PDF as the actual knowledge of the system, implying that the ultimate goal of data assimilation is to estimate the PDF. As will be shown in Chapters 3, 7, 8, 9, and 12, estimating the PDF is quite a challenging problem in realistic high-dimensional applications of data assimilation, mostly limiting practical data assimilation to estimating the first PDF moment (e.g., mean) and eventually the second PDF moment (e.g., covariance), with only an occasional capability of estimating the higher-order PDF moments.

Another critical aspect of data assimilation is the processing of information. Both prior model realizations and data contain information that can potentially contribute to improving the state of knowledge. Shannon's information theory (Shannon and Weaver, 1949), also based on using the probabilistic approach, offers the mathematical formalism for quantifying and processing information. Although still not used to its maximum, this information theory is a very handy tool for data assimilation. Implied from the above discussion of the impact of model parameters, such as ICs, MEs, and EMPs, on the prediction made by the model, and the aspiration of data assimilation to improve prediction by modifying model parameters ICs, MEs, and/or EMPs, the control theory is also an important tool of data assimilation. The implied dynamic-stochastic characterization of a prediction model also implies the important role of statistics and possibly chaotic nonlinear dynamics in data assimilation. Given that data assimilation is typically multivariate and applied to vectors and matrices, it relies heavily on using linear algebra and functional analysis.

There are several other considerations that are important for data assimilation. Realistic physical phenomena and processes, and their relation to observed variables, are all inherently nonlinear. As such, the treatment of nonlinearity in data assimilation plays an important role in choosing the adequate control theory methods and limiting the utility of linear algebra. The dynamical aspect of prediction models, generally characterized by time-dependent phenomena, implies that prediction uncertainties have to be dynamical and time-dependent as well. Given the sensitivity of PDEs to the initial (and boundary) conditions, data assimilation has to provide dynamically balanced ICs that would not cause spurious perturbations in prediction. In the case of chaotic nonlinear dynamics, as most realistic dynamical systems are, data assimilation needs to capture and eventually remove the errors of growing and neutral modes from the ICs.

With all these components, probability theory, statistics, information theory, control theory, linear algebra, and functional analysis, make data assimilation very complex and challenging.

## 1.2 Historical Background

First attempts to address what we now call data assimilation could be traced to data fitting and regression analysis applied in astronomy, most notably by Legendre (1805) and Gauss (1809). In solving the problem Gauss assumed normally distributed errors and introduced the normal probability distribution. Around that time Laplace (1814) introduced the Bayesian approach by developing a mathematical system on inductive reasoning based on probability. Starting with these discoveries, and after a considerable development of mathematical tools and theories, the modern-age data assimilation was made possible.

Early methods for data assimilation were deterministic and essentially represented a function fitting to measurements. This included the interpolation methods with distance-based interpolation weights in order to determine the relative importance of observations, such as the objective analysis schemes of Bergthórsson and Döös (1955), Gilchrist and Cressman (1954), Cressman (1959), and Barnes (1964). While useful for operational numerical weather prediction (NWP) of that time, these methods did not explicitly include probabilistic considerations. Other deterministic methods include nudging data assimilation (Hoke and Anthes, 1976; Davies and Turner, 1977), sometimes also referred to as four-dimensional data assimilation (4DDA) or a dynamic relaxation method. Later developments of the method include a generalization to accept uncertainties (e.g., Zou et al., 1992). Nudging implies a change of the original dynamical equations of a prediction model to include a forcing term. The coefficients associated with the forcing are generally determined by fitting the model state closer to the observations. Although nudging has been improved to implicitly accept uncertainties, it does not rely on using the Bayesian approach and does not attempt to estimate PDF moments as probabilistic data assimilation does.

Probably the first data assimilation method that is critically relevant for understanding modern-age data assimilation is the Kalman filter (KF) (Kalman and Bucy, 1961), initially developed for signal processing. It provides a mathematically consistent methodology based on probability and Bayesian principles that produces a minimum variance solution. The KF is also helpful in describing the role of dynamics in forecast error covariance, as well in model error covariance. Since the KF is defined for linear systems, it fully resolves the Gaussian PDF and in that sense represents a satisfactory solution to general probabilistic data assimilation problems. There are, however, major obstacles in making the KF a practical data assimilation method. For one, it is a linear filtering method and as such it cannot satisfactorily address nonlinearities in the prediction model and observations. Another major obstacle is the required matrix inversion, which becomes practically impossible to calculate in realistic high-dimensional applications. Strictly relying on the Gaussian PDF assumption is also a disadvantage of the KF, given that prediction model variables and observations could have non-Gaussian errors.

The first practical method that incorporates the basic data assimilation setup with Bayesian and probabilistic assumptions is the optimal interpolation (OI) method of Gandin (1963), sometimes referred to as statistical interpolation. This is a minimum variance estimator and as such it can be related to the KF and other probabilistic data assimilation methods. A more detailed overview of OI can be found in Daley (1991) (see chapters 3, 4, and 5 therein). The OI method is very much a simplified version of the KF. The OI employs a linear observation operator, in early versions only the identity matrix. For nonlinear observations, such as satellite radiances, an inversion algorithm (i.e., retrieval) that produces a model variable from observations is required. The forecast error covariance is modeled and includes separate vertical and horizontal correlations. By construction the forecast error, covariance is homogeneous (i.e., all grid points are treated equally) and isotropic (all directions are treated equally). In addition, the covariance is stationary, being approximated by a correlation function with statistically estimated correlation parameters. Since it is related to the KF, OI can also produce an estimate of the posterior error covariance. However, such an estimate is not reliable since the input covariances and parameters are not accurate. The OI is also local, in the sense that only observations within a certain distance from the model point impact the analysis at that point. Although theoretically and practically an important step in probabilistic data assimilation development, when measured against our motivation to produce a reliable estimate of PDFs, OI leaves much to be desired. At best it can produce a meaningful estimate of the first PDF moment only, however with serious limitations related to preferred capabilities such as the nonlinearity of observation operators and dynamical structure of forecast error covariance.

Another fundamental development that led to current variational data assimilation (VAR) methods was the introduction of variational principles in data assimilation by Sasaki (1958). While at the time it was understood as a method for objective analysis based on least squares, the new method for the first time introduced variational formalism and minimization under the geostrophic constraint and also under the more general balance constraint between winds and geopotential. Then, in a trilogy of papers (Sasaki, 1970a, 1970b, 1970c) expanded the previous approach to include the time dependency of observations and established a basis for future development of four-dimensional variational data assimilation (4DVAR) methodology.

While the use of variational principles in data assimilation have been known since the early work of Sasaki (1958), it took almost 25 years before variational methodology had another push into the field of data assimilation, mostly because of the advancements of computers in NWP. Addressing the deficiencies of OI, most importantly the local character of the analysis, nonlinearity of observations, and to some extent the specification of forecast error covariance, variational methods for data assimilation were revived in the mid 1980s (e.g., Lewis and Derber, 1985; Le Dimet and Talagrand, 1986). The subtypes of variational methods include three-dimensional variational (3DVAR) (e.g., Parrish and Derber, 1992) and 4DVAR data assimilation (e.g., Navon et al., 1992; Županski, 1993; Courtier et al., 1994). They include a global minimization (i.e., over all model points) of the cost function that can incorporate nonlinear observations and solves the inversion problem using adjoint equations. The forecast error covariance is improved over OI as it includes complex cross-correlations with additional dynamical balance constraints, but the correlations are still modeled. On the positive side, the modeling of error covariance allows the covariance to be of full rank, meaning that all degrees of freedom (DOF) required for solving the analysis problem are included. The covariance is stationary, although in 4DVAR there is limited capability to introduce time dependence during the assimilation window. Also, variational methods primarily estimate the first moment of PDFs. Although it is possible to estimate the second PDF moment, especially in 4DVAR, there is no feedback of uncertainties from one data assimilation cycle to the next implying a limited use of Bayesian inference. The main advantage of variational methods is their capability to assimilate nonlinear observations, in particular the satellite radiances that now represent the major source of information in meteorology (e.g., Derber and Wu, 1998). By introducing 4DVAR the prediction model itself could be used as a constraint in optimization. The cost of applying VAR has increased compared to previous methods, but it can still be considered efficient since potentially costly matrix inversions are avoided. The variational methods are still used in practice.

Immediately following this development of variational methods, ensemble Kalman filtering (EnKF) methods have been introduced to data assimilation (Evensen, 1994; Houtekamer and Mitchell, 1998). The EnKF successfully addressed the problem of the nonlinear prediction model in the KF by introducing the Monte Carlo approach to the KF forecast step. At the same time the forecast error covariance is dynamic, and is therefore an improvement on the stationary and modeled error covariance used in variational methods. The most important impact of the EnKF was that a realistic data assimilation could be used to produce the first two moments of the PDF, the mean and the covariance, although still under a Gaussian PDF assumption. One of the issues of the EnKF is not being able to account for nonlinearity of observations, since the same linear KF analysis equation is used. More recently (e.g., Sakov et al., 2012), an iterative EnKF was introduced in a manner similar to the iterated KF to address the nonlinearity of observation operators. Implementing the EnKF requires the assimilation of perturbed observations, which results in the calculation of numerous analyses for each ensemble member, and therefore an increase in the cost. Square-root

EnKF methods were introduced to reduce the computational cost, by directly calculating the mean of the analysis. Including a large number of ensembles required to resolve a realistic data assimilation problem proved to be practically impossible due to the computational cost and storage requirements. This prompted a need for covariance localization to increase the number of DOF of the low-rank ensemble covariance that could be feasible. This localization greatly helped the EnKF and related ensemble methods to remain of practical significance, although a modification of the dynamically based ensemble forecast error covariance via convolution with a prespecified localizing covariance matrix is required. Covariance localization implies that practical EnKF methods can be interpreted, in terms of forecast error covariance, as an intermediate approach between the full-rank EnKF (with all DOF) and the local OI method. The analysis solution in the EnKF with localization is essentially local since only observations within a certain distance can impact the analysis point.

Both EnKF and variational methods have practical and theoretical limitations. Variational methods have the capability of addressing nonlinearity through applying global numerical optimization. The EnKF is inherently designed to use the linear analysis equation of the KF. An alternative way of bridging this issue was introduced by the maximum likelihood ensemble filter (MLEF) (Županski, 2005), in which it was shown how the calculation of adjoint operators could be avoided by using nonlinear ensemble perturbations and applied in variational-like minimization of the cost function. As with other ensemble methods, MLEF includes the flow-dependent ensemble covariance and estimates the posterior uncertainty.

The implied limitation of error covariance representation in the practical EnKF due to an insufficient number of DOF and to some extent the nondynamical impact of covariance localization, even though the covariance is flow dependent, can result in an analysis that is not of the desired quality. The same could be said for variational methods, where the use of stationary and modeled error covariance is not sufficiently realistic and can produce unsatisfactory analysis. As a result, hybrid ensemble-variational methods that allow a combination of the flow-dependent ensemble, but low-rank covariance, and the stationary variational, but full-rank, error covariance were introduced (Lorenc, 2003; Buehner, 2005; Wang et al., 2007; Bonavita et al., 2012; Clayton et al., 2013).

Data assimilation can also be viewed as an application of Pontryagin's minimum principle (PMP) (e.g., Pontryagin et al., 1961; Lakshmivarahan et al., 2013) where a least squares fit of an idealized path to dynamics law follows from Hamiltonian mechanics. In this application of optimal control theory, the problem is posed as finding the best possible forcing for taking a dynamical system from one state to another, in the presence of dynamical constraints. This forcing is also related to accounting for ME in data assimilation. While the use of forcing reminds us of nudging, the PMP method is more general since it includes an optimization subject to dynamical constraints as well as uncertainties (Lakshmivarahan and Lewis, 2013). Similar to previous methods, it searches for optimal analysis that could be interpreted as the first PDF moment, but estimation of the posterior uncertainties is not an essential part of the method. It is possible to view 4DVAR as a special case of PMP.

The above historical overview also indicates the current status of practical data assimilation development. Other methods with stronger theoretical foundations have been introduced to data assimilation, such as particle filters (PFs) (e.g., van Leeuwen, 2009; Chorin et al., 2010), but they still have limitations for realistic high-dimensional applications.

However, by directly calculating arbitrary PDFs through the Bayesian framework they have a theoretical advantage in accounting for nonlinearity and non-Gaussianity and therefore offer numerous possibilities for the future development of data assimilation.

## 1.3 Terminologies and Notation

Data assimilation consists of two major elements – a model of the dynamical system and a set of data (i.e., observations), and aims to procure optimal estimates of model states by combining model forecasts and observations. We represent the model states and the observations in terms of vectors, $\mathbf{x}$ and $\mathbf{y}^o$, respectively. The *true* state, $\mathbf{x}^t$, can never be obtained but can be estimated through an adequate estimation procedure. Such an estimate, made at a given time, is called the *analysis*, $\mathbf{x}^a$. The *estimate* is also denoted by $\hat{\mathbf{x}}$ and is interchangeably used with $\mathbf{x}^a$. The *background*, $\mathbf{x}^b$, is an a priori estimate of $\mathbf{x}^t$ before the analysis is conducted. For the notations in data assimilation, we generally follow Ide et al. (1997).

Data assimilation represents a process to obtain $\mathbf{x}^a$, as close to $\mathbf{x}^t$ as possible, by correcting $\mathbf{x}^b$ using a correction, $\Delta\mathbf{x}$. Mathematically, it is formulated, in its simplest form, as:

$$\mathbf{x}^a = \mathbf{x}^b + \Delta\mathbf{x}. \tag{1.1}$$

Note that $\Delta\mathbf{x}$ is a function of both $\mathbf{y}^o$ and $\mathbf{x}^b$, and it is called an *analysis increment.*

### *1.3.1 Observation Equation*

A variety of observations, assembled in $\mathbf{y}^o$, are used for data assimilation (see Figure 1.1). As observations are much fewer than model states and are irregularly distributed, direct comparison between observations and model states is unfeasible. Thus, we define a nonlinear function, $H$, called an *observation operator*, that transforms the state vector from the state space, $\mathcal{R}^m$, to the observation vector in the observation space, $\mathcal{R}^n$. The observation is described in terms of the true state as:

$$\mathbf{y}^o = H\mathbf{x}^t + \boldsymbol{\varepsilon}^o, \tag{1.2}$$

where $\boldsymbol{\varepsilon}^o$ is the observation (measurement) error. Equation (1.2) is called the *observation equation* or the *observation model.*

### *1.3.2 Observation Error Statistics*

We assume that the measurement error $\boldsymbol{\varepsilon}^o$ in (1.2) is random and independent, and hence have zero mean, i.e.,

$$mean(\varepsilon_i^o) = E(\varepsilon_i^o) = 0 \text{ for } i = 1, \dots, n. \tag{1.3}$$

This implies that $\mathbf{y}^o$ in (1.2) depends only on $\mathbf{x}^t$ and all other variation in $\mathbf{y}^o$ is random. For the random errors, the variance and the covariance of the errors are

$$var(\varepsilon_i^o) = E(\varepsilon_i^o \varepsilon_i^o) = \sigma_i^2 \text{ for } i = 1, \dots, n \tag{1.4}$$

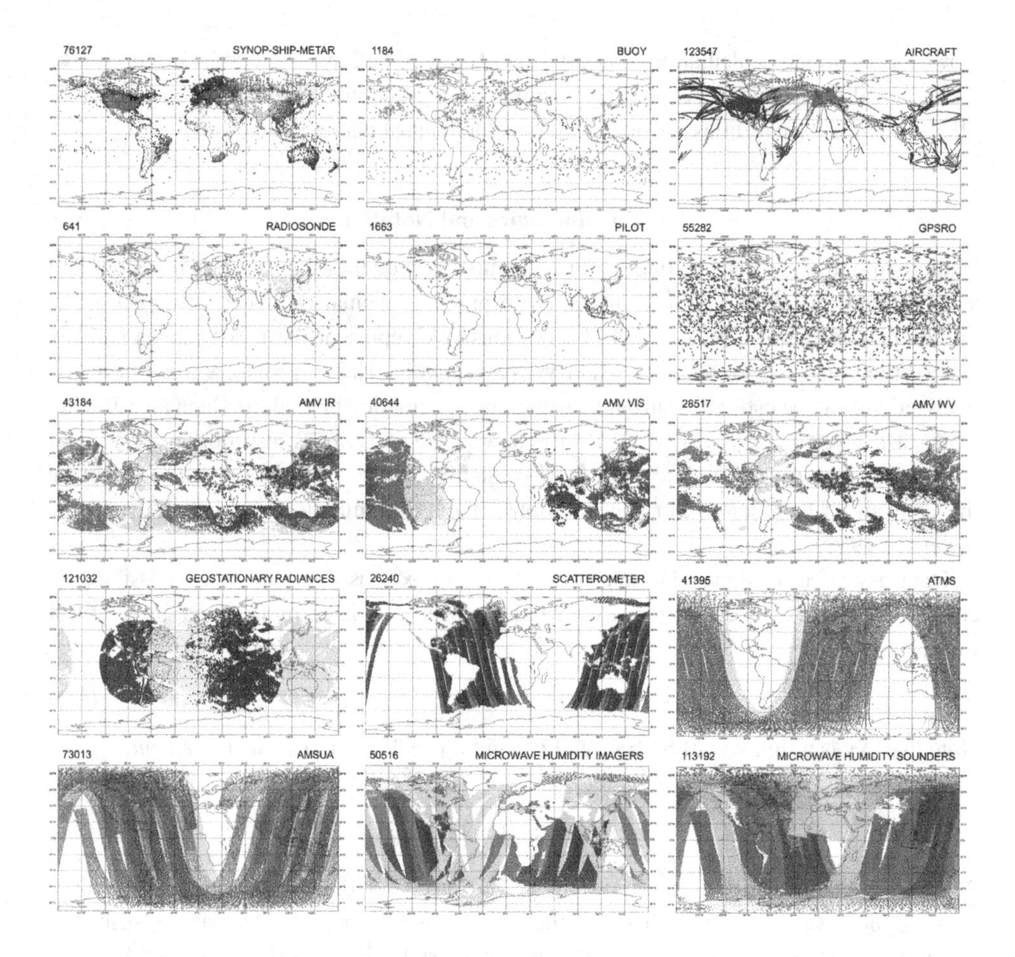


Figure 1.1 Various observation data at the global scale, available on 0000 UTC 18 August 2020, with the observation platforms (top-right corners) and the number of data used for data assimilation (top-left corners). The details of legends in the subfigures refer to the data coverage from the European Centre for Medium-Range Weather Forecasts (ECMWF, 2020). CC BY-NC-ND 4.0 License.

and

$$cov(\varepsilon_i^o, \varepsilon_j^o) = E(\varepsilon_i^o \varepsilon_j^o) = 0 \text{ for } i, j = 1, \ldots, n \text{ and } i \neq j, \tag{1.5}$$

respectively. Here, $\sigma_i^2$ is the squared standard deviation of $\varepsilon_i^o$ and (1.4) assumes that the variance of $\boldsymbol{\varepsilon}^o$ is constant; thus, not dependent on $\mathbf{x}^t$. With zero covariances in (1.5), the variables in $\boldsymbol{\varepsilon}^o$ are uncorrelated with each other. By combining the three assumptions in (1.3)–(1.5), we have

$$\begin{aligned} mean(\boldsymbol{\varepsilon}^o) &= E(\boldsymbol{\varepsilon}^o) = \mathbf{0}, \\ cov(\boldsymbol{\varepsilon}^o) &= E(\boldsymbol{\varepsilon}^o (\boldsymbol{\varepsilon}^o)^T) = \sigma^2 \mathbf{I} = \mathbf{R}. \end{aligned} \tag{1.6}$$

That is, the observation error covariance **R** is a diagonal matrix composed of $\sigma_i^2$ though **R** is in general a nondiagonal matrix.

### *1.3.3 Observation Operator*

Note that it is not only that the observation sites are not usually located at the grid points where model states are calculated, but also that the observation quantities often do not match the model variables. For instance, some remotely sensed observations do not have corresponding model states; thus, it is essential to convert the model state variables into observation variables. In practice, the operator $H$ is performed in two steps:

1. *Interpolation*, say $H^I$, from the model grid points to the observation sites where the conversion will be performed for indirect observations or directly when the state variables are the same as the observation quantities.
2. *Conversion*, say $H^C$, of the model variables to the observables when the measurements are indirect, e.g., radiances measured by sensors onboard satellites. A radiative transfer model can serve as an $H$ to calculate the radiance, using the whole state vector components, at a specific waveband (see, e.g., Figure 1.2).

The operation by $H: \mathcal{R}^m \rightarrow \mathcal{R}^n$ is composed of two mappings, that is, $H^I: \mathcal{R}^m \rightarrow \mathcal{R}^n$ and $H^C: \mathcal{R}^n \rightarrow \mathcal{R}^n$, giving

$$\hat{\mathbf{y}} = H\left(\hat{\mathbf{x}}\right) = H^C\left(\mathbf{x}^o\right) = H^C\left(H^I\left(\hat{\mathbf{x}}\right)\right) = H^C H^I\left(\hat{\mathbf{x}}\right). \tag{1.7}$$

This decomposition is illustrated in Figure 1.3. Here, $H^I(\hat{\mathbf{x}})$ interpolates an $m \times 1$ state estimate vector $\hat{\mathbf{x}}$ from the model space (i.e., grid points) to an $n \times 1$ vector $\mathbf{x}^o$ in the observation space. $H^C(\mathbf{x}^o)$ converts $n$ state variables $\mathbf{x}^o$ into a set of $n$ observations ($\hat{\mathbf{y}}$) – that is, observation estimate (or model equivalents of observations) – the radiance computed

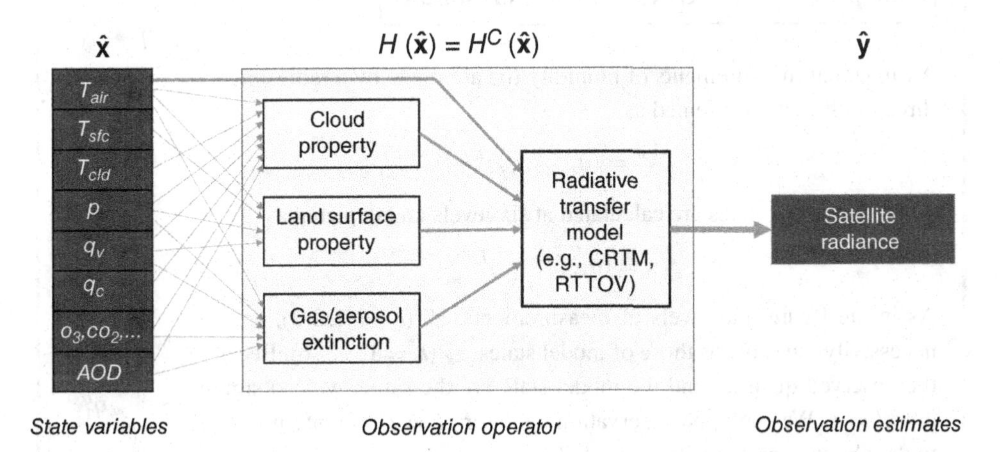


Figure 1.2 Composition of observation operator, $H$.

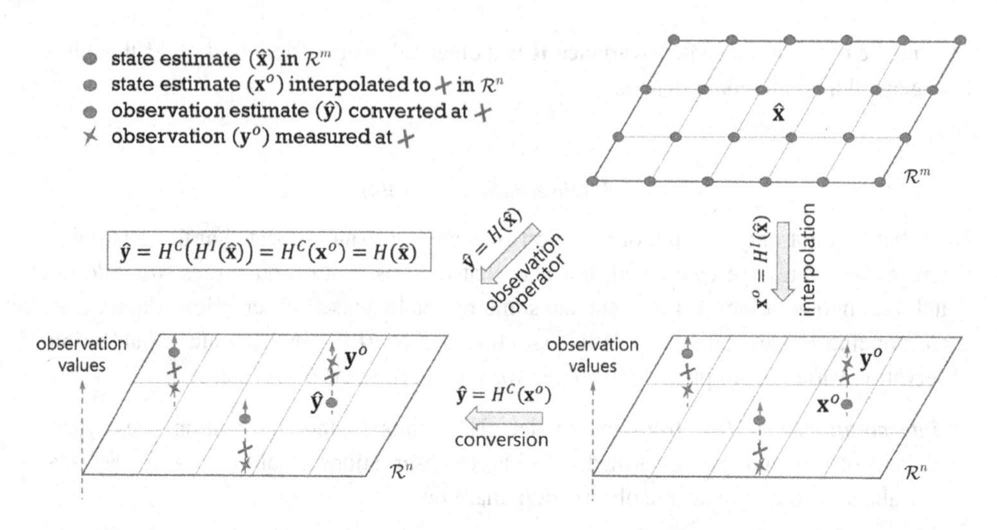


Figure 1.3 Composition of observation operator, $H$.

through a radiance transfer model using temperature, water vapor mixing ratio, cloud mixing ratio, etc. (see Figure 1.2). If the model states and the observation quantities were the same, $H(\hat{\mathbf{x}}) = H^I(\hat{\mathbf{x}})$: If the observation sites were exactly located at the grid points and the model states and observation quantities were different, $H(\hat{\mathbf{x}}) = H^C(\hat{\mathbf{x}})$.

The observation operator $H$ is generally nonlinear; however, it is more convenient to use its linearized version, denoted by $\mathbf{H}$, in explaining the concept of data assimilation. For practical applications, in (1.7), we adopt linear operators, i.e., $H = \mathbf{H}$, for direct observations that match the model variables (wind, temperature, humidity, etc.): We employ the nonlinear operator $H$ for indirect observations (satellite radiance, radar reflectivity, etc.).

**Example 1.1** ***H* for vertical sounding of moisture**

Assume that measurements of humidity ($q$) are made by a radiosonde at three levels and represented as

$$\mathbf{y}^o = (q_1^o, q_2^o, q_3^o)^T$$

while the model states are calculated at six levels and depicted as

$$\mathbf{x} = (q_1, \ldots, q_6)^T.$$

As in the figure, the levels of measurements, $z_l^o$ ($l = 1,2,3$), are not necessarily the same as those of model states, $z_k$ ($k = 1, \ldots, 6$). Because the observed quantity and the model state are the same, we can simply put $\mathbf{x}^o = \mathbf{x}$. We apply the observation operator $H = H^I$ to interpolate $\mathbf{x}^o$ to the observation space (i.e., levels $l$):

$q_6$ $q_3^o$ $q_5$ $q_4$ $q_2^o$ $q_3$ $q_2$ $q_1^o$ $q_1$

$$H\left(\mathbf{x}^o\right) = \begin{pmatrix} q_1 + \frac{q_2-q_1}{z_2-z_1}\left(z_1^o - z_1\right) \\ q_3 + \frac{q_4-q_3}{z_4-z_3}\left(z_2^o - z_3\right) \\ q_5 + \frac{q_6-q_5}{z_6-z_5}\left(z_3^o - z_5\right) \end{pmatrix} = \begin{pmatrix} \beta_1 q_1 + \beta_2 q_2 \\ \beta_3 q_3 + \beta_4 q_4 \\ \beta_5 q_5 + \beta_6 q_6 \end{pmatrix} = \hat{\mathbf{y}}.$$

This is a set of linear equations; thus, we can define a linear observation operator $\mathbf{H} \equiv \frac{\partial H}{\partial \mathbf{x}}$, given by

$$\mathbf{H} = \begin{pmatrix} \beta_1 & \beta_2 & 0 & 0 & 0 & 0 \\ 0 & 0 & \beta_3 & \beta_4 & 0 & 0 \\ 0 & 0 & 0 & 0 & \beta_5 & \beta_6 \end{pmatrix}.$$

**Example 1.2 *H* for the radar reflectivity factor**

The following relation, derived from the Marshall–Palmer distribution of raindrop size without considering ice phases, was used as an observation operator for the radar reflectivity factor (e.g., Sun and Crook, 1997; Xiao et al., 2007; Sugimoto et al., 2009):

$$Z = 43.1 + 17.5 \log\left(\rho q_r\right), \tag{1.8}$$

where $Z$ is the reflectivity factor (in dBZ), $\rho$ is the air density (in kg m$^{-3}$), and $q_r$ is the rainwater mixing ratio (in g kg$^{-1}$).

By considering two ice phases – the snow and hail mixing ratios ($q_s$ and $q_h$, respectively) – Gao (2017) devised the reflectivity observation operator as:

$$Z = 10 \log Z_e. \tag{1.9}$$

Here, the equivalent radar reflectivity factor $Z_e$ is given by

$$Z_e = \begin{cases} Z(q_r) & \text{for} \quad T_b \geq 5°\text{C} \\ \alpha Z(q_r) + (1-\alpha)\left[Z(q_s) + Z(q_h)\right] & \text{for} \quad -5°\text{C} < T_b < 5°\text{C} \\ Z(q_s) + Z(q_h) & \text{for} \quad T_b \leq -5°\text{C}, \end{cases} \tag{1.10}$$

where $T_b$ is the background temperature, and $\alpha$ varies linearly between 0 at $T_b = -5°$C and 1 at $T_b = 5°$C, for different components of the reflectivity factors as the following:

| Phase | Reflectivity factor | References/Conditions |
|---|---|---|
| Rain | $Z(q_r) = 3.63 \times 10^9 (\rho q_r)^{1.75}$ | Smith Jr. et al. (1975) |
| Snow (dry) | $Z(q_s) = 9.80 \times 10^8 (\rho q_s)^{1.75}$ | $T_b < 0°$C |
| Snow (wet) | $Z(q_s) = 4.26 \times 10^{11} (\rho q_s)^{1.75}$ | $T_b > 0°$C |
| Hail | $Z(q_h) = 4.33 \times 10^{10} (\rho q_h)^{1.75}$ | Lin et al. (1983); Gilmore et al. (2004) |

**Practice 1.1 $H$ for irradiance**

Assume that a satellite measures irradiance, $E$, emitted from an atmospheric layer. In a numerical model, $E_i$, from the $i$th grid box, can be simply calculated by the Stefan–Boltzmann law as:

$$E_i = \sigma T_i^4,$$

where $\sigma$ is the Stefan–Boltzmann constant and $T_i$ is the layer temperature in that grid box. Develop the observation operator $H$ and its linearized operator $\mathbf{H}$ $(\equiv \partial H/\partial \mathbf{x})$, for the following figure, which has measurements of $E$ from grid boxes 1 and 3 only.

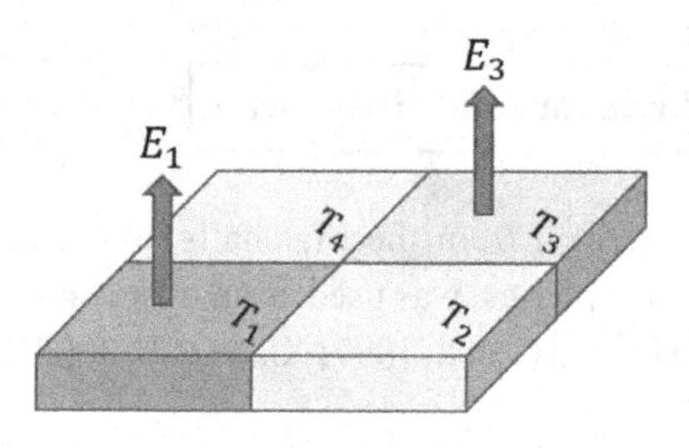


### 1.3.4 Background Field

Observations are generally much fewer than model states that are assigned to 3D grid points per time step: the total number of conventional observations is of the order of $10^4$ while that of grid-point variables to be calculated in an NWP model is of the order of $10^7$ (Kalnay, 2003). This makes data assimilation an *underdetermined* problem. Furthermore, observations are distributed irregularly over the globe, e.g., there exists much poorer data over the southern hemisphere than the northern hemisphere.

Therefore, additional information on each grid point – called a *background* or *first guess* and denoted by $\mathbf{x}^b$ – is necessary to create ICs for numerical prediction. In these days, operational centers generate $\mathbf{x}^b$ using a short-range forecast (e.g., a 6-h forecast for a global model) out of an analysis cycle (e.g., 6-h cycle in the global data assimilation system), as shown in Kalnay (2003).

### 1.3.5 Analysis Equation

The goal of data assimilation is to find the analysis ($\mathbf{x}^a$) by correcting the background ($\mathbf{x}^b$) using the observation ($\mathbf{y}^o$), which is formulated as

$$\mathbf{x}^a = \mathbf{x}^b + \mathbf{W}\left[\mathbf{y}^o - H\left(\mathbf{x}^b\right)\right]. \tag{1.11}$$

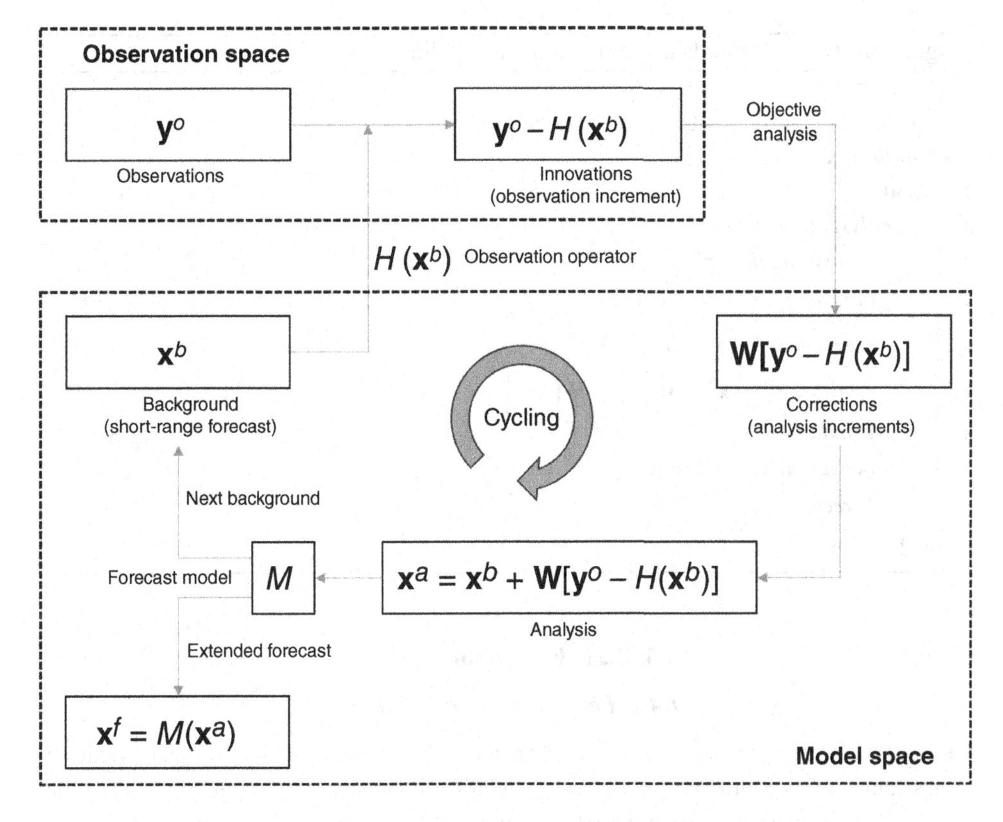


Figure 1.4 Basic concept of data assimilation.

Here, $\mathbf{y}^o - H\left(\mathbf{x}^b\right)$ is called the *observation increment* or *innovation*, representing the difference between the observations and the model states. The term $\mathbf{W}\left[\mathbf{y}^o - H\left(\mathbf{x}^b\right)\right]$ is called the *analysis increment*, where $\mathbf{W}$ is a weight – or *gain* – represented by the error characteristics (e.g., the background and observation error covariances).

Figure 1.4 describes the basic concept of data assimilation in the framework of NWP. Note that we are dealing with two separate spaces – the observation space and the model space – which are linked to each other via the observation operator $H$. The model prediction starts from $\mathbf{x}^b$, obtained from a short-range forecast (e.g., 6-h forecast in a global prediction system and 1-h forecast in a regional prediction system); when observations are available, $\mathbf{x}^b$ is transformed to the observation space through $H\left(\mathbf{x}^b\right)$ to calculate an innovation $\mathbf{y}^o - H\left(\mathbf{x}^b\right)$. Using the innovation, a correction term $\mathbf{W}\left[\mathbf{y}^o - H\left(\mathbf{x}^b\right)\right]$ is obtained in the model space through the objective analysis and added to $\mathbf{x}^b$ to get $\mathbf{x}^a$; then, a forward propagator $M$ (i.e., a forecast model), operating on $\mathbf{x}^a$, produces an extended forecast and a new background for the next cycle of data assimilation. An algorithmic view of data assimilation and numerical prediction is depicted in Algorithm 1.1.

**Algorithm 1.1** General data assimilation with cycling

/* index $k$ denotes cycle number */

/* $\mathcal{R}^m$ denotes model space; $\mathcal{R}^n$, observation space */

1 ***Initiation***: $\mathbf{x}^b$ ! Provide a background $\mathbf{x}^b$ at $k = 1$

2 **repeat** ! Loop for cycle $k = 1$ to $kmax$

3 ***Analysis*** at cycle $k$ ! Procedure to obtain analysis $\mathbf{x}^a$

4 ***Transformation***: $\mathbf{x}^b \longrightarrow H(\mathbf{x}^b)$ ! Operate $H: \mathcal{R}^m \rightarrow \mathcal{R}^n$

5 ***Innovation***: $\mathbf{y}^o - H\left(\mathbf{x}^b\right)$ ! Calculate innovation at $\mathcal{R}^n$

6 ***Correction***: $\mathbf{W}\left[\mathbf{y}^o - H\left(\mathbf{x}^b\right)\right]$ ! Calculate correction at $\mathcal{R}^m$

7 ***Analysis***: $\mathbf{x}^a = \mathbf{x}^b - \mathbf{W}\left[\mathbf{y}^o - H\left(\mathbf{x}^b\right)\right]$ ! Obtain analysis at $\mathcal{R}^m$

8 ***Forecast*** at cycle $k$ ! New background and forecast

9 ***Background***: $\mathbf{x}^b = M(\mathbf{x}^a)\big|_{t_b}$ ! Short-range forecast (e.g., $t_b = 6$ h)

10 ***Forecast***: $\mathbf{x}^f = M(\mathbf{x}^a)\big|_{t_f}$ ! Extended forecast (e.g., $t_f = 48$ h)

11 **until** *end of cycling*

## 1.4 Basic Estimation Problem

### *1.4.1 Least Squares Estimation*

The least-squares approach was invented in the 1800s independently by Carl Friedrich Gauss and Adrien-Marie Legendre for calculating planetary motion. It constitutes the foundation of modern data assimilation (Sorenson, 1970; Kalnay, 2003; Lewis et al., 2006) through the core concept of estimating the unknown parameters by minimizing the squared differences between the model and the data.

From the observation model (1.2), we can express $\mathbf{y}^o$ in terms of the state estimate $\hat{\mathbf{x}}$, rather than the true state $\mathbf{x}^t$, and a linear observation operator $\mathbf{H}$ as

$$\mathbf{y}^o = \mathbf{H}\hat{\mathbf{x}} + \boldsymbol{\varepsilon}^r = \hat{\mathbf{y}} + \boldsymbol{\varepsilon}^r, \tag{1.12}$$

where $\boldsymbol{\varepsilon}^r$ is called the residual error – the difference between the *true* measurement $\mathbf{y}^o$ and the *estimated* measurement $\hat{\mathbf{y}} = \mathbf{H}\hat{\mathbf{x}}$. Equation (1.12) is also called the linear *regression* model (see Colloquy 1.1).

In the least-squares estimation, we seek to find a specific $\hat{\mathbf{x}}$ that minimizes a functional $J = J(\hat{\mathbf{x}})$, defined as the sum of squared residual errors:

$$J = (\boldsymbol{\varepsilon}^r)^T \boldsymbol{\varepsilon}^r = \|\boldsymbol{\varepsilon}^r\|_2^2 = \left(\mathbf{y}^o - \mathbf{H}\hat{\mathbf{x}}\right)^T \left(\mathbf{y}^o - \mathbf{H}\hat{\mathbf{x}}\right), \tag{1.13}$$

where $\|\boldsymbol{\varepsilon}^r\|_2 = \left((\varepsilon_1^r)^2 + \cdots + (\varepsilon_n^r)^2\right)^{1/2}$ represents the Euclidean or $L^2$ norm. Minimization of the quadratic function $J$ should satisfy the following requirements, for the *gradient* $\nabla_{\hat{\mathbf{x}}} J$ and the *Hessian* $\nabla_{\hat{\mathbf{x}}}^2 J$:

$$\nabla_{\hat{\mathbf{x}}} J = \frac{\partial J}{\partial \hat{\mathbf{x}}} = -2\mathbf{H}^T \mathbf{y}^o + 2\left(\mathbf{H}^T\mathbf{H}\right)\hat{\mathbf{x}} = 0 \tag{1.14}$$

and

$$\nabla_{\hat{\mathbf{x}}}^2 J = \frac{\partial^2 J}{\partial \hat{\mathbf{x}} \partial \hat{\mathbf{x}}^T} = 2\left(\mathbf{H}^T \mathbf{H}\right) \text{ is positive definite.} \tag{1.15}$$

Through the necessary condition (1.14), we can calculate the minimizer $\hat{\mathbf{x}}$ – the optimal estimate that minimizes $J$ – as the solution of the *normal* equation

$$\left(\mathbf{H}^T \mathbf{H}\right) \hat{\mathbf{x}} = \mathbf{H}^T \mathbf{y}^o. \tag{1.16}$$

When $\mathbf{H}^T\mathbf{H}$ is square ($m \times m$) and *nonsingluar* – i.e., having its inverse, nonzero determinant, and $\mathbf{H}$ with full rank of $m$ – we get the solution of (1.16) as

$$\hat{\mathbf{x}} = \left(\mathbf{H}^T \mathbf{H}\right)^{-1} \mathbf{H}^T \mathbf{y}^o. \tag{1.17}$$

Here, $\left(\mathbf{H}^T\mathbf{H}\right)^{-1}\mathbf{H}^T \equiv \mathbf{H}^\iota$ is called the *generalized inverse* or *pseudoinverse* of $\mathbf{H}$. The sufficient condition (1.15) implies that, for any nonzero norm of $\mathbf{q}$ (i.e., for all $\|\mathbf{q}\| > 0$),

$$\mathbf{q}^T \left(\mathbf{H}^T \mathbf{H}\right) \mathbf{q} = (\mathbf{H}\mathbf{q})^T (\mathbf{H}\mathbf{q}) = \|\mathbf{H}\mathbf{q}\|_2^2 > 0, \tag{1.18}$$

which will hold only when $\mathbf{H}\mathbf{q} \neq 0$ for $\mathbf{q} \neq 0$: This is true when the columns of $\mathbf{H}$ are linearly independent (Lewis et al., 2006). Furthermore, the rank of $\mathbf{H}$ is $m$ (i.e., full). These confirm that $\hat{\mathbf{x}}$ in (1.17) is the minimizer of $J$ only when $\mathbf{H}^T\mathbf{H}$ is *positive definite.*

**Practice 1.2 Least squares cost function, $J$**

Show that the least squares cost function, $J$, satisfies the following equality:

$$\begin{aligned} J &= \left(\mathbf{y}^o - \mathbf{H}\hat{\mathbf{x}}\right)^T \left(\mathbf{y}^o - \mathbf{H}\hat{\mathbf{x}}\right) \\ &= (\mathbf{y}^o)^T \mathbf{y}^o - 2(\mathbf{y}^o)^T \mathbf{H}\hat{\mathbf{x}} + \hat{\mathbf{x}}^T \left(\mathbf{H}^T \mathbf{H}\right) \hat{\mathbf{x}}. \end{aligned}$$

Then, derive $\nabla_{\hat{\mathbf{x}}} J$ and $\nabla_{\hat{\mathbf{x}}}^2 J$.

**Colloquy 1.1**

### Linear regression

Linear regression – finding a line that best fits a set of data in the least-squares sense – is regarded as a useful paradigm for more complex inverse problems (e.g., atmospheric/oceanic data assimilation) (see Thacker, 1992). Assume that we use a set of data $(X_i, Y_i)$, with a total of $N$ pairs, to develop a function (i.e., regression model) relating the dependent variable $Y$ (or predictand) to the independent variable $X$ (or predictor) as in the following form:

$$Y_i = \hat{Y}_i + \varepsilon_i = aX_i + b + \varepsilon_i, \quad i = 1, \ldots, N, \tag{1.19}$$

where $a$ and $b$ are the slope and the intercept, respectively, of the regression line $\hat{Y}_i$ and $\varepsilon_i$ is the residual (or error). Finding the best fit implies estimating the values of $a$ and $b$ that minimize the sum of squares between the observations ($Y_i$) and and the model solutions ($\hat{Y}_i$):

$$SSE = \sum_i \varepsilon_i^2 = \sum_i \left(Y_i - \hat{Y}_i\right)^2 = \sum_i (Y_i - (aX_i + b))^2, \tag{1.20}$$

where $SSE$ stands for the error sum of squares. The following normal equations are derived by taking the derivatives of the $SSE$ with respect to $a$ and $b$ and setting them to 0:

$$\frac{\partial(SSE)}{\partial a} = 0 = \sum_i 2X_i\left((aX_i + b) - Y_i\right) = 2a\sum_i X_i^2 + 2b\sum_i X_i - 2\sum_i Y_iX_i$$
$$\frac{\partial(SSE)}{\partial b} = 0 = \sum_i 2\left((aX_i + b) - Y_i\right) = 2Nb + 2a\sum_i X_i - 2\sum_i Y_i,$$

producing the least squares estimates of $a$ and $b$ as

$$a = \frac{\sum_i \left(Y_i - \overline{Y}\right)\left(X_i - \overline{X}\right)}{\sum_i \left(X_i - \overline{X}\right)^2}; \quad b = -a\overline{X} + \overline{Y},$$

where $\overline{X}$ and $\overline{Y}$ denote the means of $X$ and $Y$, respectively (see the figure in Colloquy 1.2).

With multiple predictors (i.e., $X_{ik}$, $k = 1, \ldots, K$, for given $Y_i$), we can construct a multiple linear regression, $\hat{Y}_i = b + \sum_k X_{ik}a_k$, represented in vector form as:

$$\mathbf{Y} = \mathbf{Xa} + \boldsymbol{\varepsilon},$$

where $b$ is included in the vector $\mathbf{a}$. Then, the *SSE* (1.20) becomes the least squares cost function (1.13) as

$$SSE = \boldsymbol{\varepsilon}^T\boldsymbol{\varepsilon} = (\mathbf{Y} - \mathbf{Xa})^T(\mathbf{Y} - \mathbf{Xa}). \tag{1.21}$$

We can obtain the optimal parameter $a$ that minimizes the $SSE$ through the normal equation

$$\frac{\partial(SSE)}{\partial \mathbf{a}} = \mathbf{X}^T(\mathbf{Y} - \mathbf{Xa}) = \mathbf{0}, \tag{1.22}$$

producing

$$\mathbf{a} = (\mathbf{X}^T\mathbf{X})^{-1}\mathbf{X}^T\mathbf{Y}, \tag{1.23}$$

which is equivalent to the least squares estimate (1.17).

**Colloquy 1.2**

## Goodness-of-fit

We can assess the goodness-of-fit by defining the total sum of squares ($SST$):

$$SST = \sum_i (Y_i - \overline{Y})^2, \tag{1.24}$$

which measures the prediction error without using regression. In contrast, the $SSE$, defined in (1.20), reflects the prediction error using the least squares regression. How much prediction error is reduced by using the regression?

By combining the $SSE$ and $SST$, we can further define a measure – the *coefficient of determination* or $R^2$ – to test the goodness of the regression model:

$$R^2 = \frac{SST - SSE}{SST} = 1 - \frac{SSE}{SST}, \tag{1.25}$$

which measures the proportion of variation in the predictand (i.e., dependent variable) that has been explained by the regression model. For instance,

$$R^2 = \frac{50.18 - 17.76}{50.18} \approx 0.6461$$

means that 64.61% of the variance in $Y$ can be explained by the variance in $X$: The remaining variation in $Y$ may be due to random variability. It further tells us that the overall sum of squares of 50.18 without regression is reduced down to 17.76 by empoying the least squares regression. Note that $R^2 = 1$ implies the perfect linear fit; thus, the higher the $R^2$ value is, the better the regression model fits the data. See the following figure for interpreting the linear regression model in a given scatter plot and the errors involved in the linear regression.

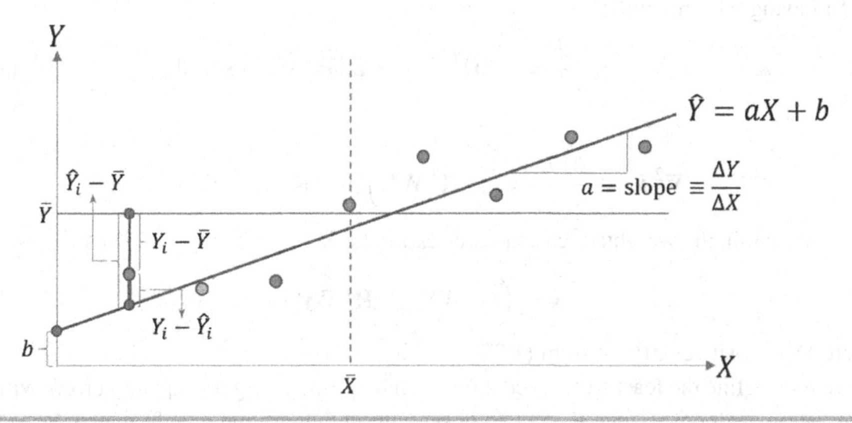

**Practice 1.3 Linear regression: Global warming vs sea level rise**

Shown in the table are 11-year average values, centered at the specified years, of the change in global mean temperature ($\Delta T$) and the change in global mean sea level ($\Delta SL$). For example, the values in 1895 are averaged over 1890–1990.

1. Plot a scatter diagram with $\Delta T$ on the $x$-axis and $\Delta SL$ on the $y$-axis. Construct a linear least squares regression model, i.e., calculate $a$ and $b$ in (1.19), and draw the regression line.
2. Estimate the corresponding values of $\Delta SL$ for the values of $\Delta T = 0.75$ and 1.5°C.
3. Discuss the accuracy of the regression model in terms of $R^2$ in (1.25).

| Year | $\Delta T$ (°C) | $\Delta SL$ (m) |
|---|---|---|
| 1885 | −0.205 | 0.004 |
| 1895 | −0.211 | 0.017 |
| 1905 | −0.315 | 0.030 |
| 1915 | −0.315 | 0.043 |
| 1925 | −0.225 | 0.048 |
| 1935 | −0.102 | 0.060 |
| 1945 | 0.018 | 0.080 |
| 1955 | −0.044 | 0.102 |
| 1965 | −0.023 | 0.115 |
| 1975 | 0.056 | 0.133 |
| 1985 | 0.250 | 0.150 |
| 1995 | 0.388 | 0.168 |
| 2005 | 0.602 | 0.198 |
| 2008 | 0.650 | 0.211 |

### *1.4.2 Weighted Least Squares Estimation*

In operational data assimilation, many observations – of different variables (temperature, humidity, pressure, wind, etc.) and from different platforms (radiosonde, radar, satellite, etc.) – are used. Previously we have acquired $\hat{\mathbf{x}}$ by minimizing $J$ (1.13) that assumed equal emphasis on all observations. However, we may have higher reliance (i.e., weight) on some measurements than others. This leads to the weighted least squares estimation, which defines the cost function as

$$J = (\boldsymbol{\varepsilon}^r)^T \mathbf{W} \boldsymbol{\varepsilon}^r = \|\boldsymbol{\varepsilon}^r\|_{\mathbf{W}}^2 = \left(\mathbf{y}^o - \mathbf{H}\hat{\mathbf{x}}\right)^T \mathbf{W} \left(\mathbf{y}^o - \mathbf{H}\hat{\mathbf{x}}\right), \tag{1.26}$$

where $\mathbf{W}$ is a symmetric weight matrix. To obtain $\hat{\mathbf{x}}$ that minimizes (1.26), we should have the following requirements:

$$\nabla_{\hat{\mathbf{x}}} J = \frac{\partial J}{\partial \hat{\mathbf{x}}} = -2\mathbf{H}^T \mathbf{W} \mathbf{y}^o + 2\left(\mathbf{H}^T \mathbf{W} \mathbf{H}\right) \hat{\mathbf{x}} = \mathbf{0} \tag{1.27}$$

and

$$\nabla_{\hat{\mathbf{x}}}^2 J = \frac{\partial^2 J}{\partial \hat{\mathbf{x}} \partial \hat{\mathbf{x}}^T} = 2\left(\mathbf{H}^T \mathbf{W} \mathbf{H}\right) \text{ is positive definite.} \tag{1.28}$$

Then, we obtain the weighted least squares estimate (WLSE) $\hat{\mathbf{x}}$ from (1.27) as:

$$\hat{\mathbf{x}} = \left(\mathbf{H}^T \mathbf{W} \mathbf{H}\right)^{-1} \mathbf{H}^T \mathbf{W} \mathbf{y}^o, \tag{1.29}$$

where $\mathbf{W}$ is positive definite from (1.28).

We now define the least squares cost function by normalizing the squared errors with the observation error covariance $\mathbf{R}$ in (1.6) or by putting the optimal weight matrix $\mathbf{W} = \mathbf{R}^{-1}$ from (1.26):

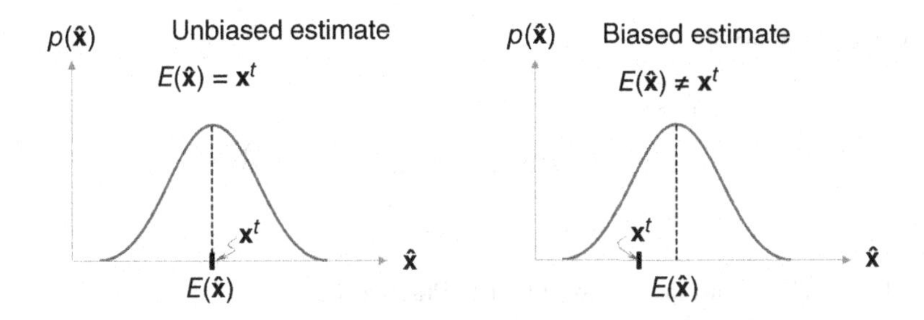


Figure 1.5 Unbiased vs biased estimate $\hat{\mathbf{x}}$ in terms of the PDF of the estimate $p(\hat{\mathbf{x}})$.

$$J = (\boldsymbol{\varepsilon}^r)^T \mathbf{R}^{-1} \boldsymbol{\varepsilon}^r = \left(\mathbf{y}^o - \mathbf{H}\hat{\mathbf{x}}\right)^T \mathbf{R}^{-1} \left(\mathbf{y}^o - \mathbf{H}\hat{\mathbf{x}}\right). \tag{1.30}$$

By minimizing $J$, we obtain

$$\hat{\mathbf{x}} = \left(\mathbf{H}^T \mathbf{R}^{-1} \mathbf{H}\right)^{-1} \mathbf{H}^T \mathbf{R}^{-1} \mathbf{y}^o. \tag{1.31}$$

Note that (1.31) becomes (1.17) when the measurement errors are uncorrelated (i.e., $\mathbf{R}$ is a diagonal matrix) and all errors have equal variance (i.e., $\mathbf{R} = \sigma^2 \mathbf{I}$) (Gelb, 1974).

### *1.4.3 Best Linear Unbiased Estimate*

Given the observation model (1.2), where $\boldsymbol{\varepsilon}^o$ has zero mean and covariance matrix $E(\boldsymbol{\varepsilon}^o (\boldsymbol{\varepsilon}^o)^T) = \mathbf{R}$, as in (1.6), the *best linear unbiased estimate* (BLUE) of the true state $\mathbf{x}^t$, based on data $\mathbf{y}^o$, should satisfy the following conditions: 1) to be *linear*; 2) to be *unbiased*; and 3) to have the *minimum variance* among all unbiased linear estimates.

The observation model becomes linear by taking $H = \mathbf{H}$, where $\mathbf{H}$ is the linearized version of $H$, i.e.,

$$\mathbf{y}^o = \mathbf{H}\mathbf{x}^t + \boldsymbol{\varepsilon}^o, \tag{1.32}$$

and by assuming the estimate is a linear function of the data, in the form of $\hat{\mathbf{x}} = \mathbf{z}^T \mathbf{y}^o$, as in (1.17) and (1.29). The *bias* of the estimate is defined as $E(\hat{\mathbf{x}}) - \mathbf{x}^t$: if $E(\hat{\mathbf{x}}) = \mathbf{x}^t$, then $\hat{\mathbf{x}}$ is an *unbiased* estimate of $\mathbf{x}^t$. Note that the zero-mean condition (1.3) also implies that $\hat{\mathbf{x}}$ is unbiased (do Practice 1.4). Figure 1.5 provides a graphical interpretation of the unbiased and biased estimate in terms of the PDF, which will be explained in more detail in Chapter 2.

**Practice 1.4 Unbiased estimate $\hat{\mathbf{x}}$**

Using Eqs. (1.17), (1.3), and (1.32), show that $E(\hat{\mathbf{x}}) = \mathbf{x}^t$.

We now discuss the minimum variance condition for BLUE. Consider a linear unbiased estimate of $\mathbf{x}^t$:

$$\hat{\mathbf{x}} = \mathbf{z}^T \mathbf{y}^o. \tag{1.33}$$

Noting that $\hat{\mathbf{x}}$ is unbiased,

$$\begin{aligned} E(\hat{\mathbf{x}}) &= \mathbf{x}^t \\ &= E\left(\mathbf{z}^T\mathbf{y}^o\right) = E\left(\mathbf{z}^T(\mathbf{H}\hat{\mathbf{x}} + \boldsymbol{\varepsilon}^r)\right) \\ &= \mathbf{z}^T\mathbf{H}\mathbf{x}^t; \end{aligned} \tag{1.34}$$

thus, $\mathbf{z}^T\mathbf{H} = \mathbf{I}$. The covariance is given by (do Practice 1.5)

$$\begin{aligned} cov(\hat{\mathbf{x}}) &= E\left((\hat{\mathbf{x}} - \mathbf{x}^t)(\hat{\mathbf{x}} - \mathbf{x}^t)^T\right) \\ &= \mathbf{z}^T\mathbf{R}\mathbf{z}. \end{aligned} \tag{1.35}$$

For BLUE, based on the Gauss–Markov Theorem (see Colloquy 1.3),

$$\mathbf{z}_{BLUE}^T = \left(\mathbf{H}^T\mathbf{R}^{-1}\mathbf{H}\right)^{-1}\mathbf{H}^T\mathbf{R}^{-1}, \tag{1.36}$$

following (1.31) and (1.43) in Colloquy 1.3. We also note that $\mathbf{z}_{BLUE}^T\mathbf{H} = \mathbf{I}$, and hence $E(\hat{\mathbf{x}}_{BLUE}) = \mathbf{x}^t$. The difference between $\mathbf{z}_{BLUE}^T$ and $\mathbf{z}^T$ is

$$\begin{aligned} \mathbf{z}_{BLUE}^T - \mathbf{z}^T &= \mathbf{d} \\ &= \left(\mathbf{H}^T\mathbf{R}^{-1}\mathbf{H}\right)^{-1}\mathbf{H}^T\mathbf{R}^{-1} - \mathbf{z}^T, \end{aligned} \tag{1.37}$$

giving $\mathbf{dH} = 0$ (do Practice 1.5); then, we can rewrite (1.35) in terms of $\mathbf{d}$ as (do Practice 1.5)

$$cov(\hat{\mathbf{x}}) = \left(\mathbf{H}^T\mathbf{R}^{-1}\mathbf{H}\right)^{-1} + \mathbf{dRd}^T. \tag{1.38}$$

The covariance of BLUE is

$$\begin{aligned} cov(\hat{\mathbf{x}}_{BLUE}) &= \mathbf{z}_{BLUE}^T\mathbf{R}\mathbf{z}_{BLUE} \\ &= \left(\mathbf{H}^T\mathbf{R}^{-1}\mathbf{H}\right)^{-1}\mathbf{H}^T\mathbf{R}^{-1}\mathbf{R}\left(\left(\mathbf{H}^T\mathbf{R}^{-1}\mathbf{H}\right)^{-1}\mathbf{H}^T\mathbf{R}^{-1}\right)^T \\ &= \left(\mathbf{H}^T\mathbf{R}^{-1}\mathbf{H}\right)^{-1}. \end{aligned} \tag{1.39}$$

By taking the difference between (1.38) and (1.39), we have

$$cov(\hat{\mathbf{x}}) - cov(\hat{\mathbf{x}}_{BLUE}) = \mathbf{dRd}^T. \tag{1.40}$$

Note that $\mathbf{dRd}^T$ is positive semidefinite (do Practice 1.5), making

$$cov(\hat{\mathbf{x}}) \geq cov(\hat{\mathbf{x}}_{BLUE}). \tag{1.41}$$

Therefore, $\hat{\mathbf{x}}_{BLUE}$ has the minimum variance (i.e., "best") among the linear unbiased estimate $\hat{\mathbf{x}}$ in (1.33).

**Colloquy 1.3**

**Gauss–Markov Theorem**

If the observations constitute a general linear model in the form of (1.32), i.e.,

$$\mathbf{y}^o = \mathbf{H}\mathbf{x}^t + \boldsymbol{\varepsilon}^o,$$

where $\mathbf{H}$ is a linear observation operator matrix of $n \times m$, $\mathbf{x}^t$ is an $m \times 1$ vector of states to be estimated, and $\boldsymbol{\varepsilon}^o$ is an $n \times 1$ measurement error vector with zero mean and covariance $\mathbf{R}$ (i.e., $E(\boldsymbol{\varepsilon}^o) = \mathbf{0}$ and $E(\boldsymbol{\varepsilon}^o(\boldsymbol{\varepsilon}^o)^T) = \mathbf{R}$), then the WLSE in (1.31) becomes the BLUE of $\mathbf{x}^t$. That is,

$$\hat{\mathbf{x}}_{WLSE} = \hat{\mathbf{x}}_{BLUE} = \left(\mathbf{H}^T\mathbf{R}^{-1}\mathbf{H}\right)^{-1}\mathbf{H}^T\mathbf{R}^{-1}\mathbf{y}^o. \tag{1.42}$$

The covariance of $\hat{\mathbf{x}}_{BLUE}$ is

$$cov\left(\hat{\mathbf{x}}_{BLUE}\right) = \left(\mathbf{H}^T\mathbf{R}^{-1}\mathbf{H}\right)^{-1}, \tag{1.43}$$

with its diagonal elements are the minimum variance:

$$var\left(\hat{x}_i\right)_{\min} = \left[\left(\mathbf{H}^T\mathbf{R}^{-1}\mathbf{H}\right)^{-1}\right]_{ii}. \tag{1.44}$$

**Practice 1.5 BLUE**

Solve the following:

1. For any linear unbiased estimate $\hat{\mathbf{x}} = \mathbf{z}^T\mathbf{y}^o$ in (1.33), show that the relation in (1.35) holds, i.e.,

$$cov(\hat{\mathbf{x}}) = \mathbf{z}^T\mathbf{R}\mathbf{z}.$$

2. For $\mathbf{d} = \mathbf{z}_{BLUE}^T - \mathbf{z}^T$ in (1.37), show that

$$\mathbf{dH} = 0.$$

(*Hint*: Note that $\mathbf{z}^T\mathbf{H} = \mathbf{I}$ from (1.33).)

3. Derive the relation in (1.38) from (1.35), i.e.,

$$\begin{aligned} cov(\hat{\mathbf{x}}) &= \mathbf{z}^T\mathbf{R}\mathbf{z} \\ &= \left(\mathbf{H}^T\mathbf{R}^{-1}\mathbf{H}\right)^{-1} + \mathbf{dRd}^T, \end{aligned}$$

using (1.37) and $\mathbf{dH} = 0$.

4. Show that $\mathbf{dRd}^T$ in (1.38) is always positive semidefinite. (*Hint*: Referring to (1.18), you may define any $\mathbf{q} \in \mathcal{R}^m$ and multiply $\mathbf{q}$ or $\mathbf{q}^T$ to both sides of $\mathbf{dRd}^T$.)

**Example 1.3 BLUE with two measurements**

Assume that we have measurements of humidity ($q$) at two places in an auditorium – say, $q_1$ near the air conditioner and $q_2$ away from the air conditioner. We define the observation model similar to (1.32), with $\mathbf{H} = \mathbf{I}$ and the observation error ($\varepsilon^o$) following the statistics in (1.6), as

$$q_i^o = q + \varepsilon_i^o \text{ for } i = 1,2. \quad (1.45)$$

The BLUE $\hat{q}$ should be 1) linear (say, $\hat{q} = c_i q_i^o$ for $i = 1,2$), 2) unbiased (i.e., $E(\hat{q}) = q$), and have 3) minimum variance. Because $\hat{q}$ is unbiased,

$$E\left(\hat{q}\right) = q = E\left(c_1 q_1^o + c_2 q_2^o\right) = E\left(c_1(q + \varepsilon_1^o) + c_2(q + \varepsilon_2^o)\right) = (c_1 + c_2)q;$$

thus, $c_1 + c_2 = 1$. The variance of $\hat{q}$ is (do Practice 1.6)

$$E\left((\hat{q} - q)^2\right) = c_1^2\sigma_1^2 + (1 - c_1)^2\sigma_2^2, \quad (1.46)$$

making its minimum occur when

$$c_1 = \frac{\sigma_2^2}{\sigma_1^2 + \sigma_2^2}; \; c_2 = \frac{\sigma_1^2}{\sigma_1^2 + \sigma_2^2}. \quad (1.47)$$

Then, the BLUE $\hat{q}$ becomes

$$\hat{q} = \frac{\sigma_2^2 q_1^o + \sigma_1^2 q_2^o}{\sigma_1^2 + \sigma_2^2}. \quad (1.48)$$

**Practice 1.6 Variance of BLUE**

Answer the following:

1. Show how (1.46) is obtained.
2. Show how to obtain $c_1$ in (1.47) that minimizes the variance (1.46).
3. Define a least squares cost function $J$ as

$$J(q) = \frac{(q - q_1^o)^2}{\sigma_1^2} + \frac{(q - q_2^o)^2}{\sigma_2^2},$$

   and find the BLUE by minimizing $J$. Is the solution similar to the one in (1.48)?
4. Assume that $q_1^o$ is a background (or first guess), i.e., $q_1^o = q^b$, and $q_2^o$ is a real observation, say $q_2^o = q$. Express $\hat{q}$ in terms of the innovation (i.e., $q - q^b$) and compare it with the analysis equation (1.11).

### 1.4.4 BLUE with a Background

Assume that we have a background $\mathbf{x}^b$ – the model-generated gridded observations – and a real observation $\mathbf{y}^o$, and we aim at getting the BLUE or the analysis, i.e., $\mathbf{x}^a = \hat{\mathbf{x}}_{BLUE}$. The observation model is the same as in (1.32) and the error statistics follow (1.6). We further assume that $\mathbf{x}^a$ is a linear combination of the background and the observation as:

$$\mathbf{x}^a = \mathbf{K}_x\mathbf{x}^b + \mathbf{K}_y\mathbf{y}^o, \tag{1.49}$$

with linear operators $\mathbf{K}_x$ and $\mathbf{K}_y$. We define the background error $\boldsymbol{\varepsilon}^b$ and the analysis error $\boldsymbol{\varepsilon}^a$ as

$$\boldsymbol{\varepsilon}^b = \mathbf{x}^b - \mathbf{x}^t \text{ and } \boldsymbol{\varepsilon}^a = \mathbf{x}^a - \mathbf{x}^t,$$

respectively, where $\boldsymbol{\varepsilon}^b$ assumed to have zero mean (i.e., $E(\boldsymbol{\varepsilon}^b) = 0$) and covariance $\mathbf{P}^b$. From (1.32) and (1.49), we have

$$\begin{aligned}
E\left(\boldsymbol{\varepsilon}^a\right) &= E\left(\mathbf{K}_x\mathbf{x}^b + \mathbf{K}_y\mathbf{y}^o - \mathbf{x}^t\right) \\
&= E\left(\mathbf{K}_x\left(\boldsymbol{\varepsilon}^b + \mathbf{x}^t\right) + \mathbf{K}_y\left(\mathbf{H}\mathbf{x}^t + \boldsymbol{\varepsilon}^o\right) - \mathbf{x}^t\right) \\
&= (\mathbf{K}_x + \mathbf{K}_y\mathbf{H} - \mathbf{I})E\left(\mathbf{x}^t\right).
\end{aligned}$$

As we are seeking the BLUE, $\mathbf{x}^a$ should be *unbiased* – i.e., $E(\mathbf{x}^a) = \mathbf{x}^t$, giving $E(\boldsymbol{\varepsilon}^a) = 0$ and hence $\mathbf{K}_x = \mathbf{I} - \mathbf{K}_y\mathbf{H}$. By inserting $\mathbf{K}_x$ into (1.49) and by simply putting $\mathbf{K}_y = \mathbf{K}$, we have

$$\mathbf{x}^a = \mathbf{x}^b + \mathbf{K}\left(\mathbf{y}^o - \mathbf{H}\mathbf{x}^b\right), \tag{1.50}$$

where $\mathbf{K}$ is called the *gain* or the *Kalman gain*, which maps from $\mathcal{R}^n$ to $\mathcal{R}^m$, and $\mathbf{y}^o - \mathbf{H}\mathbf{x}^b$ is the innovation. Note that (1.50) is equivalent to the analysis equation (1.11) except that the observation operator is linearized ($H = \mathbf{H}$) and the weight matrix $\mathbf{W}$ is replaced by $\mathbf{K}$ (i.e., $\mathbf{W} = \mathbf{K}$).

The gain $\mathbf{K}$ can be specified by finding the condition for minimum variance of $\boldsymbol{\varepsilon}^a$ – another property of BLUE. This implies that the analysis error covariance $\mathbf{P}^a$ must have the sum of diagonal elements that are the smallest among the linear estimates, that is,

$$\arg\min E\left(\boldsymbol{\varepsilon}_i^a(\boldsymbol{\varepsilon}^a)_i^T\right) = \arg\min\left(tr(\mathbf{P}^a)\right), \tag{1.51}$$

where $tr(\mathbf{P}^a)$ is the trace of a square matrix $\mathbf{P}^a$, defined for its diagonal elements $p_{ii}^a$ as

$$tr(\mathbf{P}^a) = \sum_i p_{ii}^a.$$

For the square matrices $\mathbf{A}$ and $\mathbf{B}$, and a scalar $\beta$, the trace has the following properties:

$$\begin{aligned}
&tr(\mathbf{A}+\mathbf{B}) = tr(\mathbf{A}) + tr(\mathbf{B});\ tr(\mathbf{A}) = tr(\mathbf{A}^T);\ tr(\beta\mathbf{A}) = \beta tr(\mathbf{A}), \\
&tr(\mathbf{AB}) = tr(\mathbf{BA});\ tr(\mathbf{BAB}^{-1}) = tr(\mathbf{A});\ tr(\mathbf{ABC}) = tr(\mathbf{BCA}) = tr(\mathbf{CAB}),
\end{aligned} \tag{1.52}$$

and

$$\nabla_{\mathbf{X}} tr(\mathbf{XB}) = \mathbf{B}^T;\ \nabla_{\mathbf{X}} tr(\mathbf{BX}^T) = \mathbf{B};\ \nabla_{\mathbf{X}} tr(\mathbf{BXC}) = \mathbf{B}^T\mathbf{C}^T,$$
$$\nabla_{\mathbf{X}} tr(\mathbf{XBX}^T) = \mathbf{X}(\mathbf{B}^T + \mathbf{B});\ \nabla_{\mathbf{X}} tr(\mathbf{X}^T\mathbf{BX}) = (\mathbf{B} + \mathbf{B}^T)\mathbf{X}. \tag{1.53}$$

Equation (1.50) is represented in terms of errors as (do Practice 1.7)

$$\boldsymbol{\varepsilon}^a = \boldsymbol{\varepsilon}^b + \mathbf{K}\left(\boldsymbol{\varepsilon}^o - \mathbf{H}\boldsymbol{\varepsilon}^b\right) = (\mathbf{I} - \mathbf{KH})\boldsymbol{\varepsilon}^b + \mathbf{K}\boldsymbol{\varepsilon}^o, \tag{1.54}$$

from which $\mathbf{P}^a$ is derived as (do Practice 1.7)

$$\begin{aligned}\mathbf{P}^a &= E\left(\left((\mathbf{I} - \mathbf{KH})\boldsymbol{\varepsilon}^b + \mathbf{K}\boldsymbol{\varepsilon}^o\right)\left((\mathbf{I} - \mathbf{KH})\boldsymbol{\varepsilon}^b + \mathbf{K}\boldsymbol{\varepsilon}^o\right)^T\right)\\ &= (\mathbf{I} - \mathbf{KH})\mathbf{P}^b(\mathbf{I} - \mathbf{KH})^T + \mathbf{KRK}^T,\end{aligned} \tag{1.55}$$

where $\mathbf{P}^b = E\left(\boldsymbol{\varepsilon}^b(\boldsymbol{\varepsilon}^b)^T\right)$ and $\mathbf{R} = E\left(\boldsymbol{\varepsilon}^o(\boldsymbol{\varepsilon}^o)^T\right)$. Following (1.51), we differentiate $tr(\mathbf{P}^a)$ with respect to $\mathbf{K}$ and set it to 0 (do Practice 1.7)

$$\begin{aligned}\nabla_{\mathbf{K}} tr(\mathbf{P}^a) &= 0\\ &= \nabla_{\mathbf{K}}\left[tr(\mathbf{P}^b) - 2tr(\mathbf{P}^b\mathbf{H}^T\mathbf{K}^T) + tr(\mathbf{KHP}^b\mathbf{H}^T\mathbf{K}^T) + tr(\mathbf{KRK}^T)\right]\\ &= -2tr(\mathbf{P}^b\mathbf{H}^T) + 2tr(\mathbf{KHP}^b\mathbf{H}^T) + 2tr(\mathbf{KR})\\ &= 2tr\left(\mathbf{K}(\mathbf{HP}^b\mathbf{H}^T + \mathbf{R}) - \mathbf{P}^b\mathbf{H}^T\right);\end{aligned} \tag{1.56}$$

thus,

$$\mathbf{K} = \mathbf{P}^b\mathbf{H}^T(\mathbf{HP}^b\mathbf{H}^T + \mathbf{R})^{-1}. \tag{1.57}$$

Alternatively, $\mathbf{K}$ is expressed as (do Practice 1.7)

$$\mathbf{K} = \left(\left(\mathbf{P}^b\right)^{-1} + \mathbf{H}^T\mathbf{R}^{-1}\mathbf{H}\right)^{-1}\mathbf{H}^T\mathbf{R}^{-1}. \tag{1.58}$$

Note that, with little or no background information, $\left(\mathbf{P}^b\right)^{-1}$ becomes very small and (1.50) reduces to (1.31) (Gelb, 1974).

**Practice 1.7 BLUE and Kalman gain**

Solve the following:

1. From (1.50), derive the error equation (1.54).
2. Derive (1.55) using the properties $E(\boldsymbol{\varepsilon}^o) = E(\boldsymbol{\varepsilon}^b) = 0$.
3. Derive (1.56) using the properties of trace (1.52) and (1.53) and the fact that $\mathbf{P}^b$ and $\mathbf{R}$ are symmetric (i.e., $(\mathbf{P}^b)^T = \mathbf{P}^b$ and $\mathbf{R}^T = \mathbf{R}$).
4. Derive (1.58) from (1.57). You may start by putting (1.57) as

$$\mathbf{K} = \mathbf{IP}^b\mathbf{H}^T(\mathbf{HP}^b\mathbf{H}^T + \mathbf{R})^{-1},$$

where

$$\mathbf{I} = \left(\left(\mathbf{P}^b\right)^{-1}\mathbf{H}^T\mathbf{R}^{-1}\mathbf{H}\right)^{-1}\left(\left(\mathbf{P}^b\right)^{-1}\mathbf{H}^T\mathbf{R}^{-1}\mathbf{H}\right).$$

5. Define a cost function as

$$J(\mathbf{x}) = \frac{1}{2}\left(\mathbf{x}-\mathbf{x}^b\right)^T\left(\mathbf{P}^b\right)^{-1}\left(\mathbf{x}-\mathbf{x}^b\right) + \frac{1}{2}\left(\mathbf{y}^o-\mathbf{H}\mathbf{x}\right)^T\mathbf{R}^{-1}\left(\mathbf{y}^o-\mathbf{H}\mathbf{x}\right).$$

Show that the optimal estimate $\mathbf{x}^a$ is given by

$$\mathbf{x}^a = \mathbf{x}^b + \mathbf{K}\left(\mathbf{y}^o - \mathbf{H}\mathbf{x}^b\right),$$

where $\mathbf{K}$ is shown as in (1.58). You may want to find $\mathbf{x}^a$ that minimizes $J$, i.e.,

$$\mathbf{x}^a = \arg\min J(\mathbf{x})$$

so that

$$\nabla_{\mathbf{x}} J = 0.$$

The result implies that the analysis obtained by minimizing the cost function $J$ (i.e., *variational* analysis) corresponds to the analysis through the minimum error variance approach (i.e., BLUE).

6. Show that the analysis error $\boldsymbol{\varepsilon}^a$ is orthogonal to the analysis $\mathbf{x}^a$, that is,

$$E\left(\mathbf{x}^a\left(\boldsymbol{\varepsilon}^a\right)^T\right) = \mathbf{0},$$

and make a geometric interpretation.

## 1.5 Optimal Interpolation

Optimal interpolation (OI) is equivalent to the BLUE obtained intermittently in a discrete time domain when observation is available. The term "optimal" is employed in a sense that the analysis error variance is minimized – see (1.56); thus, $\mathbf{K}$ in (1.57) is actually regarded as an *optimal gain*, denoted by $\mathbf{K}^{\mathrm{O}}$. By putting $\mathbf{K} = \mathbf{K}^{\mathrm{O}}$ and substituting (1.57) into (1.55), we obtain

$$\mathbf{P}^a = (\mathbf{I} - \mathbf{K}^{\mathrm{O}}\mathbf{H})\mathbf{P}^b. \tag{1.59}$$

Therefore, the OI solution is nothing but the BLUE, represented by (1.50), (1.57), and (1.59). Solving the analysis equation requires direct inversion and is computationally expensive with all global observations. In OI, the calculation of $\mathbf{K}^{\mathrm{O}}$ is simplified by using observations only near the grid (analysis) point. That is, OI acquires the analysis $\mathbf{x}^a$ over an analysis circle (or block), consisting of a grid point and nearby observations (i.e., *localized*) within the so-called radius of influence, $r$ (see Figure 1.6). Depending on $r$, some observations are used twice while some other observations are not used.

The OI scheme, for a given analysis cycle, is written in terms of the analysis circle index, $i$, as

$$\begin{aligned}
\mathbf{x}_i^a &= \mathbf{x}_i^b + \mathbf{K}_i^{\mathrm{O}}\left(\mathbf{y}^o - \mathbf{H}\mathbf{x}^b\right)_i \\
\mathbf{K}_i^{\mathrm{O}} &= \left(\mathbf{P}^b\mathbf{H}^T\right)_i\left((\mathbf{H}\mathbf{P}^b\mathbf{H}^T)_i + \mathbf{R}_i\right)^{-1} \\
\mathbf{P}_i^a &= \left(\mathbf{I} - \mathbf{K}^{\mathrm{O}}\mathbf{H}\right)_i \mathbf{P}_i^b,
\end{aligned} \tag{1.60}$$

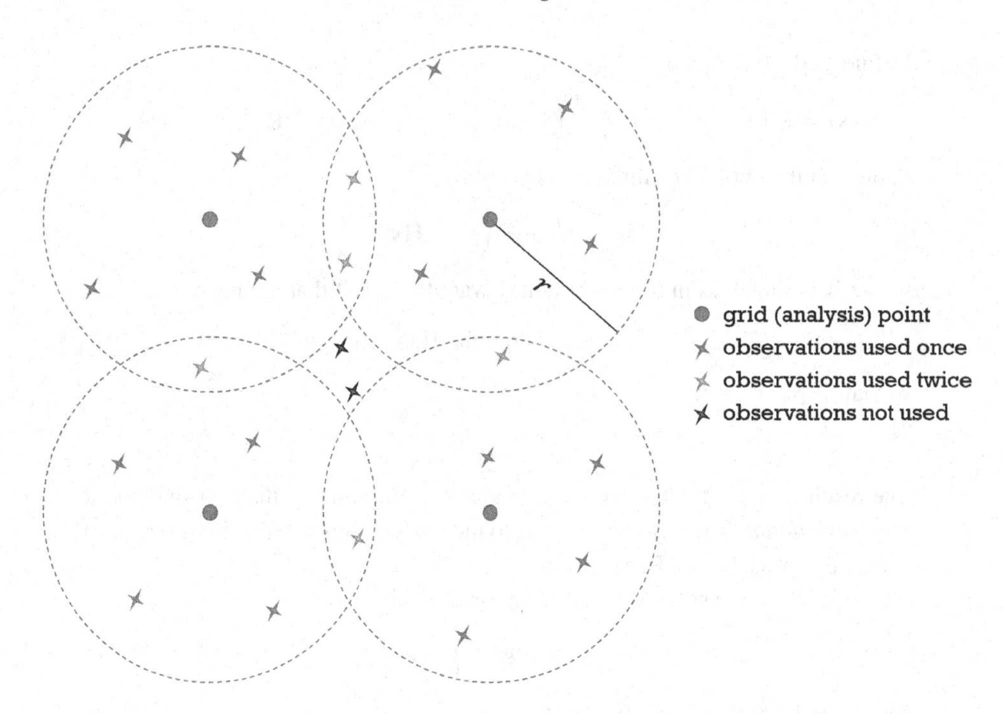


Figure 1.6 OI performed on the grid (analysis) points, centered at the analysis circles. Depending on the size of the analysis circle, which is determined by the radius of influence, $r$, the numbers of observations and their usage for analysis are different.

where $\mathbf{P}_i^b$ is specified through statistical measures (e.g., autocorrelation functions) and dynamic constraints (e.g., geostrophic balance). The analysis equation in OI is equivalent to (1.11) except that the weight $\mathbf{W}$ is replaced by the optimal gain $\mathbf{K}_i^{\mathrm{O}}$ and the observation operator $H$ is linearized. That is, the analysis is given by correcting the background with the analysis increment – the product of the optimal gain and the innovation. The optimal gain is obtained as the product between the background error covariance in the observation space and the inverse of total error covariance (Kalnay, 2003).

From (1.60), by representing a fixed $\mathbf{P}^b$ as $\mathbf{B}$, the analysis increment can be expressed as

$$\mathbf{x}_i^a - \mathbf{x}_i^b = \overbrace{\left(\mathbf{BH}^T\right)_i \underbrace{\left((\mathbf{HBH}^T)_i + \mathbf{R}_i\right)^{-1} \left(\mathbf{y}^o - \mathbf{Hx}^b\right)_i}_{\mathcal{R}^n}}^{\mathcal{R}^m}, \tag{1.61}$$

where $\mathcal{R}^m$ is the model (grid) space and $\mathcal{R}^n$ is the observation space. Note that $\mathbf{H}$ performs a transformation of $\mathcal{R}^m \longrightarrow \mathcal{R}^n$ while $\mathbf{H}^T$ does that of $\mathcal{R}^n \longrightarrow \mathcal{R}^m$. The analysis increment is calculated first by computing $\left((\mathbf{HBH}^T)_i + \mathbf{R}_i\right)^{-1} \left(\mathbf{y}^o - \mathbf{Hx}^b\right)_i$ in the observation space, then by transforming it to the model space by applying $(\mathbf{BH}^T)_i$. This implies that the OI analysis is affected not only by the relevant observations $(\mathbf{y}^o)_i$ but also by the information and structure of the background error covariance $\mathbf{B}_i$. In OI, $\mathbf{B}_i$ has a stationary (i.e., time-invariant) structure: It is generally defined by an isotropic correlation function, depending only on the radius of influence with zero correlations for very large separations between grid points and observations, and enforces most dynamical balance properties reasonably well (see Bouttier and Courtier, 2002). With complex observation operators, calculating $(\mathbf{BH}^T)_i$ is quite difficult.

**Practice 1.8 Properties of operators in analysis increment**

Take the same measurements and model states as in Example 1.1; that is, the measurements are made at three levels and the model states are calculated at six levels (see the figure in Example 1.1). Then the analysis $\mathbf{x}^a$ and the background $\mathbf{x}^b$ are given by

$$\mathbf{x}^a = \left(q_1^a, \ldots, q_6^a\right) \text{ and } \mathbf{x}^b = \left(q_1^b, \ldots, q_6^b\right),$$

respectively. To avoid any confusion, set the observation levels using an alphabetical index, say, $\mathbf{y}^o = \left(q_a^o, q_b^o, q_c^o\right)^T$.

1. Construct the matrices of the background error covariance $\mathbf{B}$ and the linearized observation operator $\mathbf{H}$. [*Hint*: The elements of $\mathbf{B}$ are the covariances between grid points ($b_{23}, b_{56}$, etc.) and those of $\mathbf{H}$ represent interpolation between the model space and observation space ($h_{a5}, h_{c2}$, etc.)]
2. Using $\mathbf{B}$ and $\mathbf{H}$ from #1, evaluate the following matrices and provide interpretation on each of them:

$$\mathbf{BH}^T \text{ and } \mathbf{HBH}^T.$$

Following Hollingsworth (1986), we can extend our interpretation of OI as a filter and an interpolator by manipulating (1.61) as follows:

$$\mathbf{x}_i^a - \mathbf{x}_i^b = \underbrace{\left(\mathbf{BH}^T\right)_i \left(\mathbf{HBH}^T\right)_i^{-1}}_{\mathcal{B}} \underbrace{\left(\mathbf{HBH}^T\right)_i \left(\left(\mathbf{HBH}^T\right)_i + \mathbf{R}_i\right)^{-1}}_{\mathcal{A}} \left(\mathbf{y}^o - \mathbf{Hx}^b\right)_i, \quad (1.62)$$

where we put $\left(\mathbf{HBH}^T\right)_i^{-1} \left(\mathbf{HBH}^T\right)_i = \mathbf{I}$. By applying $\mathcal{A}$ to the innovations $\left(\mathbf{y}^o - \mathbf{Hx}^b\right)_i$, OI filters the innovations (or observations) to generate the analysis increments $\left(\mathbf{x}^a - \mathbf{x}^b\right)_i$ in the observation space. Then, by further operating $\mathcal{B}$, OI interpolates (or propagates) the analysis increments from the observation space to the model space (i.e., grid points). Note that $\mathbf{HBH}^T$ maps the background to the observation space and the observation-error covariances are given; thus, the filtering process occurs in the observation space by converting the innovations to the analysis increments. As $\left(\mathbf{BH}^T\right)_i$ represents the background error covariances operating between the observation space and the model space, the interpolating process transforms the analysis increments from the observation points to the model grid points through normalization by $\left(\mathbf{HBH}^T\right)_i^{-1}$. When the cross-variable background error correlations are provided, the process $\mathcal{B}$ can transform the analysis increment of a certain variable to that of another variable. Overall, the spatial propagation of the corrections by observations (i.e., innovations) is performed by the background error covariances $\mathbf{B}$.

The OI method basically assumes that, for each state variable, the analysis increment is mainly determined by just a few relevant observations (Bouttier and Courtier, 2002). Since the dimension of $\left((\mathbf{HBH}^T)_i + \mathbf{R}_i\right)$ in (1.60) is equal to the number of observations, selecting relevant observations reduces its size (i.e., approximating a block diagonal matrix) and makes the direct inversion viable. For effective calculation of $(\mathbf{BH}^T)_i$ and $(\mathbf{HBH}^T)_i$, one should employ simple observation operators (e.g., interpolation of direct state observations) that are sparse. Due to the selection and use of local data within the analysis circles, the OI analysis fields include spurious noise and sometimes show incoherency between small and large scales (Lorenc, 1981; Bouttier and Courtier, 2002; Dance, 2004). The OI scheme had been widely used in operational centers in 1980s and 1990s (e.g., Lorenc, 1981; Lyne et al., 1982; DiMego, 1988; Kanamitsu, 1989) because of its simplicity and computational efficiency but had been replaced by a variational method due to the disadvantages mentioned above (e.g., Parrish and Derber, 1992; Courtier et al., 1998).

Algorithm 1.2 shows a general prediction process based on OI while Algorithm 1.3 depicts the computation process of $\mathbf{K}^{\mathrm{O}}$ in detail (e.g., Bouttier and Courtier, 2002; Dance, 2004).

**Algorithm 1.2** Prediction process based on OI

```
/* This algorithm is based on (1.60) with P^b = B                                   */
/* index n denotes analysis cycle (time)                                            */
/* index i denotes analysis circle (block) centered at grid (analysis) point        */
/* x^f denotes the future state (forecast) and x^b the background                   */
/* M denotes the model propagator                                                   */
1  Initiation: (x^b)_i^0 = (x^f)_i^0 at the initial time    ! Provide a background at n = 0
2  for n=0 to nmax do                                       ! Loop for analysis cycle (time)
3      for i=1 to imax do                                   ! Loop for analysis circle (block)
4          if (y^o exists) then                             ! When observations are available
5              (K^O)_i^n = (BH^T)_i^n ((HBH^T)_i^n + R_i^n)^-1   ! Calculate the optimal gain
6                                                           ! See Algorithm 1.3
7              (x^a)_i^n = (x^b)_i^n + (K^O)_i^n (y^o - Hx^b)_i^n   ! Calculate the analysis
8              (x^f)_i^{n+1} = M^n (x^a)_i^n                ! Obtain the future state using the analysis
9          else                                             ! When no observations are available
10             (x^f)_i^{n+1} = M^n (x^f)_i^n                ! Obtain the future state using current state
11         end
12         (x^b)_i^{n+1} = (x^f)_i^{n+1}                    ! Assign the forecast to the background
13     endfor
14 endfor
```

**Algorithm 1.3** Process of calculating $\mathbf{K}^{\mathrm{O}}$ for each state variable in OI

/* Calculate $\mathbf{K}^{\mathrm{O}}$ for a given state variable, e.g., temperature, humidity, etc. */
/* This algorithm is based on (1.60) with $\mathbf{P}^b = \mathbf{B}$ */
1 **Input**: $\mathbf{x}^b, \mathbf{y}^o, \mathbf{B}, \mathbf{R}$
2 **Output**: $\mathbf{K}^{\mathrm{O}}$
3 **begin**
4 *Step 1:* Determine the radius of influence $r$ based on empirical selection criteria to specify the analysis circle (block).
5 *Step 2:* Select $l$ observations within the the analysis circle.
6 *Step 3:* Calculate $\left(\mathbf{y}^o - \mathbf{H}\mathbf{x}^b\right)_l$ relevant to the $l$ observations.
7 *Step 4:* Calculate $l \times l$ submatrices of $\mathbf{HBH}^T$ and $\mathbf{R}$ to form $\left(\mathbf{HBH}^T + \mathbf{R}\right)_l$.
8 *Step 5:* Calculate a row vector $\left(\mathbf{BH}^T\right)_l$ for the given state variable, restricted to the $l$ observations.
9 *Step 6:* Calculate the inverse of $\left(\mathbf{HBH}^T + \mathbf{R}\right)_l$.
10 *Step 7:* Calculate $\mathbf{K}_l^{\mathrm{O}} = \left(\mathbf{BH}^T\right)_l \left(\mathbf{HBH}^T + \mathbf{R}\right)_l^{-1}$.
11 **end**

## References


Barnes SL (1964) A technique for maximizing details in numerical weather map analysis. *J Appl Meteor* 3:396–409.

Bergthórsson P, Döös BR (1955) Numerical weather map analysis. *Tellus* 7:329–340.

Bonavita M, Isaksen L, Hólm E (2012) On the use of EDA background error variances in the ECMWF 4D-Var. *Quart J Roy Meteor Soc* 138:1540–1559.

Bouttier F, Courtier P (2002) *Data Assimilation Concepts and Methods*. Meteorological Training Course Lecture Series, ECMWF, Reading, 59 pp., www.ecmwf.int/node/16928

Buehner M (2005) Ensemble-derived stationary and flow-dependent background-error covariances: Evaluation in a quasi-operational NWP setting. *Quart J Roy Meteor Soc* 131:1013–1043.

Chorin A, Morzfeld M, Tu X (2010) Implicit particle filters for data assimilation. *Commun Appl Math Comput Sci* 5:221–240.

Clayton AM, Lorenc AC, Barker DM (2013) Operational implementation of a hybrid ensemble/4D-Var global data assimilation system at the Met Office. *Quart J Roy Meteor Soc* 139:1445–1461.

Courtier P, Thépaut JN, Hollingsworth A (1994) A strategy for operational implementation of 4D-Var, using an incremental approach. *Quart J Roy Meteor Soc* 120:1367–1387.

Courtier P, Andersson E, Heckley W, et al. (1998) The ECMWF implementation of three-dimensional variational assimilation (3D-Var). I: Formulation. *Quart J Roy Meteor Soc* 124:1783–1807.

Cressman GP (1959) An operational objective analysis system. *Mon Wea Rev* 87:367–374.

Daley R (1991) *Atmospheric Data Analysis*. Cambridge University Press, New York, 457 pp.

Dance SL (2004) Issues in high resolution limited area data assimilation for quantitative precipitation forecasting. *Physica D* 196:1–27.

Davies HC, Turner RE (1977) Updating prediction models by dynamical relaxation: An examination of the technique. *Quart J Roy Meteor Soc* 103:225–245.

Derber JC, Wu WS (1998) The use of TOVS cloud-cleared radiances in the NCEP SSI analysis system. *Mon Wea Rev* 126:2287–2299.

DiMego GJ (1988) The National Meteorological Center regional analysis system. *Mon Wea Rev* 116:977–1000.

ECMWF (2020) Monitoring: Data coverage. www.ecmwf.int/en/forecasts/charts/monitoring/dcover

Evensen G (1994) Sequential data assimilation with a nonlinear quasi-geostrophic model using Monte Carlo methods to forecast error statistics. *J Geophys Res Oceans* 99:10143–10162.

Gandin LS (1963) *Objective Analysis of Meteorological Fields*. Gidromet, Leningrad, 285 pp., English translation, Israel Program for Scientific Translations, 1965.

Gao J (2017) A three-dimensional variational radar data assimilation scheme developed for convective scale NWP. In *Data Assimilation for Atmospheric, Oceanic and Hydrologic Applications* (vol. III), (eds.) Park SK, Xu L, Springer International Publishing, Cham, 285–326.

Gauss CF (1809) *Theoria Motus Corporum Coelestium in Sectionibus Conicis Solem Ambientium (Theory of the Motion of Heavenly Bodies Moving about the Sun in Conic Section)*. English translation by C.H. Davis, published by Little Brown, and Co. of Boston in 1857, reprinted in 1963 by Dover Publications, Inc., New York, 374 pp.

Gelb A (ed.) (1974) *Applied Optimal Estimation*. MIT Press, Cambridge, MA, and London, 374 pp.

Gilchrist B, Cressman GP (1954) An experiment in objective analysis. *Tellus* 6:309–318.

Gilmore MS, Straka JM, Rasmussen EN (2004) Precipitation and evolution sensitivity in simulated deep convective storms: Comparisons between liquid-only and simple ice and liquid phase microphysics. *Mon Wea Rev* 132:1897–1916.

Hoke JE, Anthes RA (1976) The initialization of numerical models by a dynamic-initialization technique. *Mon Wea Rev* 104:1551–1556.

Hollingsworth A (1986) Objective analysis for numerical weather prediction. *J Meteor Soc Japan* 64A:11–59.

Houtekamer PL, Mitchell HL (1998) Data assimilation using an ensemble Kalman filter technique. *Mon Wea Rev* 126:796–811.

Ide K, Courtier P, Ghil M, Lorenc AC (1997) Unified notation for data assimilation: Operational, sequential and variational. *J Meteor Soc Japan* 75:181–189.

Kalman RE, Bucy RS (1961) New results in linear filtering and prediction theory. *J Basic Eng* 83:95–108.

Kalnay E (2003) *Atmospheric Modeling, Data Assimilation and Predictability*. Cambridge University Press, New York, 341 pp.

Kanamitsu M (1989) Description of the NMC global data assimilation and forecast system. *Wea Forecasting* 4:335–342.

Lakshmivarahan S, Lewis JM (2013) Nudging methods: A critical overview. In *Data Assimilation for Atmospheric, Oceanic and Hydrologic Applications* (vol. II), (eds.) Park SK, Xu L, Springer, 27–57.

Lakshmivarahan S, Lewis JM, Phan D (2013) Data assimilation as a problem in optimal tracking: Application of Pontryagin's minimum principle to atmospheric science. *J Atmos Sci* 70:1257–1277.

Laplace PS (1814) *Essai philosophique sur les probabilités*. Nabu Press, French ed., 2010, 212 pp.
Le Dimet F-X, Talagrand O (1986) Variational algorithms for analysis and assimilation of meteorological observations: Theoretical aspects. *Tellus A* 38:97–110.
Legendre AM (1805) *Nouvelles méthodes pour la détermination des orbites des comètes*. Firmin Didot, Paris. Pages 72–75 of the appendix reprinted in Stigler (1986, p.56). English translation of these pages by Ruger, HA and Walker, HM in *A Source Book of Mathematics* by Smith, DE, McGraw-Hill Book Company, New York, 1929, p. 576–579.
Lewis JM, Derber JC (1985) The use of adjoint equations to solve a variational adjustment problem with advective constraints. *Tellus A* 37:309–322.
Lewis JM, Lakshmivarahan S, Dhall S (2006) *Dynamic Data Assimilation: A Least Squares Approach*. Cambridge University Press, Cambridge, 654 pp.
Lin YL, Farley RD, Orville HD (1983) Bulk parameterization of the snow field in a cloud model. *J Climate Appl Meteor* 22:1065–1092.
Lorenc AC (1981) A global three-dimensional multivariate statistical interpolation scheme. *Mon Wea Rev* 109:701–721.
Lorenc AC (2003) The potential of the ensemble Kalman filter for NWP – a comparison with 4D-Var. *Quart J Roy Meteor Soc* 129:3183–3203.
Lyne WH, Swinbank R, Birch NT (1982) A data assimilation experiment and the global circulation during the FGGE special observing periods. *Quart J Roy Meteor Soc* 108:575–594.
Navon IM, Zou X, Derber J, Sela J (1992) Variational data assimilation with an adiabatic version of the NMC spectral model. *Mon Wea Rev* 120:1433–1446.
Parrish DF, Derber JC (1992) The National Meteorological Center's spectral statistical-interpolation analysis system. *Mon Wea Rev* 120:1747–1763.
Pontryagin LS, Boltyanskii VG, Gamkrelidze RV, Mischenko EF (1961) *Matematicheskaya Teoriya Optimal'nykh Prozessov*. Fizmatgiz, Moscow, English translation *The Mathematical Theory of Optimal Control Processes* by Trirogoff, KN, published in 1962 by Interscience Publishers, John Wiley & Sons, Inc., New York, London, Sydney, 360 pp.
Sakov P, Oliver DS, Bertino L (2012) An iterative enkf for strongly nonlinear systems. *Mon Wea Rev* 140:1988–2004.
Sasaki Y (1958) An objective analysis based on the variational method. *J Meteor Soc Japan* 36:77–88.
Sasaki Y (1970a) Some basic formalisms in numerical variational analysis. *Mon Wea Rev* 98:875–883.
Sasaki Y (1970b) Numerical variational analysis formulated under the constraints as determined by longwave equations and a low-pass filter. *Mon Wea Rev* 98:884–898.
Sasaki Y (1970c) Numerical variational analysis with weak constraint and application to surface analysis of severe storm gust. *Mon Wea Rev* 98:899–910.
Shannon CE, Weaver W (1949) *The Mathematical Theory of Communication*. University of Illinois Press, Chicago, IL, 117 pp.
Smith Jr. P, Myers C, Orville H (1975) Radar reflectivity factor calculations in numerical cloud models using bulk parameterization of precipitation. *J Appl Meteor* 14:1156–1165.
Sorenson HW (1970) Least-squares estimation: From Gauss to Kalman. *IEEE Spectrum* 7:63–68.
Sugimoto S, Crook NA, Sun J, Xiao Q, Barker DM (2009) An examination of WRF 3DVAR radar data assimilation on its capability in retrieving unobserved variables and forecasting precipitation through observing system simulation experiments. *Mon Wea Rev* 137:4011–4029.

Sun J, Crook NA (1997) Dynamical and microphysical retrieval from Doppler radar observations using a cloud model and its adjoint. Part I: Model development and simulated data experiments. *J Atmos Sci* 54:1642–1661.

Thacker WC (1992) Oceanographic inverse problems. *Physica D* 60:16–37.

van Leeuwen PJ (2009) Particle filtering in geophysical systems. *Mon Wea Rev* 137:4089–4114.

Wang X, Hamill TM, Whitaker JS, Bishop CH (2007) A comparison of hybrid ensemble transform Kalman filterâŁ"OI and ensemble square-root filter analysis schemes. *Mon Wea Rev* 136:5116–5131.

Xiao Q, Kuo YH, Sun J, et al. (2007) An approach of radar reflectivity data assimilation and its assessment with the inland QPF of Typhoon Rusa (2002) at landfall. *J Appl Meteor Climatol* 46:14–22.

Zou X, Navon IM, Ledimet FX (1992) An optimal nudging data assimilation scheme using parameter estimation. *Quart J Roy Meteor Soc* 118:1163–1186.

Županski M (1993) Regional four-dimensional variational data assimilation in a quasi-operational forecasting environment. *Mon Wea Rev* 121:2396–2408.

Županski M (2005) Maximum likelihood ensemble filter: Theoretical aspects. *Mon Wea Rev* 133:1710–1726.

# 2
# Probability and the Bayesian Approach

## 2.1 Introduction

In this section the mathematical foundation of probabilistic data assimilation will be presented and discussed. Although the mathematical formulation of probability is unique, it can have various interpretations, most notably the frequentist and Bayesian views. Given that probabilistic data assimilation relies heavily on using Bayes' theorem in order to produce an estimate of the posterior probability density function (PDF), it is naturally related to the Bayesian view. At the same time, implied by using Monte Carlo–based methods, data assimilation also relies on the statistical sampling of the theory. Therefore, both views of probability are very much mixed in interpretation of data assimilation. We will try to stay within the mathematical formalism of probability theory only and leave the interpretation of probability out as much as possible.

## 2.2 Probability

Before formally defining the probability, it is important to introduce the notion of an event $\boldsymbol{\omega}$ that belongs to a collection of all logically plausible events $\boldsymbol{\Omega}$, i.e., $\boldsymbol{\omega} \in \boldsymbol{\Omega}$. For example, if tossing a coin in the air, all logically plausible events $\boldsymbol{\Omega}$ include a heads *and* tails collection, while the individual event $\boldsymbol{\omega}$ could be heads *or* tails. Given a countable set of events $A_1, A_2, \ldots \in \boldsymbol{\Omega}$, we consider such a collection $\boldsymbol{\Omega}$ that

1. includes the empty subset $\varnothing$ ($\varnothing \in \boldsymbol{\Omega}$);
2. is closed under complement ($\sum + \sum^c = \boldsymbol{\Omega}$); and
3. is closed under countable unions and intersections,

$$\bigcup_{n=1}^{\infty} A_n = \{\omega : \forall k, \omega \in A_k\}, \quad \bigcup_{n=1}^{\infty} A_n = \{\omega : \exists k, \omega \in A_k\}.$$

The above statements describe the so-called $\boldsymbol{\sigma}$-algebra, or $\boldsymbol{\sigma}$-field, which represents a measurable space, and is fundamental for defining probability. In probability theory it is fundamentally important to associate an event $\boldsymbol{\omega} \in \boldsymbol{\Omega}$ with a real number $X(\boldsymbol{\omega}) \in \mathbb{R}$. However, in order to make this association between an event and a real number, it is important to introduce the *measurability* of $X$. This is commonly done by assuming that an event is defined as a set of $X$ that belongs to an interval $[a, b)$, where the symbol "["

implies that $a$ is included in the interval, while the symbol ")" implies that $b$ is an upper bound that is not included in the interval. Therefore, we assume

$$\{\omega : a \leq X(\omega) < b\} = \{a \leq X < b\}. \tag{2.1}$$

Using the above equation one can define a *random* variable as a measurable function $X : \mathbf{\Omega} \to \mathbb{R}$ such that for any $x \in \mathbb{R}$

$$\{\omega : X(\omega) < x\} = \{X < x\}. \tag{2.2}$$

Formally, the probability $P$ is a continuous function defined over the $\sigma$-field, with the following properties:

1. $P(A) \geq 0$ for any $A \in \mathbf{\Omega}$.
2. $P(\mathbf{\Omega}) = 1$.
3. $P\left(\sum_{n=1}^{\infty} A_n\right) = \sum_{n=1}^{\infty} P\left(A_n\right)$ for independent events $A_n \in \mathbf{\Omega}$, i.e., $A_i \cap A_j = \varnothing$ for $i \neq j$.

There are a few additional rules that follow from the basic probability properties (1–3):

4. $P(\varnothing) = 0$.
5. $P(A) \leq P(B)$ for $A \subseteq B$, where $A, B \in \mathbf{\Omega}$.
6. $P(A^c) = 1 - P(A)$ for $A \cup A^c = A + A^c = \mathbf{\Omega}$, where $A^c$ denotes the complement of $A$.
7. $P(A \cup B) = P(A) + P(B) - P(AB)$, where $AB = A \cap B$.
8. $P\left(\bigcup_{n=1}^{\infty} A_n\right) \leq \sum_{n=1}^{\infty} P\left(A_n\right)$.
9. $P\left(\bigcap_{n=1}^{\infty} A_n\right) = \lim_{n\to\infty} P(A_n)$ for $A_1 \subseteq A_2 \subseteq \cdots$.

From properties (1), (2), and (4) it follows that the probability has numerical values between 0 and 1. For $P(A) = 1$ the event $A$ is *certain*, while for $P(A) = 0$ the event $A$ is *impossible*.

In order to formally deal with probability using standard calculus it is beneficial to introduce a function of *probability distributions*

$$F(x) = P\left((-\infty, x)\right) = P(X < x), \quad -\infty < x < +\infty \tag{2.3}$$

with the following properties:

1. $F(-\infty) = \lim_{x\to-\infty} F(x) = 0$.
2. $F(+\infty) = \lim_{x\to+\infty} F(x) = 1$.
3. $F(x - 0) = F(x)$ (continuous from the left).

The function $F$ is also referred to as the cumulative distribution function (CDF). If a random variable $X$ is discrete, then the probability distribution function $F$ is also discrete.

If the function $F$ is *absolutely continuous* one can define the PDF denoted by $\varphi(x)$:

$$P\left([a, b)\right) = P\{a \leq X < b\} = \int_a^b \varphi(x)dx. \tag{2.4}$$

It is clear that $\int_{-\infty}^{+\infty} \varphi(x)dx = 1$. In data assimilation we mostly deal with PDFs, rather than CDFs, so understanding the role of PDFs is quite important.

The properties of probability are valid for multidimensional variables as well. Given an $N$-dimensional random variable $X = (X_1, X_2, \ldots, X_N)$, the probability is

$$F(x_1, x_2, \ldots, x_N) = P(\{X_1 < x_1\} \cap \{X_2 < x_2\} \cap \cdots \{X_N < x_N\}). \qquad (2.5)$$

The properties (1–3) also hold for multidimensional random variables. The corresponding multidimensional PDF is given by

$$P(X_1, X_2, \ldots, X_N) = \int\int \cdots \int \varphi(x_1, x_2, \ldots, x_N)\, dx_1, dx_2, \ldots, dx_N. \qquad (2.6)$$

The PDF follows most of the properties of CDF. One of the more relevant properties of PDF is related to independent variables. A sufficient and necessary condition for two variables $X$ and $Y$ to be independent is that $\varphi(x, y) = \varphi(x)\varphi(y)$.

## 2.3 Probability Density Function Moments

A probability is sometimes described in terms of its moments. To express probability in terms of its moments one needs to define mathematical expectation, typically denoted $E$, which is a function that maps random variables to the most likely value. For a random variable $X$ with PDF $\varphi(x)$ the mathematical expectation is given by

$$EX = \int_{-\infty}^{+\infty} x\varphi(x)dx. \qquad (2.7)$$

More generally, for function $f(X)$, the mathematical expectation is

$$E(f(X)) = \int_{-\infty}^{+\infty} f(x)\varphi(x)dx. \qquad (2.8)$$

A generalization for an $N$-dimensional random variable $X = X(X_1, X_2, \ldots, X_N)$ with density $\varphi(x_1, x_2, \ldots, x_N)$ produces

$$E(X_1, X_2, \ldots, X_N) = \int_{-\infty}^{+\infty}\int_{-\infty}^{+\infty} \cdots \int_{-\infty}^{+\infty} x_1 x_2 \cdots x_N \varphi(x_1, x_2, \ldots, x_N)\, dx_1, dx_2, \ldots, dx_N. \qquad (2.9)$$

Using definitions (2.7)–(2.9), one can also easily derive the following properties of mathematical expectation:

1. $E(X_1 + X_2) = E(X_1) + E(X_2)$.
2. $E(X_1 X_2 \cdots X_N) = E(X_1)E(X_2) \cdots E(X_N)$ for independent variables $X_1, X_2, \ldots, X_N$.

The standard moment of the order $N (N = 1, 2, \ldots)$ of random variable $X$ is defined as $EX^N$, and the central moment of the order $N$ is defined as $E(X - EX)^N$. We have already defined the first moment as mathematical expectation (Eq. (2.7)). Of interest also is the second central moment ($N$=2):

$$E(X - EX)^2 = \int_{-\infty}^{+\infty} (x - m)^2 \varphi(x)dx, \qquad (2.10)$$

which is referred to as the variance, with $m = EX$ representing the first moment. For Gaussian random variable $X \sim N(m,\sigma^2)$ we have

$$E(X - EX)^2 = \frac{1}{\sqrt{2\pi\sigma^2}} \int_{-\infty}^{+\infty} (x - m)^2 e^{-\frac{(x-m)^2}{2\sigma^2}} dx = \sigma^2. \qquad (2.11)$$

Of general interest are also the third central moment, referred to as *skewness*, and the fourth central moment, referred to as *kurtosis*. Skewness measures how asymmetric the distribution is, and kurtosis measures how heavy the distribution tail is.

## 2.4 Common Probability Density Functions

Among the most commonly used probabilities are the uniform, Gaussian (normal), and lognormal probabilities. In data assimilation it is of interest to know not only the moments but also a few other pointwise characterizations of PDFs, such as the median and the mode. The median is defined as a point where $F(x) = 0.5$, while the mode is a point of global maximum of the PDF.

• *Uniform distribution*: If a random variable defined on an interval $[a,b]$ has uniform distribution, then its PDF is

$$\varphi(x) = \begin{cases} \frac{1}{b-a} & a \leq x \leq b \\ 0 & x < a,\ x > b \end{cases}. \qquad (2.12)$$

This distribution is symmetric and has an identical median and mean. The mode could be any value in the interval $[a,b]$. The PDF is shown in Fig. 2.1. Uniform PDFs can be generalized to multivariate forms. For example, a bivariate uniform distribution on an interval $([a,b],[c,d])$ is

$$\varphi(x_1,x_2) = \frac{1}{(b-a)(c-d)} \begin{cases} x_1x_2 & a \leq x_1 \leq b, c \leq x_2 \leq d \\ 0 & \text{otherwise} \end{cases}. \qquad (2.13)$$

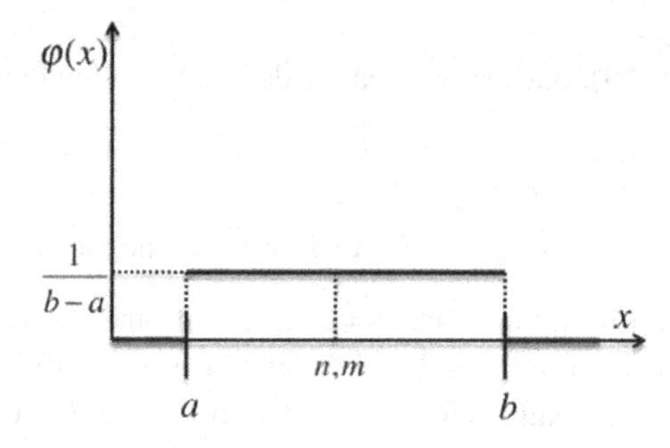


Figure 2.1 Univariate uniform PDF defined by Eq. (2.12). The mean and median are identical, equal to $\frac{1}{2}(a + b)$, while mode is any value in the interval $[a,b]$.

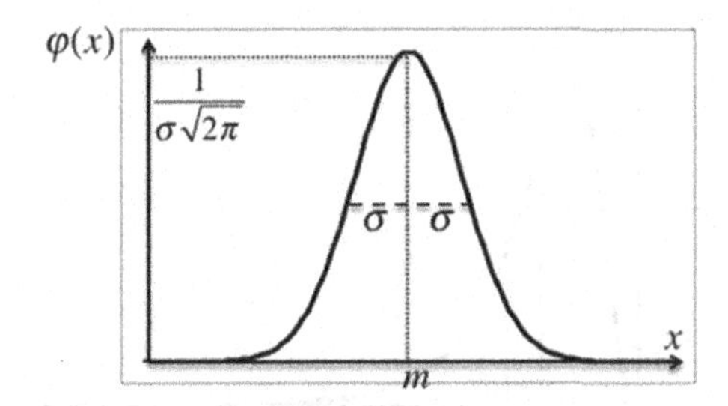


Figure 2.2 Univariate Gaussian PDF defined by Eq. (2.14). The mean, mode, and median are identical.

• *Normal (Gaussian) distribution*: If a random variable defined on an interval $-\infty < x < +\infty$ has normal distribution, then its PDF is

$$\varphi(x) = \frac{1}{\sqrt{2\pi\sigma^2}} e^{-\frac{(x-m)^2}{2\sigma^2}}, \tag{2.14}$$

where the parameters $m$ and $\sigma$ are referred to as the mean and standard deviation, respectively. Such a random variable is generally denoted $N(m, \sigma^2)$. The PDF is shown in Fig. 2.2. The multivariate Gaussian probability density is obtained by generalization of the univariate formula (2.14) as

$$\varphi(\mathbf{x}) = \frac{1}{\sqrt{(2\pi)^N \det(\mathbf{\Sigma})}} e^{-\frac{1}{2}(\mathbf{x}-\mathbf{m})^T \mathbf{\Sigma}^{-1}(\mathbf{x}-\mathbf{m})}, \tag{2.15}$$

where $\mathbf{\Sigma}$ is a covariance matrix, *det* denotes matrix determinant, and $\mathbf{x} = (x_1, x_2, \ldots, x_N)$ is multivariate. The multivariate normal variable is typically denoted $N(\mathbf{m}, \mathbf{\Sigma})$.

**Practice 2.1 PDF of univariate Gaussian distribution**

Calculate the PDF of a univariate Gaussian distribution with mean $m = 0$ and standard deviation $\sigma = 1$ at point $x = 0$.

**Practice 2.2 Interval length of univariate uniform distribution**

Find the interval length $a$ of a univariate uniform distribution on the interval $(-a, +a)$ such that its PDF value at point $x = 0$ is equal to the Gaussian PDF $N(0, 1)$ at $x = 0$.

• *Lognormal distribution*: This distribution is obtained by introducing a change of variable $x = \ln z$, where $x$ belongs to the normal distribution. If a random variable defined on an interval $0 < z < \infty$ has a lognormal distribution, then its PDF is

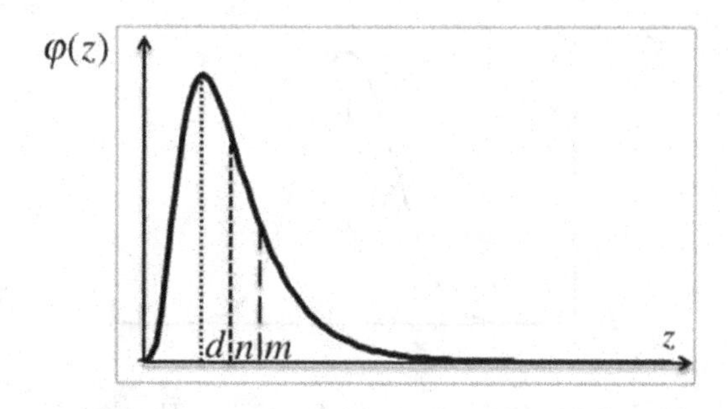


Figure 2.3 Univariate lognormal PDF defined by Eq. (2.16). Note that the distribution is skewed, and that the mode ($d$), median ($n$), and mean ($m$) are different.

$$\varphi(z) = \frac{1}{z\sqrt{2\pi\eta^2}} e^{-\frac{(\ln z-\mu)^2}{2\eta^2}}, \tag{2.16}$$

where the parameters $\mu$ and $\eta$ are the mean and standard deviation of the related Gaussian distribution, respectively. The mean of the lognormal distribution is $m = \exp\left[\mu + \frac{\eta^2}{2}\right]$ and standard deviation is $\sigma = \sqrt{\exp(\eta^2) - 1} \cdot \exp\left(\mu + \frac{\eta^2}{2}\right)$. The PDF is shown in Fig. 2.3. A lognormal distribution can also be generalized to a multivariate distribution (Fletcher and Županski, 2006):

$$\varphi(\mathbf{z}) = \frac{1}{\sqrt{(2\pi)^N \det(\mathbf{E})}} \left(\prod_{i=1}^{N} \frac{1}{z_i}\right) e^{-\frac{1}{2}(\ln \mathbf{z}-\boldsymbol{\mu})^T \mathbf{E}^{-1}(\ln \mathbf{z}-\boldsymbol{\mu})}, \tag{2.17}$$

where $\mathbf{E}$ is the covariance matrix, $\prod$ denotes the product operator, and $\mathbf{z}$ is multivariate.

The mentioned probability distributions are most often used in data assimilation, but there are many other probability distributions that may be of interest.

## 2.5 Bayes' Formula

Especially in data assimilation, it is often important to know how the probability of an event $B$ is influenced by a related event $A$ that is realized. This introduces *conditional* probability $P(B \mid A)$ that can be defined as

$$P(B \mid A) = \frac{P(AB)}{P(A)}, \tag{2.18}$$

where $P(AB)$ is referred to as *joint* probability. Assuming $P(A) > 0$, it is also clear that $P(A \mid A) = 1$. From definition (2.4) it also follows that

$$P(AB) = P(A)P(B \mid A) = P(B)P(A \mid B), \tag{2.19}$$

which essentially shows how the joint probability of two dependent events, $AB \neq \emptyset$, can be described in terms of probabilities of independent events $A$ and $B|A$, or events $B$ and $A|B$. By induction, this could be further generalized to a finite number of events

$$P(A_1A_2\cdots A_n) = P(A_1)P(A_2 \mid A_1)\cdots P(A_n \mid A_1\cdots A_{n-1}). \tag{2.20}$$

Given independent events $A_1, A_2, \ldots, A_n$ with $P(A_i) > 0$ and $\sum_{i=1}^{n} A_i = \mathbf{\Omega}$, then

$$P(B) = \sum_{i=1}^{n} P(A_i)P(B \mid A_i), \tag{2.21}$$

which follows from $B = \sum_{i=1}^{n} A_i B$. Under the same conditions

$$P(B \mid A_i) = \frac{P(B)P(A_i \mid B)}{P(A_i)}, \tag{2.22}$$

which is the *Bayes' formula* (theorem), of fundamental importance in data assimilation. In particular, in data assimilation, event $B$ represents the realization of a model state and events $A_i$ represent the measurements. The probability $P(B)$ in (2.22) is referred to as the *prior* or *marginal* probability, as it represents the probability of a state before knowing the measurements. Therefore, Bayes' formula gives the probability of a state after realization of the measurements and of the prior state.

Bayes' formula also holds for PDFs. Given variables $x$ and $y = (y_1, y_2, \ldots, y_N)$, with $\varphi(y_i) > 0$, the conditional PDF is

$$\varphi(x \mid y_i) = \frac{\varphi(x)\varphi(y_i \mid x)}{\varphi(y_i)}. \tag{2.23}$$

## 2.6 Recursive Bayes' Formula for Data Assimilation

Data assimilation typically begins with the application of Bayes' formula (2.23). However, in applications to data assimilation with *probabilistic modeling* few additional adjustments are required. Typical probabilistic modeling refers to ensemble forecasting, but in principle could be any prediction of PDFs. Note that prediction is commonly defined as a Markov process, in the sense that the state at time $t$ explicitly depends only on the state at previous time $t-1$, and only conditionally depends on all previous times. Consider a state vector defined over multiple times $(x_0, x_1, \ldots, x_k)$ and observations defined over multiple times $(y_1, y_2, \ldots, y_k)$, where $k$ is the time index. Note that the state begins with index $k = 0$, while observations begin at $k = 1$. Assuming an initial forecast PDF $\varphi(x_0)$, the forecast PDF at time $k$ is

$$\varphi(x_k) = \varphi(x_k|x_{k-1}) \cdots \varphi(x_1|x_0)\varphi(x_0). \tag{2.24}$$

We are now interested in developing a *recursive* Bayes' formula that is relevant to data assimilation, and implicitly shows how a data assimilation system "learns" from previous experiences. At the first observation time ($k = 1$) the Bayes' formula (2.23) is

$$\varphi(x_1 \mid y_1) = \frac{\varphi(x_1)\varphi(y_1 \mid x_1)}{\varphi(y_1)}, \tag{2.25}$$

which is optimized during data assimilation to get the analysis PDF

$$\varphi(x_1^{opt}) = \varphi(x_1^{opt}|y_1). \tag{2.26}$$

Bayes' formula at the second observation time ($k = 2$) is

$$\varphi(x_2|y_2) = \frac{\varphi(x_2)\varphi(y_2|x_2)}{\varphi(y_2)}. \tag{2.27}$$

Note that the forecast PDF at $k = 2$ is obtained by starting from the analysis PDF as the initial conditions

$$\varphi(x_2) = \varphi(x_2|x_1^{opt})\varphi(x_1^{opt}), \tag{2.28}$$

which produces a guess of conditional PDF as

$$\varphi(x_2|y_2) = \frac{\varphi(x_2|x_1^{opt})\varphi(x_1^{opt})\varphi(y_2|x_2)}{\varphi(y_2)}, \tag{2.29}$$

and after optimization the analysis PDF is $\varphi(x_2^{opt}) = \varphi(x_2^{opt}|y_2)$. After extending these formulas to time $k$, the forecast PDF is

$$\varphi(x_k) = \varphi(x_k|x_{k-1}^{opt})\varphi(x_{k-1}^{opt}) \tag{2.30}$$

and the guess conditional PDF is

$$\varphi(x_k|y_k) = \frac{\varphi(x_k|x_{k-1}^{opt})\varphi(x_{k-1}^{opt})\varphi(y_k|x_k)}{\varphi(y_k)}, \tag{2.31}$$

which after data assimilation produces the analysis PDF at time $k$ as

$$\varphi(x_k^{opt}) = \varphi(x_k^{opt}|y_k). \tag{2.32}$$

The formulas (2.30)–(2.32) represent a recursive application of data assimilation. Consequently, data assimilation is an algorithm that incorporates PDF information from previous and current observations, as well as from the past forecast PDFs. As a result, the information gathered results in an improved forecast and analysis of the PDFs.

## 2.7 Uncertainty and Probability: Shannon Entropy

Consider a proposition $X = (X_1, X_2, \ldots, X_N)$ with assigned probabilities $p_i = P(X_i)$. We are interested in finding a measure of the amount of information a random variable $X$ produces, i.e., knowing how much information is gained by observing each event $X_i$. There are several intuitive conditions of such *information gain* function $I(p_i)$:

1. *Information gain is monotone and inversely proportional to probability*:
   An increase in probability implies a decrease of possible additional information, since for large probability there is a reduced communication of information. A system with a high probability of occurrence can gain little new information, and vice versa.
2. *Information gain is nonnegative*: $I(p_i) \geq 0$:
   If nothing is observed, the information stays the same, i.e., the information gain is zero. On the other hand, if new observations are made, they can only increase the information of the system.
3. *Events with total certainty do not accept new information*: $I(1) = 0$.
4. *Information gain for two independent events is additive*: $I(p_i p_j) = I(p_i) + I(p_j)$.

One such information gain function can be $I(p_i) = -\log p_i$, also referred to as the *log-likelihood* function. The information entropy $H$ is defined as the mathematical expectation (i.e., the average) of the information gain function (Shannon and Weaver, 1949)

$$H_E(X) = -\sum_{i=1}^{N} P(X_i) \log [P(X_i)]. \tag{2.33}$$

Typical bases of the logarithm in (2.33) include 2, Euler's number $e$, and 10. We will use base $e$, in which case the logarithm becomes the natural logarithm $ln$ and the units are called *natural units* or *nats*. Entropy also represents a measure of uncertainty as larger information gain can be achieved only if there is a large uncertainty. On the other hand, a small information gain can be associated with relatively predictable events, i.e., events with small uncertainty. It can be shown that the information gain function and entropy are unique solutions to the problem that satisfies the above conditions, up to a multiplicative constant. One can also introduce conditional and joint entropy, in agreement with conditional and joint probabilities. Given random variables $X$ and $Y$ with joint probability $P(X, Y)$, one can also define joint entropy

$$H_E(X, Y) = -\sum_{i=1}^{N} P(X, Y) \ln[P(X, Y)]. \tag{2.34}$$

The conditional entropy can be defined from marginal and joint entropy

$$H_E(X|Y) = H_E(X, Y) - H_E(Y). \tag{2.35}$$

**Example 2.1 Marginal entropy for a Gaussian random variable**

Additional details of the presented derivation could be found in McEliece (1977). For marginal entropy we use the formulation (2.33). The marginal Gaussian PDF is

$$\varphi(x) = \frac{1}{\sigma\sqrt{2\pi}} \exp\left[-\frac{1}{2}\left(\frac{x-\mu}{\sigma}\right)^2\right]. \tag{2.36}$$

Substituting (2.36) in (2.33)

$$H_E(x) = -\int_{-\infty}^{+\infty} \left[\frac{1}{\sigma\sqrt{2\pi}} \exp\left[-\frac{1}{2}\left(\frac{x-\mu}{\sigma}\right)^2\right]\right] \ln\left[\frac{1}{\sigma\sqrt{2\pi}} \exp\left[-\frac{1}{2}\left(\frac{x-\mu}{\sigma}\right)^2\right]\right] dx.$$

After using the rules for logarithmic functions, this integral reduces to

$$\begin{aligned} H_E(x) = {} & \frac{\ln(\sigma\sqrt{2\pi})}{\sigma\sqrt{2\pi}} \int_{-\infty}^{+\infty} \exp\left[-\frac{1}{2}\left(\frac{x-\mu}{\sigma}\right)^2\right] dx \\ & + \frac{1}{\sigma\sqrt{2\pi}} \int_{-\infty}^{+\infty} \frac{1}{2}\left(\frac{x-\mu}{\sigma}\right)^2 \exp\left[-\frac{1}{2}\left(\frac{x-\mu}{\sigma}\right)^2\right] dx. \end{aligned} \tag{2.37}$$

For the first integral in (2.37) we use the solution of a general Gaussian integral

$$\int_{-\infty}^{+\infty} \exp[-a(x+b)^2]\,dx = \sqrt{\frac{\pi}{a}}$$

to obtain

$$\frac{\ln(\sigma\sqrt{2\pi})}{\sigma\sqrt{2\pi}} \int_{-\infty}^{+\infty} \exp\left[-\frac{1}{2}\left(\frac{x-\mu}{\sigma}\right)^2\right] dx = \ln(\sigma\sqrt{2\pi}). \tag{2.38}$$

For the second integral in (2.37) we can use integration by parts by noticing that

$$\frac{1}{2}\left(\frac{x-\mu}{\sigma}\right)^2 \exp\left[-\frac{1}{2}\left(\frac{x-\mu}{\sigma}\right)^2\right] dx = \left[-\frac{1}{2}(x-\mu)\right] d\left[\exp\left[-\frac{1}{2}\left(\frac{x-\mu}{\sigma}\right)^2\right]\right]$$

which produces

$$\frac{1}{\sigma\sqrt{2\pi}} \int_{-\infty}^{+\infty} \frac{1}{2}\left(\frac{x-\mu}{\sigma}\right)^2 \exp\left[-\frac{1}{2}\left(\frac{x-\mu}{\sigma}\right)^2\right] dx = \frac{1}{2}. \tag{2.39}$$

Finally, (2.38) and (2.39) give the marginal entropy of X

$$H_E(X) = \frac{1}{2}\ln[2\pi\sigma^2] + \frac{1}{2}. \tag{2.40}$$

**Example 2.2 Joint entropy for a Gaussian random variable**

The joint entropy can be calculated from (2.34). It requires the joint Gaussian PDF for variables $X_1 \sim N(\mu_1, \sigma_1)$ and $X_2 \sim N(\mu_2, \sigma_2)$

$$\varphi(x_1, x_2) = \frac{1}{\sqrt{2\pi(\sigma_1^2 + \sigma_2^2)}} \exp\left[-\frac{1}{2}\frac{[x-(\mu_1+\mu_2)]^2}{\sigma_1^2+\sigma_2^2}\right]. \tag{2.41}$$

Following the same reasoning as for the marginal entropy calculation, the joint entropy is

$$H_E(X_1, X_2) = \frac{1}{2}\ln[2\pi(\sigma_1^2+\sigma_2^2)] + \frac{1}{2}. \tag{2.42}$$

**Example 2.3 Conditional entropy for a Gaussian random variable**

Using (2.35) one can utilize (2.40) and (2.42) to obtain conditional entropy

$$H_E(X_1|X_2) = \frac{1}{2}\ln[2\pi(\sigma_1^2+\sigma_2^2)] + \frac{1}{2} - \left[\frac{1}{2}\ln[2\pi\sigma_2^2] + \frac{1}{2}\right] = \frac{1}{2}\ln\left[1 + \frac{\sigma_1^2}{\sigma_2^2}\right]. \tag{2.43}$$

It is also of interest to define mutual information, the amount of common information shared between variables $X$ and $Y$,

$$I_E(X;Y) = \sum_X \sum_Y P(X,Y) \ln\left[\frac{P(X,Y)}{P(X)P(Y)}\right]. \tag{2.44}$$

Mutual information is strictly nonnegative, since the shared information between two variables can be only 0 or positive. Mutual information can be viewed as a generalization of the correlation coefficient. For example, if the joint variable $(X,Y)$ has a bivariate normal distribution, $I_E(X;Y) = -0.5\ln(1-\rho^2)$ , where $\rho$ is the correlation coefficient.

**Example 2.4 Mutual information for Gaussian variables**

One needs to apply the formula (2.44). The marginal entropy for random variables $X_1 \sim N(\mu_1, \sigma_1)$ and $X_2 \sim N(\mu_2, \sigma_2)$ are

$$H_E(X_1) = \frac{1}{2}\ln[2\pi\sigma_1^2] + \frac{1}{2} \qquad H_E(X_2) = \frac{1}{2}\ln[2\pi\sigma_2^2] + \frac{1}{2} \tag{2.45}$$

and the joint entropy is given by (2.42). The substitution in (2.44) gives

$$I_E(X_1;X_2) = \frac{1}{2}\ln[2\pi\sigma_1^2] + \frac{1}{2} + \frac{1}{2}\ln[2\pi\sigma_2^2] + \frac{1}{2} - \left[\frac{1}{2}\ln[2\pi(\sigma_1^2+\sigma_2^2)] + \frac{1}{2}\right]$$

and the mutual information is

$$I_E(X_1;X_2) = \frac{1}{2}\ln\left[2\pi\frac{\sigma_1^2\sigma_2^2}{\sigma_1^2+\sigma_2^2}\right] + \frac{1}{2}. \tag{2.46}$$

Note that the above definitione of entropies and mutual information are strictly valid for discrete random variables. For continuous random variables one can use probability density function $\varphi$ to define their counterparts

$$H_E(X) = -\int_X \varphi(x)\ln(\varphi(x))\,dx \tag{2.47}$$

$$H_E(X,Y) = -\int_X\int_Y \varphi(x,y)\ln(\varphi(x,y))\,dxdy \tag{2.48}$$

$$H_E(X|Y) = \int_X\int_Y \varphi(x,y)\ln\left(\frac{\varphi(y)}{\varphi(x,y)}\right)dxdy \tag{2.49}$$

$$I_E(X;Y) = \int_X\int_Y \varphi(x,y)\ln\left(\frac{\varphi(x,y)}{\varphi(x)\varphi(y)}\right)dxdy. \tag{2.50}$$

One can also define Bayes' rule for entropies, in a manner similar to probabilities, except that addition is used instead of multiplication:

$$H_E(X|Y) = H_E(X) + H_E(Y|X) - H_E(Y). \tag{2.51}$$

There are other useful entropy relationships:

1. $H_E(X,Y) = H_E(Y,X)$
2. $H_E(X,Y) \leq H_E(X) + H_E(Y)$
3. $I_E(X;Y) = H_E(X) - H_E(X|Y) = H_E(Y) - H_E(Y|X)$
4. $I_E(X;Y) = H_E(X) + H_E(Y) - H_E(X,Y)$.

In data assimilation, one is mostly interested in entropy (uncertainty) reduction due to observation $Y$ of an event $X$

$$\Delta H_E = H_E(X) - H_E(X|Y). \tag{2.52}$$

These various entropy-related quantities form information measures that can be used to assess the performance of data assimilation, as will be described for Gaussian random variables.

## 2.8 Data Assimilation with Gaussian Probability Density Functions

We now pose the main data assimilation problem in terms of probabilities, and suggest common solution approaches, such as the maximum a posteriori and minimum variance. Although it is generally known that some variables used in data assimilation are non-Gaussian, it is still common to assume Gaussian distribution. Since Gaussian distribution assumption greatly simplifies the mathematics, we will use it to illustrate data assimilation equations.

It was indicated earlier that recursive Bayes' formulation (2.30)–(2.32) is used in data assimilation. In order for the prior and conditional PDFs to be defined, however, the variables $x$ and $y$ have to be defined first. "Data assimilation is addressing a difference between variables, or errors, not actual variables". In practical applications the variables $x$ and $y$ are multidimensional and are expressed as vectors $\mathbf{x}$ and $\mathbf{y}$, respectively. Continuing with vector formulation, if $\mathbf{x}_g$ and $\mathbf{y}_g$ denote guess values of the variables $\mathbf{x}$ and $\mathbf{y}$, respectively, then in Gaussian data assimilation we consider differences $\mathbf{x}-\mathbf{x}_g$ and $\mathbf{y}-\mathbf{y}_g$. In this notation we define $\mathbf{x}$ to be a state variable and $\mathbf{y} = \mathbf{y}^o$ an observation variable. Although guess values can be any physically meaningful value, there is a benefit if the guess values are close to the (unknown) true value. For example, the guess $\mathbf{x}_g$ can be a climatological average (denoted $\mathbf{x}^c$) or a short-term forecast (denoted $\mathbf{x}^f$), among many other possibilities. In data assimilation, a short-term forecast is most often used as a good approximation of the truth. This, of course, depends on the quality of the model used for prediction, and may not be the best choice in some applications. For the observation guess, a transformed state vector is commonly used as a guess, i.e., $\mathbf{y}^o_g = H(\mathbf{x})$, where $H$ is a nonlinear mapping from state space to observation space. Sometimes a linear mapping $\mathbf{H} = \frac{\partial H}{\partial \mathbf{x}}$ is used instead, i.e. $\mathbf{y}^o = \mathbf{H}\mathbf{x}$. The mappings $H$ and $\mathbf{H}$ are also referred to as *observation operators*. Therefore, we can define Gaussian random variables

$$\boldsymbol{\varepsilon} = \mathbf{x} - \mathbf{x}_g \quad \boldsymbol{\gamma} = \mathbf{y}^o - \mathbf{y}^o_g. \tag{2.53}$$

Another common assumption that further simplifies the problem is that such variables have zero expectation which implies $\boldsymbol{\varepsilon} \sim N(0, \mathbf{P})$ and $\boldsymbol{\gamma} \sim N(0, \mathbf{R})$, where $\mathbf{P}$ and $\mathbf{R}$ are state and observation error covariances, respectively. The multivariate covariances can be defined from PDF moments as

$$\mathbf{P} = E(\varepsilon\varepsilon^T) \quad \mathbf{R} = E(\gamma\gamma^T), \tag{2.54}$$

where $E$ denotes the expectation operator.

Although we will discuss specifics data assimilation algorithms in Chapters 7 and 8, it is worth mentioning here that some data assimilation algorithms, such as *particle filters* (PFs), do not require an assumption about the type of PDFs used in Bayes' formula. Most data assimilation algorithms, however, require explicit assumptions about the PDF. In Gaussian data assimilation, therefore, we assume that the prior and conditional PDF in (2.30)–(2.32) are all multivariate Gaussian. The prior PDF is

$$\varphi(\mathbf{x}) = \frac{1}{\sqrt{(2\pi)^N \det(\mathbf{P})}} e^{-\frac{1}{2}(\mathbf{x}-\mathbf{x}^f)^T \mathbf{P}^{-1}(\mathbf{x}-\mathbf{x}^f)} \tag{2.55}$$

and the conditional PDF is

$$\varphi(\mathbf{y}^o|\mathbf{x}) = \frac{1}{\sqrt{(2\pi)^N \det(\mathbf{R})}} e^{-\frac{1}{2}(\mathbf{y}^o - H(\mathbf{x}))^T \mathbf{R}^{-1}(\mathbf{y}^o - H(\mathbf{x}))}. \tag{2.56}$$

A note of caution is to remember that the above formulation has an implicit assumption of zero mean, i.e., the values $\mathbf{x}^f$ and $H(\mathbf{x})$ are guess values, not mean values of $\mathbf{x}$ and $\mathbf{y}^o$, respectively. The use of Bayes' formula further implies that the prior and conditional PDFs are independent, which for Gaussian PDFs means that the random variables $\mathbf{x} - \mathbf{x}^f$ and $\mathbf{y}^o - H(\mathbf{x})$ are uncorrelated.

The two most common solution methods for data assimilation are the minimum variance and maximum a posteriori estimates. In the *minimum variance method*, one first defines a loss function $L$, such that it is real, positive semidefinite, and $L(0) = 0$. The most common example of the loss function is a quadratic form $L(\mathbf{w}) = \mathbf{w}^T\mathbf{S}\mathbf{w}$, where $\mathbf{w}$ is an $N$-dimensional vector and $\mathbf{S}$ is a symmetric positive semidefinite matrix. The minimum variance method minimizes the expected, or average loss

$$\mathbf{x}_{opt} = \underset{\mathbf{x}}{\text{argmin}} E[L(\mathbf{x} - \mathbf{x}^f)], \tag{2.57}$$

and it can be shown that the optimal minimum variance estimate is the conditional mean $E(\mathbf{x}|\mathbf{y}^o)$ (e.g., Jazwinski, 1970). The *argmin* denotes the values at which the function is minimized. The minimization of the expectation of the loss function is only implicit in the method, and the estimate conditional mean is directly used as a solution. In the *maximum a posteriori* (*MAP*) *method*, one defines the optimal solution as the one that maximizes the posterior PDF in (2.23)

$$\mathbf{x}_{opt} = \underset{\mathbf{x}}{\text{argmax}} \left[\varphi(\mathbf{x}|\mathbf{y}^o)\right]. \tag{2.58}$$

The *argmax* denotes the values at which the function is maximized. The maximization of the function is never actually applied. Instead, the problem (2.58) is transformed into a minimization problem, which is then explicitly solved. This transformation is typically defined as a negative log-likelihood function, which transforms Bayes formula into

$$f(\mathbf{x}) = -\ln \varphi(\mathbf{x}|\mathbf{y}^o) = -\ln \varphi(\mathbf{x}) - \ln \varphi(\mathbf{y}^o|\mathbf{x}) + \ln \varphi(\mathbf{y}^o). \tag{2.59}$$

By applying a negative logarithm, we transform the maximization to minimization

$$\mathbf{x}_{opt} = \underset{\mathbf{x}}{\operatorname{argmax}} \left[\varphi(x|y)\right] = \underset{x}{\operatorname{argmin}} \left[f(x)\right]. \tag{2.60}$$

In (2.50) $f$ denotes a function of $\mathbf{x}$ only, as the observations $\mathbf{y}^o$ are the input to the system and are not the subject of *argmax* estimation in (2.58). After substituting (2.55) and (2.56), and using the properties of logarithmic and exponential functions we obtain

$$f(x) = \frac{1}{2}\left[\mathbf{x} - \mathbf{x}^f\right]^T \mathbf{P}^{-1}\left[\mathbf{x} - \mathbf{x}^f\right] + \frac{1}{2}\left[\mathbf{y}^o - H(\mathbf{x})\right]^T \mathbf{R}^{-1}\left[\mathbf{y}^o - H(\mathbf{x})\right] - c_x - c_y - \ln \varphi(\mathbf{y}^o), \tag{2.61}$$

where $c_x = \ln \sqrt{(2\pi)^N \det(\mathbf{P})}$, $c_y = \ln \sqrt{(2\pi)^N \det(\mathbf{R})}$, and $\ln \varphi(\mathbf{y}^o)$ are not functions of $\mathbf{x}$, i.e., effectively do not alter the solution of minimization problem (2.60). Therefore, we define the so-called *cost function*, $f(\mathbf{x})$:

$$f(\mathbf{x}) = \frac{1}{2}\left[\mathbf{x} - \mathbf{x}^f\right]^T \mathbf{P}^{-1}\left[\mathbf{x} - \mathbf{x}^f\right] + \frac{1}{2}\left[\mathbf{y}^o - H(\mathbf{x})\right]^T \mathbf{R}^{-1}\left[\mathbf{y}^o - H(\mathbf{x})\right], \tag{2.62}$$

and solve the minimization problem equivalent to (2.60)

$$\mathbf{x}_{opt} = \underset{\mathbf{x}}{\operatorname{argmax}} \left[\varphi(x|y)\right] = \underset{\mathbf{x}}{\operatorname{argmin}} \left[f(\mathbf{x})\right]. \tag{2.63}$$

The two solution methods, minimum variance and maximum a posteriori, are widely used in data assimilation. One should be aware that in the Gaussian PDF framework, both solutions should produce the same estimate, as the mean and mode are identical for a Gaussian PDF. In practice, however, this is generally not the case, and even the algorithms within each group can produce very different results. For non-Gaussian PDFs, these two solution methods already differ from a theoretical point of view. As noted in Figure 2.3, for example, the lognormal PDF has different values for the mode (maximum a posteriori) and the mean (minimum variance). Finding an acceptable solution in such cases is even more difficult, which explains why non-Gaussian data assimilation is not widely used in today's data assimilation practice. However, a majority of state variables (e.g., their errors) is non-Gaussian, such as atmospheric chemistry variables, and cloud and moisture variables. This suggests that non-Gaussian data assimilation is needed and may become a mainstream approach in the future.

## 2.9 Overview

Probabilistic data assimilation is introduced through notions of probability, Bayes' formula, and entropy-based information measures. These topics are fundamental for understanding data assimilation, and for eventually being able to improve data assimilation by changing its often-hidden assumptions. One should also be aware that the material presented in this chapter is theoretical. As will be seen in Chapters 7–9 and 12, there are numerous technical difficulties associated with changing basic data assimilation assumptions in practice. This, however, should only be viewed as a nuisance in further development of data assimilation and not an accepted limitation.

Further elaborating on the meaning and implications of using probability in data assimilation, there are at least two different interpretations of probability that are worth mentioning. The most common interpretation is that probability is a physical quantity that has a measurable objective value. Effectively, probability is equated with statistical frequency distribution, and the process of finding probability is the process of collecting random samples organized in the form of frequency distribution. According to this interpretation there is only one, unique probability of an event. This interpretation is closely related to the standard statistical view. Another interpretation is that probability represents the state of knowledge, or the state of information about an event. Since different people can have different knowledge about the nature of an event and the conditions impacting this event, there may be multiple probabilities associated with a single event. Of course, if the knowledge about the event is identical, the resulting probability will be the same. This interpretation is closely related to the use of Bayes' formula as a logical inference. The two probability interpretations do not need to be mutually disqualifying, especially from the practitioner's point of view. In theory and practice of data assimilation there is often a mixture of the two probability interpretations.

If one is interpreting probability in terms of knowledge (information), then Bayes' formula gives a "smart" logical system that learns from previous experiences. This learning process is very appealing since it gives an automated way of improving a system, and is widely used in data assimilation, machine learning, and related applications.

COLLOQUY 2.1

**Derivation of a Gaussian PDF without assuming randomness**

While the original derivation of normal PDFs by Gauss (1809) relied on assuming random vectors, an alternative derivation without assuming randomness is also possible. An alternative approach to normal probability distribution was originally proposed by Quetelet (1850) for 2D problem, extended by Maxwell (1860) to a 3D problem. This alternative derivation is interesting since it offers a different view of the Gaussian probability distribution based on geometric arguments. We generally follow Quetelet's derivation presented by Jaynes (2003).

The problem Herschel addressed was to find the 2D probability distribution of errors in measuring the position of a star. If $x$ and $y$ denote the errors in two Cartesian coordinate directions, then the problem was finding a function representing the joint probability density $\varphi(x, y)$. Two intuitive geometric postulates that Herschel made were: (P1) probabilities of errors are independent (knowledge of $x$ tells nothing about $y$), and (P2) probability of errors is isotropic (independent of angle).

From the first postulate (P1) one can write

$$\varphi(x,y)dxdy = f(x)dx \cdot f(y)dy, \tag{2.64}$$

where $f$ represents the marginal probability. In order to better address the angle requirement of the second postulate (P2) it is convenient to represent the above probability in polar coordinates

$$x = r\cos(\theta)$$
$$y = r\sin(\theta),$$

where $r$ is the radius and $\theta$ is the angle. From the above relationship between Cartesian and polar coordinates one can form the coordinate transformation

$$\begin{pmatrix} dx \\ dy \end{pmatrix} = \begin{pmatrix} \cos(\theta) & -r\sin(\theta) \\ \sin(\theta) & r\cos(\theta) \end{pmatrix} \begin{pmatrix} dr \\ d\theta \end{pmatrix},$$

where matrix

$$\mathbf{J} = \begin{pmatrix} \cos(\theta) & -r\sin(\theta) \\ \sin(\theta) & r\cos(\theta) \end{pmatrix}$$

is the transformation Jacobian matrix. Since an infinitesimal area transformation between coordinates is represented by $dxdy = det(\mathbf{J})drd\theta$, and since $det(\mathbf{J}) = r\cos(\theta)^2 + r\sin(\theta)^2 = r$, one has

$$dxdy = rdrd\theta. \tag{2.65}$$

Using (2.65), the joint probability density in (2.64) can be written in polar coordinates

$$\varphi(x,y)dxdy = g(r,\theta)rdrd\theta. \tag{2.66}$$

The second postulate (P2) implies an independence of the angle

$$g(r,\theta) = g(r), \tag{2.67}$$

so that (2.66) becomes

$$\varphi(x,y)dxdy = g(r)rdrd\theta. \tag{2.68}$$

After equating the right-hand side of (2.64) and (2.68)

$$f(x)f(y)dxdy = g(r)rdrd\theta. \tag{2.69}$$

Making use of (2.65) and since $r = \sqrt{x^2+y^2}$, the functional equation to solve is

$$f(x)f(y) = g\left(\sqrt{x^2+y^2}\right). \tag{2.70}$$

For $y = 0$ the last equation becomes

$$f(x)f(0) = g(x), \tag{2.71}$$

and after substituting $x$ by $\sqrt{x^2+y^2}$ in (2.71)

$$f\left(\sqrt{x^2+y^2}\right) f(0) = g\left(\sqrt{x^2+y^2}\right). \tag{2.72}$$

Finally, by combining (2.72) and (2.70), and additionally dividing both sides by $f(0)^2$ (assuming $f(0) \neq 0$)

$$\frac{f(x)}{f(0)} \cdot \frac{f(y)}{f(0)} = \frac{f\left(\sqrt{x^2+y^2}\right)}{f(0)}. \tag{2.73}$$

After taking the logarithm of (2.73)

$$\ln \frac{f(x)}{f(0)} + \ln \frac{f(y)}{f(0)} = \ln \frac{f\left(\sqrt{x^2+y^2}\right)}{f(0)}.$$

The unique solution to the above equation is $\ln \frac{f(x)}{f(0)} = ax^2$ which implies

$$f(x) = f(0) \exp(ax^2).$$

In order for the above function $f(x)$ to be defined everywhere, including $+\infty$ and $-\infty$, the coefficient $a$ has to be negative, i.e., $a = -\alpha$ where $\alpha > 0$, so that

$$f(x) = f(0) \exp(-\alpha x^2). \tag{2.74}$$

It is obvious that $f(x)$ is now well defined as $f(+\infty) = f(-\infty) = 0$. The unknown value of the function at $x = 0$ can be obtained using the probability normalization. In order for function $f(x)$ to represent the probability density it has to satisfy

$$\int_{-\infty}^{+\infty} f(x)dx = 1.$$

Therefore, taking an integral of both sides of (2.74) and noting that $f(0)$ is independent of $x$

$$f(0) \int_{-\infty}^{+\infty} \exp(-\alpha x^2)dx = 1. \tag{2.75}$$

Given that the solution of a general Gaussian integral (see Example 2.1) is

$$\int_{-\infty}^{+\infty} \exp(-\alpha x^2)dx = \sqrt{\frac{\pi}{\alpha}}$$

it follows from (2.75) that

$$f(0) = \sqrt{\frac{\alpha}{\pi}},$$

so that the normalized probability density function is

$$f(x) = \sqrt{\frac{\alpha}{\pi}} \exp(-\alpha x^2). \tag{2.76}$$

Equation (2.76) represents the unique solution subject to postulates (P1) and (P2) that Quetelet (1850) was able to obtain. Note that for $\alpha = \frac{1}{2}$ the above function represents the probability density of normal distribution $N(0, 1)$. It may be also interesting to know that arguments similar to the above Quetelet's derivation were used by Maxwell (1860) to find the probability distribution for velocities of molecules in a gas, and by Einstein (1905) to deduce the Lorentz transformation law from his two postulates of special relativity theory, as noted by Jaynes (2003).

In overview, the above derivation illustrates that even such fundamental probability as the Gaussian is, can be uniquely determined without the need to define a random variable or process. This is one of the arguments that Jaynes (2003) makes in his interpretation of probability as a measure of the *information content* or *knowledge*.

## References

Einstein A (1905) Zur elektrodynamik bewegter körper. *Ann Phys* 322:891–921, English translation On the electrodynamics of moving bodies by Perrett W, Jeffery GB, in *The Principle of Relativity*, (eds.) Lorentz HA, Einstein A, Minkowski H, Weyl H, published by Methuen and Company, Ltd. of London in 1923, reprinted in 1952 by Dover Publications, Inc., New York. www.fourmilab.ch/etexts/einstein/specrel/

Fletcher SJ, Županski M (2006) A hybrid multivariate normal and lognormal distribution for data assimilation. *Atmos Sci Lett* 7:43–46.

Gauss CF (1809) *Theoria Motus Corporum Coelestium in Sectionibus Conicis Solem Ambientium (Theory of the Motion of Heavenly Bodies Moving about the Sun in Conic Section)*. English translation by CH Davis, published by Little Brown, and Co. of Boston in 1857, reprinted in 1963 by Dover Publications, Inc., New York, 374 pp.

Jaynes ET (2003) *Probability Theory: The Logic of Science*. Cambridge University Press, New York, 753 pp.

Jazwinski AH (1970) *Stochastic Processes and Filtering Theory*. Academic Press, New York, 376 pp.

Maxwell JC (1860) V. Illustrations of the dynamical theory of gases – Part I. On the motions and collisions of perfectly elastic spheres. *London Edinburgh Dublin Philos Mag J Sci* 19:19–32.

McEliece RJ (1977) *The Theory of Information and Coding: A Mathematical Framework for Communication*. Addison-Wesley, London, Amsterdam, Don Mills, Ontario, Sydney, Tokyo, 302 pp.

Quetelet MA (1850) Letters addressed to H.R.H. the Grand Duke of Saxe-Cobourg and Gotha on the Theory of Probabilities as applied to the Moral and Political Sciences. *Edinburgh Rev* 92(185):1–57, https://hdl.handle.net/2027/mdp.39015027446098, Translated from the French by Downes OG. Collected by Herschel J under a title Probabilities, from the Edinburgh Review, July, 1850, in

*Essays from the Edinburgh and Quarterly Reviews: With Addresses and Other Pieces*, pp. 365–465, published by Longman, Brown, Green, Longmans, & Roberts of London in 1857, reprinted in 2014 by Cambridge University Press, Cambridge. `http://echo.mpiwg-berlin.mpg.de/MPIWG:DWUE9CMM`

Shannon CE, Weaver W (1949) *The Mathematical Theory of Communication*. University of Illinois Press, Chicago, IL, 117 pp.

# 3
# Filters and Smoothers

## 3.1 Introduction

In data assimilation, we deal with dynamical models, available observations, and analysis. The observations are generally sparser than the model states in both space and time; that is, the dimension of observations, $N$, is much smaller than that of states, $K$ (i.e., $N \ll K$). Thus, data assimilation includes estimation of the time-varying model states using insufficient observations. By combining the model state ($\mathbf{x}$) and the observation ($\mathbf{y}^o$), the estimation problem attempts to obtain an improved estimate ($\hat{\mathbf{x}}$), or analysis ($\mathbf{x}^a$). We classify the estimation task as either *filtering*, *smoothing*, or *prediction*, depending on when the estimation and the observation incorporation are made.

Figure 3.1 illustrates the concept of such classification. The *filtering* problem aims to provide the best estimate of the model state ($\hat{\mathbf{x}}_{k=N}$) at the analysis time ($t_k = t_N$) using the latest observation ($\mathbf{y}_N$) as well as previous observations ($\mathbf{y}_{n<N}$). At the next analysis time when new data and forecasts are provided (e.g., at $t = t_{2N}$), the filtering reproduces an improved estimate of the state (i.e., $\hat{\mathbf{x}}_{k=2N}$). That is, a filter treats observations sequentially in time. In contrast, at any analysis time $t_k$ prior to $t_N$, the *smoothing* problem combines all data – past, present and future – to yield an optimal estimate, $\hat{\mathbf{x}}_{k<N}$; that is, a smoother treats observations all together in a smoothing period (i.e., assimilation window), essentially requiring backward propagation of information. The optimal estimate through the smoothing can be obtained by giving more weight to observations near the desired analysis time, $t_k$, than those distant from $t_k$. The *prediction* problem seeks to get a future estimate of the state ($\hat{\mathbf{x}}_{k>N}$) at a forecast time $t_k$ beyond $t_N$ by integrating the optimal state vector, estimated at or prior to $t_N$, forward in time.

The filtering process, in a sense, produces a "nowcast" by combining past and present data (Müller and von Storch, 2004), which in turn can be used as an initial condition (IC) for a new forecast. New data arrive continuously; thus, filtering usually takes a *sequential* approach. The smoothing procedure usually requires backward integration from $t_N$ to $t_k < t_N$; thus, a smoother is also called a backward filter, and it produces an optimal estimate that is dynamically consistent. Smoothing is often called a *batch* process as all measurements in the smoothing period are processed simultaneously (McLaughlin, 2002). As smoothing incorporates more observations than filtering does, the accuracy of a smoother is generally better than that of a filter.

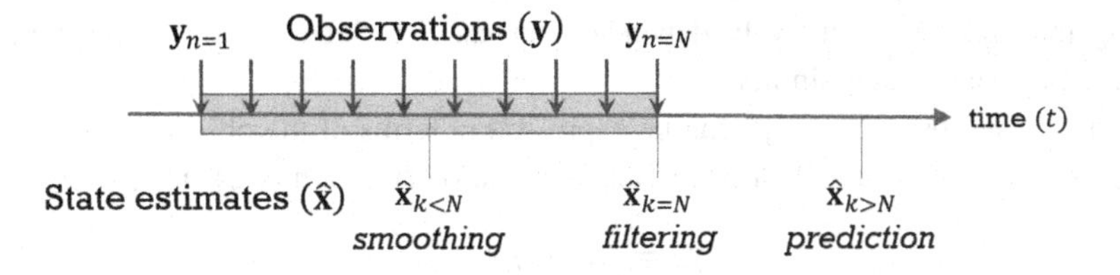


Figure 3.1 Estimation of the model state ($\hat{\mathbf{x}}_k$) at $t = t_k$ through filtering ($\hat{\mathbf{x}}_{k=N}$), smoothing ($\hat{\mathbf{x}}_{k<N}$), and prediction ($\hat{\mathbf{x}}_{k>N}$).

## 3.2 Discrete Model and Observations

In Chapter 1, we introduced *static* data assimilation, in which we sought a unique estimate in the space domain with observations in a fixed time or a very short time interval. In real-time data assimilation, observations are collected over time and analyses are made accordingly, thus making data assimilation cycled sequentially (see Figure 1.4 and Algorithm 1.1) – typically a 6-h cycle for a global prediction. We now discuss *dynamic* data assimilation where observations are distributed over both space and time and a series of model states are estimated in a dynamical prediction system.

A dynamical model is expressed, for the discrete-time index $n$, as

$$\mathbf{x}_{n+1} = M_n (\mathbf{x}_n ; \boldsymbol{\alpha}) + \boldsymbol{\varepsilon}_n^m, \tag{3.1}$$

where $M$ represents the nonlinear model (NLM) or propagator, $\mathbf{x}$ the model state, and $\boldsymbol{\alpha}$ a set of parameters. The error term $\boldsymbol{\varepsilon}_n^m$ represents inherent errors involved in the state $\mathbf{x}$ through the uncertainties, in the model equations, initial and boundary conditions, external forcing fields, parameterizations of subgrid-scale processes, discretization in both space and time, etc.

Equation (3.1) shows that the evolution of model states depends on not only the state's IC $\mathbf{x}_0$ but also the parameters $\boldsymbol{\alpha}$, which are specified either empirically or theoretically with some uncertainties. Both the states and parameters can be estimated simultaneously in the data assimilation framework (e.g., Aksoy et al., 2006; Bocquet and Sakov, 2013; Ruckstuhl and Janjić, 2020); the optimal parameter values are also obtained independently via *parameter estimation* (e.g., Severijns and Hazeleger, 2005; Lee et al., 2006; Storch et al., 2007). For simplicity, we disregard $\boldsymbol{\alpha}$ here; then, the dynamical model becomes

$$\mathbf{x}_{n+1} = M_n (\mathbf{x}_n) + \boldsymbol{\varepsilon}_n^m. \tag{3.2}$$

Note that (3.2) also represents a *recursive* system, in which output at any time, say, $\mathbf{x}_n$, depends on previous output(s), say, $\mathbf{x}_{n-1}$. In a recursive system, the model output at any time depends on the IC $\mathbf{x}_0$ through the recursive evolution by the model propagator $M$; that is, by neglecting $\boldsymbol{\varepsilon}^m$, (3.2) becomes

$$\mathbf{x}_{n+1} = M_n \overbrace{M_{n-1} \cdots M_1 \underbrace{M_0 (\mathbf{x}_0)}_{\mathbf{x}_1}}^{\mathbf{x}_n},$$

showing that $\mathbf{x}_{n+1}$ is recursively dependent on $\mathbf{x}_0$. See Section 2.6 for the concept of recursive Bayesian data assimilation.

The available observations $\mathbf{y}^o$ can be expressed in terms of the observation operator $H$ and the observation error $\boldsymbol{\varepsilon}^o$, following the observation equation (1.2), as

$$\mathbf{y}_n^o = H_n(\mathbf{x}_n) + \boldsymbol{\varepsilon}_n^o. \tag{3.3}$$

Here, $H_n$ is a nonlinear operator that transforms the state vector $\mathbf{x}_n$ from the model space to the observation space, and $\boldsymbol{\varepsilon}_n^o$ represents the errors due to instrument, sampling, etc. More details of the observation operator $H$ are described in Section 1.3.3.

## 3.3 Filters

Filters are sequential estimation algorithms that obtain an estimate of state ($\hat{\mathbf{x}}$) or an analysis ($\mathbf{x}^a$) at the analysis time ($t_{n=N_i}$) using past and present observations ($\mathbf{y}^o_{n\leq N_i}$), as depicted in Figure 3.2. The estimation process by filters is typically divided into two steps:

1. a **forecast** (or **prediction** or **propagation**) **step** that calculates changes in state estimates between observation times, and
2. an **analysis** (or **correction** or **update**) **step** that adjusts state estimates to account for information contained in a new observation.

The general analysis equation (1.11) shows that the analysis $\mathbf{x}^a$ is affected by the background error covariance via the weight matrix $\mathbf{W}$ as well as the background field $\mathbf{x}^b$. As the background fields are obtained through model forecasts based on the previous analysis, i.e., $\mathbf{x}_{n+1}^b = M_n(\mathbf{x}_n^a)$, the background errors are characterized by previous errors in both model and analysis, evolving through the analysis times. We discuss various filters below with time-evolving background errors in a recursive dynamic system.

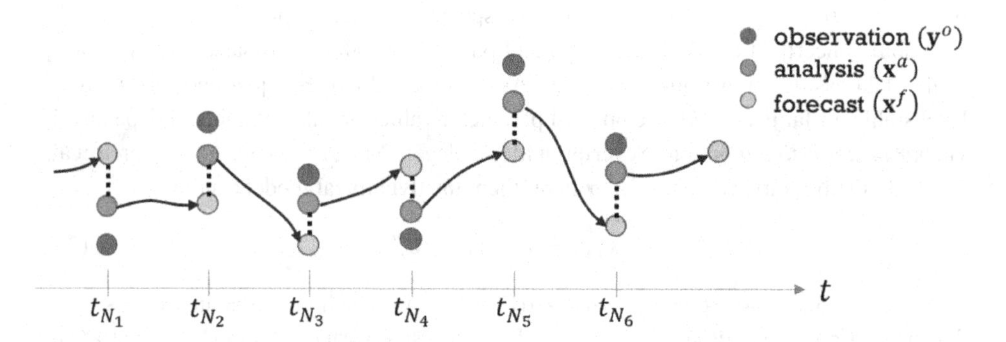


Figure 3.2 A filter obtains the analysis fields ($\mathbf{x}^a$) at the analysis times $t_{N_i}$ where the latest observations are provided. The dotted lines represent corrections to the previous forecasts ($\mathbf{x}^f$) using information from the observations ($\mathbf{y}^o$) to yield $\mathbf{x}^a$. Solid arrows indicate the model state trajectories starting from $\mathbf{x}^a$ to yield $\mathbf{x}^f$.

### 3.3.1 Kalman Filter

The Kalman filter (KF), following the seminal work by Rudolf E. Kalman (Kalman, 1960), is a dynamical extension of the static data assimilation represented by the BLUE with background – i.e., a minimum variance estimate (see Section 1.4.4). Therefore, it follows the same assumptions as in BLUE, including

1. *Background and forecast fields*: At the initial time ($n = 0$), the true state $\mathbf{x}_0^t$ is approximated by the background $\mathbf{x}_0^b$, i.e., $E(\mathbf{x}_0^t) = \mathbf{x}_0^b$; the error $\boldsymbol{\varepsilon}_0^b$ has zero mean, i.e., $E(\boldsymbol{\varepsilon}_0^b) = \mathbf{0}$, and covariance $\mathbf{P}_0^b$. For $n \geq 1$, the model state and the error covariance are expressed as $\mathbf{x}_n^f$ and $\mathbf{P}_n^f$.
2. *Unbiased errors*: Errors of the model ($\boldsymbol{\varepsilon}_n^m$) and the observation ($\boldsymbol{\varepsilon}_n^o$) are both unbiased (i.e., zero mean or $E(\boldsymbol{\varepsilon}_n^m) = E(\boldsymbol{\varepsilon}_n^o) = \mathbf{0}$) and have error covariances $\mathbf{Q}_n$ and $\mathbf{R}_n$, respectively.
3. *Uncorrelated (white) errors*: All the errors are uncorrelated in time, that is,
$$E\left(\boldsymbol{\varepsilon}_k^m(\boldsymbol{\varepsilon}_l^m)^T\right) = E\left(\boldsymbol{\varepsilon}_k^o(\boldsymbol{\varepsilon}_l^o)^T\right) = \mathbf{0} \text{ for } k \neq l.$$
Furthermore, errors of different types are not correlated, i.e.,
$$E\left(\boldsymbol{\varepsilon}_n^m(\boldsymbol{\varepsilon}_l^o)^T\right) = E\left(\boldsymbol{\varepsilon}_n^m(\boldsymbol{\varepsilon}_0^b)^T\right) = E\left(\boldsymbol{\varepsilon}_n^o(\boldsymbol{\varepsilon}_0^b)^T\right) = \mathbf{0}.$$
4. *Linear operators*: Both the model operator (propagator) and the observation operator are assumed to be linear, i.e., $M_n = \mathbf{M}_n$ and $H_n = \mathbf{H}_n$.

In the Bayesian framework, the KF is optimal for a linear, Gaussian system, providing $E\left(\mathbf{x}_n|\mathbf{y}_n^o\right)$ and $cov\left(\mathbf{x}_n|\mathbf{y}_n^o\right)$ (Snyder, 2015). The errors $\boldsymbol{\varepsilon}_n^m$ and $\boldsymbol{\varepsilon}_n^o$ have Gaussian (normal) distribution, i.e., $\boldsymbol{\varepsilon}_n^m \sim N(\mathbf{0}, \mathbf{Q}_n)$ and $\boldsymbol{\varepsilon}_n^o \sim N(\mathbf{0}, \mathbf{R}_n)$ (see Section 2.4). From the Bayesian viewpoint, $\mathbf{x}_n^f (= \mathbf{x}_n^b)$ is an a priori estimate while $\mathbf{x}_n^a$ is an a posteriori estimate, represented as

$$\mathbf{x}_n^a = E\left(\mathbf{x}_n^t|\mathbf{y}_{1:n}^o\right), \tag{3.4}$$

where $\mathbf{y}_{1:n}^o = \left\{\mathbf{y}_1^o, \ldots, \mathbf{y}_n^o\right\}$. That is, in the KF, $\mathbf{x}_n^a$ is the expected value of $\mathbf{x}_n^t$ conditioned by all the observations up to the current analysis time $n$. In Section 3.5, we discuss Bayesian perspectives on filtering.

Under the assumptions mentioned above, the KF evolves both the states and the error covariances through the analysis and forecast steps (see Figure 3.3). The state evolves through the linear propagator, $\mathbf{M}_n$, as

$$\mathbf{x}_{n+1}^f = \mathbf{M}_n\mathbf{x}_n^a, \tag{3.5}$$

that is, the forecast field ($\mathbf{x}_{n+1}^f$) is obtained using the analysis field $\mathbf{x}_n^a$. The error covariance also changes with time as

$$\mathbf{P}_{n+1}^f = \mathbf{M}_n\mathbf{P}_n^a\mathbf{M}_n^T + \mathbf{Q}_n, \tag{3.6}$$

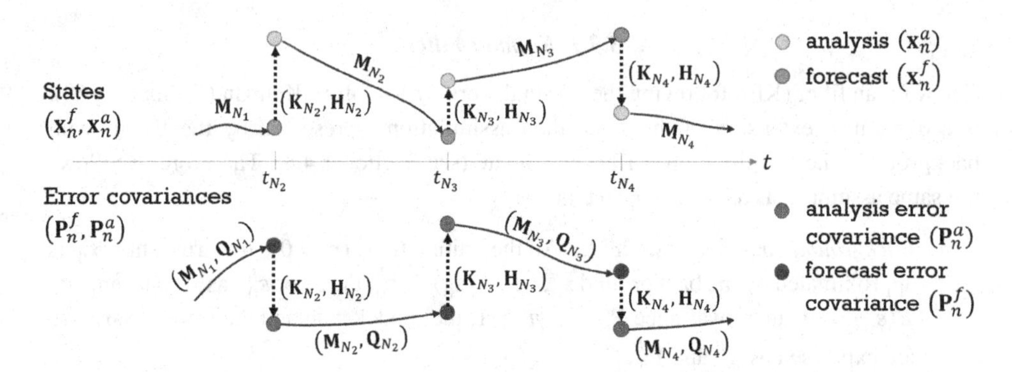


Figure 3.3 The discrete KF evolves the model states and the error covariances through the forecast steps (solid arrows) and the analysis steps (dotted arrows) at the analysis times $t_{n=N_i}$ where the latest observations are provided. Light circles represent the model states ($\mathbf{x}_n^a$ and $\mathbf{x}_n^f$) while dark circles depict the error covariances ($\mathbf{P}_n^a$ and $\mathbf{P}_n^f$).

where the forecast error covariance $\mathbf{P}_{n+1}^f$ is obtained using the analysis-error covariance $\mathbf{P}_n^a$ and the model error covariance $\mathbf{Q}_n$. Equations (3.5) and (3.6) constitute the *forecast step* of the discrete KF (see the solid arrows in Figure 3.3).

The analysis $\mathbf{x}_n^a$ and the analysis-error covariance $\mathbf{P}_n^a$ are obtained through the *analysis step*, shown as the dotted arrows in Figure 3.3, as the following:

$$\mathbf{x}_n^a = \mathbf{x}_n^f + \mathbf{K}_n \left(\mathbf{y}^o - \mathbf{H}\mathbf{x}^f\right)_n, \tag{3.7}$$

$$\mathbf{P}_n^a = (\mathbf{I} - (\mathbf{KH})_n)\,\mathbf{P}_n^f, \tag{3.8}$$

where $\mathbf{y}_n^o$ is the observation, $\mathbf{H}_n$ is the linear observation operator, and $\mathbf{x}_n^f$ and $\mathbf{P}_n^f$ are from the forecast step. The Kalman gain $\mathbf{K}_n$ is obtained by

$$\mathbf{K}_n = \left(\mathbf{P}^f\mathbf{H}^T\right)_n \left((\mathbf{H}\mathbf{P}^f\mathbf{H}^T)_n + \mathbf{R}_n\right)^{-1}, \tag{3.9}$$

where $\mathbf{R}_n$ is the observation error covariance. At the initial time ($n = 0$), $\mathbf{x}_0^f$ and $\mathbf{P}_0^f$ are replaced by their background correspondents, i.e., $\mathbf{x}_0^b$ and $\mathbf{P}_0^b$, respectively. The algorithmic view of the discrete KF is represented in Algorithm 3.1.

**Practice 3.1 Forecast error and its covariance in the KF**

Solve the following:

1. The forecast error $\boldsymbol{\varepsilon}_n^f$ and the analysis error $\boldsymbol{\varepsilon}_n^a$ are defined as

$$\boldsymbol{\varepsilon}_n^f = \mathbf{x}_n^f - \mathbf{x}_n \quad \text{and} \quad \boldsymbol{\varepsilon}_n^a = \mathbf{x}_n^a - \mathbf{x}_n,$$

respectively. Using (3.2) with $M_n = \mathbf{M}_n$, show that

$$\boldsymbol{\varepsilon}_{n+1}^f = \mathbf{M}_n \boldsymbol{\varepsilon}_n^a - \boldsymbol{\varepsilon}_{n+1}^m.$$

2. The error covariances of forecast ($\mathbf{P}_n^f$), analysis ($\mathbf{P}_n^a$), and model ($\mathbf{Q}_n$) are defined as

$$\mathbf{P}_n^f = E\left(\boldsymbol{\varepsilon}_n^f(\boldsymbol{\varepsilon}_n^f)^T\right), \quad \mathbf{P}_n^a = E\left(\boldsymbol{\varepsilon}_n^a(\boldsymbol{\varepsilon}_n^a)^T\right), \quad \text{and} \quad \mathbf{Q}_n = E\left(\boldsymbol{\varepsilon}_n^m(\boldsymbol{\varepsilon}_n^m)^T\right),$$

respectively. Using the result in #1, show that

$$\mathbf{P}_{n+1}^f = \mathbf{M}_n\mathbf{P}_n^a\mathbf{M}_n^T + \mathbf{Q}_{n+1}.$$

**Algorithm 3.1** Discrete KF

/* index $n$ denotes time step */
/* **M**: the linear model propagator, **H**: the linear observation operator */

1 ***Initiation***: $\mathbf{x}_0^f = \mathbf{x}_0^b$, $\mathbf{P}_0^f = \mathbf{P}_0^b$, $\mathbf{y}_n^o$, $\mathbf{R}_n$, $\mathbf{Q}_n$ ! ICs and inputs
2 **for** $n = 0$ **to** *nmax* **do** ! Loop for time step $n$
3 **if** ($\mathbf{y}^o$ exists) **then** ! When observations are available
4 ***Analysis***: ! Obtain $\mathbf{K}$, $\mathbf{x}^a$, and $\mathbf{P}^a$ at $n$
5 $\mathbf{K}_n = \left(\mathbf{P}^f\mathbf{H}^T\right)_n\left((\mathbf{H}\mathbf{P}^f\mathbf{H}^T)_n + \mathbf{R}_n\right)^{-1}$ ! Calculate the Kalman gain
6 $\mathbf{x}_n^a = \mathbf{x}_n^f + \mathbf{K}_n\left(\mathbf{y}^o - \mathbf{H}\mathbf{x}^f\right)_n$ ! Calculate the analysis
7 $\mathbf{P}_n^a = (\mathbf{I} - (\mathbf{K}\mathbf{H})_n)\,\mathbf{P}_n^f$ ! Calculate the analysis error covariance
8
9 ***Forecast***: ! Propagating $\mathbf{x}^f$ and $\mathbf{P}^f$ from $n$ to $n+1$
10 $\mathbf{x}_{n+1}^f = \mathbf{M}_n\mathbf{x}_n^a$ ! Evolve the state using the analysis
11 $\mathbf{P}_{n+1}^f = \mathbf{M}_n\mathbf{P}_n^a\mathbf{M}_n^T + \mathbf{Q}_n$ ! Evolve the forecast error covariance
12 **else** ! When no observations are available
13 ***Forecast only***: ! Propagating $\mathbf{x}^f$ and $\mathbf{P}^f$ from $n$ to $n+1$
14 $\mathbf{x}_{n+1}^f = \mathbf{M}_n\mathbf{x}_n^f$ ! Evolve the state using the previous forecast
15 $\mathbf{P}_{n+1}^f = \mathbf{M}_n\mathbf{P}_n^f\mathbf{M}_n^T + \mathbf{Q}_n$ ! Evolve the forecast error covariance
16 **end**
17 **endfor**

### *3.3.2 Extended Kalman Filter*

The standard KF is based on the assumption that the model and observation operators are linear, that is, $M_n = \mathbf{M}_n$ and $H_n = \mathbf{H}_n$; however, in practice, we often encounter cases in which those operators are nonlinear. For example, as for $M_n$, most realistic models in geosciences and environmental science are based on nonlinear governing equations; as for $H_n$, most remotely sensed observations, including observations from satellite, radar, lidar, etc., are indirectly linked with model variables through nonlinear functions or models (e.g., satellite radiances via a radiative transfer model).

The extended Kalman filter (EKF) is a nonlinear extension of the KF through linear approximations of $M_n$ and $H_n$ along the nonlinear solution trajectories, i.e., the current estimate (see Gelb, 1974; Gibbs, 2011; Crassidis and Junkins, 2012). Nonlinear evolution of a probability distribution usually results in a non-Gaussian output even though one starts with a Gaussian input. Therefore, in the EKF, we assume that the errors ($\boldsymbol{\varepsilon}_n^m$, $\boldsymbol{\varepsilon}_o^m$) and the state ($\mathbf{x}_n$) are *approximately* Gaussian; thus, $\mathbf{x}_n \sim N\left(\mathbf{x}_n^a, \mathbf{P}_n^a\right)$ (Snyder, 2015). To derive the EKF formulation, we start with the dynamical system (3.2) and (3.3), rewritten for $\mathbf{x}_n = \mathbf{x}_n^t$ as

$$\begin{aligned} \mathbf{x}_{n+1} &= M_n\left(\mathbf{x}_n\right) + \boldsymbol{\varepsilon}_n^m \\ \mathbf{y}_n^o &= H_n\left(\mathbf{x}_n\right) + \boldsymbol{\varepsilon}_n^o. \end{aligned} \tag{3.10}$$

We linearize $M_n$ and $H_n$ at the current estimate. By defining the linearized $\mathbf{M}_n$ and $\mathbf{H}_n$ as the first-order derivatives (i.e., Jacobians) of $M_n$ and $H_n$, respectively, with respect to the state evaluated at the current estimate, i.e., $\hat{\mathbf{x}}_n = \mathbf{x}_n^a$ for $\mathbf{M}_n$ and $\hat{\mathbf{x}}_n = \mathbf{x}_n^f$ (or $\mathbf{x}_n^b$) for $\mathbf{H}_n$, we have

$$\mathbf{M}_n = \left.\frac{\partial M}{\partial \mathbf{x}}\right|_{\hat{\mathbf{x}}_n = \mathbf{x}_n^a} \quad \text{and } \mathbf{H}_n = \left.\frac{\partial H}{\partial \mathbf{x}}\right|_{\hat{\mathbf{x}}_n = \mathbf{x}_n^f}. \tag{3.11}$$

Then, expanding $M_n(\mathbf{x}_n)$ and $H_n(\mathbf{x}_n)$ in Taylor series (see Section 4.2), for a sufficiently small variation $\delta\mathbf{x} = \mathbf{x}_n - \hat{\mathbf{x}}_n$, gives

$$\begin{aligned} M_n(\mathbf{x}_n) &= M_n(\hat{\mathbf{x}}_n) + \mathbf{M}_n\delta\mathbf{x} + O\left((\delta\mathbf{x})^2\right) \\ H_n(\mathbf{x}_n) &= H_n(\hat{\mathbf{x}}_n) + \mathbf{H}_n\delta\mathbf{x} + O\left((\delta\mathbf{x})^2\right). \end{aligned} \tag{3.12}$$

By neglecting the high-order terms $O\left((\delta\mathbf{x})^2\right)$ and by inserting (3.12) into (3.10), we have approximations to (3.10) as

$$\begin{aligned} \mathbf{x}_{n+1} &\approx M_n\left(\mathbf{x}_n^a\right) + \mathbf{M}_n\left(\mathbf{x}_n - \mathbf{x}_n^a\right) + \boldsymbol{\varepsilon}_n^m \\ \mathbf{y}_n^o &\approx H_n\left(\mathbf{x}_n^f\right) + \mathbf{H}_n\left(\mathbf{x}_n - \mathbf{x}_n^f\right) + \boldsymbol{\varepsilon}_n^o. \end{aligned} \tag{3.13}$$

As we assumed $\mathbf{x}_n \sim N\left(\mathbf{x}_n^a, \mathbf{P}_n^a\right)$, we can further assume that $\mathbf{x}_{n+1} \sim N\left(\mathbf{x}_{n+1}^f, \mathbf{P}_{n+1}^f\right)$; thus, the forecast equation of the EKF becomes

$$\mathbf{x}_{n+1}^f = M_n\left(\mathbf{x}_n^a\right), \tag{3.14}$$

where the forecast $\mathbf{x}_{n+1}^f$ is obtained by applying the nonlinear propagator $M_n$ to the analysis $\mathbf{x}_n^a$ – the best estimate of $\mathbf{x}_n^t$. We can estimate the forecast error $\boldsymbol{\varepsilon}_{n+1}^f$, using (3.13), as

$$\boldsymbol{\varepsilon}_{n+1}^f = \mathbf{x}_{n+1} - \mathbf{x}_{n+1}^f \approx \mathbf{M}_n\boldsymbol{\varepsilon}_n^a + \boldsymbol{\varepsilon}_n^m. \tag{3.15}$$

By assuming that $E(\boldsymbol{\varepsilon}_n^a) = \mathbf{0}$, $E(\boldsymbol{\varepsilon}_n^m) = \mathbf{0}$ and $E(\boldsymbol{\varepsilon}_n^a(\boldsymbol{\varepsilon}_n^m)^T) = \mathbf{0}$, the forecast error covariance $\mathbf{P}_{n+1}^f$ is given by

$$\mathbf{P}_{n+1}^f = E\left(\boldsymbol{\varepsilon}_{n+1}^f(\boldsymbol{\varepsilon}_{n+1}^f)^T\right) = \mathbf{M}_n\mathbf{P}_n^a\mathbf{M}_n^T + \mathbf{Q}_n. \tag{3.16}$$

The analysis equation for the EKF is

$$\mathbf{x}_n^a = \mathbf{x}^f + \mathbf{K}_n \left(\mathbf{y}_n^o - H_n(\mathbf{x}_n^f)\right), \tag{3.17}$$

the analysis error $\boldsymbol{\varepsilon}_n^a$, using (3.13) and (3.17), is

$$\boldsymbol{\varepsilon}_n^a = \mathbf{x}_n - \mathbf{x}_n^a \approx (\mathbf{I} - \mathbf{KH})_n \boldsymbol{\varepsilon}_n^f + \mathbf{K}_n \boldsymbol{\varepsilon}_n^o \tag{3.18}$$

and the analysis error covariance $\mathbf{P}_n^a$, following (1.55), is

$$\mathbf{P}_n^a = E\left(\boldsymbol{\varepsilon}_n^a (\boldsymbol{\varepsilon}_n^a)^T\right) = (\mathbf{I} - \mathbf{KH})_n \mathbf{P}_n^f (\mathbf{I} - \mathbf{KH})_n^T + \mathbf{K}_n \mathbf{R}_n \mathbf{K}_n^T, \tag{3.19}$$

where $\mathbf{R}_n = E\left(\boldsymbol{\varepsilon}_n^o (\boldsymbol{\varepsilon}_n^o)^T\right)$. We want to get the minimum variance estimate, i.e., find a $\mathbf{K}_n$ that satisfies $\nabla_{\mathbf{K}} tr(\mathbf{P}_n^a) = 0$, as in (1.56); thus, we obtain the Kalman gain of the EKF as the same as that of the standard KF – see (3.9). By putting the Kalman gain (3.9) into (3.19), we get $\mathbf{P}_n^a$ as in (3.8).

In summary, the EKF is the same as the standard KF except that the analysis equation (3.17) and the forecast equation (3.14) have nonlinear operators, $H_n$ and $M_n$, respectively; the other equations keep the same forms while the linearized operators $\mathbf{M}_n$ and $\mathbf{H}_n$ are defined as the Jacobians evaluated at the current estimate – see (3.11). The EKF is summarized in Algorithm 3.2.

**Practice 3.2 Errors and error covariance in the EKF**

Solve the following:

1. Using (3.13) and (3.14), show that

$$\boldsymbol{\varepsilon}_{n+1}^f \approx \mathbf{M}_n \boldsymbol{\varepsilon}_n^a + \boldsymbol{\varepsilon}_n^m$$

and

$$\mathbf{P}_{n+1}^f = \mathbf{M}_n \mathbf{P}_n^a \mathbf{M}_n^T + \mathbf{Q}_n.$$

2. Using (3.13) and (3.17), show that

$$\boldsymbol{\varepsilon}_n^a \approx (\mathbf{I} - \mathbf{KH})_n \boldsymbol{\varepsilon}_n^f + \mathbf{K}_n \boldsymbol{\varepsilon}_n^o$$

and

$$\mathbf{P}_n^a = (\mathbf{I} - \mathbf{KH})_n \mathbf{P}_n^f (\mathbf{I} - \mathbf{KH})_n^T + \mathbf{K}_n \mathbf{R}_n \mathbf{K}_n^T.$$

3. Show that, by taking $\nabla_{\mathbf{K}} tr(\mathbf{P}_n^a) = 0$, the Kalman gain $\mathbf{K}_n$ is obtained as in (3.9).
4. By inserting (3.9) into (3.19), show that $\mathbf{P}_n^a$ becomes

$$\mathbf{P}_n^a = (\mathbf{I} - (\mathbf{KH})_n)\, \mathbf{P}_n^f.$$

**Algorithm 3.2** Discrete EKF

```
   /* index n denotes time step                                                        */
   /* M and H are Jacobians of M and H, respectively – Eq. (3.11)                       */
1  Initiation: x_0^f = x_0^b, P_0^f = P_0^b, y_n^o, R_n, Q_n          ! ICs and inputs
2  for n = 0 to nmax do                                                ! Loop for time step n
3     if (y^o exists) then                                             ! When observations are available
4        Analysis:                                                     ! Obtain K, x^a, and P^a at n
5           K_n = (P^f H^T)_n ((H P^f H^T)_n + R_n)^{-1}               ! Calculate the Kalman gain
6           x_n^a = x_n^f + K_n (y^o - H x^f)_n                        ! Calculate the analysis
7           P_n^a = (I - (KH)_n) P_n^f                                 ! Calculate the analysis error covariance
8
9        Forecast:                                                     ! Propagating x^f and P^f from n to n+1
10          x_{n+1}^f = M_n x_n^a                                      ! Evolve the state using the analysis
11          P_{n+1}^f = M_n P_n^a M_n^T + Q_n                          ! Evolve the forecast error covariance
12    else                                                             ! When no observations are available
13       Forecast only:                                                ! Propagating x^f and P^f from n to n+1
14          x_{n+1}^f = M_n x_n^f                                      ! Evolve the state using the previous forecast
15          P_{n+1}^f = M_n P_n^f M_n^T + Q_n                          ! Evolve the forecast error covariance
16    end
17 endfor
```

**Example 3.1 EKF application: A lake model**

We introduce the work by Kourzeneva (2014) as an example of constructing an EKF scheme to assimilate the lake water surface temperature (LWST) observations into the lake model Freshwater Lake (FLake). Among different model regimes for FLake, only the open water period and stratified regime is considered here, which is controlled by four variables in the state vector:

$$\mathbf{x} = \left(\overline{T}\ \ \eta\ \ T_b\ \ C_T\right)^T,$$

where $\overline{T}$ and $T_b$ are the mean water and bottom temperatures (in K), $C_T$ is the shape factor, $\eta = 1 - h/D$ is the dimensionless mixed layer depth, and $h$ and $D$ (both in m) are the mixed layer depth and the lake depth, respectively. In open water conditions, the mixed layer temperature $T_{ML}$ is the same as the LWST, which is the only observed quantity; thus, the observation vector has just one component:

$$\mathbf{y}^o = \left(T_{ML}^o\right),$$

where the superscript $o$ implies observation. Then, the diagnostic equation for $T_{ML}$ is used for the observation operator $H$, i.e.,

$$H = T_{ML} = \frac{\overline{T} - C_T \eta T_b}{1 - C_T \eta}.$$

The linearized (or tangent linear) observation operator $\mathbf{H} \equiv \frac{\partial H}{\partial \mathbf{x}}$ is derived as

$$\mathbf{H} = \left( \frac{1}{1 - C_T \eta} \quad \frac{C_T(\overline{T} - T_b)}{(1 - C_T \eta)^2} \quad \frac{-C_T \eta}{1 - C_T \eta} \quad \frac{\eta(\overline{T} - T_b)}{(1 - C_T \eta)^2} \right).$$

The tangent linear model operator $\mathbf{M} \equiv \frac{\partial \mathbf{x}^n}{\partial \mathbf{x}^0}$ is given as

$$\mathbf{M} = \begin{pmatrix} \frac{\partial \overline{T}^n}{\partial \overline{T}^0} & \frac{\partial \overline{T}^n}{\partial \eta^0} & \frac{\partial \overline{T}^n}{\partial T_b^0} & \frac{\partial \overline{T}^n}{\partial C_T^0} \\ \frac{\partial \eta^n}{\partial \overline{T}^0} & \frac{\partial \eta^n}{\partial \eta^0} & \frac{\partial \eta^n}{\partial T_b^0} & \frac{\partial \eta^n}{\partial C_T^0} \\ \frac{\partial T_b^n}{\partial \overline{T}^0} & \frac{\partial T_b^n}{\partial \eta^0} & \frac{\partial T_b^n}{\partial T_b^0} & \frac{\partial T_b^n}{\partial C_T^0} \\ \frac{\partial C_T^n}{\partial \overline{T}^0} & \frac{\partial C_T^n}{\partial \eta^0} & \frac{\partial C_T^n}{\partial T_b^0} & \frac{\partial C_T^n}{\partial C_T^0} \end{pmatrix},$$

where the superscript $n$ is a time index and the superscript 0 means the initial time ($n = 0$). The Jacobians are calculated numerically by approximating

$$\mathbf{M} \approx \frac{M(\mathbf{x} + \delta \mathbf{x}) - M(\mathbf{x})}{\delta \mathbf{x}},$$

where values of $\delta \mathbf{x}$ are small enough to accurately approximate the derivative but not too small that roundoff errors occur. For the stratified regime, the following initial perturbation values are used:

$$\delta \mathbf{x}^0 = (0.2K \;\; 0.05 \;\; 0.1K \;\; 0.05)^T.$$

Algorithm 3.2 can be applied to solve this problem. For descriptions on the error covariances $\mathbf{Q}$ and $\mathbf{P}^f$, see Kourzeneva (2014).

### 3.3.3 Practical Issues with the Kalman Filter

The quality of data assimilation using the KF is affected by nonlinearity in the model and observation operators and insufficient knowledge on error statistics. The nonlinearity issue is partly solved by applying the EKF to nonlinear systems, whereas the error statistics are represented by the error covariances $\mathbf{R}$, $\mathbf{Q}$, and $\mathbf{P}^b$ with proper estimations (e.g., Dee, 1995; Bannister, 2008). Another issue that makes the KF impractical is related to the computational burden in propagating the error covariances.

A typical atmospheric operational data assimilation system involves the analysis state in $O(N \sim 10^7\text{–}10^8)$, whereas the KF and EKF stores and evolves temporally the covariance matrices $\mathbf{P}^a$ and $\mathbf{P}^f$ in $O(N \times N)$. Therefore, the computational demand of the KF is huge for large dimensional systems. In practice, approximate (or *suboptimal*) KFs are used (see, e.g., Dee, 1991; Todling and Cohn, 1994; Cohn and Todling, 1996). The error covariances advancing with the state estimates are often represented in a reduced-dimension subspace defined by some basis functions, using, e.g., empirical orthogonal functions (Buehner and Malanotte-Rizzoli, 2003), singular vectors (Cohn and Todling, 1996), etc. Simon (2007) took the full-model order into account while achieving the reduced-order filter by minimizing the trace of the estimation error covariance.

Several methods have been suggested for developing the suboptimal KFs: 1) various *covariance modeling* approaches with constant forecast error covariances, such as OI (Lorenc, 1981), three-dimensional variational analysis (3DVAR) (Barker et al., 2004), etc., 2) *simplified dynamical model* or *reduced-order model* to propagate the error covariances (Dee, 1991; Farrell and Ioannou, 2001; Rozier et al., 2007); 3) *localized approximation* that advances error covariances only for adjacent grid points (Todling and Cohn, 1994); 4) computing the error covariances at *lower resolution* than the model states (Fukumori, 1995); 5) *reducing the state space order* (Cane et al., 1996); 6) computing an asymptotic error covariance and using a *steady-state filter* with a fixed gain matrix (Heemink, 1988); and 7) *reducing the error space order* (Verlaan and Heemink, 1997; Pham et al., 1998; Lermusiaux and Robinson, 1999), including the Monte Carlo approach (Evensen, 1994).

The KF also has a numerical stability issue related to uncertainties in computing the a posteriori error covariances $\mathbf{P}_n^a$ (see (3.8)), which is assumed to be a positive definite. Note that $\mathbf{P}_n^a$ is given by the difference between $\mathbf{P}_n^f$ and $\mathbf{K}_n\mathbf{H}_n\mathbf{P}_n^f$; thus, unless the numerical algorithm is highly accurate, $\mathbf{P}_n^a$ may become a negative definite. This induces unstable behavior in the KF, which is called *filter divergence*: After an extended operation period, the errors in the estimates eventually diverge out, making the filter useless (Fitzgerald, 1971). Filter divergence can be caused by imprecise representation of the model/error dynamics, biased observations, numerical roundoff errors, etc.: It is also evident with overconfidence on the filter estimates (i.e., $\mathbf{P}_n^a$ too small), which prevents a correct update by subsequent observations and brings about a large estimate error leading to ill-conditioned error covariances.

The divergence phenomenon can be treated by adopting *square root filtering* through the Cholesky factorization (Kaminski et al., 1971): A matrix $\mathbf{P}$ evolves in square root form as

$$\mathbf{P}_n = \mathbf{S}_n\mathbf{S}_n^T, \qquad (3.20)$$

where $\mathbf{S}_n = \mathbf{P}_n^{\frac{1}{2}}$ is a lower-triangular square matrix, representing the square root of $\mathbf{P}_n$. Note that the product of any square matrix and its transpose is always positive definite; thus, $\mathbf{S}_n\mathbf{S}_n^T$ never becomes indefinite even with roundoff errors. The numerical conditioning of $\mathbf{S}_n$ is generally much better than that of $\mathbf{P}_n$ itself. In the square root filter, the covariance matrices are replaced by their square root counterparts, then $\mathbf{S}_n$ is propagated for computing $\mathbf{P}_n^a$ and $\mathbf{P}_{n+1}^f$ (see Algorithm 3.1).

**Colloquy 3.1**

**Cholesky factorization**

A symmetric matrix **P** is positive definite if

$$\mathbf{x}^T \mathbf{P} \mathbf{x} > 0$$

for all nonzero vectors **x**. The Cholesky factorization (or decomposition) implies that a positive definite matrix **P** is factored as $\mathbf{P} = \mathbf{L}\mathbf{L}^T$ or $\mathbf{P} = \mathbf{U}^T\mathbf{U}$, where **L** is lower triangular and **U** is upper triangular. When **P** depicts a factorization, for a square matrix **S**, as

$$\mathbf{P} = \mathbf{S}\mathbf{S}^T, \tag{3.21}$$

**S** is called the *square root* (or *Cholesky square root*) of **P**. When (3.21) is expressed, for an $n \times n$ matrix **P**, as

$$\begin{pmatrix} P_{11} & P_{12} & \cdots & P_{1n} \\ P_{21} & P_{22} & \cdots & P_{2n} \\ \vdots & & \ddots & \vdots \\ P_{n1} & P_{n2} & \cdots & P_{nn} \end{pmatrix} = \begin{pmatrix} S_{11} & 0 & \cdots & 0 \\ S_{21} & S_{22} & \cdots & 0 \\ \vdots & & \ddots & \vdots \\ S_{n1} & S_{n2} & \cdots & S_{nn} \end{pmatrix} \begin{pmatrix} S_{11} & S_{21} & \cdots & S_{n1} \\ 0 & S_{22} & \cdots & S_{n2} \\ \vdots & & \ddots & \vdots \\ 0 & 0 & \cdots & S_{nn} \end{pmatrix}, \tag{3.22}$$

one can obtain **S** recursively for the entries $i = 1, \ldots, n$, following Kaminski et al. (1971), as

$$S_{ii} = \sqrt{P_{ii} - \sum_{j=1}^{i-1} S_{ij}^2}, \tag{3.23}$$

$$S_{ji} = \frac{1}{S_{ii}} \left( P_{ji} - \sum_{k=1}^{i-1} S_{jk} S_{ik} \right), \quad j = i+1, \ldots, n, \tag{3.24}$$

and $S_{ji} =$ for $j < i$.

**Colloquy 3.2**

**Low-rank KFs in square root form**

We introduce the reduced-order error space (or error subspace) approach, i.e., method 7) mentioned in p. 64, in which the Cholesky factorization is adopted.

A real and symmetric (positive definite) matrix **P** of order $n \times n$, with real eigenvalues and orthogonal eigenvectors can be written in a factored form as

$$\mathbf{P} = \mathbf{E}\mathbf{D}\mathbf{E}^T, \tag{3.25}$$

where $\mathbf{D}$ is a diagonal matrix of order $n$ including the eigenvalues of $\mathbf{P}$ and $\mathbf{E}$ is a matrix containing $r$ eigenvectors of $\mathbf{P}$ ($r \ll n$). Note that $n \sim 10^6$–$10^8$ and $r \sim 10$–$100$; thus, the order reduction is accomplished through $\mathbf{E}$ having the order $n \times r$ rather than $n \times n$. If we define an error subspace $\mathbf{S} = \mathbf{E}\mathbf{D}^{\frac{1}{2}}$ of order $n \times r$, limited to the dominant eigenmodes that best represent the covariance $\mathbf{P}$; then, $\mathbf{P}$ is written as

$$\mathbf{P} = \mathbf{S}\mathbf{S}^T \tag{3.26}$$

and is specified as a *low-rank* matrix. The error subspace can be built based on various basis functions, empirical orthogonal functions, singular vectors, Lyapunov vectors, breeding vectors, etc., and on the Monte Carlo method.

The EKF is now reconstructed by replacing $\mathbf{P}$ with $\mathbf{S}$ (see Algorithm 3.2). The Kalman gain $\mathbf{K}_n$ is expressed in terms of $\mathbf{S}_n^f$ as

$$\begin{aligned}\mathbf{K}_n &= \mathbf{S}_n^f \left(\mathbf{H}\mathbf{S}^f\right)_n^T \left((\mathbf{H}\mathbf{S}^f)(\mathbf{H}\mathbf{S}^f)^T + \mathbf{R}\right)_n^{-1} \\ &= \mathbf{S}_n^f \left(\mathbf{I} + (\mathbf{H}\mathbf{S}^f)^T\mathbf{R}^{-1}(\mathbf{H}\mathbf{S}^f)\right)_n^{-1} (\mathbf{H}\mathbf{S}^f)_n^T \mathbf{R}_n^{-1}.\end{aligned} \tag{3.27}$$

The analysis error covariance $\mathbf{P}_n^a$ is described in terms of $\mathbf{S}_n^f$, using (3.27), as

$$\begin{aligned}\mathbf{P}_n^a &= \left(\mathbf{I} - \mathbf{S}_n^f \left(\mathbf{I} + (\mathbf{H}\mathbf{S}^f)^T\mathbf{R}^{-1}(\mathbf{H}\mathbf{S}^f)\right)_n^{-1} (\mathbf{H}\mathbf{S}^f)_n^T \mathbf{R}_n^{-1}\mathbf{H}_n\right) \mathbf{S}_n^f (\mathbf{S}^f)_n^T \\ &= \mathbf{S}_n^f \left(\mathbf{I} + (\mathbf{H}\mathbf{S}^f)^T\mathbf{R}^{-1}(\mathbf{H}\mathbf{S}^f)\right)_n^{-1} (\mathbf{S}^f)_n^T.\end{aligned} \tag{3.28}$$

Using (3.26), we apply the order reduction $\mathbf{P}_n^a = \mathbf{S}_n^a(\mathbf{S}^a)_n^T$ to obtain $\mathbf{P}_{n+1}^f$:

$$\mathbf{P}_{n+1}^f = \left(\mathbf{M}_n\mathbf{S}_n^a\right)\left(\mathbf{M}_n\mathbf{S}_n^a\right)^T + \mathbf{Q}_n, \tag{3.29}$$

where $\mathbf{S}_n^a$ is given by

$$\mathbf{S}_n^a = \mathbf{S}_n^f \left(\mathbf{I} + (\mathbf{H}\mathbf{S}^f)^T\mathbf{R}^{-1}(\mathbf{H}\mathbf{S}^f)\right)_n^{-\frac{1}{2}}. \tag{3.30}$$

To propagate $\mathbf{S}_n^a$, we calculate $\mathbf{S}_n^f$ for each eigenvector column $l$, i.e., $\{\mathbf{S}_n^f\}_l$ for $l = 1, \ldots, r$. In the *singular evolutive extended Kalman (SEEK)* filter (Pham et al., 1998), $\{\mathbf{S}_{n+1}^f\}_l$ is obtained by applying the tangent linear model $\mathbf{M}_n$, i.e.,

$$\{\mathbf{S}_{n+1}^f\}_l = \mathbf{M}_n\{\mathbf{S}_n^a\}_l. \tag{3.31}$$

Finite difference approximation, through the NLM $M_n$, can be also used to compute $\{\mathbf{S}_{n+1}^f\}_l$, i.e.,

$$\{\mathbf{S}_{n+1}^f\}_l = \mathbf{\Lambda}_n^{-1}\left(M_n\left(\mathbf{x}_n^a + \mathbf{\Lambda}_n\{\mathbf{S}_n^a\}_l\right) - M_n\left(\mathbf{x}_n^a\right)\right), \tag{3.32}$$

assuming the Cholesky decomposition $\mathbf{D}_n = \mathbf{\Lambda}_n\mathbf{\Lambda}_n^T$ (e.g., Pham, 1996; Brasseur et al., 1999). This approach is a variant of the SEEK filter and is named the *singular evolutive interpolated Kalman (SEIK)* filter (Pham, 1996).

### 3.4 Smoothers

Smoothers estimate the current state using future data as well as past data at the analysis time, $t_k$, located between the start and the end of a given smoothing period (or assimilation window); thus, they are regarded as non-real-time data processing schemes (Gelb, 1974) or batch estimators (Crassidis and Junkins, 2012). Figure 3.4 illustrates a smoother where $t_k$ is located at the start of each smoothing period, while it is also allowed to lie anywhere inside the smoothing period. Smoothers seek an optimal estimate of the state that produces an optimized solution trajectory within the smoothing period (see Figure 3.4). Another implication of *smoothers* is that they typically *smooth* out the effect of measurement noise (Crassidis and Junkins, 2012).

As smoothers include future observations (say, at $t_{k+l}$) while the state estimates are made at an earlier time (say, at $t_k$), conditioned by the observations, it is convenient to define the states in terms of the conditioning (e.g., Cohn et al., 1994; Cosme et al., 2012); for example, an a priori state estimate at $t_k$ conditioned by all the observations up to $t_{k-1}$ is denoted as $\hat{\mathbf{x}}_{k|k-1}$ (or $\mathbf{x}^f_{k|k-1}$ or $\mathbf{x}^b_{k|k-1}$), whereas an a posteriori state estimate at $t_k$ given all the observations up to $t_k$ is denoted as $\hat{\mathbf{x}}_{k|k}$ (or $\mathbf{x}^a_{k|k}$). With this notation, the *smoother estimate* $\hat{\mathbf{x}}_{j|k}$ (or $\mathbf{x}^s_{j|k}$), with $j < k$, is the expectation of the state at $t_j$ based on the observations $\{\mathbf{y}^o_1, \ldots, \mathbf{y}^o_k\}$ from $t_1$ to $t_k$, represented as

$$\hat{\mathbf{x}}_{j|k} = E\left(\mathbf{x}_j \mid \mathbf{y}^o_{1:k}\right). \tag{3.33}$$

The forecast and analysis errors at $t_k$ are denoted as $\boldsymbol{\varepsilon}^f_{k|k-1}$ with observations up to $t_{k-1}$ and $\boldsymbol{\varepsilon}^a_{k|k}$ with observations up to $t_k$, respectively; then, the corresponding error covariances are defined as

$$\mathbf{P}^f_{k|k-1} = E\left(\boldsymbol{\varepsilon}^f_{k|k-1}\left(\boldsymbol{\varepsilon}^f_{k|k-1}\right)^T\right); \ \mathbf{P}^a_{k|k} = E\left(\boldsymbol{\varepsilon}^a_{k|k}\left(\boldsymbol{\varepsilon}^a_{k|k}\right)^T\right). \tag{3.34}$$

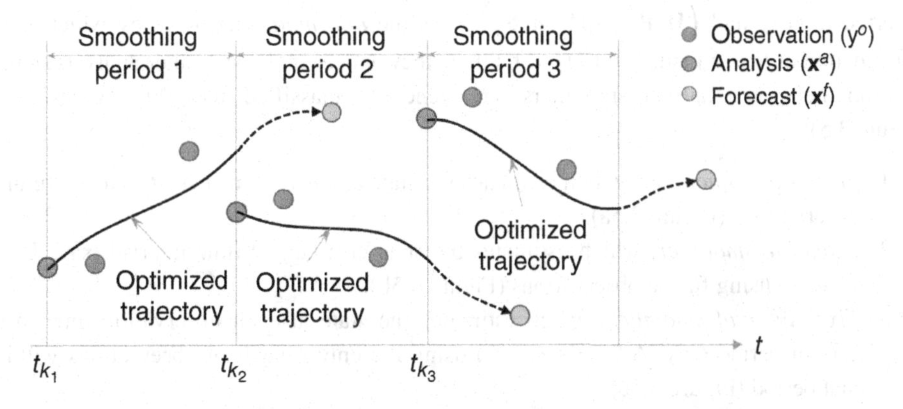


Figure 3.4 A smoother obtains the analysis fields ($\mathbf{x}^a$) at the analysis times $t_{k_i}$, which minimizes the cost function, using information from the observations ($\mathbf{y}^o$) in the given smoothing periods (i.e., assimilation windows). Solid lines represent model state trajectories optimized in the given smoothing periods, which start from the optimized ICs (i.e., $\mathbf{x}^a$). Dotted arrows depict model state trajectories beyond the given smoothing periods to yield the extended forecasts ($\mathbf{x}^f$).

We also define the analysis-forecast error cross-covariances for different times, $t_j$ (analysis) and $t_k$ (forecast), as

$$\mathbf{P}^{af}_{j,k|k-1} \equiv E\left(\boldsymbol{\varepsilon}^a_{j|k-1}\left(\boldsymbol{\varepsilon}^f_{k|k-1}\right)^T\right) \tag{3.35}$$

and the analysis error cross-covariances as

$$\mathbf{P}^{aa}_{j,k|k} \equiv E\left(\boldsymbol{\varepsilon}^a_{j|k}\left(\boldsymbol{\varepsilon}^a_{k|k}\right)^T\right). \tag{3.36}$$

Using the conditioning notation above, the discrete KF (see Algorithm 3.1) is rewritten as

- *Forecast step:*

$$\begin{aligned}\mathbf{x}^f_{k|k-1} &= \mathbf{M}_{k-1}\mathbf{x}^a_{k-1|k-1},\\ \mathbf{P}^f_{k|k-1} &= \mathbf{M}_{k-1}\mathbf{P}^a_{k-1|k-1}\mathbf{M}^T_{k-1} + \mathbf{Q}_{k-1} \text{ and}\end{aligned} \tag{3.37}$$

- *Analysis step:*

$$\begin{aligned}\mathbf{K}_{k|k} &= \mathbf{P}^f_{k|k-1}\mathbf{H}^T_k\left(\mathbf{H}_k\mathbf{P}^f_{k|k-1}\mathbf{H}^T_k + \mathbf{R}_k\right)^{-1},\\ \mathbf{x}^a_{k|k} &= \mathbf{x}^f_{k|k-1} + \mathbf{K}_{k|k}\left(\mathbf{y}^o_k - \mathbf{H}_k\mathbf{x}^f_{k|k-1}\right),\\ \mathbf{P}^a_{k|k} &= \left(\mathbf{I} - \mathbf{K}_{k|k}\mathbf{H}_k\right)\mathbf{P}^f_{k|k-1},\end{aligned} \tag{3.38}$$

with ICs $\mathbf{x}^a_0$ and $\mathbf{P}^a_0$. Note that optimal linear smoothers are based on the assumptions and equations in the KF and differ from filters only in terms of the error cross-covariance in time (i.e., *cross-time* covariances); thus, a series of filter estimates, e.g., from (3.37)–(3.38), are used to obtain the smoother estimates (Gibbs, 2011; Cosme et al., 2012). Cohn et al. (1994) referred to $\mathbf{K}_{k|k}$ and $\left(\mathbf{H}_k\mathbf{P}^f_{k|k-1}\mathbf{H}^T_k + \mathbf{R}_k\right)^{-1}$ as the *information carriers* by which new information at $t_k$ is transmitted to $t_j < t_k$; thus, they appear in the smoother analysis step.

Smoothers, or Kalman smoothers, are generally classified into three types (see Figure 3.5):

1. *Fixed-point smoother*: which estimates the state at a *fixed* time (say, $t_{k_0}$) using future observations (Figure 3.5a).
2. *Fixed-lag smoother*: which estimates the state at a lagged time interval (say, $L = t_N - t_k$) using future observations (Figure 3.5b).
3. *Fixed-interval smoother*: which estimates the state at each observation time in a certain period (say, $\Delta T = t_N - t_1$) using the entire batch of observations within that period (Figure 3.5c).

Smoothers often take a sequential approach as model integrations proceed alternately with observation updates and the KF innovation is used to correct the past state estimates (see, e.g., Cohn et al., 1994; Cosme et al., 2010; Cosme, 2015). The sequential smoother problem can be converted into a special case of the standard filtering problem by adopting

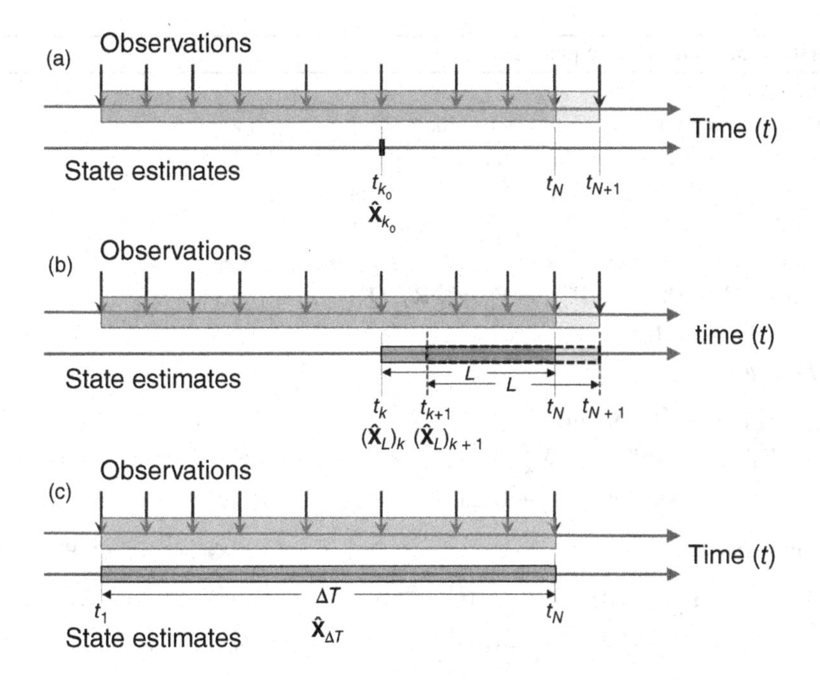


Figure 3.5 Classification of smoothers a) fixed-point smoother evaluating $\hat{\mathbf{x}}_{k_0}$ at a fixed time $t_{k_0}$; b) fixed-lag smoother evaluating $\hat{\mathbf{x}}_L$ at any lagged time intervals, e.g., $L = t_N - t_k$ and $L = t_{N+1} - t_{k+1}$; and c) fixed-interval smoother evaluating $\hat{\mathbf{x}}_{\Delta T}$ at a fixed time interval that includes all available observations, i.e., $\Delta T = t_N - t_1$.

the *state augmentation* approach (Willman, 1969). The sequential smoothers are represented in Algorithm 3.3: A different choice of $j$ results in a different type of smoother.

### 3.4.1 Fixed-Point Smoother

The fixed-point smoother estimates $\hat{\mathbf{x}}_{k_0|k} = E\left(\mathbf{x}_{k_0}|\mathbf{y}_{1:k-1}\right)$ for a fixed $k_0 \leq k$ by taking $j = k_0$ from Algorithm 3.3. In this smoother, the observation time index $k$ continues to increase as more measurements arrive. Figure 3.6 shows an example of the fixed-point smoother estimate at $t_{k_0=3}$.

We define a new augmented state vector, $\mathcal{X}_k$, which contains all the necessary past information of the system (Willman, 1969) and its corresponding error covariances, $\mathcal{P}_k$, as

$$\mathcal{X}_k = \begin{pmatrix} \mathbf{x}_k \\ \mathbf{x}_j \end{pmatrix} \text{ and } \mathcal{P}_k = \begin{pmatrix} \mathbf{P}_{k,k|k} & \mathbf{P}_{j,k|k}^T \\ \mathbf{P}_{j,k|k} & \mathbf{P}_{j,j|k} \end{pmatrix}, \tag{3.39}$$

where $j$ is fixed (say, $j = k_0$) and $k$ $(> j)$ is variable. Then, the augmented state equation can be written as

$$\mathcal{X}_{k+1} = \mathcal{M}_k \mathcal{X}_k + \mathcal{E}_k^m, \tag{3.40}$$

**Algorithm 3.3** Discrete sequential smoothers

```
/* index n denotes time step                                                        */
/* M: the linear model propagator, H: the linear observation operator              */
/* Fixed-point smoother for j = k0 (for fixed k0)                                   */
/* Fixed-lag smoother for j = {k − L, ..., k − 1} (for fixed L)                     */
/* Fixed-interval smoother for j = {0, 1, ..., N − 1, N} (for fixed N)              */
/* Information carriers (see Cohn et al., 1994) are represented in gray            */
```

1 ***Initiation***: $\mathbf{x}_0^s = \mathbf{x}_0^a = \mathbf{x}_0^b$, $\mathbf{P}_{j,0}^{aa} = \mathbf{P}_0^b$, $\mathbf{y}_n^o$, $\mathbf{R}_n$, $\mathbf{Q}_n$ — ! ICs and inputs

2 **for** $k = 1$ **to** *kmax* **do** — ! Loop for time step $k$

3   ***Forecast***: — ! Propagating $\mathbf{x}^f$, $\mathbf{P}_j^{af}$, and $\mathbf{P}^f$

4     $\mathbf{x}_{k|k-1}^f = \mathbf{M}_{k-1}\mathbf{x}_{k-1|k-1}^s$ — ! Evolve the state using the analysis

5     $\mathbf{P}_{j,k|k-1}^{af} = \mathbf{P}_{j,k-1|k-1}^{aa}\mathbf{M}_{k-1}^T$ — ! Evolve the error cross-covariance

6     $\mathbf{P}_{k|k-1}^f = \mathbf{M}_{k-1}\mathbf{P}_{j,k|k-1}^{af} + \mathbf{Q}_{k-1}$ — ! Evolve the forecast error covariance

7   ***Smoother analysis***: — ! Obtain $\mathbf{K}_j$, $\mathbf{x}_j^a$, $\mathbf{P}_j^{aa}$, and $\mathbf{P}_j^a$

8     $\mathbf{K}_{j|k} = \mathbf{P}_{j,k|k-1}^{af}\mathbf{H}_k^T\left(\mathbf{H}_k\mathbf{P}_{k|k-1}^f\mathbf{H}_k^T + \mathbf{R}_k\right)^{-1}$ — ! Calculate the smoother gain

9     $\mathbf{x}_{j|k}^s = \mathbf{x}_{j|k-1}^s + \mathbf{K}_{j|k}\left(\mathbf{y}_k^o - \mathbf{H}_k\mathbf{x}_{k|k-1}^f\right)$ — ! Calculate the smoother analysis

10     $\mathbf{P}_{j,k|k}^{aa} = \mathbf{P}_{j,k|k-1}^{af}\left(\mathbf{I} - \mathbf{K}_{k|k}\mathbf{H}_k\right)^T$ — ! Calculate the analysis error covariance

11     $\mathbf{P}_{j|k}^a = \mathbf{P}_{j|k-1}^a - \mathbf{P}_{j,k|k-1}^{af}\left(\mathbf{K}_{j|k}\mathbf{H}_k\right)^T$ — ! Diagnose the analysis error covariance

12 **endfor**

where

$$\mathcal{M}_k = \begin{pmatrix} \mathbf{M}_k & \mathbf{0} \\ \mathbf{0} & \mathbf{I} \end{pmatrix} \quad \text{and} \quad \mathcal{E}_k^m = \begin{pmatrix} \boldsymbol{\varepsilon}_k^m \\ \mathbf{0} \end{pmatrix}.$$

The observation equation, in terms of $\mathcal{X}_k$, becomes

$$\mathbf{y}_k^o = \mathcal{H}_k\mathcal{X}_k + \boldsymbol{\varepsilon}_k^o, \tag{3.41}$$

where

$$\mathcal{H}_k = \begin{pmatrix} \mathbf{H}_k & \mathbf{0} \end{pmatrix}.$$

The augmented system (3.40) and (3.41) can be solved for a minimum variance filtering problem to get the estimate $\hat{\mathcal{X}}_k$; then, the fixed-point smoother estimate $\hat{\mathbf{x}}_{j|k}$ is obtained from the partitioned form of $\hat{\mathcal{X}}_k$, by separately processing the filtering for all $k > 0$ and the smoothing for $k > j$. The augmented gain matrix, $\mathcal{K}_k$, is expressed as

$$\mathcal{K}_k = \begin{pmatrix} \mathbf{K}_k \\ \mathbf{K}_k^s \end{pmatrix},$$

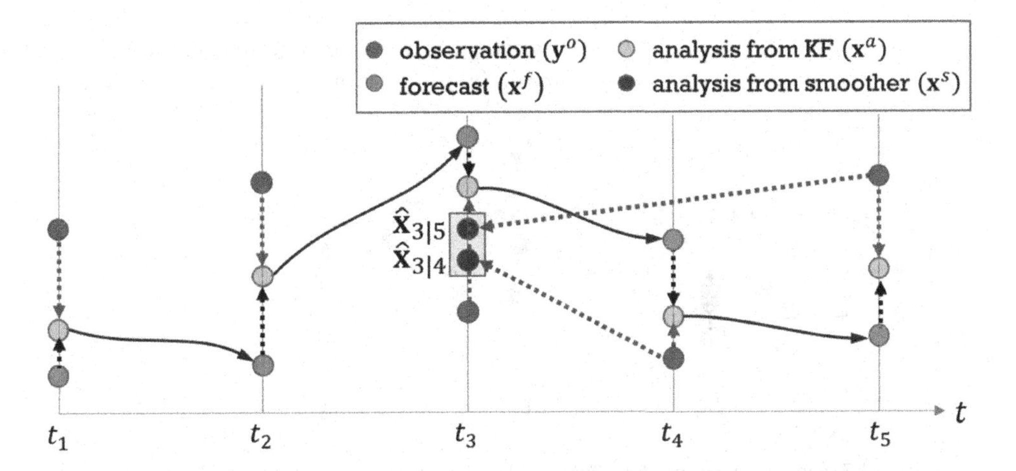


Figure 3.6 Fixed-point smoother evaluating $\hat{\mathbf{x}}_{k_0|k}$ where $k_0 = 3$ and $k = 4$ and 5 (see the shaded zone). Solid arrows are filter forecast paths; dotted arrows from forecasts ($\mathbf{x}^f$) and from observations ($\mathbf{y}^o$) at $t_k$ represent corrections of filter forecasts and observation impacts on the filter analyses ($\mathbf{x}^a$), respectively, and dotted arrows from the observations at $t_4$ and $t_5$ to the smoother analyses ($\mathbf{x}^s$) at $t_3$ (i.e., $\hat{\mathbf{x}}_{3|4}$ and $\hat{\mathbf{x}}_{3|5}$ depict retrospective effect of future observations to the smoother analyses.

where $\mathbf{K}_k$ and $\mathbf{K}_k^s$ represent the filter gain and the smoother gain, respectively. The KF, shown in Algorithm 3.1, can be formulated for the augmented system as

$$\begin{aligned}
\mathcal{X}_{k+1} &= \mathcal{M}_k \mathcal{X}_k + \mathcal{M}_k \mathcal{K}_k \left(\mathbf{y}_k^o - \mathcal{H}_k \mathcal{X}_k\right), \\
\mathcal{K}_k &= \mathcal{P}_k \mathcal{H}_k^T \left(\mathcal{H}_k \mathcal{P}_k \mathcal{H}_k^T + \mathbf{R}_k\right)^{-1}, \\
\mathcal{P}_{k+1} &= \mathcal{M}_k \left(\mathbf{I} - \mathcal{K}_k \mathcal{H}_k\right) \mathcal{P}_k \mathcal{M}_k^T + \mathcal{Q}_k,
\end{aligned} \tag{3.42}$$

where

$$\mathcal{Q}_k = E\left(\mathcal{E}_k^m \left(\mathcal{E}_k^m\right)^T\right) = \begin{pmatrix} \boldsymbol{\varepsilon}_k^m \left(\boldsymbol{\varepsilon}_k^m\right)^T & \mathbf{0} \\ \mathbf{0} & \mathbf{0} \end{pmatrix} = \begin{pmatrix} \mathbf{Q}_k & \mathbf{0} \\ \mathbf{0} & \mathbf{0} \end{pmatrix}.$$

By solving the augmented KF equations (3.42) and following the notation in (3.37)–(3.38), we can obtain the fixed-point smoother solutions, for $j = k_0 < k$, in terms of the smoother state $\hat{\mathbf{x}}_{j|k}$, the smoother error covariances $\mathbf{P}_{j,j|k}$ and the smoother gain $\mathbf{K}_k^s$, as follows:

$$\begin{aligned}
\hat{\mathbf{x}}_{j|k} &= \hat{\mathbf{x}}_{j|k-1} + \mathbf{K}_k^s \left(\mathbf{y}_k^o - \mathbf{H}_k \mathbf{M}_{k-1} \hat{\mathbf{x}}_{k-1|k-1}\right), \\
\mathbf{K}_k^s &= \mathbf{P}_{j,k|k-1} \mathbf{H}_k^T \left(\mathbf{H}_k \mathbf{P}_{k,k|k-1} \mathbf{H}_k^T + \mathbf{R}_k\right)^{-1}, \\
\mathbf{P}_{j,j|k} &= \mathbf{P}_{j,j|k-1} - \mathbf{K}_k^s \mathbf{H}_k \mathbf{P}_{j,k|k-1}^T, \\
\mathbf{P}_{j,k|k} &= \mathbf{P}_{j,k|k-1} - \mathbf{K}_k^s \mathbf{H}_k \mathbf{P}_{k,k|k-1}, \\
\mathbf{P}_{j,k|k-1} &= \mathbf{P}_{j,k-1|k-1} \mathbf{M}_{k-1}^T.
\end{aligned} \tag{3.43}$$

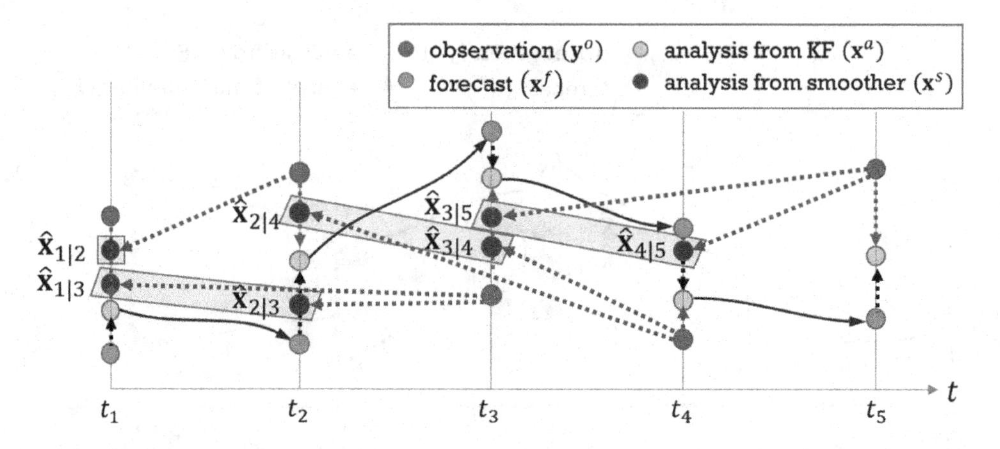


Figure 3.7 Same as in Figure 3.6 but for fixed-lag smoother, evaluating $\hat{\mathbf{x}}_{k-l|k}$ where $l = 1, 2$ (i.e., $L = 2$).

The equations in (3.43) are essentially the same as those in the smoother analysis step in Algorithm 3.3 by putting $\hat{\mathbf{x}}_{j|k} = \mathbf{x}^s_{j|k}$, $\mathbf{K}^s_k = \mathbf{K}_{j|k}$, $\mathbf{P}_{k,k|k-1} = \mathbf{P}^f_{k|k-1}$, $\mathbf{P}_{j,j|k} = \mathbf{P}^a_{j|k}$, $\mathbf{P}_{j,k|k} = \mathbf{P}^{aa}_{j,k|k}$, and $\mathbf{P}_{j,k|k-1} = \mathbf{P}^{af}_{j,k|k-1}$.

**Practice 3.3 Fixed-point smoother from the augmented system**

Derive the fixed-point smoother equation set (3.43) from the filter equation set (3.42) of the augmented system.

### 3.4.2 Fixed-Lag Smoother

The fixed-lag smoother estimates $\hat{\mathbf{x}}_{k-l|k} = E\left(\mathbf{x}_{k-l}|\mathbf{y}_{1:k}\right)$ for $l = 1, \ldots, L$ with a fixed lag $L$ by taking $j = k - L, \ldots, k - 1$ (or simply $j = k - l$) from Algorithm 3.3. In this smoother, the states are estimated at $t_{k-l}$ using measurements up to $t_k$. Figure 3.7 shows an example of the fixed-lag smoother estimate at $t_{k-l}$ using observations up to $t_k$ for $l = 1, \ldots, L$ with the lag $L = 2$.

We construct the augmented KF in the same fashion as (3.42):

$$\mathcal{X}_k = \begin{pmatrix} \mathbf{x}_k & \mathbf{x}_{k-1} & \cdots & \mathbf{x}_{k-L} \end{pmatrix}^T, \tag{3.44}$$

$$\mathcal{M}_k = \begin{pmatrix} \mathbf{M}_k & \mathbf{0} & \cdots & \mathbf{0} \\ \mathbf{I} & \mathbf{0} & \cdots & \mathbf{0} \\ \vdots & \ddots & \ddots & \vdots \\ \mathbf{0} & \cdots & \mathbf{I} & \mathbf{0} \end{pmatrix}, \tag{3.45}$$

$$\mathcal{K}_k = \begin{pmatrix} \mathbf{K}_{k|k} & \mathbf{K}_{k-1|k} & \cdots & \mathbf{K}_{k-L|k} \end{pmatrix}^T, \tag{3.46}$$

$$\mathcal{H}_k = \begin{pmatrix} \mathbf{H}_k & \mathbf{0} & \cdots & \mathbf{0} \end{pmatrix}, \tag{3.47}$$

and

$$\mathcal{P}_k = \begin{pmatrix} \mathbf{P}_{k,k|k} & \mathbf{P}_{k,k-1|k} & \cdots & \mathbf{P}_{k,k-L|k} \\ \mathbf{P}_{k-1,k|k} & \mathbf{P}_{k-1,k-1|k} & \cdots & \mathbf{P}_{k-1,k-L|k} \\ \vdots & \vdots & \ddots & \vdots \\ \mathbf{P}_{k-L,k|k} & \mathbf{P}_{k-L,k-1|k} & \cdots & \mathbf{P}_{k-L,k-L|k} \end{pmatrix}, \tag{3.48}$$

where $\mathbf{P}_{k,k-l|k} = \mathbf{P}^T_{k-l,k|k}$ for $l = 1, \ldots, L$. Then, we have the fixed-lag smoother equations as

$$\begin{aligned}
\hat{\mathbf{x}}_{k-l|k} &= \hat{\mathbf{x}}_{k-l|k-1} + \mathbf{K}_{k-l|k}\left(\mathbf{y}^o_k - \mathbf{H}_k\mathbf{M}_{k-1}\hat{\mathbf{x}}_{k-1|k-1}\right), \\
\mathbf{K}_{k-l|k} &= \mathbf{P}_{k-l,k|k-1}\mathbf{H}^T_k\left(\mathbf{H}_k\mathbf{P}_{k,k|k-1}\mathbf{H}^T_k + \mathbf{R}_k\right)^{-1}, \\
\mathbf{P}_{k-l,k-l|k} &= \mathbf{P}_{k-l,k-l|k-1} - \mathbf{K}_{k-l|k}\mathbf{H}_k\mathbf{P}^T_{k-l,k|k-1}, \\
\mathbf{P}_{k-l,k|k} &= \mathbf{P}_{k-l,k|k-1} - \mathbf{K}_{k|k}\mathbf{H}_k\mathbf{P}_{k-l,k|k-1}, \\
\mathbf{P}_{k-l,k|k-1} &= \mathbf{P}_{k-l,k-1|k-1}\mathbf{M}^T_{k-1},
\end{aligned} \tag{3.49}$$

for $l = 1, \ldots, L$. This is the same as Algorithm 3.3 (the smoother analysis step) by putting for $j = k - l$, $\hat{\mathbf{x}}_{k-l|k} = \mathbf{x}^s_{j|k}$, $\mathbf{P}_{k,k|k-1} = \mathbf{P}^f_{k|k-1}$, $\mathbf{P}_{k-l,k-l|k} = \mathbf{P}^a_{j|k}$, $\mathbf{P}_{k-l,k|k} = \mathbf{P}^{aa}_{j,k|k}$, and $\mathbf{P}_{k-l,k|k-1} = \mathbf{P}^{af}_{j,k|k-1}$.

**Practice 3.4 Fixed-lag smoother from the augmented system**

Derive the fixed-lag smoother equation set (3.49) from the augmented filter equation set (3.42) using definitions of the augmented system variables in (3.44)–(3.48).

### *3.4.3 Fixed-Interval Smoother*

The fixed-interval smoother estimates $\hat{\mathbf{x}}_{k|N} = E\left(\mathbf{x}_k|\mathbf{y}_{1:N}\right)$ for a fixed interval $\Delta T = t_N - t_1$ by taking $j = 0, \ldots, N$ from Algorithm 3.3. In this smoother, the states are estimated at varying times $t_k$ using total measurements available only up to $t_N$ (fixed). Figure 3.8 shows an example of the fixed-interval smoother estimate at $t_k$ using observations up to $t_{N=5}$.

One can easily develop the sequential fixed-interval smoother equations in the same vein with the fixed-lag smoother by employing the augmented KF system (3.42) and by defining the augmented system variables as in (3.44)–(3.48) – by replacing $j = k - l$ ($l = 1, \ldots, L$ for a fixed $L$) with $j = 0, 1, \ldots, N$ for a fixed $N$ (see Algorithm 3.3).

Another common method for solving the fixed-interval smoothing problem is the *forward–backward* smoother or the *Rauch–Tung–Striebel (RTS)* smoother, which was introduced by Rauch et al. (1965) in the linear Gaussian framework.

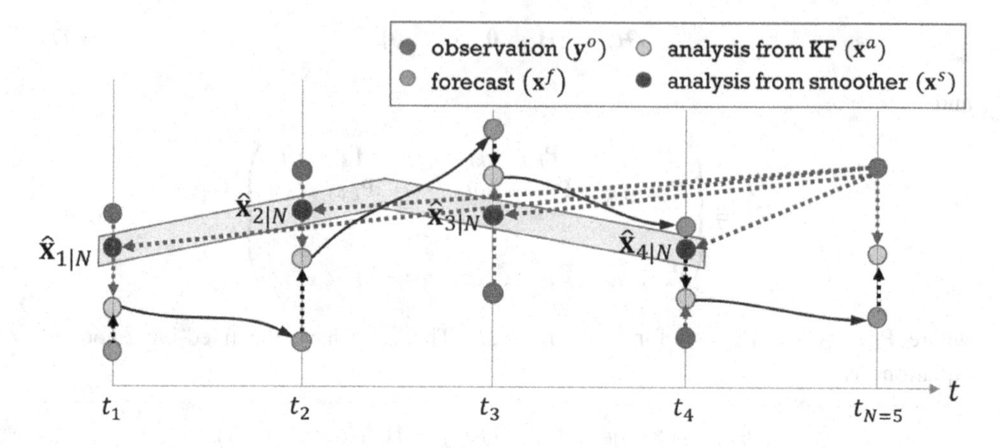


Figure 3.8 Same as in Figure 3.6 but for fixed-interval smoother, evaluating $\hat{\mathbf{x}}_{k|N}$ where $N = 5$.

## 3.5 Bayesian Perspectives

Data assimilation, dealing with uncertainties in both models and observations, is considered to be a process of Bayesian inference that fits a probability model to a set of data and summarizes the result by a probability distribution on the parameters of the model and on unobserved quantities such as predictions for new observations (Gelman et al., 2014). In Chapter 2, we introduced Bayes' formula and its recursive application of data assimilation (see Sections 2.5 and 2.6, respectively). Here, we introduce the Bayesian viewpoints of filtering and smoothing. More details on this issue are referred to in Wikle and Berliner (2007), Särkkä (2013), and Carrassi et al. (2018).

### 3.5.1 Bayesian Inference and the Markov Process

We define vector time series of model states (or unobservable quantities) $\mathbf{x}_{0:N} = \{\mathbf{x}_0, \mathbf{x}_1, \ldots, \mathbf{x}_N\}$ and a set of observed noisy measurements $\mathbf{y}_{1:N} = \{\mathbf{y}_1, \ldots, \mathbf{y}_N\}$ where $N$ indicates the analysis period. Let $p(\cdot|\cdot)$ denote a conditional probability density (or distribution) and $p(\cdot)$ a marginal distribution. In the Bayesian sense, data assimilation aims at obtaining the posterior distribution of the states ($\mathbf{x}_{0:N}$) given the measurements ($\mathbf{y}_{1:N}$):

$$p\left(\mathbf{x}_{0:N}|\mathbf{y}_{1:N}\right) = \frac{p\left(\mathbf{y}_{1:N}|\mathbf{x}_{0:N}\right) p\left(\mathbf{x}_{0:N}\right)}{p\left(\mathbf{y}_{1:N}\right)} = \frac{p\left(\mathbf{x}_{0:N}, \mathbf{y}_{1:N}\right)}{p\left(\mathbf{y}_{1:N}\right)}, \tag{3.50}$$

which consists of the following components:

- $p\left(\mathbf{x}_{0:N}|\mathbf{y}_{1:N}\right)$: *Posterior distribution* of the states $\mathbf{x}_{0:N}$ given knowledge of the data $\mathbf{y}_{1:N}$, implying the update of prior knowledge about the states, $p\left(\mathbf{x}_{0:N}\right)$, given the actual observations.
- $p\left(\mathbf{x}_{0:N}\right)$: *Prior distribution* of the states without knowledge of the data, quantifying a priori knowledge of the states (e.g., climatology or background fields from a 6-h cycle forecast).

- $p(\mathbf{y}_{1:N}|\mathbf{x}_{0:N})$: *Sampling distribution* or *data distribution* that simply depicts distribution of the data given the unobservables, $\mathbf{x}_{0:N}$. When regarded as a function of $\mathbf{x}_{0:N}$, for fixed $\mathbf{y}_k$, it is called the *likelihood function* through which the data $\mathbf{y}_k$ modifies prior knowledge of $\mathbf{x}_{0:N}$; thus, representing the state information transferring from the data.
- $p(\mathbf{y}_{1:N})$: *Marginal distribution* of the data or *prior predictive distribution* in the sense that it is the distribution of observable quantities while not conditioned on previous observations. It is independent of the model states and serves as a *normalizing constant*, defined as:

$$p(\mathbf{y}_{1:N}) = \sum_{\mathbf{x}_{0:N}} p(\mathbf{y}_{1:N}|\mathbf{x}_{0:N})\, p(\mathbf{x}_{0:N}), \tag{3.51}$$

ensuring that the posterior distribution sums (or integrates) to the unity.

A stochastic process is called a Markov process when it has the Markov property:

$$p(\mathbf{x}_k|\mathbf{x}_{0:k-1}) = p(\mathbf{x}_k|\mathbf{x}_{k-1}), \tag{3.52}$$

that is, the conditional probability distribution of future states given all the past states depend only on the current state. In probabilistic Markov sequences, major components of data assimilation are defined, in terms of probability distributions, as the following:

- $p(\mathbf{x}_0)$: Prior probabilistic distribution of the model state $\mathbf{x}_0$ at the initial time step $k = 0$, specifying the ***initial state distribution***.
- $p(\mathbf{x}_k|\mathbf{x}_{k-1})$: Transition (or evolution) probability distribution, defining the *dynamic model* that describes the system dynamics and its uncertainties as a Markov sequence.
- $p(\mathbf{y}_k|\mathbf{x}_k)$: Conditional probability distribution of the measurement given the state, designating the ***measurement model*** that describes how the measurement $\mathbf{y}_k$ depends on the current state $\mathbf{x}_k$.

Based on the Markov assumption that the state at time $t = k$ depends only on the state at $t = k - 1$, the prior distribution is given by:

$$p(\mathbf{x}_{0:N}) = p(\mathbf{x}_0) \prod_{k=1}^{N} p(\mathbf{x}_k|\mathbf{x}_{k-1}), \tag{3.53}$$

which is called the *transition model*. The observations are assumed to be independent given that the true state is known:

$$p(\mathbf{y}_{1:N}|\mathbf{x}_{0:N}) = \prod_{k=1}^{N} p(\mathbf{y}_k|\mathbf{x}_k). \tag{3.54}$$

Then the Bayesian inference (3.50) is expressed as

$$p(\mathbf{x}_{0:N}|\mathbf{y}_{1:N}) \propto p(\mathbf{x}_0) \prod_{k=1}^{N} p(\mathbf{y}_k|\mathbf{x}_k)\, p(\mathbf{x}_k|\mathbf{x}_{k-1}), \tag{3.55}$$

implying that a sequential update on the previous estimate of the model state is made by employing new available observations.

### 3.5.2 Bayesian Filtering and Smoothing

In the Bayesian viewpoint, filtering and smoothing are expressed in terms of marginal distribution as the following (see also Figure 3.1):

- *Filtering* obtains the marginal distributions of the states at time $t = k$, $\mathbf{x}_k$, by using measurements accumulated up to and including $k$, $\mathbf{y}_{1:k}$:

$$p\left(\mathbf{x}_k|\mathbf{y}_{1:k}\right), \quad k = 1, \ldots, N. \tag{3.56}$$

- *Smoothing* obtains the marginal distributions of the state $\mathbf{x}_k$ given the measurements over a certain time interval $[0, N]$, $\mathbf{y}_{1:N}$, for $t = k < N$:

$$p\left(\mathbf{x}_k|\mathbf{y}_{1:N}\right), \quad k = 1, \ldots, N. \tag{3.57}$$

  Using the data measured after the analysis time, smoothing is considered as an a posteriori estimation.
- *Prediction* evaluates the marginal distributions of the future state at $t = k + l$ $(l > 0)$, $\mathbf{x}_{k+l}$, given the measurements accumulated up to and including $k$, $\mathbf{y}_{1:k}$:

$$p\left(\mathbf{x}_{k+l}|\mathbf{y}_{1:k}\right), \quad k = 1, \ldots, N, \quad l = 1, 2, \ldots. \tag{3.58}$$

  Therefore, prediction distributions are obtained from the Bayesian filter through the prediction step. Using data from the past and the present, prediction is regarded as an a priori estimation.

Filtering gains the analysis distribution at $t = k$ in two steps – the prediction step and the analysis step – using $p\left(\mathbf{x}_{k-1}|\mathbf{y}_{1:k-1}\right)$ at $t = k - 1$:

1. *Prediction step*: Based on the Markov assumption, the prediction distribution (or posterior predictive distribution), $p\left(\mathbf{x}_k|\mathbf{y}_{1:k-1}\right)$, is given by

$$p\left(\mathbf{x}_k|\mathbf{y}_{1:k-1}\right) = \int p\left(\mathbf{x}_k|\mathbf{x}_{k-1}\right) p\left(\mathbf{x}_{k-1}|\mathbf{y}_{1:k-1}\right) d\mathbf{x}_{k-1}. \tag{3.59}$$

2. *Analysis step*: Bayesian inference provides the analysis distribution, $p\left(\mathbf{x}_k|\mathbf{y}_{1:k}\right)$, as

$$p\left(\mathbf{x}_k|\mathbf{y}_{1:k}\right) \propto p\left(\mathbf{y}_k|\mathbf{x}_k\right) p\left(\mathbf{x}_k|\mathbf{y}_{1:k-1}\right). \tag{3.60}$$

Smoothing is categorized into *joint* smoothing and *marginal* smoothing (see Cosme et al., 2012; Cosme, 2015): The former identifies the probability distribution of a set of system states while the latter seeks that of a system state, given observations of past, present, and future. These smoothers are discussed in detail below.

#### 3.5.2.1 Joint Smoother: Fixed-Interval

The *fixed-interval* smoother obtains the smoothing distribution $p\left(\mathbf{x}_{0:N}|\mathbf{y}_{1:N}\right)$ that estimates all the states $\mathbf{x}_{0:N}$ within a time interval conditioned to observations from $t_1$ to $t_N$. Using Bayes' rule and the Markov property, we can decompose $p\left(\mathbf{x}_{0:N}|\mathbf{y}_{1:N}\right)$, for $k \in [0, N]$, as

$$p\left(\mathbf{x}_{0:k} \mid \mathbf{y}_{1:k}\right) \propto p\left(\mathbf{x}_{0:k-1} \mid \mathbf{y}_{1:k-1}\right) p\left(\mathbf{x}_k \mid \mathbf{x}_{k-1}\right) p\left(\mathbf{y}_k \mid \mathbf{x}_k\right). \tag{3.61}$$

This smoother obtains the smoothed estimates sequentially via the prediction and analysis steps (see Cosme et al., 2012):

$$p(\mathbf{x}_{0:k} \mid \mathbf{y}_{1:k-1}) = p(\mathbf{x}_{0:k-1} \mid \mathbf{y}_{1:k-1})\, p(\mathbf{x}_k \mid \mathbf{x}_{k-1}) \quad \text{(prediction)},$$
$$p(\mathbf{x}_{0:k} \mid \mathbf{y}_{1:k}) \propto p(\mathbf{x}_{0:k} \mid \mathbf{y}_{1:k-1})\, p(\mathbf{y}_k \mid \mathbf{x}_k) \quad \text{(analysis)}.$$

#### 3.5.2.2 Joint Smoother: Fixed-Lag

The *fixed-lag* smoother obtains $p(\mathbf{x}_{k-L:k}|\mathbf{y}_{1:k})$ that estimates some consecutive states using the $L$-lagged future observations as well as past and present observations. It is similar to the fixed-interval smoother while the estimated states keep the same size through time (see Figure 3.7). The prediction and analysis steps are given, following Cosme et al. (2012), as

$$p(\mathbf{x}_{k-L:k} \mid \mathbf{y}_{1:k-1}) = p(\mathbf{x}_k \mid \mathbf{x}_{k-1}) \int p(\mathbf{x}_{k-L-1:k-1} \mid \mathbf{y}_{1:k-1})\, d\mathbf{x}_{k-L-1} \quad \text{(prediction)},$$
$$p(\mathbf{x}_{k-L:k} \mid \mathbf{y}_{1:k}) \propto p(\mathbf{x}_{k-L:k} \mid \mathbf{y}_{1:k-1})\, p(\mathbf{y}_k \mid \mathbf{x}_k) \quad \text{(analysis)}.$$

#### 3.5.2.3 Marginal Smoother: Fixed-Point

The *marginal* or *fixed-point* smoother obtains $p(\mathbf{x}_k|\mathbf{y}_{1:N})$, for $0 \le k < N$, where all observations, from $t_1$ to $t_N$, are used to estimate only one system state at $t_k$. We use the subsequent state at $t_{k+1}$ to estimate the state at $t_k$ as

$$p(\mathbf{x}_k \mid \mathbf{y}_{1:N}) = \int p(\mathbf{x}_k \mid \mathbf{x}_{k+1}, \mathbf{y}_{1:N})\, p(\mathbf{x}_{k+1} \mid \mathbf{y}_{1:N})\, d\mathbf{x}_{k+1}. \tag{3.62}$$

Noting that observations $\mathbf{y}_{k+1:N}$ are independent of $\mathbf{x}_k$, given $\mathbf{x}_{k+1}$, we have the following Bayesian inference

$$p(\mathbf{x}_k \mid \mathbf{x}_{k+1}, \mathbf{y}_{1:N}) = p(\mathbf{x}_k \mid \mathbf{x}_{k+1}, \mathbf{y}_{1:k}) = p(\mathbf{x}_k \mid \mathbf{y}_{1:k}) \frac{p(\mathbf{x}_{k+1} \mid \mathbf{x}_k)}{p(\mathbf{x}_{k+1} \mid \mathbf{y}_{1:k})}, \tag{3.63}$$

leading to

$$\underbrace{p(\mathbf{x}_k \mid \mathbf{y}_{1:N})}_{A} = \underbrace{p(\mathbf{x}_k \mid \mathbf{y}_{1:k})}_{B} \underbrace{\int \frac{p(\mathbf{x}_{k+1} \mid \mathbf{x}_k)}{p(\mathbf{x}_{k+1} \mid \mathbf{y}_{1:k})} p(\mathbf{x}_{k+1} \mid \mathbf{y}_{1:N})\, d\mathbf{x}_{k+1}}_{C}, \tag{3.64}$$

as in Cosme et al. (2012). This implies that the smoother estimate at $t_k$ (term $A$) is obtained through the filter analysis at $t_k$ (term $B$) with a correction by the smoother analysis at $t_{k+1}$ and the ratio between the transition probability density and the filter forecast at $t_{k+1}$ (term $C$). As the analysis distribution $p(\mathbf{x}_k|\mathbf{y}_{1:k})$ is given by the filter, the smoothing distribution can be attained by a forward–backward recursive algorithm as

1. *Forward filtering*: Starting from the initial state distribution $p(\mathbf{x}_0)$, the analysis (filter) distributions $p(\mathbf{x}_k|\mathbf{y}_{1:k})$ are stored for $t = 1$ to $N$.
2. *Backward smoothing*: Eq. (3.63) provides $p(\mathbf{x}_k|\mathbf{x}_{k+1}, \mathbf{y}_{1:k})$, for $t = N - 1$ to 1, using the stored filter solutions $p(\mathbf{x}_k|\mathbf{y}_{1:k})$. Then, Eq. (3.62) yields the smoothing distributions $p(\mathbf{x}_k|\mathbf{y}_{1:N})$ recursively, using the smoothing distributions from the previous iteration (i.e., $t = k + 1$), $p(\mathbf{x}_{k+1}|\mathbf{y}_{1:N})$.

This algorithm is also called the *forward–backward* smoother or the RTS smoother (see Rauch et al., 1965).

One can obtain the probability distributions in (3.59)–(3.64) explicitly when the dynamical and measurement models are linear and all the distributions are Gaussian. The KF and the Kalman smoother are examples of such recursions of the linear system with a Gaussian distribution.

## References

Aksoy A, Zhang F, Nielsen-Gammon JW (2006) Ensemble-based simultaneous state and parameter estimation in a two-dimensional sea-breeze model. *Mon Wea Rev* 134:2951–2970.

Bannister RN (2008) A review of forecast error covariance statistics in atmospheric variational data assimilation. II: Modelling the forecast error covariance statistics. *Quart J Roy Meteor Soc* 134:1971–1996.

Barker DM, Huang W, Guo Y-R, Bourgeois AJ, Xiao QN (2004) A three-dimensional variational data assimilation system for MM5: Implementation and initial results. *Mon Wea Rev* 132:897–914.

Bocquet M, Sakov P (2013) Joint state and parameter estimation with an iterative ensemble kalman smoother. *Nonlinear Processes Geophys* 20:803–818.

Brasseur P, Ballabrera-Poy J, Verron J (1999) Assimilation of altimetric data in the mid-latitude oceans using the singular evolutive extended Kalman filter with an eddy-resolving, primitive equation model. *J Mar Syst* 22:269–294.

Buehner M, Malanotte-Rizzoli P (2003) Reduced-rank Kalman filters applied to an idealized model of the wind-driven ocean circulation. *J Geophys Res* 108, doi:10.1029/2001jc000873

Cane MA, Kaplan A, Miller RN, Tang B, Hackert EC, Busalacchi AJ (1996) Mapping tropical Pacific sea level: Data assimilation via a reduced state space Kalman filter. *J Geophys Res Oceans* 101:22599–22617.

Carrassi A, Bocquet M, Bertino L, Evensen G (2018) Data assimilation in the geosciences: An overview of methods, issues, and perspectives. *WIREs Clim Change* 9:e535, doi:10.1002/wcc.535

Cohn SE, Todling R (1996) Approximate data assimilation schemes for stable and unstable dynamics. *J Meteor Soc Japan* 74:63–75.

Cohn SE, Sivakumaran NS, Todling R (1994) A fixed-lag Kalman smoother for retrospective data assimilation. *Mon Wea Rev* 122:2838–2867.

Cosme E (2015) Smoothers. In *Advanced Data Assimilation for Geosciences: Lecture Notes of the Les Houches School of Physics: Special Issue, June 2012*, (eds.) Blayo E, Bocquet M, Cosme E, Cugliandolo LF, chap. 4, Oxford University Press, New York, 121–136.

Cosme E, Brankart JM, Verron J, Brasseur P, Krysta M (2010) Implementation of a reduced rank square-root smoother for high resolution ocean data assimilation. *Ocean Modell* 33:87–100.

Cosme E, Verron J, Brasseur P, Blum J, Auroux D (2012) Smoothing problems in a bayesian framework and their linear gaussian solutions. *Mon Wea Rev* 140:683–695.

Crassidis JL, Junkins JL (2012) *Optimal Estimation of Dynamic Systems*. 2nd ed., CRC Press, Boca Raton, FL, 733 pp.

Dee DP (1991) Simplification of the Kalman filter for meteorological data assimilation. *Quart J Roy Meteor Soc* 117:365–384.

Dee DP (1995) On-line estimation of error covariance parameters for atmospheric data assimilation. *Mon Wea Rev* 123:1128–1145.

Evensen G (1994) Sequential data assimilation with a nonlinear quasi-geostrophic model using Monte Carlo methods to forecast error statistics. *J Geophys Res Oceans* 99:10143–10162.

Farrell BF, Ioannou PJ (2001) State estimation using a reduced-order Kalman filter. *J Atmos Sci* 58:3666–3680.

Fitzgerald R (1971) Divergence of the Kalman filter. *IEEE Trans Autom Control* 16:736–747.

Fukumori I (1995) Assimilation of TOPEX sea level measurements with a reduced-gravity, shallow water model of the tropical Pacific Ocean. *J Geophys Res* 100:25027, doi: 10.1029/95jc02083

Gelb A (Ed.) (1974) *Applied Optimal Estimation*. MIT Press, Cambridge, MA, and London, 374 pp.

Gelman A, Carlin JB, Stern HS, et al. (2014) *Bayesian Data Analysis*. 3rd ed., CRC Press, Boca Raton, FL, 639 pp.

Gibbs BP (2011) *Advanced Kalman Filtering, Least-Squares and Modeling: A Practical Handbook*. John Wiley & Sons, Hoboken, NJ, 605 pp.

Heemink A (1988) Two-dimensional shallow water flow identification. *Appl Math Modell* 12:109–118.

Kalman RE (1960) A new approach to linear filtering and prediction problems. *J Basic Eng* 82:35–45.

Kaminski P, Bryson A, Schmidt S (1971) Discrete square root filtering: A survey of current techniques. *IEEE Trans Autom Control* 16:727–736.

Kourzeneva E (2014) Assimilation of lake water surface temperature observations using an extended Kalman filter. *Tellus A* 66:21510, doi:10.3402/tellusa.v66.21510

Lee YH, Park SK, Chang DE (2006) Parameter estimation using the genetic algorithm and its impact on quantitative precipitation forecast. *Ann Geophys* 24:3185–3189.

Lermusiaux PFJ, Robinson AR (1999) Data assimilation via error subspace statistical estimation. Part I: Theory and schemes. *Mon Wea Rev* 127:1385–1407.

Lorenc A (1981) A global three-dimensional multivariate statistical interpolation scheme. *Mon Wea Rev* 109:701–721.

McLaughlin D (2002) An integrated approach to hydrologic data assimilation: Interpolation, smoothing, and filtering. *Adv Water Resour* 25:1275–1286.

Müller PK, von Storch H (eds.) (2004) Models and data. In *Computer Modelling in Atmospheric and Oceanic Sciences: Building Knowledge*, chap. 3, Springer-Verlag, Berlin, Heidelberg, 55–68.

Pham DT (1996) *A Singular Evolutive Interpolated Kalman Filter for Data Assimilation in Oceanography*. Tech. Rep. IDOPT project INRIA-CNRS-UJF-INGP RT 163, Laboratoire de Modeélisation et Calcul, France, Grenoble Cédex, France.

Pham DT, Verron J, Roubaud MC (1998) A singular evolutive extended Kalman filter for data assimilation in oceanography. *J Mar Syst* 16:323–340.

Rauch HE, Tung F, Striebel CT (1965) Maximum likelihood estimates of linear dynamic systems. *AIAA J* 3:1445–1450.

Rozier D, Birol F, Cosme E, et al. (2007) A reduced-order Kalman filter for data assimilation in physical oceanography. *SIAM Rev* 49:449–465.

Ruckstuhl Y, Janjić T (2020) Combined state-parameter estimation with the LETKF for convective-scale weather forecasting. *Mon Wea Rev* 148:1607–1628.

Särkkä S (2013) *Bayesian Filtering and Smoothing*. Institute of Mathematical Statistics Textbooks, Cambridge University Press, Cambridge, 256 pp.

Severijns CA, Hazeleger W (2005) Optimizing parameters in an atmospheric general circulation model. *J Clim* 18:3527–3535.

Simon D (2007) Reduced order Kalman filtering without model reduction. *Control Intell Syst* 35:169–174.

Snyder C (2015) Introduction to the Kalman filter. In *Advanced Data Assimilation for Geosciences*, (eds.) Blayo É, Bocquet M, Cosme E, Cugliandolo LF, chap. 3, Oxford University Press, New York, 75–120.

Storch RB, Pimentel LCG, Orlande HRB (2007) Identification of atmospheric boundary layer parameters by inverse problem. *Atmos Environ* 41:1417–1425.

Todling R, Cohn SE (1994) Suboptimal schemes for atmospheric data assimilation based on the Kalman filter. *Mon Wea Rev* 122:2530–2557.

Verlaan M, Heemink AW (1997) Tidal flow forecasting using reduced rank square root filters. *Stoch Hydrol Hydraul* 11:349–368.

Wikle CK, Berliner LM (2007) A Bayesian tutorial for data assimilation. *Physica D* 230:1–16.

Willman WW (1969) On the linear smoothing problem. *IEEE Trans Autom Control* 14:116–117.

# Part II

## Practical Tools

# 4

# Tangent Linear and Adjoint Models

## 4.1 Introduction

The performance of a numerical model depends on the spatial and temporal evolution of errors or uncertainties in the model (from initial and boundary conditions, physical parameterizations, etc.). Error growth in a nonlinear system (e.g., atmosphere) leads to a predictability limit for a given phenomenon, which can be quantified using either a brute-force approach or a gradient method. In the brute-force (or Monte Carlo) approach, the error evolution is examined by running the nonlinear model (NLM) repeatedly for various initial perturbations on an input parameter, which provides the true nonlinear evolution of the perturbations but is limited to only a small subset of perturbations in the practical sense. In the gradient approach, the evolution of all possible perturbations is obtained through the model solution sensitivity (i.e., gradient), in the context of linear dynamics, using the tangent linear model (TLM) and the adjoint model (ADJM).

Both the TLM and the ADJM are important tools in various fields in meteorology, including sensitivity analysis (e.g., Errico and Vukićević, 1992; Park and Droegemeier, 1999, 2000), variational data assimilation (VAR) (e.g., Lewis and Derber, 1985; Navon et al., 1992; Park and Županski, 2003), and predictabilty assessment (e.g., Urban, 1993; Ehrendorfer and Errico, 1995; Park, 1999). They are derived from the NLM, based on the tangent linear approximation; thus, their validity strongly depends on the nonlinearity of processes involved in the NLM (e.g., Errico et al., 1993; Park and Droegemeier, 1997). That is, their results are valid when the perturbations remain "reasonably small." Establishing this validity and its generality is thus of vital importance in the application of this approach.

In this chapter, we provide a theoretical background in developing the TLM and ADJM and show examples of their applications to sensitivity studies. As both the TLM and ADJM are based on a linear approximation, we first explore the Taylor series expansion to approximate a linear function for a small displacement from a given point.

### 4.2 Taylor Series Approximation

A continuous function $f(x)$ can be expanded around a given point $x_0$ with a small displacement $\delta x$ by the Taylor series:

$$f(x_0+\delta x)=f(x_0)+\delta x\left.\frac{\partial f}{\partial x}\right|_{x_0}+\frac{\delta x^2}{2!}\left.\frac{\partial^2 f}{\partial x^2}\right|_{x_0}+\frac{\delta x^3}{3!}\left.\frac{\partial^3 f}{\partial x^3}\right|_{x_0}+\cdots. \tag{4.1}$$

We can express higher-order terms with a symbol $O(\cdot)$, in terms of $\delta x$; for example, the third- and higher-order terms can be expressed as $O\left(\delta x^3\right)$. As $\delta x$ is small, we can neglect $O\left(\delta x^3\right)$ to approximate a quadratic function:

$$\begin{aligned} f(x_0+\delta x)&=f(x_0)+\delta x\left.\frac{\partial f}{\partial x}\right|_{x_0}+\frac{\delta x^2}{2!}\left.\frac{\partial^2 f}{\partial x^2}\right|_{x_0}+\underbrace{\frac{\delta x^3}{3!}\left.\frac{\partial^3 f}{\partial x^3}\right|_{x_0}+\cdots}_{O(\delta x^3)}\\ &\approx f(x_0)+\delta x\left.\frac{\partial f}{\partial x}\right|_{x_0}+\frac{\delta x^2}{2!}\left.\frac{\partial^2 f}{\partial x^2}\right|_{x_0}, \end{aligned} \tag{4.2}$$

whereas we can neglect $O\left(\delta x^2\right)$ to approximate a linear function:

$$\begin{aligned} f(x_0+\delta x)&=f(x_0)+\delta x\left.\frac{\partial f}{\partial x}\right|_{x_0}+\underbrace{\frac{\delta x^2}{2!}\left.\frac{\partial^2 f}{\partial x^2}\right|_{x_0}+\frac{\delta x^3}{3!}\left.\frac{\partial^3 f}{\partial x^3}\right|_{x_0}+\cdots}_{O(\delta x^2)}\\ &\approx f(x_0)+\delta x\left.\frac{\partial f}{\partial x}\right|_{x_0}. \end{aligned} \tag{4.3}$$

For a multivariable function $f(\mathbf{x})=f(x_1,x_2,\ldots,x_N)$, the Taylor series around $\mathbf{x}$ with a small displacement vector $\delta\mathbf{x}=[\delta x_1,\delta x_2,\ldots,\delta x_N]^T$ can be expressed as:

$$f(\mathbf{x}+\delta\mathbf{x})=f(\mathbf{x})+\sum_{j=1}^{N}\frac{\partial f(\mathbf{x})}{\partial x_j}\delta x_j+\sum_{i=1}^{N}\sum_{j=1}^{N}\frac{\partial^2 f(\mathbf{x})}{\partial x_i\partial x_j}\delta x_i\delta x_j+\cdots. \tag{4.4}$$

In vector form, Eq. (4.4) becomes

$$f(\mathbf{x}+\delta\mathbf{x})=f(\mathbf{x})+\mathbf{g}^T\delta\mathbf{x}+\frac{1}{2}\delta\mathbf{x}^T\mathbf{H}\delta\mathbf{x}+\cdots, \tag{4.5}$$

where $\mathbf{g}$ and $\mathbf{H}$, respectively, are the gradient vector (first-order derivatives) and Hessian matrix (second-order derivatives) of $f(\mathbf{x})$, which are defined as:

$$\mathbf{g}=\nabla f(\mathbf{x})=\frac{d}{d\mathbf{x}}f(\mathbf{x})=\begin{bmatrix}\frac{\partial f(\mathbf{x})}{\partial x_1}\\ \vdots\\ \frac{\partial f(\mathbf{x})}{\partial x_N}\end{bmatrix}_{N\times 1} \tag{4.6}$$

and

$$\mathbf{H} = \nabla \mathbf{g} = \frac{d}{d\mathbf{x}}\mathbf{g} = \begin{bmatrix} \frac{\partial^2 f(\mathbf{x})}{\partial x_1^2} & \cdots & \frac{\partial^2 f(\mathbf{x})}{\partial x_1 \partial x_N} \\ \vdots & \ddots & \vdots \\ \frac{\partial^2 f(\mathbf{x})}{\partial x_N \partial x_1} & \cdots & \frac{\partial^2 f(\mathbf{x})}{\partial x_N^2} \end{bmatrix}_{N\times N}. \tag{4.7}$$

If $f(\mathbf{x})$ has continuous second-order partial derivatives in a neighborhood $P$ of a given point $(p_1, \ldots, p_n)$, the off-diagonal mixed partial derivative components of $\mathbf{H}$ have the following relationship:

$$\frac{\partial^2}{\partial x_j \partial x_i} f(p_1, \ldots, p_n) = \frac{\partial^2}{\partial x_i \partial x_j} f(p_1, \ldots, p_n),$$

implying that the Hessian of $f$ is symmetric (i.e., $\mathbf{H} = \mathbf{H}^T$) throughout $P$.

Equation (4.5) can be expressed, in terms of higher-order terms, as:

$$\begin{aligned} f(\mathbf{x} + \delta\mathbf{x}) &= f(\mathbf{x}) + \mathbf{g}^T \delta\mathbf{x} + \frac{1}{2}\delta\mathbf{x}^T \mathbf{H} \delta\mathbf{x} + O\left(\|\delta\mathbf{x}\|^3\right) \\ &= f(\mathbf{x}) + \mathbf{g}^T \delta\mathbf{x} + O\left(\|\delta\mathbf{x}\|^2\right), \end{aligned} \tag{4.8}$$

where $\|\delta\mathbf{x}\|$ represents the magnitude (norm) of $\delta\mathbf{x}$. As $\|\delta\mathbf{x}\|$ is small, we can neglect $O\left(\|\delta\mathbf{x}\|^3\right)$ to approximate a quadratic function of $f(\mathbf{x} + \delta\mathbf{x})$ as:

$$f(\mathbf{x} + \delta\mathbf{x}) \approx f(\mathbf{x}) + \mathbf{g}^T \delta\mathbf{x} + \frac{1}{2}\delta\mathbf{x}^T \mathbf{H} \delta\mathbf{x}, \tag{4.9}$$

whereas we can neglect $O\left(\|\delta\mathbf{x}\|^2\right)$ to approximate a linear function as:

$$f(\mathbf{x} + \delta\mathbf{x}) \approx f(\mathbf{x}) + \mathbf{g}^T \delta\mathbf{x}. \tag{4.10}$$

**Example 4.1 Taylor series approximation: $f(x) = \cos x$**

The Taylor series of a cosine function, centered at 0, is given by:

$$\cos x = \overbrace{\underbrace{\overbrace{\underbrace{1 - \frac{x^2}{2!}}_{P2} + \frac{x^4}{4!}}^{P4} - \frac{x^6}{6!}}_{P6} + \frac{x^8}{8!}}^{P8} - \cdots. \tag{4.11}$$

Here, $\cos x$ can be approximated by a polynomial of degree 2 as $\cos x \approx 1 - \frac{x^2}{2!} = P2$, by that of degree 4 as $\cos x \approx 1 - \frac{x^2}{2!} + \frac{x^4}{4!} = P4$, and so on. The following figure shows approximations of $\cos x$ from different degrees of polynomials – the higher the polynomial degree, the higher the accuracy.

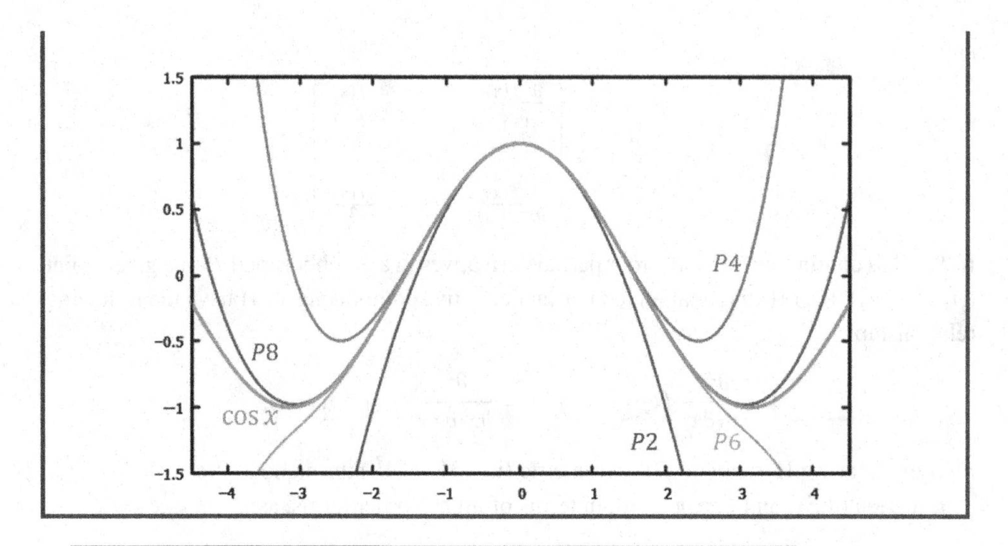


**Example 4.2 Taylor series approximation: $f(x, y) = \ln(1 + x)e^y$**

A second-order Taylor series expansion around point (0, 0) of a function

$$f(x, y) = \ln(1 + x)e^y \tag{4.12}$$

will have the following form:

$$f(x,y) = f(0,0) + (x-0)\left.\frac{\partial f}{\partial x}\right|_{(0,0)} + (y-0)\left.\frac{\partial f}{\partial y}\right|_{(0,0)} + \frac{1}{2!}\left[(x-0)^2\left.\frac{\partial^2 f}{\partial x^2}\right|_{(0,0)}\right.$$

$$\left. + 2(x-0)(y-0)\left.\frac{\partial^2 f}{\partial x \partial y}\right|_{(0,0)} + (y-0)^2\left.\frac{\partial^2 f}{\partial y^2}\right|_{(0,0)}\right] + \cdots.$$

By evaluating the first-order and second-order derivatives at (0,0), and noting that $f(0,0) = 0$, we can approximate $f(x, y)$ to the second order as

$$f(x, y) = \ln(1 + x)e^y \approx x + xy - \frac{1}{2}x^2. \tag{4.13}$$

The following figure shows $f(x, y) = \ln(1+x)e^y$ and the second-order approximation $f(x, y) \approx x + xy - \frac{1}{2}x^2$.

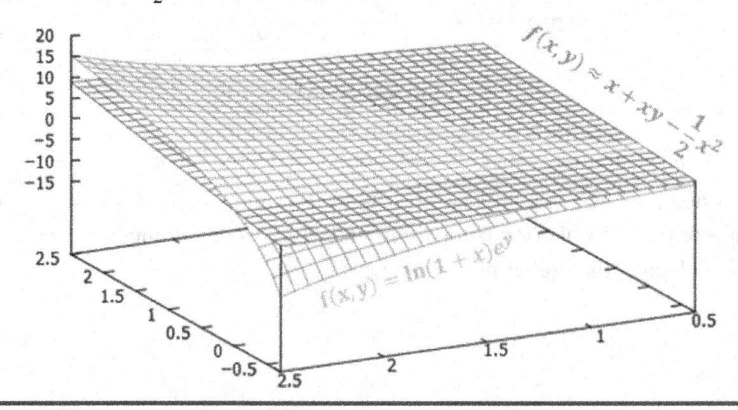

**Practice 4.1 Taylor series calculation**

Derive (4.13) by evaluating the derivative terms in the Taylor series in Example 4.2, and using $f(0,0) = 0$.

## 4.3 Tangent Linear Model

### *4.3.1 Formulation of the TLM*

The TLM is based on the first-order derivative term of the Taylor expansion and describes the linear evolution of perturbations along the solution trajectories of the NLM (Park and Droegemeier, 1997). The TLM is useful for analyzing forward sensitivity (e.g., Park and Droegemeier, 1999, 2000).

A simple dynamical system, or NLM, that represents the temporal evolution of model states can be expressed as:

$$\frac{d}{dt}\mathbf{X} = \mathbf{F}(\mathbf{X}, t), \tag{4.14}$$

where $\mathbf{X}$ is an $m$-dimensional vector of model states and $\mathbf{F}$ is a tendency vector that is a nonlinear function of $X_i$ for $i = 1$ to $m$. The Taylor series expansion of (4.14) for small perturbations $\delta\mathbf{X}(t)$ in the vicinity of $\mathbf{X}(t)$ yields

$$\begin{aligned}\frac{d}{dt}\left[\mathbf{X}(t) + \delta\mathbf{X}(t)\right] &= \mathbf{F}\left[\mathbf{X}(t) + \delta\mathbf{X}(t)\right] \\ &= \mathbf{F}\left[\mathbf{X}(t)\right] + \left.\frac{\partial\mathbf{F}}{\partial\mathbf{X}}\right|_{\mathbf{X}(t)} \delta\mathbf{X}(t) + O\left(\|\delta\mathbf{X}(t)\|^2\right).\end{aligned}$$

By taking the first-order (linear) term, i.e., by neglecting $O\left(\|\delta\mathbf{X}(t)\|^2\right)$, we have

$$\frac{d}{dt}\delta\mathbf{X}(t) = \left.\frac{\partial\mathbf{F}}{\partial\mathbf{X}}\right|_{\mathbf{X}(t)} \delta\mathbf{X}(t) = \mathbf{G}(t)\delta\mathbf{X}(t), \tag{4.15}$$

where the elements of the $m \times m$ Jacobian $\mathbf{G}$ are given by $g_{ik} = \partial F_i / \partial X_k$. Equation (4.15) denotes a linear model whose coefficients are determined by the slopes of the tangents to the nonlinear state trajectories – i.e., $\left.\frac{\partial\mathbf{F}}{\partial\mathbf{X}}\right|_{\mathbf{X}(t)}$; thus, defining the TLM.

In a discrete form, (4.14) can be expressed recursively as

$$\mathbf{X}^{n+1} = \mathbf{F}\left(\mathbf{X}^n\right), \tag{4.16}$$

where the subscript $n$ denotes a time index, where $t_n = n\Delta t + t_0$ with $t_0$ the initial time. The TLM of (4.16) is expressed as

$$\begin{aligned}\delta\mathbf{X}^{n+1} &= \frac{\partial\mathbf{F}\left(\mathbf{X}^n\right)}{\partial\mathbf{X}^n}\delta\mathbf{X}^n \\ &= \mathbf{G}^n\delta\mathbf{X}^n, \end{aligned} \tag{4.17}$$

where $\mathbf{G}^n$ is a Jacobian matrix, whose components are $g_{ij}^n = \partial F_i^n / \partial X_j^n$. Then, a perturbation at any time $t_n$, $\delta\mathbf{X}^n$, can be expressed in terms of initial perturbations, $\delta\mathbf{X}^0$, as a recursive system (e.g., Errico and Vukićević, 1992; Park and Droegemeier, 1997):

$$\delta\mathbf{X}^n = \mathbf{G}^{n-1}\mathbf{G}^{n-2}\cdots\mathbf{G}^1\mathbf{G}^0\delta\mathbf{X}^0 = \mathbf{L}^n\delta\mathbf{X}^0, \tag{4.18}$$

where the elements of $\mathbf{L}^n$ are expressed as $l_{ij}^n = \partial X_i^n / \partial X_j^0$.

Noting that the numerical prediction is an initial value problem, we can regard the solution of Eq. (4.14), $\mathbf{X}(t)$, as a trajectory projected from an initial value $\mathbf{X}(t_0)$ through time integration of Eq. (4.14) from $t_0$ to $t$. Then, we can express the state (prediction) vector $\mathbf{X}(t)$ as a mapping of the initial condition vector $\mathbf{X}(t_0)$ through a nonlinear propagator $\mathbf{M}$ as:

$$\mathbf{X}(t) = \mathbf{M}\mathbf{X}(t_0). \tag{4.19}$$

In the same vein, Eq. (4.18) is expressed as:

$$\delta\mathbf{X}(t) = \mathbf{L}\delta\mathbf{X}(t_0), \tag{4.20}$$

where $\mathbf{L}$ is a linear (or tangent linear) propagator that maps the initial perturbation $\delta\mathbf{X}(t_0)$ onto the projected perturbation $\delta\mathbf{X}(t)$. We note from Eqs. (4.19) and (4.20) that $\mathbf{L}$ is related to $\mathbf{M}$ as:

$$\mathbf{L}(t) = \left.\frac{\partial\mathbf{M}}{\partial\mathbf{X}}\right|_{\mathbf{X}(t)}, \tag{4.21}$$

indicating that $\mathbf{L}(t)$ is the linear first-order derivative of the NLM $\mathbf{M}$, evaluated at a fixed model trajectory $\mathbf{X}(t)$. In other words, $\mathbf{L}(t)$ represents a slope of the tangent to the nonlinear state $\mathbf{X}(t)$ and evolves along the trajectory of $\mathbf{X}(t)$.

**Example 4.3 Deriving the TLM from the NLM**

A simple nonlinear equation or NLM is given as:

$$z = x^2 + y^3, \tag{4.22}$$

and can be represented in a matrix form, by adding $x = x$ and $y = y$, as:

$$\begin{pmatrix} x \\ y \\ z \end{pmatrix} = \begin{pmatrix} 1 & 0 & 0 \\ 0 & 1 & 0 \\ x & y^2 & 0 \end{pmatrix} \begin{pmatrix} x \\ y \\ z \end{pmatrix}.$$

The TLM representation of (4.22) becomes

$$\delta z = 2x\delta x + 3y^2\delta y, \tag{4.23}$$

which can be represented in a matrix form, by adding $\delta x = \delta x$ and $\delta y = \delta y$, as:

$$\begin{pmatrix} \delta x \\ \delta y \\ \delta z \end{pmatrix} = \begin{pmatrix} 1 & 0 & 0 \\ 0 & 1 & 0 \\ 2x & 3y^2 & 0 \end{pmatrix} \begin{pmatrix} \delta x \\ \delta y \\ \delta z \end{pmatrix}.$$

### 4.3.2 Correctness Check of the TLM

To check if the TLM is coded correctly, we do the correctness test by comparing the difference in two nonlinear solutions with the TLM solution. Consider two state vectors $\mathbf{X}_c^n$ (say, control run) and $\mathbf{X}_p^n$ (say, perturbed run) evolved from $\mathbf{X}^0$ and $\mathbf{X}^0 + \delta\mathbf{X}^0$, respectively, through integration of the NLM $\mathbf{M}$ from $t_0$ to $t_n$. Here, $\delta\mathbf{X}^0$ is a small displacement from $\mathbf{X}^0$. Then the nonlinear perturbation (NLP), $\Delta\mathbf{X}$, is defined as:

$$\begin{aligned}\Delta\mathbf{X}^n &= \mathbf{X}_p^n - \mathbf{X}_c^n \\ &= \mathbf{M}\left(\mathbf{X}^0 + \delta\mathbf{X}^0\right) - \mathbf{M}\left(\mathbf{X}^0\right) \\ &= \underbrace{\mathbf{M}\left(\mathbf{X}^0\right) + \mathbf{L}\delta\mathbf{X}^0 + O\left(\|\delta\mathbf{X}\|^2\right)}_{\text{Taylor series expansion of } \mathbf{M}(\mathbf{X}^0+\delta\mathbf{X}^0)} - \mathbf{M}\left(\mathbf{X}^0\right) \\ &= \mathbf{L}\delta\mathbf{X}^0 + O\left(\|\delta\mathbf{X}\|^2\right),\end{aligned} \tag{4.24}$$

where $\mathbf{L}$ is the TLM, i.e., the first derivative of $\mathbf{M}$, as defined in Eq. (4.23). By neglecting $O\left(\|\delta\mathbf{X}\|^2\right)$, we can approximate the NLP, using Eq. (4.22), as:

$$\Delta\mathbf{X}^n \approx \mathbf{L}\delta\mathbf{X}^0 = \delta\mathbf{X}^n,$$

where $\delta\mathbf{X}^n$ can be regarded as the tangent linear perturbation (TLP). This implies that the TLM is accurate (or valid), for small $\delta\mathbf{X}^0$, when $\delta\mathbf{X}^n$ stays as close as possible to $\Delta\mathbf{X}^n$. By introducing a perturbation scaling factor $\alpha$, we can rewrite Eq. (4.24) as:

$$\begin{aligned}\Delta\mathbf{X}^n &= \mathbf{M}\left(\mathbf{X}^0 + \alpha\delta\mathbf{X}^0\right) - \mathbf{M}\left(\mathbf{X}^0\right) \\ &= \alpha\mathbf{L}\delta\mathbf{X}^0 + O\left(\alpha^2\right) \\ &\approx \delta\mathbf{X}^n.\end{aligned}$$

Then we can evaluate the correctness of TLM by checking if the ratio between the NLP and the TLP tends to unity as $\alpha$ tends to 0, that is,

$$R_\alpha \equiv \lim_{\alpha\to 0} \frac{\left\|\mathbf{M}\left(\mathbf{X}^0 + \alpha\delta\mathbf{X}^0\right) - \mathbf{M}\left(\mathbf{X}^0\right)\right\|}{\left\|\alpha\mathbf{L}\delta\mathbf{X}^0\right\|} \longrightarrow 1. \tag{4.25}$$

In practice, if $R_\alpha$ approaches unity like $\alpha$ and approaches 0 decreasing from unity, we can ensure that the TLM is constructed correctly. For example, Vidard et al. (2015) made a TLM correctness test for a subroutine and obtained the $R_\alpha$ values for various values of $\alpha$ as in Table 4.1. The table clearly shows that $R_\alpha$ approaches unity as $\alpha$ approaches 0, ensuring that the subroutine's TLM is correct.

Relative errors are also used to assess the TLM correctness alternatively, for instance, as in Rabier and Courtier (1992):

$$\varepsilon_R = \frac{\|\Delta\mathbf{X}^n - \delta\mathbf{X}^n\|}{\|\delta\mathbf{X}^n\|} = \frac{\left\|\mathbf{M}\left(\mathbf{X}^0 + \alpha\delta\mathbf{X}^0\right) - \mathbf{M}\left(\mathbf{X}^0\right) - \alpha\mathbf{L}\delta\mathbf{X}^0\right\|}{\left\|\alpha\mathbf{L}\delta\mathbf{X}^0\right\|} \tag{4.26}$$

Table 4.1. *An example of the TLM correctness test. Modified from Vidard et al. (2015).* 

| $\alpha$ | $R_\alpha$ |
|---|---|
| $1 \times 10^0$ | 0.999961862090 |
| $1 \times 10^{-1}$ | 0.999995878199 |
| $1 \times 10^{-2}$ | 0.999999584740 |
| $1 \times 10^{-3}$ | 0.999999958442 |
| $1 \times 10^{-4}$ | 0.999999995846 |

or as in Vukićević and Bao (1998):

$$\varepsilon_R = \frac{\|\Delta \mathbf{X}^n - \delta \mathbf{X}^n\|}{\|\Delta \mathbf{X}^n\|} = \frac{\left\|\mathbf{M}\left(\mathbf{X}^0 + \alpha\delta\mathbf{X}^0\right) - \mathbf{M}\left(\mathbf{X}^0\right) - \alpha\mathbf{L}\delta\mathbf{X}^0\right\|}{\left\|\mathbf{M}\left(\mathbf{X}^0 + \alpha\delta\mathbf{X}^0\right) - \mathbf{M}\left(\mathbf{X}^0\right)\right\|}. \tag{4.27}$$

## 4.4 Adjoint Model

### 4.4.1 Formulation of the ADJM

The ADJM is derived as an exact transpose of the TLM; thus, the accuracy of ADJM applications strongly depends on the TLM validity. The ADJM provides gradient information for both adjoint sensitivity analysis (e.g., Errico and Vukićević, 1992; Rabier et al., 1992; Langland et al., 1995) and VAR (e.g., Lewis and Derber, 1985; Le Dimet and Talagrand, 1986; Navon et al., 1992; Li and Droegemeier, 1993; Zou et al., 1993; Li et al., 1994).

#### 4.4.1.1 Adjoint Operator Method

A linear operator $\mathbf{L}^*$ is termed the *adjoint* of another linear operator $\mathbf{L}$ if the following relation is satisfied for all elements of a linear vector space, $\mathbf{x}$ and $\mathbf{y}$:

$$\langle \mathbf{y}, \mathbf{L}\mathbf{x} \rangle = \left\langle \mathbf{L}^*\mathbf{y}, \mathbf{x} \right\rangle, \tag{4.28}$$

where the scalar product $\langle \mathbf{A}, \mathbf{B} \rangle$ represents the sum of the products of corresponding components of $\mathbf{A}$ and $\mathbf{B}$. Because of the symmetry of the scalar product, it follows that $\mathbf{L}$ is also the adjoint of $\mathbf{L}^*$, that is, $(\mathbf{L}^*)^* = \mathbf{L}$ (Friedman, 1956).

Let $J = P(\mathbf{X})$ denote a scalar function, which is related to the dependent variables $\mathbf{X}$, and $\delta J$ its TLM solution, i.e.,

$$\delta J^n = \mathbf{Q}^n \delta \mathbf{X}^n = \mathbf{Q}^n \mathbf{L}^n \delta \mathbf{X}^0,$$

where the elements of $\mathbf{Q}^n$ are $q_j^n = \partial P^n / \partial X_j^n$ for $j = 1, \ldots, m$. By applying the adjoint operator (4.28), we have

$$\delta J^n = \left\langle \mathbf{Q}^n, \mathbf{L}^n \delta \mathbf{X}^0 \right\rangle$$
$$= \left\langle \left(\mathbf{L}^n\right)^* \mathbf{Q}^n, \delta \mathbf{X}^0 \right\rangle.$$

We also notice that

$$\delta J^n = \left\langle \nabla J^n, \delta \mathbf{X}^0 \right\rangle, \tag{4.29}$$

where $\nabla J^n$ denotes $\partial J^n / \partial \mathbf{X}^0$. Therefore, we can describe the ADJM as

$$\frac{\partial J^n}{\partial \mathbf{X}^0} = \left(\mathbf{L}^n\right)^* \mathbf{Q}^n = \left(\mathbf{L}^n\right)^* \frac{\partial P^n}{\partial \mathbf{X}^n} = \left(\mathbf{L}^n\right)^* \frac{\partial J^n}{\partial \mathbf{X}^n}. \tag{4.30}$$

Note that $(\mathbf{L}^n)^*$ is just the transpose of $\mathbf{L}^n$ because the elements of $(\mathbf{L}^n)^*$ is $\left(l_{ij}^n\right)^* = \partial X_j^n / \partial X_i^0 = l_{ji}^n$. By defining the adjoint variables $\hat{\mathbf{X}}^0 = \partial J^n / \partial \mathbf{X}^0$ and $\hat{\mathbf{X}}^n = \partial J^n / \partial \mathbf{X}^n$, Eq. (4.30) is represented as

$$\hat{\mathbf{X}}^0 = \left(\mathbf{L}^n\right)^* \hat{\mathbf{X}}^n. \tag{4.31}$$

*4.4.1.2 Lagrangian-λ Method*

A Lagrange function, $\mathcal{L}$, is defined as the sum of the original dependent variables, usually the cost function, and the products of the state equations and the adjoint variables (i.e., the Lagrange multiplier, $\boldsymbol{\lambda}$). Components in the state equation (4.16) are used as auxiliary conditions (Lanczos, 1970). Lagrange multipliers are usually considered as a means of incorporating constraints into an optimization problem (Thacker, 1991, see also Colloquy 7.4 in Chapter 7).

Let us define a dependent variable $J$ which is a scalar function of model states at a given time (i.e., $\mathbf{X}^n$). Then, we can express the Lagrange function, using (4.16), as

$$\mathcal{L} = J + \sum_{n=1}^{N} \boldsymbol{\lambda}^n \left(\mathbf{X}^n - \mathbf{F}\left(\mathbf{X}^{n-1}\right)\right). \tag{4.32}$$

Then, by taking partial derivatives of $\mathcal{L}$ with respect to each independent variable, we have

$$\begin{aligned}
\frac{\partial \mathcal{L}}{\partial \mathbf{X}^N} &= \frac{\partial J}{\partial \mathbf{X}^N} + \boldsymbol{\lambda}^N \\
\frac{\partial \mathcal{L}}{\partial \mathbf{X}^n} &= \frac{\partial J}{\partial \mathbf{X}^n} + \boldsymbol{\lambda}^n - \boldsymbol{\lambda}^{n+1} \frac{\partial \mathbf{F}\left(\mathbf{X}^n\right)}{\partial \mathbf{X}^n}, \quad n = N-1, \ldots, 1 \\
\frac{\partial \mathcal{L}}{\partial X^0} &= \frac{\partial J}{\partial X^0} - \boldsymbol{\lambda}^1 \frac{\partial F\left(\mathbf{X}^0\right)}{\partial \mathbf{X}^0}.
\end{aligned} \tag{4.33}$$

The unknown adjoint variables ($\boldsymbol{\lambda}$) are determined by requiring that the variations in $\mathcal{L}$ with respect to the independent variables vanish. Thus, one can obtain the ADJM from (4.33) as

$$\begin{aligned}
\boldsymbol{\lambda}^{N+1} &= 0 \\
\boldsymbol{\lambda}^n &= -\frac{\partial J}{\partial \mathbf{X}^n} + \boldsymbol{\lambda}^{n+1} \frac{\partial \mathbf{F}\left(\mathbf{X}^n\right)}{\partial \mathbf{X}^n}, \quad n = N, \ldots, 0.
\end{aligned} \tag{4.34}$$

Notice that $\boldsymbol{\lambda}^0 = -\partial \mathcal{L}/\partial \mathbf{X}^0$. Using this method, we can derive the ADJM directly, without developing the corresponding TLM explicitly.

**Practice 4.2 Derivation of the ADJM**

Assume that you have the following NLM:

$$\frac{dx_1}{dt} = -\frac{1}{2}x_1x_2; \quad \frac{dx_2}{dt} = \frac{1}{2}x_1^2,$$

where the model state vector is given by $\mathbf{x} = (x_1 \ x_2)^T$. We define a functional $J$ as

$$J = \int_0^T \left\langle \mathbf{x} - \mathbf{x}^{obs}, \mathbf{x} - \mathbf{x}^{obs} \right\rangle dt,$$

where $\mathbf{x}^{obs} = \left(x_1^{obs} \ x_2^{obs}\right)^T$ is the observation vector. Noting that the NLM above can be generally expressed as

$$\frac{d\mathbf{x}}{dt} = \mathbf{F}(\mathbf{x}),$$

and by defining the Lagrange multiplier as $\boldsymbol{\Lambda} = (\lambda_1 \ \lambda_2)^T$, we can form a Lagrangian:

$$\mathcal{L} = J + \int_0^T \left\langle \boldsymbol{\Lambda}, \frac{d\mathbf{x}}{dt} - \mathbf{F}(\mathbf{x}) \right\rangle dt.$$

By assuming the following conditions

$$\lambda_1(T) = \lambda_1(0) = \lambda_2(T) = \lambda_2(0) = 0,$$

derive the ADJM, using the Langrangian–$\lambda$ method.

#### 4.4.1.3 Chain Rule Method

A more straightforward and simple method for developing the ADJM was proposed by Talagrand (1991) and Errico and Vukićević (1992), using the chain rule. Noting that $J$ is a scalar function of model states, i.e., $J = J(\mathbf{X}^n)$, the gradient of $J$ with respect to the model initial states $\mathbf{X}^0$ is given, by applying the chain rule, as:

$$\frac{\partial J}{\partial X_i^0} = \sum_j \frac{\partial X_j^n}{\partial X_i^0} \frac{\partial J}{\partial X_j^n}. \tag{4.35}$$

Compared with Eq. (4.18), $\partial X_j^n / \partial X_i^0$ is regarded as $l_{ji}$ elements of $\mathbf{L}^n$, i.e., representing $(\mathbf{L}^n)^T$. By defining the adjoint variables $\partial J/\partial X_i^0 = \hat{\mathbf{X}}^0$ and $\partial J/\partial X_i^n = \hat{\mathbf{X}}^n$, (4.35) can be written as

$$\hat{\mathbf{X}}^0 = \left(\mathbf{L}^n\right)^T \hat{\mathbf{X}}^n, \tag{4.36}$$

which is essentially the same as (4.31). As in (4.17), $\hat{\mathbf{X}}^0$ can be determined recursively as

$$\hat{\mathbf{X}}^{n-1} = \left(\mathbf{G}^{n-1}\right)^T \hat{\mathbf{X}}^n, \tag{4.37}$$

where $\mathbf{G}$ is defined in (4.15). Comparing with the TLM (4.17), we notice that, in the ADJM (4.37), the time index $n$ decreases – i.e., marching backward in time, the linear perturbation $\delta\mathbf{X}$ is replaced by the gradient of $J$ (i.e., $\hat{\mathbf{X}} = \nabla_{\mathbf{X}} J$), and the Jacobian $\mathbf{G}$ is replaced by its transpose $\mathbf{G}^T$ (see Errico and Vukićević, 1992).

**Example 4.4 Deriving the ADJM from the TLM**

We start from the TLM (4.23), rewritten here in a matrix form

$$\begin{pmatrix} \delta x \\ \delta y \\ \delta z \end{pmatrix} = \begin{pmatrix} 1 & 0 & 0 \\ 0 & 1 & 0 \\ 2x & 3y^2 & 0 \end{pmatrix} \begin{pmatrix} \delta x \\ \delta y \\ \delta z \end{pmatrix},$$

which yields a set of tangent linear equations

$$\begin{aligned} \delta x &= \delta x \\ \delta y &= \delta y \\ \delta z &= 2x\delta x + 3y^2 \delta y. \end{aligned}$$

In reality, the first two statements above are not coded; however, they are represented here for convenience in developing the ADJM by exploiting the fact that the ADJM is a transpose of the TLM.

To develop the ADJM from the TLM above, we transpose the matrix and replace the tangent linear variables $\delta x$, $\delta y$, and $\delta z$ with the adjoint variables $\delta x^*$, $\delta y^*$, and $\delta z^*$, respectively; then the ADJM is represented in a matrix form as:

$$\begin{pmatrix} \delta x^* \\ \delta y^* \\ \delta z^* \end{pmatrix} = \begin{pmatrix} 1 & 0 & 2x \\ 0 & 1 & 3y^2 \\ 0 & 0 & 0 \end{pmatrix} \begin{pmatrix} \delta x^* \\ \delta y^* \\ \delta z^* \end{pmatrix}.$$

This yields a set of adjoint equations:

$$\begin{aligned} \delta x^* &= \delta x^* + 2x\delta z^* \\ \delta y^* &= \delta y^* + 3y^2 \delta z^* \\ \delta z^* &= 0. \end{aligned}$$

Here, the last equation should not be neglected because it is a part of the ADJM.

### 4.4.2 Correctness and Gradient Check of the ADJM

The correctness of the ADJM can be checked by testing the adjoint identity, Eq. (4.31), for random perturbations $\delta\mathbf{X}$ and $\delta\mathbf{Y}$ as:

$$\langle \mathbf{L}\delta\mathbf{X},\ \delta\mathbf{Y}\rangle = \left\langle \delta\mathbf{X},\ \mathbf{L}^{*}\delta\mathbf{Y}\right\rangle. \tag{4.38}$$

By noting that $\langle \mathbf{A},\ \mathbf{B}\rangle = \mathbf{A}^T\mathbf{B}$ and $\mathbf{L}^* = \mathbf{L}^T$ and by putting $\delta\mathbf{Y} = \mathbf{L}\delta\mathbf{X}$, Eq. (4.38) becomes

$$\underbrace{(\mathbf{L}\delta\mathbf{X})^T\ (\mathbf{L}\delta\mathbf{X})}_{\delta\mathbf{Y}^T\delta\mathbf{Y}} = \underbrace{\delta\mathbf{X}^T\mathbf{L}^T\ (\mathbf{L}\delta\mathbf{X})}_{\delta\mathbf{X}^T\delta\mathbf{X}^*}, \tag{4.39}$$

where $\mathbf{L}\delta\mathbf{X} = \delta\mathbf{Y}$ implies the output of a forward TLM run with the input $\delta\mathbf{X}$, whereas $\mathbf{L}^T\ (\mathbf{L}\delta\mathbf{X}) = \delta\mathbf{X}^*$ denotes that of a backward ADJM run starting from the output of the TLM. This should be verified for each subroutine up to the machine accuracy; for example, 13-digit accuracy on the 64-bit machines (e.g., Zhang et al., 2014).

From the viewpoint of data assimilation, especially using the ADJM, the gradient of the cost function ($J$) with respect to the input vector, i.e., $\nabla_{\mathbf{X}_0} J$ is of utmost importance; thus, we also need to check the gradient accuracy for the developed ADJM. For the gradient check, the following formula is used (e.g., Navon et al., 1992):

$$\Phi_\alpha \equiv \lim_{\alpha\to 0} \frac{J\left(\mathbf{X} + \alpha\mathbf{h}\right) - J(\mathbf{X})}{\alpha\mathbf{h}^T\nabla J(\mathbf{X})} \longrightarrow 1, \tag{4.40}$$

where $\mathbf{h} = \nabla J/\|\nabla J\|$.

Figure 4.1 shows the gradient check results, from Navon et al. (1992), using (4.40) along with calculation of $\log_{10}|\Phi_\alpha - 1|$. We notice that $\Phi_\alpha$ tends to unity for sufficiently small $\alpha$ values (i.e., $10^{-14} \leq \alpha \leq 10^{-7}$), whereas $\log_{10}|\Phi_\alpha - 1|$ shows a "V" shape with the minimum at $\alpha = 10^{-10}$. This ensures that the gradient $\nabla_{\mathbf{X}_0} J$ is calculated correctly from the given ADJM. The gradient checks using other models produced results similar to Figure 4.1 (e.g., Fang et al., 2006; Chen et al., 2014). Therefore, it is essential to check both the correctness and the gradient calculation before using the ADJM for any purpose (sensitivity analysis, VAR, etc.).

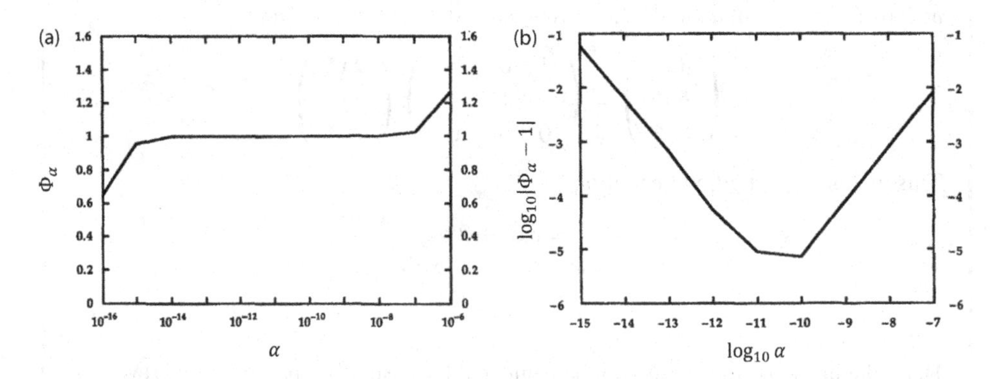


Figure 4.1 Verification of gradient calculation: (a) variation of $\Phi_\alpha$ with respect to $\alpha$; and (b) variation of $\log_{10}|\Phi_\alpha - 1|$ with respect to $\log_{10}\alpha$. Modified from Navon et al. (1992). ©American Meteorological Society. Used with permission.

## 4.5 Uncertainty and Validity of the TLM/ADJM

Most sensitivity and data assimilation problems require the gradient information, especially from the ADJM, whose accuracy definitely depends on the accuracy or uncertainty of the TLM. Therefore, it is of utmost importance to understand the nature of uncertainties in the TLM and their impact on the validity of the TLM/ADJM.

### *4.5.1 Sources of Uncertainty in the TLM*

As the TLM represents derivatives of the corresponding NLM, the basic requirement is that the original NLM formulas should be differentiable. Furthermore, the TLM solution is the first-order linear approximation to the NLP; thus, the TLM is valid for sufficiently small perturbations. Most meteorological models include various on–off switching (i.e., discontinuous) processes – e.g., phase changes of hydrometors – that bring up several issues in developing the TLMs.

Several factors making the TLM invalid and uncertain include the followings (e.g., Errico et al., 1993; Vukićević and Errico, 1993; Park and Droegemeier, 1997; Janisková and Lopez, 2013): 1) discontinuity and nondifferentiability; 2) variation in the on–off switching time; 3) nonlinearity and chaotic systems; 4) exponential forcing terms; 5) frequency of updating nonlinear trajectories; 6) specific formulations of parameterizations; and 7) linearized physical processes. Each of these are discussed below.

#### *4.5.1.1 Discontinuity and Nondifferentiability*

As realistic meteorological models include moist and diabatic processes that contain subgrid-scale physical parameterizations (cumulus convection and turbulent flow), the model solutions, especially hydrometeors such as cloud water, rain water, etc., can have on–off switching processes due to phase changes of water substances, which are discontinuous and nondifferentiable. These on–off processes also affect other model states and their associated diagnostic functions (e.g., cost function). This nature of discontinuity and nondifferentiability in the NLM solutions conflicts with the basic requirement of the TLM that includes the first-order derivatives.

#### *4.5.1.2 Variation in the On–Off Switching Time*

The subgrid-scale physical processes occur in a grid-box scale; thus, the on–off switches in physical parameterizations are strongly controlled by local atmospheric conditions (Zhang et al., 2001). Most parameterization schemes include parameters that determine the on or off time in the switching processes via pre-specified threshold values, which are governed by the model state (control) variables and their derived quantities. Therefore, the switching time varies depending on the parameterization schemes in the NLM, which influences both the TLM and ADJM. Even with a specific parameterization scheme, the switching time varies with different perturbation sizes (e.g., Park and Droegemeier, 1997; Zhang et al., 2001), thus affecting the validity of TLM.

#### 4.5.1.3 Nonlinearity and Chaotic Systems

Numerical models that have highly nonlinear or chaotic properties in solutions invalidate the linear approximation, especially for extended time integration, large perturbations, and regime shifts. Nonlinearity increases as numerical solutions evolve in the time domain; thus, even with a very small perturbation, solutions show substantially different behaviors. Li (1991) showed that a cost function, having a global minimum, changed its shape to have multiple local minima with a longer assimilation time due to the increase in nonlinearity. Chaotic systems involve bifurcation points, indicating regime shifts where solutions from nearby initial conditions with a tiny difference depict totally diverging trajectories (e.g., Ehrendorfer, 1997; Scheffer et al., 2001; Thompson and Sieber, 2011; Müller et al., 2014). Because the TLM is linearized along nonlinear trajectories, its solutions (i.e., TLPs) near a bifurcation point or a regime shift boundary fail to describe the true evolution of NLPs even with infinitesimal initial perturbations.

#### 4.5.1.4 Exponential Forcing Terms

Certain physical processes are presented by exponential forcing terms, such as $Qq^{\beta}$, where $q$ is a water-related variable and $Q$ and $\beta$ are constants. When $q$ is substantially small and $0 < \beta < 1$, the TLM may have unreasonably large derivatives. In cloud models, some parameterizations (e.g., terminal velocity and cloud/rain evaporation) include such a term (e.g., Park and Droegemeier, 1997), which produces a derivative term in the TLM as:

$$Q\beta q^{\beta-1}\delta q = Q\beta\left(\frac{q^{\beta}}{q}\right)\delta q. \tag{4.41}$$

The magnitude of $q$ can sometimes be exceptionally small due to machine rounding off (e.g., $10^{-30}$); thus, (4.41) becomes unreasonably large, making the TLM invalid. We can regulate this undesirable effect by imposing an artificial lower bound on $q$ (e.g., $10^{-8}$) though the exact linearization is violated (Vukićević and Errico, 1993; Park and Droegemeier, 1997).

#### 4.5.1.5 Frequency of Updating Nonlinear Trajectories

The infrequent update of nonlinear trajectories in the TLM is a significant source of error – see, e.g., Errico et al. (1993) for a dry model and Park and Droegemeier (1997) for a cloud model – especially in convective storms where physical processes are active, and solutions change their behaviors remarkably in a short time.

Following Errico et al. (1993), the nonlinear (exact) effect of perturbations of variables $q$ and $r$ on a quadratic term $qr$ at time $t$ is

$$N = \left(q_0^t + \delta q^t\right)\left(r_0^t + \delta r^t\right) - q_0^t r_0^t,$$

where $q_0^t$ and $r_0^t$ are nonlinear trajectories and $\delta q^t$ and $\delta r^t$ are perturbations. Then, the TLM approximation, using nonlinear trajectories at $t_n$ (i.e., $q_0^n$ and $r_0^n$), is given by

$$L = q_0^n \delta r^t + r_0^n \delta q^t.$$

Thus, the linearization error becomes

$$N - L = \underbrace{\left(q_0^t - q_0^n\right)\delta r^t}_{A} + \underbrace{\left(r_0^t - r_0^n\right)\delta q^t}_{B} + \underbrace{\delta q^t \delta r^t}_{C},$$

where terms $A$ and $B$ are related to the update periods of nonlinear trajectories and term $C$ to perturbation sizes. Term $C$ dominates for large perturbations and frequent updates (i.e., small $q_0^t - q_0^n$ and $r_0^t - r_0^n$). For infrequent updates, terms $A$ and $B$ govern the error, prohibiting immediate matches between NLPs and TLPs, even with small perturbations.

#### *4.5.1.6 Specific Formulations of Parameterizations*

Certain formulations of subgrid-scale parameterizations, in both microphysical and dynamical processes, can influence the TLM solutions. Park and Droegemeier (1997) showed that, in the cloud-process parameterizations (terminal velocity, lateral eddy mixing, etc.), a slight change in specific parameters resulted in substantial differences between the NLPs and TLPs. For example, in a cloud model, the terminal velocity $V_T$ can be parameterized as

$$V_T = a\,(\rho q_r)^b,$$

where $a$ and $b$ are dimensionless coefficients, $\rho$ the air density, and $q_r$ the rainwater mixing ratio. For different choices of the values $a$ and $b$, given the same perturbation in a variable, e.g., vertical velocity, the NLM solutions may show different behaviors in the cloud. For instance, more prominent discontinuity may occur in the NLM solutions around a specific parameter value than other values; thus, making the TLM solutions sensitive to that parameter.

#### *4.5.1.7 Linearized Physical Processes*

In order to include moist physics in the TLM/ADJM and to practically implement them in operational data assimilation, one may employ linearized physical parameterization schemes (e.g., Janisková and Lopez, 2013); however, without proper regularization (i.e., smoothing), the linearized models may have problems. As shown in Figure 4.2, a physical process may include an on/off switching function, $y = f_0(x)$ (stepwise black solid line), whose derivative $\left(\frac{\partial y}{\partial x}\right)_0$ (black dashed line) is undefined (i.e., nondifferentiable). To alleviate this problem, one can modify the function to make it differentiable and less steep. For example, the autoconversion of cloud water to rainwater is generally described as

$$P = cLT, \tag{4.42}$$

where $P$ is the autoconversion rate (in g cm$^{-3}$ s$^{-1}$), $c$ is an empirical coefficient (in s$^{-1}$), $L$ is the cloud liquid water (in g cm$^{-3}$), and $T$ is the threshold function. For a Kessler-type parameterization (Kessler, 1969), $T$ is given by

$$T = \mathcal{H}\,(L - L_c), \tag{4.43}$$

where $\mathcal{H}$ is the Heaviside step function that forces $P$ to be 0 when the driving cloud water liquid content is less than its critical value ($L_c$). In contrast, a Sundqvist-type parameterization (Sundqvist, 1978) includes a smoother threshold function as

$$T = 1 - e^{-\left(\frac{L}{L_c}\right)^{\mu}}, \tag{4.44}$$

in general form, where $\mu \geq 0$ is an empirical constant (Liu et al., 2006). As $\mu \to \infty$, the Sundqvist-type parameterization approaches the Kessler-type parameterization.

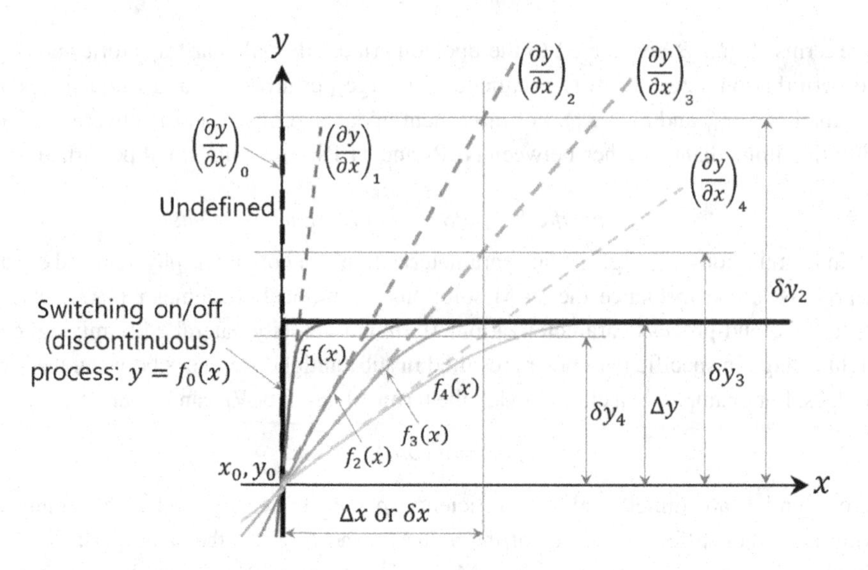


Figure 4.2 Many physical processes (e.g., condensation) occur when certain conditions are satisfied, producing a solution with on–off switching (discontinuous) function; thus, making its first-order derivative (i.e., tangent line) undefined. Here, $x$ and $y$ are any quantities related to a given physical process, and the derivatives are evaluated at $(x_0, y_0)$. The discontinuity problem can be alleviated by modifying the function to be less steep (e.g., Janisková and Lopez, 2013). A change $\Delta x$ results in an NLP $(\Delta y)$. The corresponding linear perturbations $(\delta y)$ are different for different functions; for example, $\delta y_4$ has better approximation to $\Delta y$ than $\delta y_3$ and $\delta y_2$.

In Figure 4.2, the original function $y = f_0(x)$ may be regarded as the Kessler-type threshold function (4.43) while the modified functions $f_1(x), \ldots, f_4(x)$ (gray solid lines) may be similar to the Sundqvist-type threshold functions (4.44) for various $\mu$. Note that the nonlinear variation $\Delta y$ is kept small for a given small change in $\Delta x$; however, the corresponding tangent linear variations, e.g., $\delta y_2$, of the modified functions are much larger than $\Delta y$. In particular, the derivative of $y = f_1(x)$, i.e., $\left(\frac{\partial y}{\partial x}\right)_1$ is very large and its linear perturbation is exceptionally large (i.e., $\delta y_1 \gg \Delta y$; not shown) though it is the closest to the original function $y = f_0(x)$. The less steep functions have better approximations of $y$-variations for the given $\Delta x$, especially $\delta y_4 \approx \Delta y$; however, they require significant modification of the original function, which can deteriorate the quality of original physical parameterization. Therefore, the linearity and the actual (nonlinear) realization in the parameterization schemes should be well balanced (Janisková and Lopez, 2013).

### *4.5.2 Validity of the TLM/ADJM*

The validity assessment of the TLM and hence the ADJM aims at checking whether the TLM solutions (i.e., TLPs) provide good approximations to NLPs for a given *realistic* perturbation comparable to analysis errors. If the TLM is valid only for unrealistically small perturbations below the accuracies of current observing systems, the TLM/ADJM-based sensitivity analysis and data assimilation would be of little practical use (Park and Droegemeier, 1997). Furthermore, it is essential to detail the range of TLM validity regarding scales of atmospheric systems and prediction, integration time, perturbation magnitude, model parameters and states, dry versus moist atmosphere, physical processes, flow characteristics, etc.

Hohenegger and Schar (2007) assessed, in operational ensemble prediction systems, the time scale of the TLM validity of the medium-range synoptic-scale model (ECMWF) and the short-range cloud-resolving model (Lokal Modell; LM) with horizontal resolutions of T255 (80 km) and 0.02° (2.2 km), respectively. They showed that the TLM validity extends to 54 h for ECMWF and 1.5 h for LM. In idealized convective storm simulations with a horizontal resolution of 1 km, Park and Droegemeier (2000) showed that the TLM was valid up to 30 min for a realistic perturbation (e.g., 10% errors in water vapor): Relative humidity observations are typically biased negatively by several percentage points to about 20%. They also demonstrated that the TLM validity strongly depended on the perturbation insertion time during the storm life cycle – the TLM becomes invalid much earlier when perturbation is inserted at a mature stage with multiple convective cells than at a developing stage with a single cell. In a mesoscale model with diabatic moist physics, the TLM errors showed the maximum values near the cyclone center where the NLPs were large due to parameterized moist processes (Vukićević and Errico, 1993). These imply that the TLM validity degrades with vigorous moist convective processes represented by strong nonlinearity and discontinuity.

## 4.6 Example: Continuous ADJM from Burgers' Equation

The nonlinear viscous Burgers' equation is

$$\frac{\partial u}{\partial t} = -u\frac{\partial u}{\partial x} + \nu\frac{\partial^2 u}{\partial x^2} = F(u), \tag{4.45}$$

where $u$ is velocity, $x$ and $t$ are space and time, respectively, and $\nu$ is a viscous coefficient. Let us define the cost function $J$, which measures the discrepancy between the model solution and observation, as

$$J = \frac{1}{2}\int_0^T\int_0^X \left(u - u^o\right)^2 \, dxdt, \tag{4.46}$$

where $u$ is model solution and $u^o$ the observations. In order to obtain the gradient information of $J$, we need to develop the ADJM of this system. In practice, the adjoint is developed directly from the computer code.

### 4.6.1 Adjoint Operator Method

Here we apply the adjoint operator method to develop the ADJM. We first develop the TLM for (4.45) by taking the first-order term in the Taylor expansion,

$$\frac{\partial}{\partial t}\delta u = -\frac{\partial}{\partial x}u\delta u + \nu\frac{\partial^2}{\partial x^2}\delta u = \frac{\partial F}{\partial u}\delta u, \tag{4.47}$$

where $\delta u$ is the tangent linear variable and $u$ is the nonlinear trajectory (basic state). By defining a scalar product $\langle A, B\rangle$ as

$$\langle A, B\rangle = \int_0^T\int_0^X AB\, dxdt, \tag{4.48}$$

we can express the cost function (4.46) in terms of a scalar product as

$$J = \frac{1}{2}\left\langle u - u^o, u - u^o\right\rangle. \tag{4.49}$$

Then the variation $\delta J$ due to variation $\delta u$ is given by

$$\begin{aligned}\delta J &= J(u + \delta u) - J(u)\\ &= \frac{1}{2}\left\langle u + \delta u - u^o, u + \delta u - u^o\right\rangle - \frac{1}{2}\langle u - u^o, u - u^o\rangle\\ &= \frac{1}{2}\int_0^T\int_0^X \left[\left(u - u^o\right)^2 + 2\delta u\left(u - u^o\right) + (\delta u)^2 - \left(u - u^o\right)^2\right] dxdt.\end{aligned}$$

By neglecting the higher-order variation terms,

$$\begin{aligned}\delta J &= \int_0^T\int_0^X \left(u - u^o\right)\delta u\, dxdt\\ &= \left\langle u - u^o, \delta u\right\rangle.\end{aligned}$$

Since $\delta J = \langle\nabla J, \delta u_0\rangle$ by definition, we have

$$\left\langle u - u^o, \delta u\right\rangle = \langle\nabla J, \delta u_0\rangle. \tag{4.50}$$

Taking the scalar product of the tangent linear equation (4.47) with an arbitrary vector $\mathbf{G}$, we have

$$\left\langle\frac{\partial}{\partial t}\delta u, \mathbf{G}\right\rangle = \left\langle\frac{\partial F}{\partial u}\delta u, \mathbf{G}\right\rangle = \left\langle\delta u, \frac{\partial F^*}{\partial u}\mathbf{G}\right\rangle, \tag{4.51}$$

where $\frac{\partial F^*}{\partial u}$ is the adjoint of $\frac{\partial F}{\partial u}$.

Noting that $\langle\mathbf{A}, \mathbf{B}\rangle$ represents the sum of the products of corresponding components of $\mathbf{A}$ and $\mathbf{B}$, and using the boundary condition $\mathbf{G}(x, T) = 0$, Eq. (4.51) becomes

$$\left\langle\frac{\partial}{\partial t}\delta u, \mathbf{G}\right\rangle = -\langle\delta u_0, \mathbf{G}_0\rangle - \int_0^X\int_0^T \delta u\frac{\partial\mathbf{G}}{\partial t}\, dtdx \tag{4.52}$$

and

$$\left\langle\delta u, \frac{\partial F^*}{\partial u}\mathbf{G}\right\rangle = \int_0^X\int_0^T \delta u\frac{\partial F^*}{\partial u}\mathbf{G}\, dtdx. \tag{4.53}$$

By combining (4.52) and (4.53), we have

$$-\langle \delta u_0, \mathbf{G}_0 \rangle = \int_0^X \int_0^T \delta u \left[ \frac{\partial \mathbf{G}}{\partial t} + \frac{\partial F^*}{\partial u} \mathbf{G} \right] dt dx$$
$$= \left\langle \delta u, \frac{\partial \mathbf{G}}{\partial t} + \frac{\partial F^*}{\partial u} \mathbf{G} \right\rangle.$$

If $\mathbf{G}$ satisfies

$$\frac{\partial \mathbf{G}}{\partial t} + \frac{\partial F^*}{\partial u} \mathbf{G} = u - u^o, \tag{4.54}$$

then from (4.50)

$$-\langle \delta u_0, \mathbf{G}_0 \rangle = \langle \delta u, u - u^o \rangle = \langle \nabla J, \delta u_0 \rangle.$$

Note that

$$\mathbf{G}_0 = \mathbf{G}(x, t = 0) = -\nabla_{u_0} J.$$

$\mathbf{G}$ is indeed the adjoint variable and (4.54) represents the adjoint equation of (4.47), which provides the gradient of $J$.

**Practice 4.3 Scalar product**

Derive (4.52) using $\mathbf{G}(x, T) = 0$.

### *4.6.2 Lagrangian–λ Method*

The Lagrange function $\mathcal{L}$ is formulated by introducing the Lagrange multiplier $\lambda$ to Eq. (4.45) as:

$$\mathcal{L} = J + \int_0^T \int_0^X \lambda \left[ \frac{\partial u}{\partial t} + u \frac{\partial u}{\partial x} - \nu \frac{\partial^2 u}{\partial x^2} \right] dx dt,$$

where $[\cdot]$ is called the strong constraint – here, the dynamical model. Taking the first variation of $\mathcal{L}$ and setting it to 0 yields,

$$0 \equiv \delta \mathcal{L} = \delta J + \int_0^T \int_0^X \lambda \left[ \frac{\partial}{\partial t} \delta u + \frac{\partial}{\partial x} \delta \left( \frac{u^2}{2} \right) - \nu \frac{\partial^2}{\partial x^2} \delta u \right] dx dt, \tag{4.55}$$

where

$$\delta J = \frac{1}{2} \int_0^T \int_0^X \delta \left( u - u^o \right)^2 dx dt$$
$$= \int_0^T \int_0^X \left( u - u^o \right) \delta u \, dx dt.$$

**Practice 4.4 Integration by parts (I)**

Using integration by parts and specifying the boundary conditions as

$$\lambda(x,0) = \lambda(x,T) = \lambda(0,t) = \lambda(X,t) = 0;$$
$$\left.\frac{\partial \lambda}{\partial x}\right|_{x=X} = \left.\frac{\partial \lambda}{\partial x}\right|_{x=0} = 0 \text{ for all } t,$$

show that

$$\int_0^T \int_0^X \lambda \frac{\partial}{\partial t} \delta u \, dxdt = -\int_0^X \int_0^T \delta u \frac{\partial \lambda}{\partial t} \, dtdx$$
$$\int_0^T \int_0^X \lambda \frac{\partial}{\partial x} \delta \left(\frac{u^2}{2}\right) dxdt = -\int_0^T \int_0^X u \delta u \frac{\partial \lambda}{\partial x} \, dxdt$$
$$\int_0^T \int_0^X \lambda \nu \frac{\partial^2}{\partial x^2} \delta u \, dxdt = \int_0^T \int_0^X \delta u \nu \frac{\partial^2 \lambda}{\partial x^2} \, dxdt.$$

Then, Eq. (4.55) becomes

$$0 \equiv \delta \mathcal{L} = \int_0^T \int_0^X \left[ \left(u - u^o\right) \delta u - \delta u \frac{\partial \lambda}{\partial t} - u \delta u \frac{\partial \lambda}{\partial x} - \delta u \nu \frac{\partial^2 \lambda}{\partial x^2} \right] dxdt$$
$$= \int_0^T \int_0^X \delta u \left[ \left(u - u^o\right) - \frac{\partial \lambda}{\partial t} - u \frac{\partial \lambda}{\partial x} - \nu \frac{\partial^2 \lambda}{\partial x^2} \right] dxdt.$$

Because $\delta u$ is arbitrary, the quantity inside the square brackets must be identically 0. Thus, we obtain the adjoint equation

$$-\frac{\partial \lambda}{\partial t} - u \frac{\partial \lambda}{\partial x} - \nu \frac{\partial^2 \lambda}{\partial x^2} = -\left(u - u^o\right), \tag{4.56}$$

where the Lagrange multiplier $\lambda$ is a sensitivity or adjoint variable. Note that the viscous term did not change in the adjoint formulation, and thus is called *self-adjoint*.

### *4.6.3 Equivalence of the ADJMs from the Two Methods*

Let us examine if the ADJMs (4.54) and (4.56) are equivalent to each other. To do this, we must try to find the adjoint of each term on the right side of (4.47) by applying the adjoint operator method.

**Practice 4.5 Integration by parts (II)**

By employing the definition of the scalar product, (4.48), and using integration by parts with the boundary conditions for **G**

$$\mathbf{G}(0,t) = \mathbf{G}(X,t) = 0;$$

$$\left.\frac{\partial \mathbf{G}}{\partial x}\right|_{x=X} = \left.\frac{\partial \mathbf{G}}{\partial x}\right|_{x=0} = 0 \text{ for all } t,$$

show that

$$\left\langle -\frac{\partial}{\partial x} u\delta u,\ \mathbf{G}\right\rangle = \left\langle \delta u,\ u\frac{\partial \mathbf{G}}{\partial x}\right\rangle$$

$$\left\langle \nu\frac{\partial^2}{\partial x^2}\delta u,\ \mathbf{G}\right\rangle = \left\langle \delta u,\ \nu\frac{\partial^2 \mathbf{G}}{\partial x^2}\right\rangle.$$

Therefore, from (4.51)

$$\frac{\partial F^*}{\partial u}\mathbf{G} = u\frac{\partial \mathbf{G}}{\partial x} + \nu\frac{\partial^2 \mathbf{G}}{\partial x^2},$$

and from (4.54)

$$-\frac{\partial \mathbf{G}}{\partial t} - u\frac{\partial \mathbf{G}}{\partial x} - \nu\frac{\partial^2 \mathbf{G}}{\partial x^2} = -\left(u - u^o\right), \tag{4.57}$$

which is equivalent to (4.56).

## 4.7 Continuous versus Discrete ADJM

In Section 4.6, we investigated two methods for developing the adjoint of a continuous system. One should note that developing the adjoint of a discrete system is not generally the same as discretizing the adjoint of a continuous system. If we discretize the continuous adjoint system, it does not give a unique discrete system since we can apply numerous computational methods to solve it.

In Figure 4.3, a continuous NLM is discretized by selecting a numerical scheme, say, Scheme 3; then an ADJM is constructed directly from the discrete NLM – producing discrete adjoint gradients. An ADJM can be also developed directly from the continuous NLM; then continuous adjoint gradients are calculated by choosing a numerical scheme, e.g., Scheme 3 – the same as the discrete NLM and ADJM. The discrete adjoint gradients are generally different from the continuous adjoint gradients.

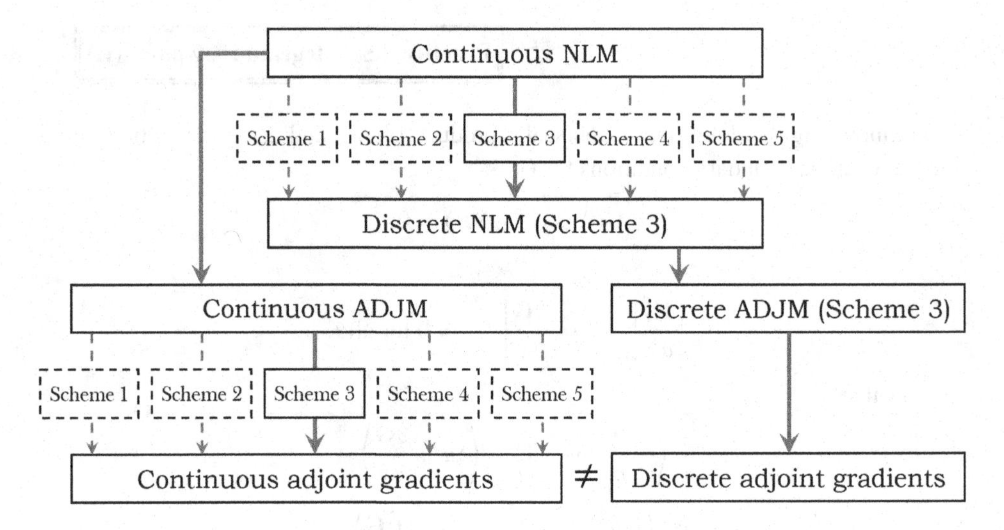


Figure 4.3 Processes to develop the continuous adjoint gradients vs the discrete adjoint gradients, starting from a continuous NLM. An example is shown with a choice of the specific numerical scheme, i.e., Scheme 3.

The ADJM constructed from the discrete system experiences similar computational problems as the original discrete model, mainly due to the specific choice of numerical scheme. For example, when the discrete system employed the leapfrog scheme that inherently has computational modes, the ADJM developed from this discrete system also includes computational modes in the solution (e.g., Sirkes and Tziperman, 1997). Generally, the discrete adjoint gradients match better with the finite difference gradients than with the continuous adjoint gradients (e.g., Nadarajah and Jameson, 2000; Gou and Sandu, 2011). However, Gou and Sandu (2011) showed that, in comparison experiments using an air quality model that uses a highly nonlinear advection scheme, the four-dimensional variational data assimilation (4DVAR) produced superior performance (i.e., faster convergence and higher analysis quality) with the continuous adjoint gradients. They also reported that while the continuous adjoint approach produced smooth gradient fields, the discrete adjoint approach led to some spurious gradients due to nonphysical derivatives of numerical solutions.

In conclusion, the choice of ADJM, constructed either from a discrete system or a continuous system, impacts the results of sensitivity analysis and data assimilation because each ADJM produces different gradient information – discrete adjoint gradients verus continuous adjoint gradients, respectively. The discrepancy becomes more serious when the numerical model employs a highly nonlinear advection scheme or a scheme that includes computational modes. In order to avoid numerical artifacts in the adjoint solution, Sirkes and Tziperman (1997) suggested the following: 1) to use the finite difference of adjoint formulation (i.e., the ADJM from a continuous system), where the numerical stability of the adjoint solution is explicitly taken care of; and 2) to develop numerical schemes that are stable and whose adjoint of finite difference (i.e., the ADJM from a discrete system) is numerically stable as well.

Note that, currently, we often use the so-called automatic differentiation (AD) tools to generate the adjoint code directly from the computer code of the forward NLM. As the AD tools generate the discrete ADJM, we may encounter potential numerical problems as discussed above; thus, requiring us to use the ADJM with caution under a deep understanding of the implemented numerical scheme.

**Practice 4.6 Continuous ADJM**

Derive a continuous ADJM of the linear advection equation:

$$\frac{\partial u}{\partial t} + c\frac{\partial u}{\partial x} = 0, \tag{4.58}$$

using the Lagrangian–$\lambda$ method with the cost function

$$J = \frac{1}{2}\int_0^X \left[u(x,T) - u^o(x,T)\right]^2 dx,$$

where $X$ is the domain length, $T$ is the final forecast time, and $u^o$ is the observation. Then develop the finite difference of the ADJM using the leapfrog scheme (see Appendix B).

**Practice 4.7 Discrete ADJM**

Develop a finite difference equation from (4.58), using the leapfrog scheme (see Appendix B). Then develop a discrete ADJM directly from the discrete NLM (see Section 4.8). Compare your result with the one you derived in Practice 4.6.

## 4.8 General Rules of Constructing the TLM/ADJM

A TLM is constructed by *linearizing* the corresponding NLM by either taking the first-order variation in the Taylor series expansion or differentiating variables directly, whereas an ADJM is obtained by taking the transpose of the corresponding TLM. Most linearized models include nonlinear trajectories (or basic states) as well as linearized (or perturbation) variables: For instance, in (4.23) of Example 4.3, $\delta z$, $\delta x$, and $\delta y$ are tangent linear (i.e., linearized) variables while $x$ and $y$ are nonlinear trajectories. To execute the linearized model codes, those nonlinear trajectories should be either stored in the memory or recomputed. By implementing the TLM line by line in the original NLM and running both simultaneously, one can update the basic state at every time step, thus reducing the uncertainty related to the update of the basic fields.

We introduce general rules in linearizing the NLM and constructing the TLM/ADJM through various examples, in addition to Examples 4.3 and 4.4. Here, we assume that the nonlinear trajectories are prestored. The very first step in linearization is to identify the input and output variables and input constants. For example, assume that we have the following NLM:

$$\begin{aligned} X &= YZ \\ W &= ZX^2 + YX. \end{aligned} \tag{4.59}$$

For simplicity, we represent a constant in uppercase and a variable in lowercase. By assuming that $Y$ is a constant and $Z$ is a variable, the TLM of (4.59) becomes

$$\begin{aligned} \delta x &= Y\delta z \\ \delta w &= x^2\delta z + 2zx\delta x + Y\delta x. \end{aligned} \tag{4.60}$$

If both $Y$ and $Z$ are variables, the TLM of (4.59) is

$$\begin{aligned} \delta x &= z\delta y + y\delta z \\ \delta w &= x^2\delta z + 2zx\delta x + x\delta y + y\delta x. \end{aligned} \tag{4.61}$$

The TLM (4.60) can be represented in matrix form as:

$$\begin{pmatrix} \delta x \\ \delta z \\ \delta w \end{pmatrix} = \begin{pmatrix} 0 & Y & 0 \\ 0 & 1 & 0 \\ 2zx + Y & x^2 & 0 \end{pmatrix} \begin{pmatrix} \delta x \\ \delta z \\ \delta w \end{pmatrix},$$

resulting in the ADJM:

$$\begin{pmatrix} \delta x^* \\ \delta z^* \\ \delta w^* \end{pmatrix} = \begin{pmatrix} 0 & 0 & 2zx + Y \\ Y & 1 & x^2 \\ 0 & 0 & 0 \end{pmatrix} \begin{pmatrix} \delta x^* \\ \delta z^* \\ \delta w^* \end{pmatrix},$$

thus, having three adjoint equations:

$$\begin{aligned} \delta x^* &= 2zx\delta w^* + Y\delta w^* \\ \delta z^* &= \delta z^* + Y\delta x^* + x^2\delta w^* \\ \delta w^* &= 0. \end{aligned} \tag{4.62}$$

**Practice 4.8 Developing an ADJM**

Show that the ADJM from the TLM (4.61) has the following equation set:

$$\begin{aligned} \delta x^* &= 2zx\delta w^* + y\delta w^* \\ \delta y^* &= \delta y^* + z\delta x^* + x\delta w^* \\ \delta z^* &= \delta z^* + y\delta x^* + x^2\delta w^* \\ \delta w^* &= 0. \end{aligned}$$

Based on these exercises, we notice that the TLM (4.60) has three variables $(\delta x, \delta z, \delta w)$ and the corresponding ADJM (4.62) has three equations. Similarly, the TLM (4.61) has four

Table 4.2. *Examples of constructing TLM/ADJM with uppercases representing constants and lowercases variables*

| NLM | TLM | ADJM |
|---|---|---|
| $x = X$ | $\delta x = 0$ | $\delta x^* = 0$ |
| $x = Yy + Zz$ | $\delta x = Y\delta y + Z\delta z$ (3 variables) | $\delta z^* = \delta z^* + Z\delta x^*$<br>$\delta y^* = \delta y^* + Y\delta x^*$<br>$\delta x^* = 0$ |
| $x = Xx + Yy + Zz$ | $\delta x = X\delta x + Y\delta y + Z\delta z$ (3 variables) | $\delta z^* = \delta z^* + Z\delta x^*$<br>$\delta y^* = \delta y^* + Y\delta x^*$<br>$\delta x^* = X\delta x^*$ |
| $x = Yy + Zz + Ww$ | $\delta x = Y\delta y + Z\delta z + W\delta w$ (4 variables) | $\delta w^* = \delta w^* + W\delta x^*$<br>$\delta z^* = \delta z^* + Z\delta x^*$<br>$\delta y^* = \delta y^* + Y\delta x^*$<br>$\delta x^* = 0$ |

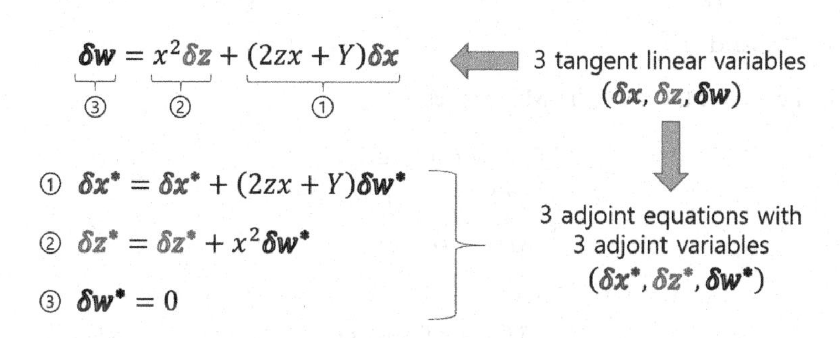


Figure 4.4 Basic rule to develop an ADJM from one equation of the TLM. The tangent linear variables are $(\delta x, \delta z, \delta w)$, and the corresponding adjoint variables are $(\delta x^*, \delta z^*, \delta w^*)$. The circled numbers depict the order of converting the TLM to the ADJM and that of the ADJM equations.

variables $(\delta x, \delta y, \delta z, \delta w)$ and the corresponding ADJM has four equations (see Practice 4.2). Figure 4.4 shows the basic rule to construct the ADJM from one equation of the TLM with three variables $(\delta x, \delta z, \delta w)$, which are converted to the adjoint variables $(\delta x^*, \delta z^*, \delta w^*)$ – each adjoint variable constituting each equation of the ADJM. Some arithmetic operation examples for coding the TLM/ADJM are summarized in Table 4.2.

We now discuss the case with iterative (loop) calculation. Consider a TLM code with the following do loop:

$$\begin{aligned} &\text{do } j = 1, J \\ &\qquad \delta x(j) = X\delta x(j-1) + Y\delta y(j) \\ &\text{end do.} \end{aligned}$$

Following the rules in Table 4.2, and making the loop reversed (i.e., decreasing from $J$ to 1), we have the ADJM as:

$$\begin{aligned}&\texttt{do } j = J, 1, -1\\&\quad \delta y^*(j) = \delta y^*(j) + Y\delta x^*(j)\\&\quad \delta x^*(j-1) = \delta x^*(j-1) + X\delta x^*(j)\\&\quad \delta x^*(j) = 0\\&\texttt{end do.}\end{aligned}$$

Many physical processes in meteorology include on–off switches that require conditional statements in the code (e.g., "if-then-else-endif" statement in Fortran). For example, we may have the following NLM and its corresponding TLM codes:

$$\begin{array}{lcl}\texttt{if}(T > T_0)\texttt{ then} & & \texttt{if}(T > T_0)\texttt{ then}\\ \quad Q = T^2 & & \quad \delta Q = 2T\delta T\\ \texttt{else} & \xrightarrow{\text{linearization}} & \texttt{else}\\ \quad Q = T & & \quad \delta Q = \delta T\\ \texttt{endif} & & \texttt{endif.}\end{array}$$

The ADJM code for the above TLM becomes

$$\begin{aligned}&\texttt{if}(T > T_0)\texttt{ then}\\&\quad \delta T^* = \delta T^* + 2T\delta Q^*\\&\quad \delta Q^* = 0\\&\texttt{else}\\&\quad \delta T^* = \delta T^* + \delta Q^*\\&\quad \delta Q^* = 0\\&\texttt{endif.}\end{aligned}$$

Note that the nonlinear trajectory $T$ should be either prestored or recalculated.

## References

Chen Y, Zhang R, Gao Y, Lu C (2014) Analysis of the properties of adjoint equations and accuracy verification of adjoint model based on fvm. *Abstract Appl Anal* 2014:407468, doi:10.1155/2014/407468

Ehrendorfer M (1997) Predicting the uncertainty of numerical weather forecasts: A review. *Meteorol Z* 6:147–183.

Ehrendorfer M, Errico RM (1995) Mesoscale predictability and the spectrum of optimal perturbations. *J Atmos Sci* 52:3475–3500.

Errico RM, Vukićević T (1992) Sensitivity analysis using an adjoint of the PSU-NCAR mesoscale model. *Mon Wea Rev* 120:1644–1660.

Errico RM, Vukićević T, Raeder K (1993) Examination of the accuracy of a tangent linear model. *Tellus A* 45:462–477.

Fang F, Piggott MD, Pain CC, Gorman GJ, Goddard AJH (2006) An adaptive mesh adjoint data assimilation method. *Ocean Modell* 15:39–55.

Friedman B (1956) *Principles and Techniques of Applied Mathematics*. John & Wiley Sons, Inc., New York, 336 pp.

Gou T, Sandu A (2011) Continuous versus discrete advection adjoints in chemical data assimilation with cmaq. *Atmos Environ* 45:4868–4881.

Hohenegger C, Schar C (2007) Atmospheric predictability at synoptic versus cloud-resolving scales. *Bull Amer Meteor Soc* 88:1783–1794.

Janisková M, Lopez P (2013) Linearized physics for data assimilation at ECMWF. In *Data Assimilation for Atmospheric, Oceanic and Hydrologic Applications (vol. II)*, (eds.) Park SK, Xu L, Springer-Verlag, Berlin, Heidelberg, 251–286.

Kessler E (1969) *On the Distribution and Continuity of Water Substance in Atmospheric Circulation*. Meteor. Monogr., No. 32, American Meteorological Society, 84 pp.

Lanczos C (1970) *The Variational Principles of Mechanics*. University of Toronto Press, Toronto, 464 pp.

Langland RH, Elsberry RL, Errico RM (1995) Evaluation of physical processes in an idealized extratropical cyclone using adjoint sensitivity. *Quart J Roy Meteor Soc* 121:1349–1386.

Le Dimet F-X, Talagrand O (1986) Variational algorithms for analysis and assimilation of meteorological observations: Theoretical aspects. *Tellus A* 38:97–110.

Lewis JM, Derber JC (1985) The use of adjoint equations to solve a variational adjustment problem with advective constraints. *Tellus A* 37:309–322.

Li Y (1991) A note on the uniqueness problem of variational adjustment approach to four-dimensional data assimilation. *J Meteor Soc Japan* 69:581–585.

Li Y, Droegemeier KK (1993) The influence of diffusion and associated errors on the adjoint data assimilation technique. *Tellus A* 45:435–448.

Li Y, Navon IM, Yang W, et al. (1994) Four-dimensional variational data assimilation experiments with a multilevel semi-Lagrangian semi-implicit general circulation model. *Mon Wea Rev* 122:966–983.

Liu Y, Daum PH, McGraw R, Wood R (2006) Parameterization of the autoconversion process. Part II: Generalization of Sundqvist-type parameterizations. *J Atmos Sci* 63:1103–1109.

Müller D, Sun Z, Vongvisouk T, et al. (2014) Regime shifts limit the predictability of land-system change. *Global Environ Change* 28:75–83.

Nadarajah S, Jameson A (2000) A comparison of the continuous and discrete adjoint approach to automatic aerodynamic optimization. In *Proceedings of 38th Aerospace Sciences Meeting and Exhibit*, Reno, NV, 667, doi:10.2514/6.2000-667

Navon IM, Zou X, Derber J, Sela J (1992) Variational data assimilation with an adiabatic version of the NMC spectral model. *Mon Wea Rev* 120:1433–1446.

Park SK (1999) Nonlinearity and predictability of convective rainfall associated with water vapor perturbations in a numerically simulated storm. *J Geophys Res Atmos* 104:31575–31587.

Park SK, Droegemeier KK (1997) Validity of the tangent linear approximation in a moist convective cloud model. *Mon Wea Rev* 125:3320–3340.

Park SK, Droegemeier KK (1999) Sensitivity analysis of a moist 1D Eulerian cloud model using automatic differentiation. *Mon Wea Rev* 127:2180–2196.

Park SK, Droegemeier KK (2000) Sensitivity analysis of a 3D convective storm: Implications for variational data assimilation and forecast error. *Mon Wea Rev* 128:140–159.

Park SK, Županski D (2003) Four-dimensional variational data assimilation for mesoscale and storm-scale applications. *Meteor Atmos Phys* 82:173–208.

Rabier F, Courtier P (1992) Four-dimensional assimilation in the presence of baroclinic instability. *Quart J Roy Meteor Soc* 118:649–672.

Rabier F, Courtier P, Talagrand O (1992) An application of adjoint models to sensitivity analysis. *Beitr Phys Atmos* 65:177–192.

Scheffer M, Carpenter S, Foley JA, Folke C, Walker B (2001) Catastrophic shifts in ecosystems. *Nature* 413:591–596.

Sirkes Z, Tziperman E (1997) Finite difference of adjoint or adjoint of finite difference? *Mon Wea Rev* 125:3373–3378.

Sundqvist H (1978) A parameterization scheme for non-convective condensation including prediction of cloud water content. *Quart J Roy Meteor Soc* 104:677–690.

Talagrand O (1991) The use of adjoint equations in numerical modelling of the atmospheric circulation. In *Automatic Differentiation of Algorithms: Theory, Implementation, and Application*, (eds.) Griewank A, Corliss GF, SIAM, Philadelphia, PA, 169–180.

Thacker WC (1991) Automatic differentiation from an oceanographer's perspective. In *Automatic Differentiation of Algorithms: Theory, Implementation, and Application*, (eds.) Griewank A, Corliss GF, SIAM, Philadelphia, PA, 191–201.

Thompson JMT, Sieber J (2011) Predicting climate tipping as a noisy bifurcation: A review. *Int J Bifurc Chaos* 21:399–423.

Urban B (1993) A method to determine the theoretical maximum error growth in atmospheric models. *Tellus A* 45:270–280.

Vidard A, Bouttier P-A, Vigilant F (2015) NEMOTAM: Tangent and adjoint models for the ocean modelling platform NEMO. *Geosci Model Dev* 8:1245–1257.

Vukićević T, Bao J-W (1998) The effect of linearization errors on 4DVAR data assimilation. *Mon Wea Rev* 126:1695–1706.

Vukićević T, Errico RM (1993) Linearization and adjoint of parameterized moist diabatic processes. *Tellus A* 45:493–510.

Zhang S, Zou X, Ahlquist JE (2001) Examination of numerical results from tangent linear and adjoint of discontinuous nonlinear models. *Mon Wea Rev* 129:2791–2804.

Zhang X, Huang X-Y, Liu J, et al. (2014) Development of an efficient regional four-dimensional variational data assimilation system for WRF. *J Atmos Ocean Technol* 31:2777–2794.

Zou X, Navon IM, Sela JG (1993) Variational data assimilation with moist threshold processes using the NMC spectral model. *Tellus A* 45:370–387.

# 5
# Automatic Differentiation

## 5.1 Introduction

In many scientific computations, calculating derivatives (or gradients) of model variables is essential. In particular, variational data assimilation (VAR; see Chapter 7) involves a minimization step, where gradient-based algorithms (see Chapter 6) are mostly employed. Tangent linear and adjoint models (see Chapter 4) provide the gradient information of prognostic/diagnostic variables from a nonlinear model (NLM) and are constructed based on differentiation of the corresponding NLM.

Derivatives can be approximated by *finite differencing* based on the Taylor series expansion (see Section 4.2). For example, the first-order derivative of $f(\mathbf{x})$ with respect to the $i$-th component of $\mathbf{x}$, at a given point $\mathbf{x}_0$, can be approximated using the forward difference as

$$\left.\frac{\partial f(\mathbf{x})}{\partial x_i}\right|_{\mathbf{x}=\mathbf{x}_0} = \frac{f\left(\mathbf{x}_0 + \mathbf{d}\right)}{d}$$

for a sufficiently small $d$, where $\mathbf{d} = d\mathbf{e}_i$ with the $i$-th Cartesian basis vector $\mathbf{e}_i$ (see Appendix B). This method is simple; however, it includes intrinsic numerical errors (e.g., truncation errors) and requires as many computations as the number of input variables.

*Symbolic differentiation* is an alternative method to obtain derivatives, especially for mathematical expressions, through symbolic manipulations. For instance, a function $f(x) = x_1 * \cdots * x_n$ has its derivatives, e.g., for $n = 4$, as

$$\frac{\partial f}{\partial x_1} = x_2 * x_3 * x_4, \quad \frac{\partial f}{\partial x_2} = x_1 * x_3 * x_4, \quad \frac{\partial f}{\partial x_3} = x_1 * x_2 * x_4, \quad \text{and} \quad \frac{\partial f}{\partial x_4} = x_1 * x_2 * x_3.$$

This approach provides exact derivatives; however, it is inefficient due to many duplicate subexpressions in representing different derivative expressions, which may demand high computational resources for functions with complex description (Bischof et al., 1992; Gebremedhin and Walther, 2020).

*Automatic* (or *algorithmic*) *differentiation* (AD) evaluates exact derivative information of a function in terms of a program rather than a formula (Griewank and Walther, 2008). The AD tools can alleviate the difficulty in developing the derivative codes by hand, which is a tedious, time-consuming, and error-prone task, especially for large codebase models (Park et al., 1996; Griewank and Walther, 2008; Margossian, 2019, etc.). By applying the chain rule systematically to elementary operations or functions to generate derivative codes of NLMs, the AD tools can evaluate the gradients efficiently and accurately (Bischof et al., 2008b), which can be used for sensitivity analysis (e.g., Park et al., 1996; Hwang et al., 1997; Zhang et al., 1998; Park and Droegemeier, 1999, 2000; Kim et al., 2006; Molkenthin et al., 2017), optimization (e.g., Courty et al., 2003; Kourounis et al., 2014; Li, 2020), and data assimilation (e.g., Qin et al., 2007; Knorr et al., 2010; Wang et al., 2018).

In this chapter, we provide an overview of AD with regard to mathematical and computational background and its various tools. More comprehensive discussions on AD are provided in textbooks (e.g., Griewank and Walther, 2008; Naumann, 2012; Henrard, 2017), and technical trends and advances in AD are compiled in editorial volumes (e.g., Bischof et al., 2008a; Forth et al., 2012). Recent reviews on AD can be found in some journal articles (e.g., Baydin et al., 2018; Margossian, 2019; Gebremedhin and Walther, 2020). A community portal for AD, available at `www.autodiff.org/`, updates information on available AD tools.

## 5.2 Background on Automatic Differentiation

Automatic differentiation techniques count on a computer program that executes functions as a sequence of elementary operations (additions, multiplications, etc.) and elementary functions (sin, cos, log, etc.), to which the chain rule of differential calculus is applied systematically (Bischof et al., 1992). For example, AD tools compute the derivative of a function $g(f(x)) = g(u)$ at $x_0$, by applying the chain rule, as

$$\left.\frac{\partial g(f(x))}{\partial x}\right|_{x=x_0} = \left.\frac{\partial g(u)}{\partial u}\right|_{u=f(x_0)} \left.\frac{\partial f(x)}{\partial x}\right|_{x=x_0} \tag{5.1}$$

(see Colloquy 5.1).

Some commonly used terminologies in AD are listed below, some of which follow Bischof et al. (2008b) and Griewank and Walther (2008):

**Independent variables:** A subset of the input variables for a program or subprogram and the set of variables with respect to which differentiation is conducted: They are also called *independents*.

**Dependent variables:** A subset of the output variables for a program or subprogram and the set of variables whose derivatives require to be evaluated: They are also called *dependents*.

**Intermediate variables:** A set of variables that serve as a link between the dependent variables and independent variables in applying the chain rule (see Colloquy 5.1), which are also called *active* variables. They vary by changes in the independent variables while their own changes cause variations in the dependent variables.

**Evaluation trace:** A record of running a given program, with particular specified values for the input variables, showing the sequence of floating-point values calculated by a processor and the operations that computed them (see, e.g., Table 5.2).

**Tree diagram:** A general graphical device that systematically represents a sequence of all possible occurrences. In particular, a differential tree diagram depicts the paths of all intermediate derivative processes in the chain rule (see, e.g., the figure in Colloquy 5.1).

**Computational graph:** The directed acyclic graph (DAG) for a statement, basic block, or execution trace. Vertex labels are operators or functions and optionally variable names (see, e.g., Figure 5.1).

**Linearized computational graph:** The computational graph with symbolic or numeric edge weights equal to the partial derivative of the target with respect to the source vertex. The derivative of a root vertex with respect to a leaf vertex is the sum over all paths of the product of the edge weights along that path (see, e.g., Figure 5.3).

**Derivative accumulation:** Application of the chain rule, typically using either forward or reverse mode.

**Preaccumulation:** Computing the partial derivatives for a statement, basic block, or other program subunit. The local partial derivatives are then used in the overall derivative accumulation.

**Activity analysis:** Identifying the set of variables lying along a dataflow path from the independent variables to the dependent variables. Variables along a path are termed *active*, and variables not along any path are termed *passive*.

COLLOQUY 5.1

**Chain rule for multivariate functions**

The chain rule describes how to differentiate composite functions that can be represented, e.g., as $g(f(x))$. Suppose that $h = g(u)$ and $u = f(x)$ are differentiable functions. Then, the composite function $h = g(f(x))$ is a differentiable function of $x$, and its derviative is given by

$$h'(x) = g'(f(x)) \cdot f'(x)$$

or alternatively expressed, in Leibniz notation (Christianson, 2012), as

$$\frac{dh}{dx} = \frac{dh}{du} \cdot \frac{du}{dx} = \frac{dg(u)}{du} \cdot \frac{df(x)}{dx}.$$

Here, $u$ denotes an intermediate variable, which serves as a dependent variable for $x$ but as an independent variable for $h$. When the derivative is evaluated at $x_0$, the chain rule is specified as

$$\left.\frac{dh}{dx}\right|_{x_0} = \left.\frac{dh}{du}\right|_{u(x_0)} \cdot \left.\frac{du}{dx}\right|_{x_0} = \left.\frac{dg(u)}{du}\right|_{u=f(x_0)} \cdot \left.\frac{df(x)}{dx}\right|_{x_0},$$

as in (5.1).

For a function $w = h(u, v)$, differentiable of $u$ and $v$, where $u = f(t)$ and $v = g(t)$ are differentiable functions of $t$, the derivative of $w = h(u(t), v(t))$ is written as

$$\frac{dw}{dt} = \frac{\partial w}{\partial u} \cdot \frac{du}{dt} + \frac{\partial w}{\partial v} \cdot \frac{dv}{dt}.$$

Note that $u$ and $v$ are functions of $t$ only while $w$ is a function of both $u$ and $v$; thus, differentiation of $u$ and $v$ is evaluated at $t$ with the ordinary derivatives while that of $w$ is evaluated at $(u, v)$ with the partial derivatives.

In general, for a multivariate function $h = g(u_1, \ldots, u_i, \ldots, u_m)$, differentiable of $m$ independent variables, where $u_i = u_i(s_1, \ldots, s_j, \ldots, s_n)$ is differentiable of $n$ independent variables, the derivative of $h$ with respect to $s_j$ is given by

$$\frac{\partial h}{\partial s_j} = \frac{\partial h}{\partial u_1}\frac{\partial u_1}{\partial s_j} + \cdots + \frac{\partial h}{\partial u_i}\frac{\partial u_i}{\partial s_j} + \cdots + \frac{\partial h}{\partial u_m}\frac{\partial u_m}{\partial s_j} \tag{5.2}$$

for any $j \in 1, \ldots, n$.

The chain rule can be represented using the so-called *tree diagram.* For example, for a function $w = h(u, v)$, differentiable of $u$ and $v$, where $u = f(s,t)$ and $v = g(s,t)$ are differentiable function of $s$ and $t$, one can draw a tree diagram as follows:

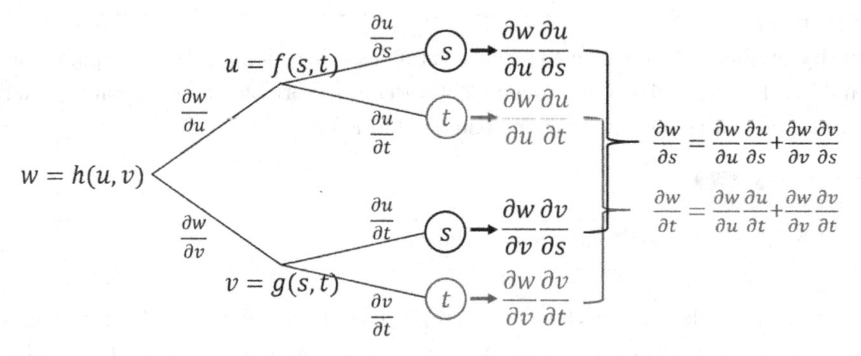


By collecting the derivative information at the branches ending with $s$ and $t$, respectively, one can evaluate the derivatives of $w$ as

$$\frac{\partial w}{\partial s} = \frac{\partial w}{\partial u}\frac{\partial u}{\partial s} + \frac{\partial w}{\partial v}\frac{\partial v}{\partial s}; \quad \frac{\partial w}{\partial t} = \frac{\partial w}{\partial u}\frac{\partial u}{\partial t} + \frac{\partial w}{\partial v}\frac{\partial v}{\partial t}.$$

**Practice 5.1 Chain rule**

Evaluate partial derivatives of the following function $w$ using a tree diagram:

$$w = g(u, v), \quad u = u(r, s, t), \quad v = v(r, s, t).$$

Application of AD essentially requires a program that computes functions/formulas, e.g., $f : \mathcal{R}^n \to \mathcal{R}^m$, mapping independent (input) variables ($\mathbf{x} \in \mathcal{R}^n$) onto dependent (output) variables ($\mathbf{y} \in \mathcal{R}^m$) as $\mathbf{y} = f(\mathbf{x})$. As a sample formula to show various aspects related to AD, we define a function $f : \mathcal{R}^3 \to \mathcal{R}^2$ as

$$\mathbf{y} = f(\mathbf{x}) = \begin{pmatrix} e^{x_1}e^{x_2} - (x_1 + x_2)/\sin(x_2) \\ (x_1 + x_2)/x_3^2 \end{pmatrix}, \tag{5.3}$$

Table 5.1. *Evaluation procedure for Equation (5.3) represented by independent variables ($x_i$), initial internal variables ($w_i^0$), intermediate variables ($w_j$), elementary functions or arithmetic operations ($\varphi_j$), and dependent variables ($y_k$). An exponentiation $b^n$ is represented, using a power operator "**," as $b**n$; for example, $b^2 = b**2$*

| Input section | Calculation section | | Output section |
|---|---|---|---|
| $w_i^0$ | $w_j$ | $\varphi_j$ | $y_k$ |
| | $w_1 = w_{0,1} + w_{0,2}$ | $\varphi_1 = +$ | |
| | $w_2 = \sin(w_{0,2})$ | $\varphi_2 = \sin$ | |
| $w_{0,1} = x_1$ | $w_3 = (w_{0,3}) ** 2$ | $\varphi_3 = ** 2$ | $y_1 = w_7$ |
| $w_{0,2} = x_2$ | $w_4 = \exp(w_1)$ | $\varphi_4 = \exp$ | |
| $w_{0,3} = x_3$ | $w_5 = w_1/w_2$ | $\varphi_5 = /$ | $y_2 = w_6$ |
| | $w_6 = w_1/w_3$ | $\varphi_6 = /$ | |
| | $w_7 = w_4 - w_5$ | $\varphi_7 = -$ | |

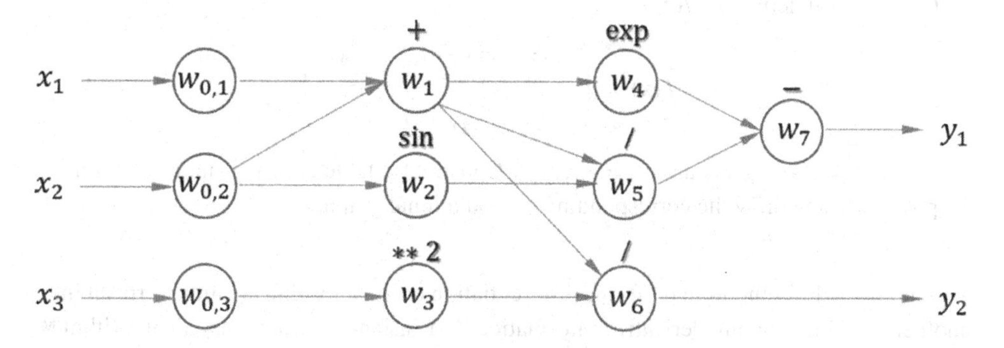


Figure 5.1 Computational graph for Eq. (5.3) and Table 5.1.

where $\mathbf{y} = (y_1, y_2)$ and $\mathbf{x} = (x_1, x_2, x_3)$. To calculate this function, a compiler may produce statements in a sequence as shown in Table 5.1. Calculation of the function is decomposed into smaller operation sections (see Table 5.1), as suggested by Griewank and Walther (2008): 1) The input section assigns the current values of the independent variables ($x_i$) to initial internal variables ($w_{0,i}$), i.e.,

$$w_{0,i} = x_i, \quad i = 1,2,3.$$

2) The calculation section carries out actual computations by applying elementary functions or arithmetic operations ($\varphi_j, j = 1, \ldots, 7$) to the intermediate variables ($w_j, j = 1, \ldots, 7$); and 3) The output section allocates the resulting values ($w_7, w_6$) to the dependent variables ($y_1, y_2$). The procedure to calculate (5.3), shown in Table 5.1, is represented as a computational graph in Figure 5.1.

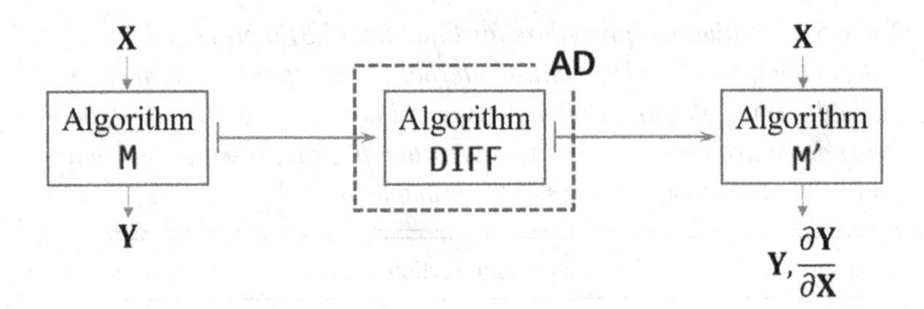


Figure 5.2 Automatic differentiation transforms an algorithm M into another algorithm M' via algorithm DIFF (i.e., differential device). For a given input $\mathbf{X}$, the transformed algorithm M' produces output $\mathbf{Y}$ and the derivatives $\frac{\partial \mathbf{Y}}{\partial \mathbf{X}}$, whereas the primal algorithm M produces only output $\mathbf{Y}$.

**Practice 5.2 Evaluation procedure and computational graph**

Consider a function $f\colon \mathcal{R}^2 \to \mathcal{R}^3$

$$\mathbf{y} = f(\mathbf{x}) = \begin{pmatrix} x_1x_2 + (x_1 - x_2)/\ln(x_1) \\ e^{x_1}/e^{x_2} - x_1/(x_2)^2 \\ (x_1 - x_2)\cos(x_1x_2) \end{pmatrix}, \tag{5.4}$$

where $\mathbf{y} = (y_1, y_2, y_3)$ and $\mathbf{x} = (x_1, x_2)$. Construct a table showing the evaluation procedure and draw the corresponding computational graph.

Automatic differentiation performs differentiation on an algorithm and transforms it into another algorithm having derivative information. For instance, assume that an algorithm M transforms an input $\mathbf{X} = (X_1, \ldots, X_n)$ into an output $\mathbf{Y} = (Y_1, \ldots, Y_m)$ and any diagnostic quantity (say, $J$; e.g., cost function) that can be computed as a function of $\mathbf{Y}$. Automatic differentiation accomplishes transformation of the original (or *primal*) algorithm M into an algorithm M' that is *augmented* with derivatives of the output (i.e., dependent variables) with respect to the input (i.e., independent variables), $\frac{\partial \mathbf{Y}}{\partial \mathbf{X}}$, as shown in Figure 5.2. In other words, for a function $F\colon \mathcal{R}^n \to \mathcal{R}^m$ represented as $\mathbf{Y} = F(\mathbf{X})$, AD evaluates the Jacobian matrix $\nabla F(\mathbf{X}) \in \mathcal{R}^{m\times n}$ that contains numerical values of partial derivatives:

$$\nabla F(\mathbf{X}) = \frac{\partial \mathbf{Y}}{\partial \mathbf{X}} = \left(\frac{\partial Y_j}{\partial X_i}\right)_{i=1,\ldots,n}^{j=1,\ldots,m} = \begin{pmatrix} \frac{\partial Y_1}{\partial X_1} & \cdots & \frac{\partial Y_1}{\partial X_n} \\ \vdots & \ddots & \vdots \\ \frac{\partial Y_m}{\partial X_1} & \cdots & \frac{\partial Y_m}{\partial X_n} \end{pmatrix}_{m\times n}, \tag{5.5}$$

following the notation in Naumann (2012).

## 5.3 Forward versus Reverse Mode of Differentiation

For evaluating the derivatives $\frac{\partial \mathbf{Y}}{\partial \mathbf{X}}$, AD has two modes of differentiation in applying the chain rule – *forward* (or tangent linear) and *reverse* (or backward or adjoint) modes.

### 5.3.1 Forward Mode

The forward mode computes the derivatives following the sequence of the primal algorithm and accumulates derivatives of intermediate (dependent) variables with respect to a specific independent variable:

$$\frac{\partial \mathbf{Y}}{\partial X_i} = \left(\frac{\partial Y_j}{\partial X_i}\right)_{i \text{ fixed}}^{j=1,\ldots,m},$$

which formulates the forward sensitivity coefficients (see Section 11.2.1). With a single forward accumulation, the derivatives of all output variables with respect to a particular input variable are calculated; thus, it is efficient, for a function $F: \mathcal{R}^n \to \mathcal{R}^m$, when $n \ll m$.

The sample function (5.3) is differentiated through the forward mode of AD as shown in Table 5.2, in which the intermediate variable $w_k$ is associated with a (tangent linear) derivative

$$w_k' = \frac{\partial w_k}{\partial x_1},$$

to compute $\partial f/\partial x_1$. In general, the forward mode of AD applies the following chain rule, repeatedly for $i$ (index for independent variables),

$$\frac{\partial f}{\partial x_i} = \sum_k \frac{\partial f}{\partial w_k}\frac{\partial w_k}{\partial x_i}, \tag{5.6}$$

initialized with $\mathbf{w}_0' = \mathbf{x}' = \mathbf{e}_i$ where $\mathbf{e}_i$ is the unit vector of the $i$-th component; that is, for $n = 3$, $\mathbf{x}' = (x_1', x_2', x_3')$ has three initial data sets – also called *seeds* – (1, 0, 0), (0, 1, 0), and (0, 0, 1). Table 5.2 also shows the evaluation trace starting from a specific input $\mathbf{x} = (x_1, x_2, x_3) = (2, 5, 2)$ and a seed $\mathbf{x}' = (x_1', x_2', x_3') = (1, 0, 0)$, which essentially produces

$$\frac{\partial f}{\partial x_1} = \left(\frac{\partial y_1}{\partial x_1}, \frac{\partial y_2}{\partial x_1}\right)^T.$$

The corresponding linear computational graph is depicted in Figure 5.3.

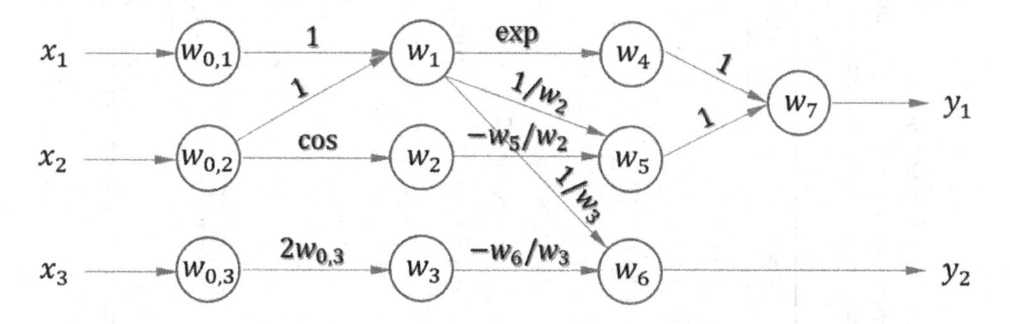


Figure 5.3 Linear computational graph for Eq. (5.3) and Table 5.2.

Table 5.2. *Evaluation trace for the primal code and the derivative code, evaluated at* $(x_1, x_2, x_3) = (2, 5, 2)$*, for the forward mode. Arrows indicate the directions of derivative accumulation*

| Primal trace | | | Derivative trace for $\partial f / \partial x_1$ | | |
|---|---|---|---|---|---|
| Code list | Evaluation | Value | Code list | Evaluation | Value |
| $w_{0,1} = x_1$ | $x_1$ | 2 | $w'_{0,1} = x'_1$ | $x'_1$ | 1 |
| $w_{0,2} = x_2$ | $x_2$ | 5 | $w'_{0,2} = x'_2$ | $x'_2$ | 0 |
| $w_{0,3} = x_3$ | $x_3$ | 2 | $w'_{0,3} = x'_3$ | $x'_3$ | 0 |
| $w_1 = w_{0,1} + w_{0,2}$ | $x_1 + x_2$ | $2 + 5$ | $w'_1 = w'_{0,1} + w'_{0,2}$ | 1 | 1 |
| $w_2 = \sin(w_{0,2})$ | $\sin(x_2)$ | $\sin(5)$ | $w'_2 = \cos(w_{0,2}) * w'_{0,2}$ | $\cos(x_2) * x'_2$ | $\cos(5) * 0$ |
| $w_3 = (w_{0,3}) * *2$ | $(x_3)^2$ | $2^2$ | $w'_3 = 2 * (w_{0,3}) * w'_{0,3}$ | $2 * x_3 * x'_3$ | $2 * 2 * 0$ |
| $w_4 = \exp(w_1)$ | $\exp(x_1 + x_2)$ | $\exp(7)$ | $w'_4 = \exp(w_1) * w'_1$ | $\exp(x_1 + x_2)$ | $\exp(7)$ |
| $w_5 = w_1 / w_2$ | $(x_1 + x_2) / \sin(x_2)$ | $7/(-0.959)$ | $w'_5 = 1/w_2 * w'_1 - w_5/w_2 * w'_2$ | $1/\sin(x_2)$ | $1/(-0.959)$ |
| $w_6 = w_1 / w_3$ | $(x_1 + x_2)/(x_3)^2$ | $7/4$ | $w'_6 = 1/w_3 * w'_1 - w_6/w_3 * w'_3$ | $1/(x_3)^2$ | $1/4$ |
| $w_7 = w_4 - w_5$ | $\exp(x_1 + x_2) - (x_1 + x_2)/\sin(x_2)$ | $1096.633 + 7.299$ | $w'_7 = w'_4 - w'_5$ | $\exp(x_1 + x_2) - 1/\sin(x_2)$ | $1096.633 + 1.043$ |
| $y_1 = w_7$ | $\exp(x_1 + x_2) - (x_1 + x_2)/\sin(x_2)$ | 1103.932 | $\mathbf{y'_1 = w'_7}$ | $\mathbf{\exp(x_1 + x_2) - 1/\sin(x_2)}$ | **1097.676** |
| $y_2 = w_6$ | $(x_1 + x_2)/(x_3)^2$ | 1.75 | $\mathbf{y'_2 = w'_6}$ | $\mathbf{1/(x_3)^2}$ | **0.25** |

**Practice 5.3 Evaluation trace (forward)**

For the sample function (5.3), construct a table showing the evaluation trace as in Table 5.2 for derivatives $\partial f/\partial x_2$ and $\partial f/\partial x_3$, evaluated at $(x_1, x_2, x_3) = (2, 5, 2)$.

**Practice 5.4 Evaluation trace and linear computational graph (forward)**

For the function in (5.4), construct a table showing the evaluation trace as in Table 5.2 for derivatives $\partial f/\partial x_1$ and $\partial f/\partial x_2$, evaluated at $(x_1, x_2) = (5, 2)$. Then, draw the corresponding linear computational graph.

### *5.3.2 Reverse Mode*

The reverse mode computes the derivatives by reversing the flow of the primal algorithm and accumulates the derivatives backward from a specific quantity of dependent variables toward intermediate (independent) variables:

$$\frac{\partial Y_j}{\partial \mathbf{X}} = \left(\frac{\partial Y_j}{\partial X_i}\right)_{i=1,\ldots,n}^{j \text{ fixed}},$$

which formulates the adjoint sensitivity coefficients (see Section 11.2.2). With a single reverse propagation, one can obtain the derivatives of a particular output with respect to all input variables, and hence it is efficient, for a function $F: \mathcal{R}^n \to \mathcal{R}^m$, when $n \gg m$.

In terms of the sample function (5.3), the reverse derivative accumulation is performed by relating each intermediate variable $w_k$ with an adjoint derivative of an output variable $y_j$ as

$$\overline{w}_k = \frac{\partial y_j}{\partial w_k}.$$

In the reverse mode of AD, the following chain rule is applied repeatedly for $j$ (index for dependent variables):

$$\frac{\partial y_j}{\partial w_k} = \sum_{l \in k} \frac{\partial y_j}{\partial w_l} \frac{\partial w_l}{\partial w_k}. \tag{5.7}$$

Thus, one just needs to know the derivatives of the parents and the formula to calculate the derivative of $w_l = g(w_k)$.

To initialize the reverse mode, e.g., for the sample function (5.3), we use the seeds $\overline{\mathbf{y}} = \mathbf{e}_j$ where $\mathbf{e}_j$ is the $j$-th component of the dependent variables; that is, for $m = 2$, $\overline{\mathbf{y}} = (\overline{y}_1, \overline{y}_2)$ has two seeds – (1,0) and (0, 1). Here,

$$\overline{y}_1 = \frac{\partial y_1}{\partial w_7} = \overline{w}_7; \;\; \overline{y}_2 = \frac{\partial y_2}{\partial w_6} = \overline{w}_6.$$

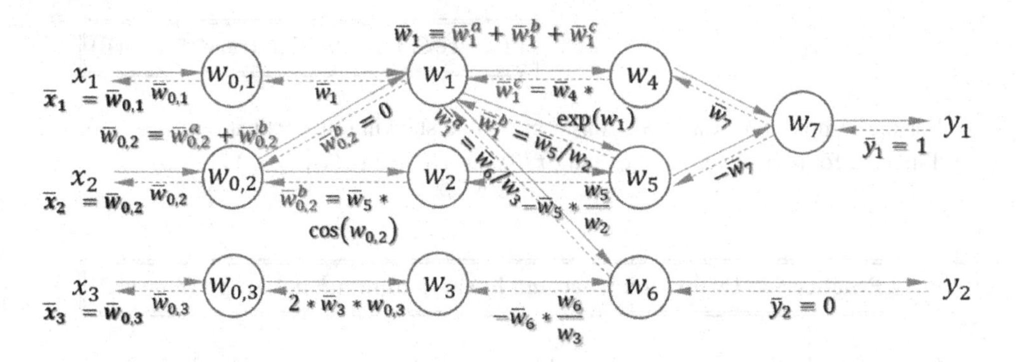


Figure 5.4 Linear computational graph for Eq. (5.3) and Table 5.3.

The reverse evaluation trace is shown in Table 5.3, starting from a seed $\bar{\mathbf{y}} = (\bar{y}_1, \bar{y}_2) = (1, 0)$, which results in

$$\frac{\partial y_1}{\partial \mathbf{x}} = \left( \frac{\partial y_1}{\partial x_1}, \frac{\partial y_1}{\partial x_2}, \frac{\partial y_1}{\partial x_3} \right).$$

The corresponding linear computational graph is depicted in Figure 5.4.

**Practice 5.5 Evaluation trace (reverse)**

For the sample function (5.3), construct a table showing the evaluation trace as in Table 5.3 for derivatives

$$\frac{\partial y_2}{\partial \mathbf{x}} = \left( \frac{\partial y_2}{\partial x_1}, \frac{\partial y_2}{\partial x_2}, \frac{\partial y_2}{\partial x_3} \right),$$

evaluated at $(x_1, x_2, x_3) = (2, 5, 2)$.

**Practice 5.6 Evaluation trace and linear computational graph (reverse)**

For the function in (5.4), construct a table showing the evaluation trace as in Table 5.3 for derivatives $\partial y_1/\partial \mathbf{x}$, $\partial y_2/\partial \mathbf{x}$, and $\partial y_3/\partial \mathbf{x}$, evaluated at $(x_1, x_2) = (5, 2)$. Then, draw the corresponding linear computational graph.

## 5.4 Automatic Differentiation Approaches and Tools

In implementing AD for differentiating program codes, there are two distinct approaches – *source transformation* and *operator overloading*:

Table 5.3. *Evaluation trace for the primal code and the derivative code, evaluated at $(x_1, x_2, x_3) = (2, 5, 2)$, for the reverse mode. The operation $a \mathrel{+}= b$ implies $a = a + b$ or $a \leftarrow a + b$; that is, it adds the right operand to the left operand and assigns the result to the left operand. Similarly, the operation $a \mathrel{-}= b$ represents $a = a - b$. Arrows indicate the directions of derivative accumulation*

| Primal trace | | | Derivative trace for $\partial y_1/\partial \mathbf{x}$ | | |
|---|---|---|---|---|---|
| Code list | Evaluation | Value | Code list | Evaluation | Value |
| $w_{0,1} = x_1$ | $x_1$ | 2 | $\boldsymbol{\overline{x}_1 = \overline{w}_{0,1}}$ | $\boldsymbol{-1/\sin(x_2) + \exp(x_1 + x_2)}$ | **1097.676** |
| $w_{0,2} = x_2$ | $x_2$ | 5 | $\overline{x}_2 = \overline{w}_{0,2}$ | $\cos(x_2) * ((x_1 + x_2)/\sin^2(x_2)) - 1/\sin(x_2) + \exp(x_1 + x_2)$ | 1099.838 |
| $w_{0,3} = x_3$ | $x_3$ | 2 | $\overline{x}_3 = \overline{w}_{0,3}$ | 0 | 0 |
| $w_1 = w_{0,1} + w_{0,2}$ | $x_1 + x_2$ | $2 + 5$ | $\overline{w}_{0,1} \mathrel{+}= \overline{w}_1 \frac{\partial w_1}{\partial w_{0,1}} = \overline{w}_1$ | $0 - 1/\sin(x_2) + \exp(x_1 + x_2)$ | $1.043 + 1096.633$ |
| | | | $\overline{w}_{0,2} \mathrel{+}= \overline{w}_1 \frac{\partial w_1}{\partial w_{0,2}} = \overline{w}_1$ | $\cos(x_2) * ((x_1 + x_2)/\sin^2(x_2)) - 1/\sin(x_2) + \exp(x_1 + x_2)$ | $0.284 * 7.613 + 1.043 + 1096.633$ |
| $w_2 = \sin(w_{0,2})$ | $\sin(x_2)$ | $\sin(5)$ | $\overline{w}_{0,2} \mathrel{+}= \overline{w}_2 \frac{\partial w_2}{\partial w_{0,2}} = \cos(w_{0,2}) * \overline{w}_2$ | $0 + \cos(x_2) * ((x_1 + x_2)/\sin^2(x_2))$ | $\cos(5) * 7.613$ |
| $w_3 = (w_{0,3}) ** 2$ | $(x_3)^2$ | $2^2$ | $\overline{w}_{0,3} \mathrel{+}= \overline{w}_3 \frac{\partial w_3}{\partial w_{0,3}} = 2 * (w_{0,3}) * \overline{w}_3$ | $0 + 2 * x_3 * 0$ | 0 |
| $w_4 = \exp(w_1)$ | $\exp(x_1 + x_2)$ | $\exp(7)$ | $\overline{w}_1 \mathrel{+}= \overline{w}_4 \frac{\partial w_4}{\partial w_1} \mathrel{+}= \exp(w_1) * \overline{w}_4$ | $-1/\sin(x_2) + \exp(x_1 + x_2) * 1$ | $1/0.959 + \exp(7)$ |
| $w_5 = w_1/w_2$ | $(x_1 + x_2)/\sin(x_2)$ | $7/(-0.959)$ | $\overline{w}_1 \mathrel{+}= \overline{w}_5 \frac{\partial w_5}{\partial w_1} \mathrel{+}= 1/w_2 * \overline{w}_5$ | $0 + 1/\sin(x_2) * (-1)$ | $-1/\sin(5)$ |
| | | | $\overline{w}_2 \mathrel{+}= \overline{w}_5 \frac{\partial w_5}{\partial w_2} = -w_5/w_2 * \overline{w}_5$ | $0 - (x_1 + x_2)/\sin^2(x_2) * (-1)$ | $7/\sin^2(5)$ |
| $w_6 = w_1/w_3$ | $(x_1 + x_2)/(x_3)^2$ | $7/4$ | $\overline{w}_1 \mathrel{+}= \overline{w}_6 \frac{\partial w_6}{\partial w_1} = 1/w_3 * \overline{w}_6$ | $0 + 1/(x_3)^2 * 0$ | 0 |
| | | | $\overline{w}_3 \mathrel{+}= \overline{w}_6 \frac{\partial w_6}{\partial w_3} = -w_6/w_3 * \overline{w}_6$ | $0 - (x_1 + x_2)/(x_3)^4 * 0$ | 0 |
| $w_7 = w_4 - w_5$ | $\exp(x_1 + x_2) - (x_1 + x_2)/\sin(x_2)$ | $1096.633 + 7.299$ | $\overline{w}_4 \mathrel{+}= \overline{w}_7 \frac{\partial w_7}{\partial w_4} = \overline{w}_7$ | $0 + 1$ | 1 |
| | | | $\overline{w}_5 \mathrel{+}= \overline{w}_7 \frac{\partial w_7}{\partial w_5} = -\overline{w}_7$ | $0 - 1$ | $-1$ |
| $y_1 = w_7$ | $\exp(x_1 + x_2) - (x_1 + x_2)/\sin(x_2)$ | 1103.932 | $\overline{w}_7 = \overline{y}_1$ | 1 | 1 |
| $y_2 = w_6$ | $(x_1 + x_2)/(x_3)^2$ | 1.75 | $\overline{w}_6 = \overline{y}_2$ | 0 | 0 |

**Source transformation:** This approach examines the original source code and generates new derivative statements and adds them to the original statements. It is considered a code translator that produces a new source code for calculating the derivatives directly from the given function evaluation source code. Source transformation is feasible for generating derivatives from existing, large, complex code: Software packages using this approach include ADIFOR,[1] TAMC,[2] and its successor TAF,[3] Tapenade (Hascoët and Pascual, 2013) and its ancestor, Odyssée (Rostaing et al., 1993), etc. Also called the *preprocessor* approach (Huiskes, 2002), source transformation can easily optimize the generated code, either by hand or with a compiler, to increase performance, but it can only differentiate functions which are fully defined at the compilation time (Margossian, 2019).

**Operator overloading:** This approach creates a new variable type that includes a real number and its derivatives and replaces the original numerical type with this new type: It does not modify the program but makes minor changes in the variable declaration, operator definition, and output format. Besides calculating the function values, common arithmetic operations and intrinsic functions are allotted to routines that evaluate the derivatives of the output operators. These routines are called for differentiating the function source code while each operation or function in the original program is conducted simultaneously. Its application is limited to specific programming languages, including Fortran 90, Ada, Pascal-XSC, C++, and Matlab, as in the software packages Adept (Hogan, 2014), ADOL-C[4], etc. This approach is generally easy to implement and handles most programming statements, but it requires some care in memory handling (Margossian, 2019).

Traditionally, source transformation has largely been used to develop the AD packages for Fortran 77 while operator overloading for C/C++; however, such a distinction is no longer valid as Fortran 95 has been developed and Matlab and scripting languages (e.g., Python) have been increasingly used (Gebremedhin and Walther, 2020). Currently, many software packages include a built-in AD capability (e.g., AMPL[5]; `https://ampl.com/`). As many machine learning techniques require the derivative information, an AD facility has also been implemented in various machine learning libraries – e.g., "`torch.autograd`" in PyTorch (see Paszke et al., 2017, `pytorch.org/docs/stable/autograd.html`). Recent surveys on various AD tools, including those in the machine learning libraries, are provided in Baydin et al. (2018), Margossian (2019), and Gebremedhin and Walther (2020). Table 5.4 lists selected AD tools based on the source transformation and operator overloading approaches. A thorough compilation of AD tools is provided in `www.autodiff.org/`.

[1] Automatic **DI**fferentiation of **FOR**tran (ADIFOR) (Bischof et al., 1992, 1996).
[2] **T**angent linear and **A**djoint **M**odel **C**ompiler (TAMC) (Giering, 1999).
[3] **T**ransformation of **A**lgorithms in **F**ortran (TAF) (Giering et al., 2005).
[4] **A**utomatic **D**ifferentiation by **O**ver**L**oading in **C++** (ADOL-C) (Griewank et al., 1996).
[5] A Modeling Language for Mathematical Programming.

Table 5.4. *Selected AD tools based on the source transformation and operator overloading approaches for the forward (F) and reverse (R) modes (see* www.autodiff.org/ *for further details)*

| | Software | Language | Mode | References | Webpage |
|---|---|---|---|---|---|
| Source transformation | ADG | Fortran 77/90/95 | R | Cheng et al. (2009a) | https://swmath.org/software/19713 |
| | ADIC | C/C++ | F | Bischof et al. (1997) | www.mcs.anl.gov/adic/ |
| | ADIFOR | Fortran 77 | F | Bischof et al. (1992), Bischof et al. (1996) | www.mcs.anl.gov/adifor/ |
| | ADiGator | MATLAB | F | Patterson et al. (2013) | https://sourceforge.net/projects/adigator/ |
| | ADiMat | MATLAB | F | Bischof et al. (2002) | www.adimat.de/ |
| | DFT | Fortran 77/90/95 | F | Cheng et al. (2009b) | N/A |
| | GRESS | Fortran 77 | F, R | Horwedel (1991) | www-rsicc.ornl.gov/ |
| | NAGWare Fortran 95 | Fortran 77/95 | F | Naumann and Riehme (2005) | www.nag.com/content/algorithmic-differentiation-solutions |
| | OpenAD | C/C++, Fortran 77/95 | F, R | Utke et al. (2014) | www.mcs.anl.gov/OpenAD/ |
| | R/ADR | R | F, R | N/A | https://r-adr.de/ |
| | TAC++ | C/C++ | F, R | Voßbeck et al. (2008) | http://fastopt.de/products/tac/ |
| | TAF | Fortran 77/90/95/2003/2008 | F, R | Giering et al. (2005) | http://fastopt.de/products/taf/ |
| | TAMC | Fortran 77 | F | Giering (1999) | http://autodiff.com/tamc/ |
| | Tapenade | C/C++, Fortran 77/95 | F, R | Hascoët and Pascual (2013) | http://tapenade.inria.fr:8080/tapenade/ |
| Operator overloading | Adept | C/C++ | F, R | Hogan (2014) | www.met.rdg.ac.uk/clouds/adept/ |
| | ADF | Fortran 77/95 | F | Straka (2005) | https://github.com/desperadoshi/ADF95 |
| | ADMAT/ ADMIT | MATLAB | F, R | Coleman and Verma (1998), Coleman and Verma (2000) | www.cayugaresearch.com/admat.html |
| | ADOL-C | C/C++, R, Python | F, R | Walther and Griewank (2012) | https://github.com/coin-or/ADOL-C |
| | AUTO_ DERIV | Fortran 77/95 | F | Stamatiadis et al. (2000) | https://tccc.iesl.forth.gr/~farantos/po_cpc/auto_deriv.html |
| | CasADI | C/C++, MATLAB, Python | F, R | Andersson et al. (2019) | https://web.casadi.org/ |
| | CoDiPack | C/C++ | F, R | Sagebaum et al. (2019) | www.scicomp.uni-kl.de/software/codi/ |
| | COSY INFINITY | C/C++, Fortran 77/95 | F | Berz and Makino (2002) | http://cosy.pa.msu.edu/ |
| | dco/c++ | C/C++ | F, R | Lotz (2016) | www.stce.rwth-aachen.de/research/software/dco/cpp |
| | FADBAD++ | C/C++ | F | Bendtsen and Stauning (1996) | www.fadbad.com/fadbad.html |
| | FFADLib | C/C++ | F | Tsukanov and Hall (2000) | https://spatial.engr.wisc.edu/ |
| | Forward Diff.ji | Julia | F | Revels et al. (2016) | https://github.com/JuliaDiff/ForwardDiff.jl |
| | TOMLAB/ MAD | MATLAB | F | Forth (2006) | https://tomopt.com/tomlab/products/mad/ |

An example of generating the tangent linear and adjoint codes for the logistic equation is provided in Appendix C (Section C.4), using an AD tool – Tapenade (see Hascoët and Pascual, 2013).

**Practice 5.7 Tangent linear code generation using AD (forward mode)**

Generate the tangent linear codes, using the Tapenade AD engine at `http://tapenade.inria.fr:8080/tapenade/`, for

1. the sample function (5.3) and
2. the function in (5.4) in Practice 5.2.

**Practice 5.8 Adjoint code generation using AD (reverse mode)**

Repeat Practice 5.7 to generate the adjoint codes.

## References

Andersson JAE, Gillis J, Horn G, Rawlings JB, Diehl M (2019) CasADi – A software framework for nonlinear optimization and optimal control. *Math Program Comput* 11:1–36.

Baydin AG, Pearlmutter BA, Radul AA, Siskind JM (2018) Automatic differentiation in machine learning: A survey. *J Mach Learn Res* 18(153):1–43.

Bendtsen C, Stauning O (1996) *FADBAD, a Flexible C++ Package for Automatic Differentiation.* Tech. Rep. IMM–REP–1996–17, Department of Mathematical Modelling, Technical University of Denmark, Lyngby, Denmark.

Berz M, Makino K (2002) *COSY INFINITY Version 8.1 – User's Guide and Reference Manual.* Department of Physics and Astronomy, Michigan State University, East Lansing, MI, 77 pp.

Bischof CH, Hovland PD, Norris B (2008b) On the implementation of automatic differentiation tools. *Higher-Order Symb Comput* 21:311–331.

Bischof CH, Roh L, Mauer A (1997) ADIC: An extensible automatic differentiation tool for ANSI-C. *Software: Pract Exper* 27:1427–1456.

Bischof CH, Khademi P, Mauer A, Carle A (1996) Adifor 2.0: Automatic differentiation of Fortran 77 programs. *IEEE Comput Sci Eng* 3:18–32.

Bischof CH, Bücker HM, Hovland P, Naumann U, Utke J (eds.) (2008a) *Advances in Automatic Differentiation.* Springer-Verlag, Berlin, Heidelberg, 386 pp.

Bischof CH, Bücker HM, Lang B, Rasch A, Vehreschild A (2002) Combining source transformation and operator overloading techniques to compute derivatives for MATLAB programs. In *Proceedings of the Second IEEE International Workshop on Source Code Analysis and Manipulation (SCAM 2002)*, IEEE Computer Society, Los Alamitos, CA, 65–72.

Bischof CH, Carle A, Corliss G, Griewank A, Hovland P (1992) ADIFOR-generating derivative codes from Fortran programs. *Sci Prog* 1:11–29.

Cheng Q, Zhang H, Wang B (2009b) Differentiation transforming system. *Prog Nat Sci* 19:397–406.

Cheng Q, Cao J, Wang B, Zhang H (2009a) Adjoint code generator. *Sci China Ser F: Inf Sci* 52:926–941.

Christianson B (2012) A Leibniz notation for automatic differentiation. In *Recent Advances in Algorithmic Differentiation*, (eds.) Forth S, Hovland P, Phipps E, Utke J, Walther A, Lecture Notes in Computational Science and Engineering, vol. 87, Springer, Berlin, Heidelberg, 1–9.

Coleman TF, Verma A (1998) ADMAT: An automatic differentiation toolbox for MATLAB. In *Proceedings of the SIAM Workshop on Object Oriented Methods for Inter-Operable Scientific and Engineering Computing*, SIAM, Philadelphia, PA.

Coleman TF, Verma A (2000) ADMIT-1: Automatic differentiation and MATLAB interface toolbox. *ACM Trans Math Software* 26:150–175.

Courty F, Dervieux A, Koobus B, Hascoët L (2003) Reverse automatic differentiation for optimum design: From adjoint state assembly to gradient computation. *Optim Methods Softw* 18:615–627.

Forth S, Hovland P, Phipps E, Utke J, Walthe A (eds.) (2012) *Recent Advances in Algorithmic Differentiation*. Springer, Berlin, Heidelberg, 362 pp.

Forth SA (2006) An efficient overloaded implementation of forward mode automatic differentiation in MATLAB. *ACM Trans Math Software* 32:195–222.

Gebremedhin AH, Walther A (2020) An introduction to algorithmic differentiation. *WIREs Data Min Knowl Discov* 10:e1334, doi:10.1002/widm.1334

Giering R (1999) *Tangent Linear and Adjoint Model Compiler, Users Manual 1.4*. 64 pp., `www.autodiff.com/tamc`

Giering R, Kaminski T, Slawig T (2005) Generating efficient derivative code with TAF: Adjoint and tangent linear Euler flow around an airfoil. *Future Gener Comput Syst* 21:1345–1355.

Griewank A, Walther A (2008) *Evaluating Derivatives: Principles and Techniques of Algorithmic Differentiation*. SIAM, Philadelphia, PA, 460 pp.

Griewank A, Juedes D, Utke J (1996) Algorithm 755: ADOL-C: A package for the automatic differentiation of algorithms written in C/C++. *ACM Trans Math Software* 22:131–167.

Hascoët L, Pascual V (2013) The Tapenade automatic differentiation tool: Principles, model, and specification. *ACM Trans Math Software* 39:20, doi:10.1145/2450153.2450158

Henrard M (2017) *Algorithmic Differentiation in Finance Explained*. Palgrave Macmillan, Cham, 103 pp.

Hogan RJ (2014) Fast reverse-mode automatic differentiation using expression templates in C++. *ACM Trans Math Software* 40:26, doi:10.1145/2560359

Horwedel JE (1991) GRESS, a preprocessor for sensitivity studies of Fortran programs. In *Automatic Differentiation of Algorithms: Theory, Implementation, and Application*, (eds.) Griewank A, Corliss GF, SIAM, Philadelphia, PA, 243–250.

Huiskes M (2002) *Automatic Differentiation Algorithms in Model Analysis*. PhD thesis, Wageningen University, Wageningen, Netherlands, 153 pp.

Hwang D, Byun DW, Odman MT (1997) An automatic differentiation technique for sensitivity analysis of numerical advection schemes in air quality models. *Atmos Environ* 31:879–888.

Kim JG, Hunke EC, Lipscomb WH (2006) Sensitivity analysis and parameter tuning scheme for global sea-ice modeling. *Ocean Modell* 14:61–80.

Knorr W, Kaminski T, Scholze M, et al. (2010) Carbon cycle data assimilation with a generic phenology model. *J Geophys Res* 115:G04017, doi:10.1029/2009jg001119

Kourounis D, Durlofsky LJ, Jansen JD, Aziz K (2014) Adjoint formulation and constraint handling for gradient-based optimization of compositional reservoir flow. *Comput Geosci* 18:117–137.

Li L (2020) Optimal inversion of conversion parameters from satellite AOD to ground aerosol extinction coefficient using automatic differentiation. *Remote Sens* 12:492, doi:10.3390/rs12030492

Lotz J (2016) *Hybrid approaches to adjoint code generation with dco/c++*. Dissertation, Department of Computer Science, RWTH Aachen University, 127 pp., `http://publications.rwth-aachen.de/record/667318`

Margossian CC (2019) A review of automatic differentiation and its efficient implementation. *WIREs Data Min Knowl Discov* 9:e1305, doi:10.1002/widm.1305

Molkenthin C, Scherbaum F, Griewank A, et al. (2017) Derivative-based global sensitivity analysis: Upper bounding of sensitivities in seismic-hazard assessment using automatic differentiation. *Bull Seismol Soc Am* 107:984–1004.

Naumann U (2012) *The Art of Differentiating Computer Programs: An Introduction to Algorithmic Differentiation.* SIAM, Philadelphia, PA, 340 pp.

Naumann U, Riehme J (2005) A differentiation-enabled Fortran 95 compiler. *ACM Trans Math Software* 31:458–474.

Park SK, Droegemeier KK (1999) Sensitivity analysis of a moist 1D Eulerian cloud model using automatic differentiation. *Mon Wea Rev* 127:2180–2196.

Park SK, Droegemeier KK (2000) Sensitivity analysis of a 3D convective storm: Implications for variational data assimilation and forecast error. *Mon Wea Rev* 128:140–159.

Park SK, Droegemeier KK, Bischof CH (1996) Automatic differentiation as a tool for sensitivity analysis of a convective storm in a 3-D cloud model. In *Computational Differentiation: Techniques, Applications, and Tools*, (eds.) Berz M, Bischof C, Corliss G, Griewank A, chap. 18, SIAM, Philadelphia, PA, 75–120.

Paszke A, Gross S, Chintala S, et al. (2017) Automatic differentiation in PyTorch. In *NIPS 2017 Autodiff Workshop: The future of gradient-based machine learning software and techniques*, `https://openreview.net/forum?id=BJJsrmfCZ`

Patterson MA, Weinstein M, Rao AV (2013) An efficient overloaded method for computing derivatives of mathematical functions in MATLAB. *ACM Trans Math Software* 39:17, doi:10.1145/2450153.2450155

Qin J, Liang S, Liu R, Zhang H, Hu B (2007) A weak-constraint-based data assimilation scheme for estimating surface turbulent fluxes. *IEEE Geosci Remote Sens Lett* 4:649–653.

Revels J, Lubin M, Papamarkou T (2016) Forward-mode automatic differentiation in Julia. *arXiv:160707892* `https://arxiv.org/abs/1607.07892`

Rostaing N, Dalmas S, Galligo A (1993) Automatic differentiation in Odyssée. *Tellus A* 45:558–568.

Sagebaum M, Albring T, Gauger N (2019) High-performance derivative computations using CoDiPack. *ACM Trans Math Software* 45:38, doi:10.1145/3356900

Stamatiadis S, Prosmiti R, Farantos SC (2000) AUTO_DERIV: Tool for automatic differentiation of a FORTRAN code. *Comput Phys Commun* 127:343–355.

Straka CW (2005) ADF95: Tool for automatic differentiation of a FORTRAN code designed for large numbers of independent variables. *Comput Phys Commun* 168:123–139.

Tsukanov I, Hall M (2000) *Fast Forward Automatic Differentiation Library (FFADLib): A User Manual.* Spatial Automation Laboratory, University of Wisconsin–Madison, Madison, WI, 60 pp.

Utke J, Naumann U, Lyons A (2014) *OpenAD/F: User Manual*. Argonne National Laboratory, 69 pp., `www.mcs.anl.gov/OpenAD/openad.pdf`

Voßbeck M, Giering R, Kaminski T (2008) Development and first applications of TAC++. In *Advances in Automatic Differentiation*, (eds.) Bischof CH, Bücker HM, Hovland P, Naumann U, Utke J, Springer, Berlin, Heidelberg, 187–197.

Walther A, Griewank A (2012) Getting started with ADOL-C. In *Combinatorial Scientific Computing*, (eds.) Naumann U, Schenk O, Chapman-Hall CRC Computational Science, Boca Raton, FL, 181–202.

Wang G, Cao X, Cai X, et al. (2018) A new data assimilation method for high-dimensional models. *PLoS ONE* 13:e0191714.

Zhang Y, Bischof CH, Easter RC, Wu P-T (1998) Sensitivity analysis of a mixed-phase chemical mechanism using automatic differentiation. *J Geophys Res Atmos* 103:18953–18979.

# 6
# Numerical Minimization Process

## 6.1 Introduction

One of the main challenges of realistic data assimilation is addressing the nonlinearity of dynamical/physical processes and observation operators, which is generally achieved by applying numerical minimization. Therefore, minimization (or optimization) is an essential process in data assimilation, which typically defines a cost (error) function and seeks the initial conditions that minimizes the function. Minimization methods are generally classified into three groups: the gradient-based methods (e.g., Kim, 2006; Arsham, 2008; Golfetto and Fernandes, 2012), the metaheuristic (gradient-free or population-based) methods (e.g., Yang and He, 2016; Maier et al., 2019; Stork et al., 2020), and the hybrid methods (e.g., Salajegheh and Salajegheh, 2019; Ahmadianfar et al., 2020; Albani et al., 2020).

The *gradient-based* methods require derivative information of the function to be minimized, which include algorithms such as the gradient descent, conjugate gradient, quasi-Newton, Gauss–Newton, Newton, etc. (e.g., Navon and Legler, 1987; Liu and Nocedal, 1989; Zou et al., 1993; Ruder, 2017; Esentürk et al., 2018; Haji and Abdulazeez, 2021). These methods may lead to a local minim and the minimization process gets stuck therein though they generally have faster convergence rate than the metaheuristic methods.

The *metaheuristic* methods perform optimization via a randomly generated population, which include algorithms such as the simulated annealing (SA), particle swarm optimization (PSO), genetic algorithm (GA), differential evolution, ant colony (AC), cuckoo search (CS), etc. (e.g., Kruger, 1993; Lee et al., 2006; Wu et al., 2006; Chaudhuri et al., 2014; Hong et al., 2014; Yu et al., 2018; Upadhyaya and Upadhyaya, 2021). These methods seek the global minimum, but they take a relatively longer time to achieve the optimization than the gradient-based methods. Feng et al. (2021) developed a population-based method, called the cooperation search algorithm, to effectively explore the decision space.

Recently, some *hybrid* approaches that combine the gradient-based and metaheuristic methods or the metaheuristic methods themselves were developed to take advantage of different methods; for instance, the combination of PSO–quasi-Newton (Salajegheh and Salajegheh, 2019), GA/PSO–Gauss–Newton (Albani et al., 2020), PSO–GA (Zhao et al., 2021), CS–SA–orthogonal design (Wang et al., 2016), AC–SA (Stodola et al., 2020), etc.

Ahmadianfar et al. (2020) developed a new population-based algorithm, called the gradient-based optimizer, by adopting Newton's method.

Certain data assimilation schemes serve as the minimization algorithms by themselves. For example, Županski et al. (2008) showed that the maximum likelihood ensemble filter (MLEF) (Županski, 2005) could be used as a stand-alone nondifferentiable minimization algorithm. Le Dimet et al. (2002) indicated the equivalence of the inverse 3D variational data assimilation (3DVAR) (Kalnay et al., 2000; Park and Kalnay, 2004) to a perfect Newton iterative method as well as the adjoint Newton algorithm (Wang et al., 1997) to solve the minimization problem at a given time.

In meteorology and other geosciences disciplines, gradient-based methods are most commonly used (e.g., Navon and Legler, 1987; Thacker, 1989; Zou et al., 1993; Wang et al., 1997; Fisher, 1998), and we focus on those methods in this chapter.

## 6.2 Mathematical Background on Minimization

### *6.2.1 Gradient and Hessian*

We aim to minimize a function $F(\mathbf{x})$ and denote $\mathbf{x}^{\dagger}$ the point (or state vector) that achieves that minimization. To understand the property of a specific minimization, it is essential to examine the solution behavior around $\mathbf{x}^{\dagger}$, in particular for a small variation (i.e., $\Delta\mathbf{x}$). Through the Taylor series expansion (see Section 4.2) about $\mathbf{x}^{\dagger}$ and by taking the first-order derivative term, we obtain the gradient vector $\mathbf{g}$ (see (4.6)) as

$$\mathbf{g} = \nabla F(\mathbf{x}) = \begin{pmatrix} \frac{\partial F(\mathbf{x})}{\partial x_1} & \cdots & \frac{\partial F(\mathbf{x})}{\partial x_n} & \cdots & \frac{\partial F(\mathbf{x})}{\partial x_N} \end{pmatrix}^T, \tag{6.1}$$

and by taking the second-order derivative term, we get the Hessian matrix $\mathbf{S}$ (see (4.7)) as

$$\mathbf{S} = \nabla \mathbf{g} = \nabla^2 F(\mathbf{x}) = \begin{pmatrix} \frac{\partial^2 F(\mathbf{x})}{\partial^2 x_1} & \cdots & \frac{\partial^2 F(\mathbf{x})}{\partial x_1 \partial x_n} & \cdots & \frac{\partial^2 F(\mathbf{x})}{\partial x_1 \partial x_N} \\ \vdots & & \vdots & & \vdots \\ \frac{\partial^2 F(\mathbf{x})}{\partial x_n \partial x_1} & \cdots & \frac{\partial^2 F(\mathbf{x})}{\partial^2 x_n} & \cdots & \frac{\partial^2 F(\mathbf{x})}{\partial x_n \partial x_N} \\ \vdots & & \vdots & & \vdots \\ \frac{\partial^2 F(\mathbf{x})}{\partial x_N \partial x_1} & \cdots & \frac{\partial^2 F(\mathbf{x})}{\partial x_N \partial x_n} & \cdots & \frac{\partial^2 F(\mathbf{x})}{\partial^2 x_N} \end{pmatrix}, \tag{6.2}$$

where $n$ is an index for the elements of $\mathbf{x}$. The gradient and Hessian provide important information on the function surface for minimization.

### *6.2.2 Directional Derivatives*

The directional derivative of $F(\mathbf{x})$ along a unit vector $\mathbf{u}$ at a specific point $\mathbf{x}$, denoted by $\nabla_{\mathbf{u}} F(\mathbf{x})$, represents the rate of change (i.e., derivative) of $F(\mathbf{x})$ in the direction of $\mathbf{u}$ through $\mathbf{x}$, which is formulated as

$$\nabla_{\mathbf{u}} F(\mathbf{x}) = \mathbf{u}^T \nabla F(\mathbf{x}) = \mathbf{u}^T \mathbf{g}. \tag{6.3}$$

The second-order derivative along $\mathbf{u}$ is given by

$$\nabla_{\mathbf{u}}^2 F(\mathbf{x}) = \mathbf{u}^T \nabla^2 F(\mathbf{x})\mathbf{u} = \mathbf{u}^T \mathbf{S}\mathbf{u}. \tag{6.4}$$

For a nonunit vector $\mathbf{v}$, its unit vector is given by $\frac{\mathbf{v}}{\|\mathbf{v}\|}$; thus, (6.3) and (6.4) become

$$\nabla_{\mathbf{v}} F(\mathbf{x}) = \frac{\mathbf{v}^T \mathbf{g}}{\|\mathbf{v}\|} \quad \text{and} \quad \nabla_{\mathbf{v}}^2 F(\mathbf{x}) = \frac{\mathbf{v}^T \mathbf{S}\mathbf{v}}{\|\mathbf{v}\|^2}, \tag{6.5}$$

respectively.

Assume that $\theta$ denotes the angle between $\mathbf{u}$ and $\mathbf{g}_0 = \nabla F(\mathbf{x}_0)$ at point $\mathbf{x}_0$; then, (6.3) becomes

$$\nabla_{\mathbf{u}} F(\mathbf{x}_0) = \mathbf{u}^T \mathbf{g}_0 = \|\mathbf{g}_0\| \, \|\mathbf{u}\| \cos\theta = \|\mathbf{g}_0\| \cos\theta. \tag{6.6}$$

Noting that $\cos\theta = [-1, 1]$, we have the following properties:

- $\nabla_{\mathbf{u}} F(\mathbf{x}_0)$ is maximized when $\theta = 0$ (i.e., $\mathbf{u}$ and $\mathbf{g}_0$ are in the same direction), with the maximum value of $\|\mathbf{g}_0\|$;
- $\nabla_{\mathbf{u}} F(\mathbf{x}_0)$ is minimized (or negatively maximized) when $\theta = \pi$ (i.e., $\mathbf{u}$ and $\mathbf{g}_0$ are in opposite directions), with the minimum (or negative maximum) value of $-\|\mathbf{g}_0\|$; and
- $\nabla_{\mathbf{u}} F(\mathbf{x}_0)$ becomes 0 when $\mathbf{g}_0 = \mathbf{0}$ or $\theta = \frac{\pi}{2}$ (i.e., $\mathbf{u}$ and $\mathbf{g}_0$ are orthogonal).

**Example 6.1 Directional derivatives**

We can calculate the directional derivative of the following function

$$F(\mathbf{x}) = 2x_1^2 + x_2^2$$

at $\mathbf{x}_0 = (-0.5, 0.5)^T$ in the direction $\mathbf{v} = (1, 2)^T$. We first evaluate the gradient at $\mathbf{x}_0$ as

$$\mathbf{g}_0 = \nabla F(\mathbf{x}_0) = \begin{pmatrix} \frac{\partial F(\mathbf{x})}{\partial x_1} \\ \frac{\partial F(\mathbf{x})}{\partial x_2} \end{pmatrix}_{\mathbf{x}_0} = \begin{pmatrix} 4x_1 \\ 2x_2 \end{pmatrix}_{\mathbf{x}_0} = \begin{pmatrix} -2 \\ 1 \end{pmatrix}.$$

Then, noting that $\|\mathbf{v}\| = \sqrt{1^2 + 2^2} = \sqrt{5}$, the derivative in the direction $\mathbf{v}$ is

$$\frac{\mathbf{v}^T \mathbf{g}_0}{\|\mathbf{v}\|} = \frac{\begin{pmatrix} 1 & 2 \end{pmatrix} \begin{pmatrix} -2 \\ 1 \end{pmatrix}}{\sqrt{5}} = \frac{0}{\sqrt{5}} = 0.$$

That is, $F(\mathbf{x})$ has zero slope in the direction $\mathbf{v}$ from the point $\mathbf{x}_0$. The function has its maximum slope at $\mathbf{x}_0$ when $\mathbf{v}$ is the same as $\mathbf{g}_0$: for $\mathbf{v} = (-2, 1)^T$, the slope is $\frac{5}{\sqrt{5}} = 2.24$. The derivatives for the above two directions are illustrated in Figure 6.1.

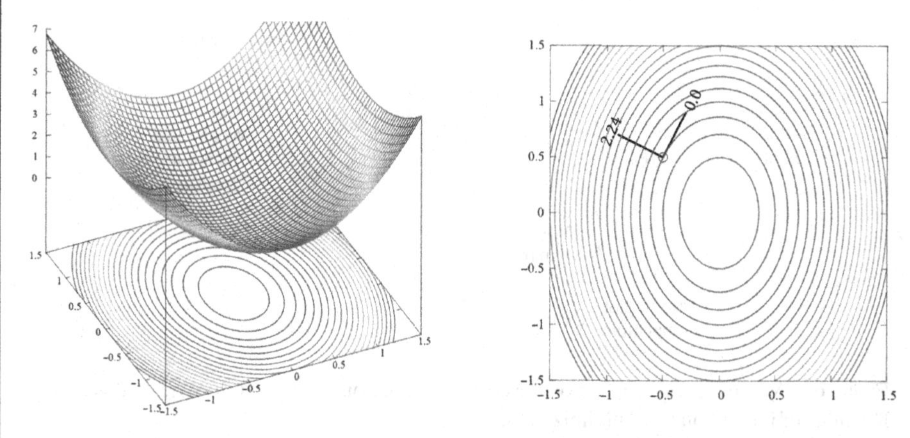


Figure 6.1 Contour plots of $F(\mathbf{x})$ with an interval of 0.25 up to the value 5, including the directional derivatives at $\mathbf{x}_0 = (-0.5, 0.5)^T$ in the directions $\mathbf{v} = (1,2)^T$ and $\mathbf{v} = (-2,1)^T$, respectively.

**Practice 6.1 Directional derivatives (I)**

Answer the following for the function

$$F(\mathbf{x}) = 2x_1^2 + 2x_1x_2 + x_2^2.$$

1. Find the directional derivative at $\mathbf{x}_0 = (-0.5, 0)^T$ in the direction $\mathbf{v} = (-1,2)^T$.
2. Find a unit vector $\mathbf{u}$ along which $\nabla_{\mathbf{u}} F(\mathbf{x})$ at $\mathbf{x}_0 = (-0.5, 0)^T$ is a maximum. What is the maximum value?
3. Discuss what you have found from the above items in terms of the gradient and the direction (e.g., tangent or orthogonal to the gradient).

**Practice 6.2 Directional derivatives (II)**

Consider the following function and answer the following:

$$F(\mathbf{x}) = x_1^2 - 2x_1x_2 + 4x_2^2.$$

1. Find the tangent vector to $F(\mathbf{x})$ at $\mathbf{x}_0 = (2, -1)^T$.
2. Make a 2D contour plot of $F(\mathbf{x})$, including the contour of $F(\mathbf{x}) = 12$. Draw the tangent vector from #1 over the contour plot.
3. Find the vector orthogonal to $F(\mathbf{x})$ at $\mathbf{x}_0 = (2, -1)^T$ and draw the vector over the contour plot in #2.

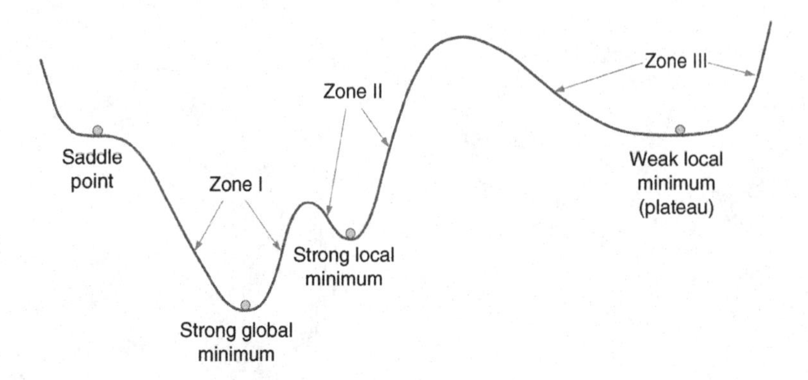


Figure 6.2 Minimization problem on the cost function. Zones I, II, and III indicate potential locations of initial point for minimization.

### *6.2.3 Properties of Minimum Points*

We desire our minimization process to lead to the global minimum, but in many cases the minimization may end up with a local minimum. This often occurs when the cost (or objective) function has many local minima. Here, we define different properties of the minima, as shown in Figure 6.2.

*Global* versus *local* minimum: A global minimum is the point where a function attains its minimum value; however, most minimization algorithms just find a local minimum at which the smallest value of the function is observed in its local neighborhood. More specifically, for a function $F(\mathbf{x})$ to be minimized,

the point $\mathbf{x}^{\dagger}$ is a unique global minimum if $F(\mathbf{x}^{\dagger}) < F(\mathbf{x})$ for all $\mathbf{x}$ over the given domain

while

the point $\mathbf{x}^{\dagger}$ is a local minimum if $F(\mathbf{x}^{\dagger}) < F(\mathbf{x})$ for all $\mathbf{x}$ that belongs to a neighborhood ($\mathcal{N}$) of $\mathbf{x}^{\dagger}$ (i.e., $\mathbf{x} = \mathbf{x}^{\dagger} + \Delta\mathbf{x} \in \mathcal{N}$ for small nonzero $\Delta\mathbf{x}$).

*Strong* versus *weak* minimum: A strong minimum is the point where the function increases with a small distance variation in any direction; in other words,

the point $\mathbf{x}^{\dagger}$ is a strong minimum if $F(\mathbf{x}^{\dagger}) < F(\mathbf{x})$ for all $\mathbf{x} \in \mathcal{N}$.

A weak minimum is the point where the function may not change in some directions or does not decrease in any direction; that is,

the point $\mathbf{x}^{\dagger}$ is a weak minimum if $F(\mathbf{x}^{\dagger}) \leq F(\mathbf{x})$ for all $\mathbf{x} \in \mathcal{N}$.

A *plateau* is an extreme case of the weak minimum, representing a nearly flat surface, where $F(\mathbf{x}^{\dagger}) \simeq F(\mathbf{x})$ for all $\mathbf{x} \in \mathcal{N}$.

*Saddle point*: A stationary point (i.e., $\nabla F(\mathbf{x}) = 0$) that is neither a maximum nor a minimum is called a saddle point. Passing through this point, the function increases in some directions and decreases in other directions.

Based on the above definitions of minima, the function in Figure 6.2 possesses one saddle point, two local minima (one strong and one weak), and a strong global minimum. It also indicates the importance of choosing the initial point in the minimization process: Starting from any point in Zone 1 will achieve the global minimum while starting from a point in Zones II and III will end up with being trapped in a local minimum. If one started at the saddle point, the minimization algorithm immediately stops. Minimization around the weak minimum or plateau usually takes a much longer time than that around the strong minimum.

### 6.2.4 Optimality Conditions

The minimum (or optimum) points should satisfy some conditions. To identify such conditions, we apply the Taylor series expansion to a twice-continuously differentiable function $F(\mathbf{x})$ with a minimum at $\mathbf{x}^\dagger$, for $\mathbf{g}_{\mathbf{x}^\dagger} = \nabla F(\mathbf{x})|_{\mathbf{x}^\dagger}$ and $\mathbf{S}_{\mathbf{x}^\dagger} = \nabla^2 F(\mathbf{x})\big|_{\mathbf{x}^\dagger}$:

$$F\left(\mathbf{x}^\dagger + \Delta\mathbf{x}\right) = F\left(\mathbf{x}^\dagger\right) + \mathbf{g}_{\mathbf{x}^\dagger}^T \Delta\mathbf{x} + \underbrace{\frac{1}{2}\Delta\mathbf{x}^T \mathbf{S}_{\mathbf{x}^\dagger} \Delta\mathbf{x} + O\left(\Delta\mathbf{x}^3\right)}_{O(\Delta\mathbf{x}^2)}, \tag{6.7}$$

for a sufficiently small perturbation $\Delta\mathbf{x}$, with negligible higher-order terms $O\left(\Delta\mathbf{x}^2\right)$ and $O\left(\Delta\mathbf{x}^3\right)$. We consider $\mathbf{x} = \mathbf{x}^\dagger + \Delta\mathbf{x}$ in the neighborhood of $\mathbf{x}^\dagger$ (i.e., $\mathbf{x} \in \mathcal{N}$).

By neglecting $O\left(\Delta\mathbf{x}^2\right)$ we notice that, for $\mathbf{x}^\dagger$ to be a local minimum, the following should be satisfied:

$$\mathbf{g}_{\mathbf{x}^\dagger}^T \Delta\mathbf{x} \geq 0. \tag{6.8}$$

Because the sign of $\Delta\mathbf{x}$ is arbitrary and (6.8) is true for any $\Delta\mathbf{x}$, we require

$$\mathbf{g}_{\mathbf{x}^\dagger} = \nabla F(\mathbf{x}^\dagger) = \mathbf{0}; \tag{6.9}$$

that is, the gradient should be 0 at a minimum point $\mathbf{x}^\dagger$. This is the *first-order necessary condition* of optimality. A stationary point denotes any point that satisfies this condition.

By neglecting $O\left(\Delta\mathbf{x}^3\right)$ and assuming that a stationary point $\mathbf{x}^\dagger$ exists (i.e., $\mathbf{g}_{\mathbf{x}^\dagger} = \mathbf{0}$), we have

$$F\left(\mathbf{x}^\dagger + \Delta\mathbf{x}\right) = F\left(\mathbf{x}^\dagger\right) + \frac{1}{2}\Delta\mathbf{x}^T \mathbf{S}_{\mathbf{x}^\dagger} \Delta\mathbf{x}. \tag{6.10}$$

To have a minimum at $\mathbf{x}^\dagger$, we require

$$\Delta\mathbf{x}^T \mathbf{S}_{\mathbf{x}^\dagger} \Delta\mathbf{x} \geq 0. \tag{6.11}$$

This implies that, for arbitrary nonzero $\Delta\mathbf{x}$, the Hessian matrix should be positive semidefinite (see Appendix A). The *second-order necessary condition* for optimality states as follows:

> If $\mathbf{S} = \nabla^2 F(\mathbf{x})$ exists and is continuous in $\mathcal{N}$ and $\mathbf{x}^\dagger$ is a local minimum, either strong or weak, then $\mathbf{g}_{\mathbf{x}^\dagger} = \mathbf{0}$ and $\mathbf{S}_{\mathbf{x}^\dagger}$ is positive semidefinite.

We also have the *second-order sufficient condition* for optimality, stating that

If $\mathbf{S} = \nabla^2 F(\mathbf{x})$ exists and continuous in $\mathcal{N}$ and $\mathbf{x}^\dagger$ is a strong local minimum, then $\mathbf{g}_{\mathbf{x}^\dagger} = \mathbf{0}$ and $\mathbf{S}_{\mathbf{x}^\dagger}$ is positive definite.

Note also that for a convex function $F(\mathbf{x})$ any local minimum $\mathbf{x}^\dagger$ becomes a global minimum. Furthermore, if $F(\mathbf{x})$ is convex and differentiable, then any stationary point $\mathbf{x}^\dagger$ is a global minimum of the function (see Nocedal and Wright, 2006).

**Practice 6.3 Optimality conditions**

Consider a function,

$$F(\mathbf{x}) = 2x_1^3 + 3x_2^2,$$

and evaluate the optimal conditions. Discuss whether you can guarantee that the stationary point, if any, is a minimum point.

### *6.2.5 Hessian Eigensystem in Quadratic Functions*

Many important features of the minimization properties in quadratic functions, the shape and curvature of the cost function, the convergence rate, etc., can be characterized by the Hessian $\mathbf{S}$ and its eigensystem (e.g., Thacker, 1989; Le Dimet et al., 2002).

Consider a quadratic function that includes a symmetric matrix $\mathbf{Q}$

$$F(\mathbf{x}) = \frac{1}{2}\mathbf{x}^T\mathbf{Q}\mathbf{x} + \mathbf{r}^T\mathbf{x} + c, \tag{6.12}$$

which brings about $\mathbf{S} = \nabla^2 F(\mathbf{x}) = \mathbf{Q}$. Through the eigenvalue decomposition (or spectral decomposition; see (A.33) in Appendix A):

$$\mathbf{Q} = \mathbf{W}\boldsymbol{\Lambda}\mathbf{W}^T, \tag{6.13}$$

where $\mathbf{W}$ is an orthonormal matrix whose column vectors ($\mathbf{w}_i$) are the *eigenvectors* and $\boldsymbol{\Lambda}$ is a diagonal matrix whose diagonal entries ($\lambda_i$) are the *eigenvalues*. Then, by performing a change of basis using the Hessian eigenvectors as the new basis (i.e., principal axes), we have a new matrix

$$\mathbf{Q}' = \mathbf{W}^T\mathbf{Q}\mathbf{W} = \boldsymbol{\Lambda}. \tag{6.14}$$

We now discuss the relationship between the Hessian eigensystem and the directional derivative (see also Hagan et al., 2014). For a direction vector $\mathbf{v}$, we have the second-order derivative (i.e., curvature) along $\mathbf{v}$, from (6.4), as

$$\nabla_{\mathbf{v}}^2 F(\mathbf{x}) = \frac{\mathbf{v}^T\mathbf{Q}\mathbf{v}}{\|\mathbf{v}\|^2}. \tag{6.15}$$

By introducing a vector $\mathbf{p}$ that represents $\mathbf{v}$ in terms of the eigenvectors of $\mathbf{Q}$ (i.e., $\mathbf{v} = \mathbf{W}\mathbf{p}$) from (6.13) and using the property $\mathbf{W}^T\mathbf{W} = \mathbf{I}$, we have

$$\nabla_{\mathbf{v}}^2 F(\mathbf{x}) = \frac{\mathbf{v}^T \mathbf{Q} \mathbf{v}}{\|\mathbf{v}\|^2} = \frac{\mathbf{p}^T \mathbf{\Lambda} \mathbf{p}}{\mathbf{p}^T \mathbf{p}}, \tag{6.16}$$

implying that $\nabla_{\mathbf{v}}^2 F(\mathbf{x})$ represents the eigenvalues along the corresponding eigenvectors (Hagan et al., 2014); thus, the maximum and minimum curvatures ($\nabla_{\mathbf{v}}^2 F(\mathbf{x})$) occur in the directions of the eigenvectors with $\lambda_{max}$ and $\lambda_{min}$, respectively.

**Practice 6.4 Hessian eigenvalues and the directional derivative**

Derive (6.16) and show that

$$\nabla_{\mathbf{v}}^2 F(\mathbf{x}) = \frac{\sum_{i=1}^n \lambda_i p_i^2}{\sum_{i=1}^n p_i^2},$$

where $\lambda_i$ are the eigenvalues of $\mathbf{Q}$ and $p_i$ are the elements of $\mathbf{p}$.

From Practice 6.4, we notice that the second-order directional derivative, $\nabla_{\mathbf{v}}^2 F(\mathbf{x})$, is a weighted average of the eigenvalues of the Hessian $\mathbf{Q}$. Thus, $\nabla_{\mathbf{v}}^2 F(\mathbf{x})$ is bounded by the maximum and minimum of the eigenvalues of $\mathbf{Q}$, that is,

$$\lambda_{min} \leq \nabla_{\mathbf{v}}^2 F(\mathbf{x}) \leq \lambda_{max}. \tag{6.17}$$

This implies that, combined with (6.16), the function varies the most quickly (slowly) along the eigenvector with $\lambda_{max}$ ($\lambda_{min}$).

**Example 6.2 Eigenvalues and eigenvectors of the Hessian**

The function in Practice 6.1 can be expressed in matrix form as

$$F(\mathbf{x}) = 2x_1^2 + 2x_1x_2 + x_2^2 = \frac{1}{2}\mathbf{x}^T \begin{pmatrix} 4 & 2 \\ 2 & 2 \end{pmatrix} \mathbf{x}.$$

Then, the eigenvalues of the Hessian can be obtained through

$$|\mathbf{S} - \lambda \mathbf{I}| = 0 \;\Rightarrow\; \begin{vmatrix} 4-\lambda & 2 \\ 2 & 2-\lambda \end{vmatrix} = 0 \;\Rightarrow\; \lambda^2 - 6\lambda + 4 = 0 \;\Rightarrow\; \begin{array}{l} \lambda_1 = 0.764 \\ \lambda_2 = 5.236. \end{array}$$

We now compute the eigenvectors $\mathbf{w}_1 = (w_{11}, w_{21})^T$ and $\mathbf{w}_2 = (w_{12}, w_{22})^T$. Using $\lambda_1 = \lambda_{min} = 0.764$, we have

$$\begin{aligned} \begin{pmatrix} 4-\lambda_1 & 2 \\ 2 & 2-\lambda_1 \end{pmatrix} \begin{pmatrix} w_{11} \\ w_{21} \end{pmatrix} &= \begin{pmatrix} 3.236w_{11} + 2w_{21} \\ 2w_{11} + 1.236w_{21} \end{pmatrix} \\ &= \left( w_{11} + \frac{1.236}{2} w_{21} \right) \begin{pmatrix} 3.236 \\ 2 \end{pmatrix} = \begin{pmatrix} 0 \\ 0 \end{pmatrix}, \end{aligned}$$

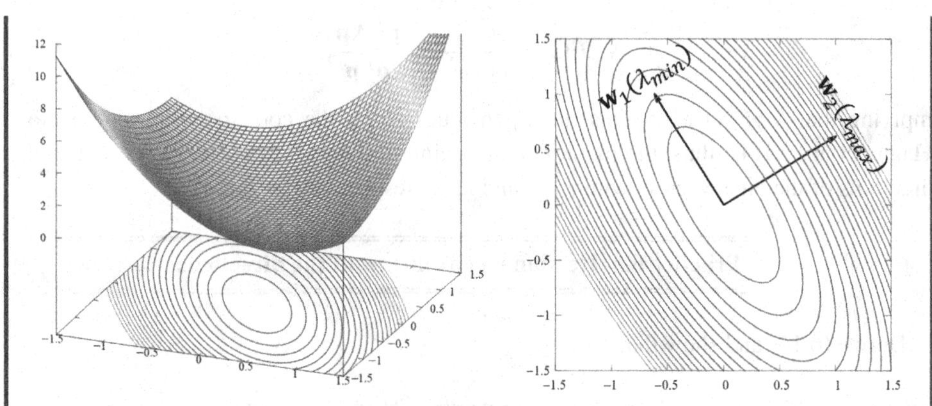


Figure 6.3 Contour plots of $F(\mathbf{x})$ with an interval of 0.25 up to the value 5, including the eigenvectors with corresponding eigenvalues.

and

$$w_{11} = -0.618w_{21} \;\Rightarrow\; \begin{pmatrix} w_{11} \\ w_{21} \end{pmatrix} = \begin{pmatrix} -0.618C_1 \\ C_1 \end{pmatrix} = C_1 \begin{pmatrix} -0.618 \\ 1 \end{pmatrix}.$$

Similarly, using $\lambda_2 = \lambda_{max} = 5.236$, we have

$$\begin{pmatrix} 4-\lambda_2 & 2 \\ 2 & 2-\lambda_2 \end{pmatrix} \begin{pmatrix} w_{12} \\ w_{22} \end{pmatrix} = \begin{pmatrix} 0 \\ 0 \end{pmatrix}$$

$$\Rightarrow\; w_{22} = 0.618w_{12} \;\Rightarrow\; \begin{pmatrix} w_{12} \\ w_{22} \end{pmatrix} = C_2 \begin{pmatrix} 1 \\ 0.618 \end{pmatrix}.$$

Thus, the eigenvectors are $\mathbf{w}_1 = (-0.618, 1)^T$ and $\mathbf{w}_2 = (1, 0.618)^T$, which are in the directions of $\lambda_{min}$ and $\lambda_{max}$, respectively (see Figure 6.3).

The eigenvalues of the Hessian $\mathbf{Q}$ also determine the shape of the function surface $\mathbf{x}^T\mathbf{Q}\mathbf{x} = \text{constant}$. For a positive definite Hessian, the contours of $\mathbf{x}^T\mathbf{Q}\mathbf{x}$ are ellipsoids whose axes are inversely proportional to $\sqrt{\lambda_i}$ (see Example 6.2). Thus, for a 2D problem, a large difference between $\lambda_1$ and $\lambda_2$ implies that the function surface is highly elongated, and the convergence can be slow, especially near the minimum.

**Practice 6.5 Eigenvalues and eigenvectors of the Hessian**

Evaluate the eigenvalues and eigenvectors of the Hessian for the following quadratic functions:

1. $F(\mathbf{x}) = x_1^2 - 2x_1x_2 + x_2^2$;
2. $F(\mathbf{x}) = x_1^2 - 4x_1x_2 + x_2^2$;

and make 2D contour plots of the functions for both $x_1$ and $x_2$ in the range $[-1.5, 1.5]$ along with the eigenvectors for $\lambda_{min}$ and $\lambda_{max}$. Discuss your results in terms of the function shape, the property of the minimum, and the sign of the eigenvalue.

Based on the results from Example 6.2 and Practice 6.5, we note the following properties of the minimum in the quadratic functions with regard to the Hessian eigenvalues: 1) when all eigenvalues are positive (i.e., **S** is positive definite), a strong minimum exists (e.g., Example 6.2); 2) when some of the nonnegative eigenvalues are 0 (i.e., **S** is positive semidefinite), a weak minimum exists (e.g., Practice 6.5 #1); and 3) when some eigenvalues are positive while others are negative (i.e., **S** is indefinite), a saddle point exists (e.g., Practice 6.5 #2).

### *6.2.6 Condition Number of the Hessian*

The performance of the iterative gradient-based minimization process strongly depends on the shape of the quadratic objective function: It takes much longer to reach the minimum when the function is more like a long ellipsoid rather than a circle. The function shape is determined by the relative magnitudes of the function's principal axes, which can be obtained by the ratio between the largest and smallest eigenvalues of the Hessian **Q** in the quadratic function (6.12), defined as the *condition number* ($\kappa$):

$$\kappa(\mathbf{Q}) = \frac{\lambda_{max}}{\lambda_{min}}, \tag{6.18}$$

where $\lambda_{max}$ and $\lambda_{min}$ represent the maximum and minimum eigenvalues of the Hessian **Q**. As the Hessian condition number $\kappa(\mathbf{Q})$ increases, the contours (i.e., the level curves) of the quadratic function become more elongated, thus resulting in slow convergence rate toward a stationary point.

Certain minimization algorithms (e.g., Newton) are required to solve the inverse of the Hessian. Computation of the inverse Hessian is prone to large errors when the Hessian is ill-conditioned, where the condition number is very large (i.e., $\kappa(\mathbf{Q}) \gg 1$). Considering the spectral decomposition in (6.13), we have the inverse of nonsingular **Q** as

$$\mathbf{Q}^{-1} = \mathbf{W}\mathbf{\Lambda}^{-1}\mathbf{W}^T, \tag{6.19}$$

where $\mathbf{\Lambda}$ is a diagonal matrix with eigenvalues $\lambda_n$ on the diagonal. When $\lambda_{min}$ is very small (i.e., the Hessian is ill-conditioned), significant instability may occur in computing the inverse Hessian. In practice, an ill-conditioned Hessian is almost singular, and a noninvertible matrix has infinite condition number (i.e., $\lambda_{min}$ is extremely small). We usually alter the Hessian from ill-conditioned to well-conditioned through *preconditioning* (e.g., Gill et al., 1981; Axelsson and Barker, 1984; Županski, 1993, 1996) before we perform minimization.

### *6.2.7 Conjugate Vectors*

For a quadratic function in (6.12), where **Q** is a symmetric and positive definite matrix, a set of nonzero vectors $\{\mathbf{v}_i\}$ is mutually *conjugate* with respect to the Hessian **Q** (or **Q**-conjugate or **Q**-orthogonal) if and only if

$$\mathbf{v}_i^T\mathbf{Q}\mathbf{v}_j = \mathbf{0} \;\text{ for }\; i \neq j, \tag{6.20}$$

and is linearly independent. Any two vectors are conjugate if $\mathbf{Q} = \mathbf{0}$, whereas the conjugacy becomes equivalent to the orthogonality if $\mathbf{Q} = \mathbf{I}$.

The eigenvectors $\mathbf{w}_i$ of the Hessian $\mathbf{Q}$ are also conjugate because $\mathbf{w}_i$ are orthogonal to each other, thus

$$\mathbf{w}_i^T \mathbf{Q} \mathbf{w}_j = \lambda_j \mathbf{w}_i^T \mathbf{w}_j = 0 \tag{6.21}$$

for the eigenvalues $\lambda_j$. We address the importance of $\mathbf{Q}$-conjugacy, which enables us to minimize the objective function without matrix inversion and use of the Hessian, in Section 6.3.3.

## 6.3 Algorithms for Minimization

The minimization (optimization) problem is generally formulated as

$$\min_{\mathbf{x} \in \mathcal{R}^n} F(\mathbf{x}) \text{ subject to } \begin{cases} c_E(\mathbf{x}) = 0 \\ c_I(\mathbf{x}) \geq 0, \end{cases} \tag{6.22}$$

where $\mathcal{R}^n$ represents the $n$-dimensional Euclidean space and $c_E$ and $c_I$ are the equality and inequality constraints, respectively. This is called a *constrained optimization problem.* A problem without any equality or inequality constraints from (6.22) is referred to as an *unconstrained optimization problem.* Here, $F(\mathbf{x})$ is usually called the *objective* (or *cost* or *loss* or *error*) function, a vector $\mathbf{x} = \mathbf{x}^\dagger$ that minimizes $F(\mathbf{x})$ is referred to as the *minimum* (or *optimum* or *minimizer*), and the corresponding output $F(\mathbf{x}^\dagger)$ is said to be the minimum (or optimum) value of $F(\mathbf{x})$.

In general, minimization is achieved through an iterative process, in which $\mathbf{x}$ is updated iteratively toward a local/global minimum of $F(\mathbf{x})$, starting from an initial guess $\mathbf{x}^0$. This process is formulated as

$$\mathbf{x}^{k+1} = \mathbf{x}^k + \Delta\mathbf{x}^k, \tag{6.23}$$

where $k$ is the index for iteration and $\Delta\mathbf{x}^k$ is the increment that determines how much $\mathbf{x}^k$ is updated at the next step. In an algorithm that adopts *line-search* methods, (6.23) can be represented by replacing $\Delta\mathbf{x}^k$ with $\alpha^k \mathbf{d}^k$ as

$$\mathbf{x}^{k+1} = \mathbf{x}^k + \alpha^k \mathbf{d}^k, \tag{6.24}$$

where $\alpha^k$ is a positive scalar representing the step length and $\mathbf{d}^k$ is a vector showing the search direction. In other words, at every iteration, $\mathbf{x}^k$ is updated as much as $\Delta\mathbf{x}^k$ in the direction of $\mathbf{d}^k$ by the length of $\alpha^k$. This iterative process terminates when either $k$ exceeds the maximum iteration (say, $K$) or one (or any combination) of the following convergence criteria is satisfied:

**Algorithm 6.1** General line-search iterative minimization process

```
/* k: iteration number; K: maximum iteration; ε_a: minimization tolerance                    */
/* x†: minimizer of F(x); F† = F(x†): minimum value of F(x)                                  */
/* d^k: search direction; α^k: step length                                                    */
1  Input: x^0, F^0 = F(x^0), ε_a                         ! Initial guess x^0 at k = 0
2  while (k ≤ K or |ΔF^{k+1}| ≥ ε_a) do                  ! Loop for termination condition
3      Determine d^k and α^k                             ! d^k and α^k using appropriate methods
4      Increment:                                        ! Calculate an increment Δx^k
5          Δx^k = α^k d^k                                ! Δx^k in search direction d^k by a length α^k
6      Update:                                           ! Move one step forward to the minimum
7          x^{k+1} = x^k + Δx^k                          ! New x at k + 1
8          F^{k+1} = F(x^{k+1})                          ! Evaluate F(x) at k + 1
9      Convergence:                                      ! Compare F(x) between k and k + 1
10         ΔF^{k+1} = F^{k+1} − F^k                      ! Difference between F^{k+1} and F^k
11         |ΔF^{k+1}| < ε_a?                             ! Convergence check using (6.25a)
12     Next iteration:                                   ! Move to the next iteration k + 1
13         Set k = k + 1
14 endwhile
15 Output: x† ← x^{k+1}, F† ← F^{k+1}                    ! Assign x^{k+1} to x† and F^{k+1} to F†
```

$$\left|F\left(\mathbf{x}^{k+1}\right)-F\left(\mathbf{x}^{k}\right)\right|<\varepsilon_a, \tag{6.25a}$$

$$\left\|\mathbf{x}^{k+1}-\mathbf{x}^{k}\right\|<\varepsilon_m, \tag{6.25b}$$

$$\frac{\left|F\left(\mathbf{x}^{k+1}\right)-F\left(\mathbf{x}^{k}\right)\right|}{\left|F\left(\mathbf{x}^{k}\right)\right|} \quad \text{or} \quad \frac{\left\|\mathbf{x}^{k+1}-\mathbf{x}^{k}\right\|}{\left\|\mathbf{x}^{k}\right\|}<\varepsilon_r, \tag{6.25c}$$

$$\left\|\nabla F\left(\mathbf{x}^{k+1}\right)\right\|<\varepsilon_g, \tag{6.25d}$$

where $\varepsilon_a$, $\varepsilon_m$, $\varepsilon_r$, and $\varepsilon_g$ represent the minimization tolerances of absolute improvement, iterate improvement, relative improvement, and gradient magnitude, respectively. The procedure in general line-search iterative minimization is illustrated in Algorithm 6.1, based on (6.23), (6.24), and (6.25a).

In this section (Section 6.3), we introduce some common minimization algorithms for the unconstrained problems in the category of the gradient descent methods, which require the first-order derivatives (e.g., steepest descent) or both the first-order and second-order derivatives (e.g., Newton) of the objective function.

### 6.3.1 Steepest Descent Method

#### 6.3.1.1 Search Direction

As we pursue minimization, we expect that the objective function reduces at each iteration as the minimization process proceeds. This implies that our algorithm should be a *descent* one, having a descent search direction $\mathbf{d}^k$, to satisfy $F(\mathbf{x}^{k+1}) - F(\mathbf{x}^k) < 0$. Then, from the first-order Taylor series approximation (see (4.5)), we have

$$\left(\mathbf{g}^k\right)^T \Delta\mathbf{x}^k < 0,$$

which further suggests from (6.24) that

$$\left(\mathbf{g}^k\right)^T \mathbf{d}^k < 0. \tag{6.26}$$

As discussed in Section 6.2.2, the directional derivative is maximized (i.e., the maximum change occurs in $F(\mathbf{x})$) along the gradient direction while it is *negatively maximized* (i.e., steepest descent) along the *negative gradient* direction (i.e., $-\mathbf{g}$). Thus, we define the search direction at each iteration, $\mathbf{d}^k$, as the *steepest descent direction* with

$$\mathbf{d}^k = -\mathbf{g}^k. \tag{6.27}$$

Therefore, in the steepest descent algorithm, (6.24) becomes

$$\mathbf{x}^{k+1} = \mathbf{x}^k - \alpha^k \mathbf{g}^k. \tag{6.28}$$

When the search direction is represented by a unit vector (say, $\hat{\mathbf{d}}^k$), (6.28) is expressed as

$$\mathbf{x}^{k+1} = \mathbf{x}^k - \alpha^k \frac{\mathbf{g}^k}{\left\|\mathbf{g}^k\right\|}, \tag{6.29}$$

where $\hat{\mathbf{d}}^k = -\frac{\mathbf{g}^k}{\left\|\mathbf{g}^k\right\|}$ is called the *normalized steepest descent direction* at $\mathbf{x}^k$.

*6.3.1.2 Step Length*

The step length $\alpha^k$ determines how far a minimzation step is allowed to move in the search direction for a given iteration: It can be assigned a *fixed*, *optimum* (exact), or *approximate* value.

A fixed value of $\alpha$ can be arbitrarily determined, but it should not be too small (to avoid slow convergence) or too large (to avoid numerical instability). For stable minimization of a quadratic function (6.12), the maximum allowed step length $\alpha_{max}$ is supposed to satisfy

$$\alpha_{max} < \frac{2}{\lambda_{max}}, \tag{6.30}$$

where $\lambda_{max}$ is the largest eigenvalue of the Hessian $\mathbf{Q}$ in (6.12) – see Hagan et al. (2014) and do Practice 6.6.

**Practice 6.6 Stable step length**

Answer the following:

1. From a quadratic function (6.12) and for a fixed step length (i.e., $\alpha^k = \alpha$), show that (6.24) can be expressed as

$$\mathbf{x}^{k+1} = (\mathbf{I} - \alpha\mathbf{Q})\,\mathbf{x}^k - \alpha\mathbf{r}$$

and that the eigenvectors of $(\mathbf{I} - \alpha \mathbf{Q})$ and $\mathbf{Q}$ are the same. To keep the steepest descent minimization stable, the magnitude of eigenvalues of $(\mathbf{I} - \alpha \mathbf{Q})$ should be less than 1. Using this condition, derive (6.30).

2. For the functions in Examples 6.1 and 6.2, evaluate the conditions for the maximum allowable step length ($\alpha_{max}$).

An optimum or exact value of $\alpha^k$ at each iteration, for a quadratic function (6.12), can be obtained by performing the following minimization:

$$\min_{\alpha^k > 0} F\left(\mathbf{x}^k - \alpha^k \mathbf{g}^k\right) = \min_{\alpha^k > 0} \frac{1}{2} \left(\mathbf{x}^k - \alpha^k \mathbf{g}^k\right)^T \mathbf{Q} \left(\mathbf{x}^k - \alpha^k \mathbf{g}^k\right) + \mathbf{r}^T \left(\mathbf{x}^k - \alpha^k \mathbf{g}^k\right). \quad (6.31)$$

By taking the derivative of $F\left(\mathbf{x}^k - \alpha^k \mathbf{g}^k\right)$ with respect to $\alpha^k$ and setting it to 0, we can solve for $\alpha^k$ to obtain

$$\alpha^k_{opt} = \frac{\left(\mathbf{g}^k\right)^T \mathbf{g}^k}{\left(\mathbf{g}^k\right)^T \mathbf{Q} \mathbf{g}^k}, \quad (6.32)$$

where $\alpha^k_{opt}$ represents the optimum $\alpha$ at each iteration and $\mathbf{g}^k = \mathbf{Q}\mathbf{x}^k + \mathbf{r}$.

**Practice 6.7 Optimum step length**

For a quadratic function (6.12), obtain an optimum step length $\alpha^k$ that minimizes

$$F\left(\mathbf{x}^k + \alpha^k \mathbf{d}^k\right)$$

and check if this $\alpha^k$ is the same as that in (6.32).

For an arbitrary (nonquadratic) function, the optimum step length can be calculated through 1D line-search algorithms such as bisection, golden section, Fibonacci, Newton–Raphson, secant, cubic polynomial fit, etc. (see, e.g., Antoniou and Lu, 2007; Chong and Żak, 2013; Arora, 2015).

An approximate value of $\alpha^k$ can be calculated through a line search that is *inexact* (i.e., approximate) but makes sufficient improvement along a search direction. One can find a *suitable* $\alpha^k$ (i.e., not too small or too large) that gives a sufficient decrease in the function, following the *sufficient decrease* or *Armijo* condition:

$$F\left(\mathbf{x}^{k+1}\right) \leq F\left(\mathbf{x}^k\right) + c_1 \alpha^k \nabla_{\mathbf{d}^k} F\left(\mathbf{x}^k\right), \quad (6.33)$$

where $c_1$ is normally set to $1 \times 10^{-4}$ and $\nabla_{\mathbf{d}^k} F\left(\mathbf{x}^k\right) = \left(\mathbf{d}^k\right)^T \nabla F\left(\mathbf{x}^k\right)$ is the directional derivative at $\mathbf{x}^k$ (see (6.3)). This condition alone is insufficient to guarantee reasonable movement toward a local minimum because of the significantly small step lengths that satisfy the condition; thus, it is complemented by the following conditions:

$$\nabla_{\mathbf{d}^k} F\left(\mathbf{x}^{k+1}\right) \geq c_2 \nabla_{\mathbf{d}^k} F\left(\mathbf{x}^k\right), \tag{6.34a}$$

$$\left|\nabla_{\mathbf{d}^k} F\left(\mathbf{x}^{k+1}\right)\right| \leq c_2 \left|\nabla_{\mathbf{d}^k} F\left(\mathbf{x}^k\right)\right|, \tag{6.34b}$$

with $0 < c_1 < c_2 < 1$. The *curvature condition* (6.34a) ensures the directional derivative (i.e., slope) at $k+1$ to be less steep. The *strong curvature condition* (6.34b) enforces $\alpha^k$ to stay around the local minimizer by not allowing the directional derivative to be too positive. A combination of the sufficient decrease and curvature conditions forms the *Wolfe conditions*, whereas that of the sufficient decrease and strong curvature conditions constitutes *strong Wolfe conditions*. These conditions can serve as termination conditions in finding the approximate step length.

**Practice 6.8 Approximate step length**

Starting from $\mathbf{x}^0 = (-1, 1.5)^T$, find the next point $\mathbf{x}^1$ using an approximate value of $\alpha$ for a function

$$F(\mathbf{x}) = 2x_1^2 + 2x_1x_2 + x_2^2$$

(see Figure 6.3 in Example 6.2), by performing an inexact line search in the direction $\mathbf{d} = (1, -1)^T$. Use a maximum $\alpha = 8$, a reduction factor of 0.5 per iteration, $c_1 = 1 \times 10^{-4}$, and $c_2 = 0.9$. Terminate the process based on the Wolfe conditions, i.e., both the sufficient decrease condition and the curvature condition should be satisfied.

Some minimization experiments using the steepest descent algorithm, in terms of different choice of step length, are shown in Example 6.3.

**Example 6.3 Steepest descent minimization and step length**

Consider a function

$$F(\mathbf{x}) = 20x_1^2 + x_2^2 = \frac{1}{2}\mathbf{x}^T \underbrace{\begin{pmatrix} 40 & 0 \\ 0 & 2 \end{pmatrix}}_{\mathbf{Q}} \mathbf{x}.$$

The eigenvalues of $\mathbf{Q}$ are $\lambda_1 = 40$ and $\lambda_2 = 2$: then, $\alpha_{max}$ should be less than 0.05 from (6.30). Starting from $\mathbf{x}^0 = (20, 20)^T$ and $F(\mathbf{x}^0) = 8400$, with a tolerance $\varepsilon_a = 1 \times 10^{-4}$ (see (6.25a)), the convergence rate generally increases as $\alpha(<0.05)$ gets larger. However, the convergence becomes very slow as $\alpha$ approaches $\alpha_{max}$: minimization with $\alpha = 0.0499$, started from $\mathbf{x}^0 = (10, 20)^T$ and $F(\mathbf{x}^0) = 2400$, satisfied the termination condition at $k = 1426$. This slow convergence is related to

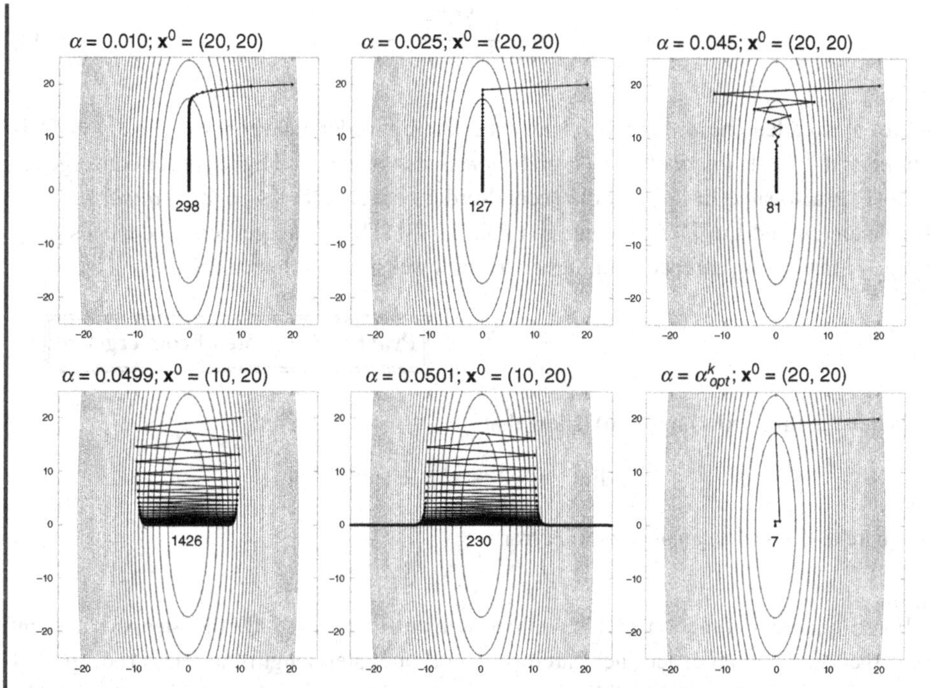


Figure 6.4 Steepest descent minimization for different step lengths ($\alpha$) starting from $\mathbf{x}^0$: $\alpha_{opt}^k$ means the optimum $\alpha$ value at each iteration. Dots indicate $\mathbf{x}^k$ at each iteration. The iteration numbers at termination are shown near the minimum. Contours are plotted with an interval of 300 up to 9000.

the large difference between $\lambda_1$ and $\lambda_2$ (see Section 6.2.5). With $\alpha = 0.0501$, the steepest descent minimization eventually diverges out of the plotting boundaries of $x_1$ after $k = 230$. Using $\alpha_{opt}^k$, based on (6.32), then the termination condition is satisfied within six iterations.

### *6.3.1.3 Orthogonality of Descent Directions*

From (6.31), the minimization of $F\left(\mathbf{x}^k - \alpha^k \mathbf{g}^k\right)$ with respect to $\alpha^k$ should satisfy the first-order necessary condition (see (6.8)), that is,

$$\frac{d}{d\alpha^k} F\left(\mathbf{x}^k - \alpha^k \mathbf{g}^k\right) = \mathbf{0} = \nabla F\left(\mathbf{x}^k - \alpha^k \mathbf{g}^k\right)^T \left(-\mathbf{g}^k\right) = -\left(\mathbf{g}^{k+1}\right)^T \mathbf{g}^k. \qquad (6.35)$$

Thus, $\mathbf{g}^{k+1}$ and $\mathbf{g}^k$ are orthogonal: This implies that, in the steepest descent algorithm using an optimum $\alpha^k$, the successive search directions are orthogonal along the minimization path (see Figure 6.4 for $\alpha = \alpha_{opt}^k$).

### *6.3.1.4 Convergence Rate*

In a minimization problem, the sequence of $\mathbf{x}^k$ is supposed to converge to a local/global minimum, $\mathbf{x}^\dagger$, i.e., $\lim_{k\to\infty} \left\| \mathbf{x}^k - \mathbf{x}^\dagger \right\| = 0$. The *convergence ratio*, $\gamma$, is defined as

$$0 < \gamma = \lim_{k\to\infty} \frac{\left\|\mathbf{x}^{k+1} - \mathbf{x}^{\dagger}\right\|}{\left\|\mathbf{x}^{k} - \mathbf{x}^{\dagger}\right\|^{p}} < \infty, \tag{6.36}$$

where $p$ is called the *order of convergence*. A first-order convergence (i.e., $p = 1$) is *sublinear* if $\gamma = 1$, *linear* if $\gamma < 1$, and *superlinear* if $\gamma = 0$. Higher-order convergence ($p > 1$) is also superlinear, and the second-order convergence ($p = 2$) is *quadratic* (see also Antoniou and Lu, 2007; Chong and Żak, 2013). The convergence rate depends on $p$ and $\gamma$ – a faster convergence with higher $p$ and lower $\gamma$ (see Practice 6.9).

**Practice 6.9 Rate of convergence**

When the limit exists in (6.36), show that

$$\lim_{k\to\infty} \left\|\mathbf{x}^{k+1} - \mathbf{x}^{\dagger}\right\| = \gamma \epsilon^{p},$$

where $0 < \varepsilon < 1$ is an asymptotic error.

For a quadratic function (6.12) with a symmetric positive $\mathbf{Q}$, the steepest descent convergence, using an exact line search (i.e., optimum step length), is linear (i.e., $p = 1$ and $\gamma < 1$). By including the sublinear convergence, the steepest descent algorithm has the following convergence rate:

$$F\left(\mathbf{x}^{k+1} - \mathbf{x}^{\dagger}\right) \leq \gamma F\left(\mathbf{x}^{k} - \mathbf{x}^{\dagger}\right), \tag{6.37}$$

where the convergence ratio, $\gamma$, is given by

$$\gamma = \left(\frac{\lambda_{max} - \lambda_{min}}{\lambda_{max} + \lambda_{min}}\right)^{2} = \left(\frac{1-\kappa}{1+\kappa}\right)^{2}. \tag{6.38}$$

Here, $\kappa = \kappa(\mathbf{Q})$ is the condition number of the Hessian $\mathbf{Q}$, representing the ratio of the largest eigenvalue, $\lambda_{max}$, to the smallest eigenvalue, $\lambda_{min}$ (see (6.18)). In the steepest descent minimization, if $\kappa \approx 1$, the function contours are nearly circular, making the convergence fast (i.e., $\gamma$ is small): If $\kappa$ is quite large, the contours form elongated ellipses, resulting in a slow convergence (i.e., $\gamma$ is large).

**Practice 6.10 Condition number and convergence rate**

Consider the following functions:

$$F(\mathbf{x}) = 2x_1^2 + x_2^2, \;\; F(\mathbf{x}) = 100x_1^2 + x_2^2 \text{ and } F(\mathbf{x}) = 400x_1^2 + x_2^2.$$

Compute the condition number ($\kappa(\mathbf{Q})$) of each function and discuss the differences in the convergence rate of the steepest descent minimization.

The process in the steepest descent algorithm is the same as in Algorithm 6.1 except that the search direction, $\mathbf{d}^k$, is replaced by the negative gradient, $-\mathbf{g}^k = -\nabla F(\mathbf{x}^k)$ (see (6.27)), and that the step length, $\alpha^k$, is determined by one of the methods in Section 6.3.1.2.

### *6.3.2 Newton's Method*

This method requires both the first-order and second-order derivative information to define the increment as

$$\Delta \mathbf{x}^k = -\left(\mathbf{S}^k\right)^{-1} \mathbf{g}^k, \tag{6.39}$$

where $\mathbf{S}$ is the Hessian and $\mathbf{g}$ is the gradient (do Practice 6.11). Thus, for the quadratic function in (6.12), Newton's method defines the following iterative process:

$$\mathbf{x}^{k+1} = \mathbf{x}^k - \left(\mathbf{Q}^k\right)^{-1} \mathbf{g}^k, \tag{6.40}$$

where $\mathbf{Q}$ is the Hessian of (6.12). For a nonsingluar $\mathbf{Q}$, by choosing the optimum step length that minimizes $F(\mathbf{x}^k + \alpha^k \mathbf{d}^k)$, $\Delta \mathbf{x}^k = \mathbf{d}^k$ at the first iteration where $\alpha^k = 1$. Then,

$$\mathbf{d}^k = -\left(\mathbf{Q}^k\right)^{-1} \mathbf{g}^k \tag{6.41}$$

is called the *Newton direction* (see Antoniou and Lu, 2007).

**Practice 6.11 Increment in Newton's method**

Derive (6.39) by taking the second-order Taylor series expansion for a function $F\left(\mathbf{x}^{k+1}\right) = F\left(\mathbf{x}^k + \Delta \mathbf{x}^k\right)$ and by minimizing it with respect to $\Delta \mathbf{x}^k$.

Newton's method is perfect for the minimization of a quadratic function with a strong minimum, especially when the initial guess $\mathbf{x}^0$ is near the minimum $\mathbf{x}^\dagger$, by achieving minimization in only one step, because it approximates a function as a quadratic that has just one minimum. However, there is no guarantee that Newton's algorithm always proceeds in the descent direction because the Hessian is not always positive definite. Moreover, Newton's method may not be a descent algorithm when $\mathbf{x}^0$ is far away from $\mathbf{x}^\dagger$. The convergence order of Newton's method is at least 2 (see Chong and Żak, 2013). An example is provided in Example 6.4, and the process is described in Algorithm 6.2.

**Example 6.4 Minimization using Newton's method**

Newton's method is used for minimization of a nonquadratic function

$$\frac{1}{5}(x_1 - x_2)^4 + 12x_1x_2 + 2(x_1 - x_2),$$

which has three stationary points – a local minimum, a global minimum, and a saddle point. The gradient ($\mathbf{g}$) and the Hessian ($\mathbf{S}$) are given as

$$\mathbf{g} = \begin{pmatrix} \frac{4}{5}(x_1 - x_2)^3 + 12x_2 + 2 \\ -\frac{4}{5}(x_1 - x_2)^3 + 12x_1 - 2 \end{pmatrix} \quad \text{and}$$

$$\mathbf{S} = \begin{pmatrix} \frac{12}{5}(x_1 - x_2)^2 & -\frac{12}{5}(x_1 - x_2)^2 + 12 \\ -\frac{12}{5}(x_1 - x_2)^2 + 12 & \frac{12}{5}(x_1 - x_2)^2 \end{pmatrix},$$

respectively. For $\mathbf{x}^0 = (0.0, 2.8)^T$, $\mathbf{g}^0$ and $\mathbf{S}^0$ are

$$\mathbf{g}^0 = \begin{pmatrix} 18.038 \\ 15.562 \end{pmatrix} \quad \text{and} \quad \mathbf{S}^0 = \begin{pmatrix} 18.816 & -6.816 \\ -6.816 & 18.816 \end{pmatrix},$$

respectively, then the first iteration gives

$$\mathbf{x}^1 = \mathbf{x}^0 - \left(\mathbf{S}^0\right)^{-1} \mathbf{g}^0 = \begin{pmatrix} 0.0 \\ 2.8 \end{pmatrix} - \begin{pmatrix} 0.061 & 0.022 \\ 0.022 & 0.061 \end{pmatrix} \begin{pmatrix} 18.038 \\ 15.562 \end{pmatrix} = \begin{pmatrix} -1.448 \\ 1.448 \end{pmatrix},$$

which is very close to the global minimum, $\mathbf{x}^\dagger = (-1.446, 1.446)^T$.

Figure 6.5 clearly shows that Newton's method strongly depends on the location of initial points and the structure and local properties of the function, such as gradient, stationary points, etc. When the initial points are properly located, Newton's method achieves minimization in just one shot. We can also observe that convergence is quite slow along the strongest gradient direction, e.g., when $\mathbf{x}^0$ is located at $(-2.5, 2.5)$ or $(2.5, -2.5)$. Once $\mathbf{x}^1$ reaches near the stationary points, the convergence is fast. In Figure 6.5, the dashed paths converge to the stationary points only in a few more steps.

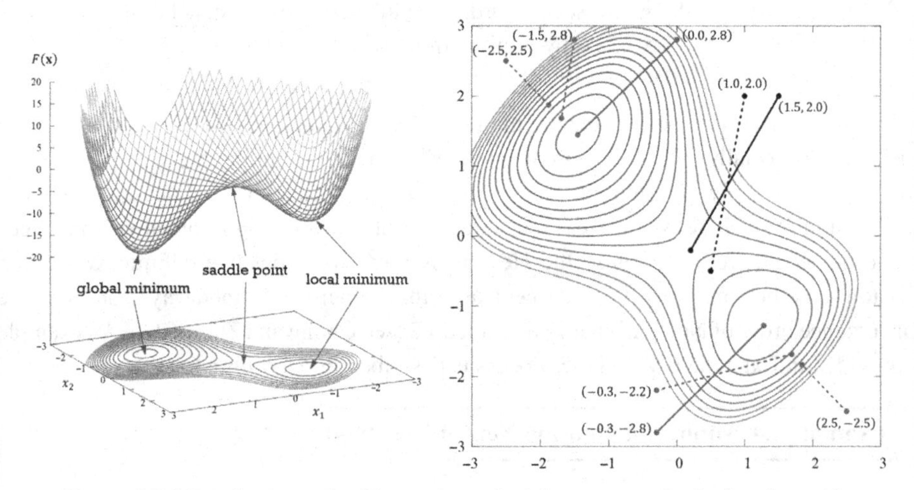


Figure 6.5 Minimization using Newton's method for a nonquadratic function with three stationary points, showing the first iteration paths from different initial points, $\mathbf{x}^0 = (x_1^0, x_2^0)$, toward the global minimum, the local minimum, and the saddle point. Some initial points converge to the stationary points in just one iteration (solid lines) while others still need more iterations (dashed lines). Contours are plotted for $[-16, 10]$ at an interval of 2.

**Algorithm 6.2** Minimization process using Newton's method

```
/* k: iteration number; K: maximum iteration; ε_a: minimization tolerance            */
/* x†: minimizer of F(x); F† = F(x†): minimum value of F(x)                          */
/* g^k: gradient (= ∇F(x^k)); S^k: Hessian (= ∇²F(x^k))                               */
1  Input: x^0, F^0 = F(x^0), ε_a                      ! Initial guess x^0 at k = 0
2  while (k ≤ K or |ΔF^{k+1}| ≥ ε_a) do               ! Loop for termination condition
3      Calculate g^k and S^k                          ! g^k and S^k using appropriate methods
4      Calculate (S^k)^{-1}                           ! Inverse of S^k using appropriate methods
5      Increment:                                     ! Calculate an increment Δx^k
6          Δx^k = −(S^k)^{-1} g^k                     ! Δx^k in Newton direction
7      Update:                                        ! Move one step forward to the minimum
8          x^{k+1} = x^k + Δx^k                       ! New x at k + 1
9          F^{k+1} = F(x^{k+1})                       ! Evaluate F(x) at k + 1
10     Convergence:                                   ! Compare F(x) between k and k + 1
11         ΔF^{k+1} = F^{k+1} − F^k                   ! Difference between F^{k+1} and F^k
12         |ΔF^{k+1}| < ε_a?                          ! Convergence check using (6.25a)
13 endwhile
14 Output: x† ← x^{k+1}, F† ← F^{k+1}                 ! Assign x^{k+1} to x† and F^{k+1} to F†
```

Although Newton's method has its own advantage, especially for the well-conditioned quadratic minimization problem, a huge computational resource is often required to exactly calculate the Hessian and its inverse. Therefore, it is almost impractical to use Newton's method for large-scale optimization that involves numerous parameters and variables. In practice, some alternative quadratic minimization methods such as the conjugate gradient, quasi-Newton, etc., are used more often.

### *6.3.3 Conjugate Gradient Method*

The orthogonal property in successive search directions in the steepest descent algorithm often produces many zigzags in small steps, which impedes convergence near the minimum, especially when the function contours are highly elongated. The *conjugate gradient* method is another first-order line-search descent algorithm that converges within a finite number of iterations in minimizing the quadratic function in (6.12), whose Hessian ($\mathbf{Q}$) is a symmetric and positive definite matrix. It shows a quadratic convergence, as in Newton's method, but avoids calculation of the Hessian and its inverse.

The essence of this method is employing a set of *conjugate vectors* (see Section 6.2.7) and successively minimizing the function along the individual directions in a conjugate set (see, e.g., Nocedal and Wright, 2006; Arora, 2015). Consider a quadratic function with $n$ variables, as in (6.12), i.e.,

$$F(\mathbf{x}) = \frac{1}{2}\mathbf{x}^T\mathbf{Q}\mathbf{x} + \mathbf{x}^T\mathbf{r} + c, \tag{6.42}$$

where $\mathbf{Q}$ is an $n \times n$ positive definite Hessian matrix. Suppose we have a set of $\mathbf{Q}$-conjugate vectors $\{\mathbf{d}^0, \ldots, \mathbf{d}^{n-1}\}$ that can be used for our search directions, as in the conjugate direction algorithm (see Colloquy 6.1).

**COLLOQUY 6.1**

**Conjugate direction algorithm**

Consider the quadratic function in (6.42) with $n$ variables and a positive definite Hessian $\mathbf{Q}$. If we have a set of nonzero $\mathbf{Q}$-conjugate directions $\{\mathbf{d}^0, \ldots, \mathbf{d}^{n-1}\}$, then we have the following sequences, for a given initial point $\mathbf{x}^0$:

$$\mathbf{g}^k = \mathbf{Q}\mathbf{x}^k + \mathbf{r},$$
$$\mathbf{x}^{k+1} = \mathbf{x}^k + \alpha^k \mathbf{d}^k,$$

which converges to the unique minimizer $\mathbf{x}^\dagger = \mathbf{x}^n$ in $n$ iterations (see Antoniou and Lu, 2007; Chong and Żak, 2013). In this basic *conjugate direction* algorithm, $\alpha^k$ is given by

$$\alpha^k = -\frac{\left(\mathbf{g}^k\right)^T \mathbf{d}^k}{\left(\mathbf{d}^k\right)^T \mathbf{Q}\mathbf{d}^k},$$

which is the minimizer of $F\left(\mathbf{x}^k + \alpha \mathbf{d}^k\right)$ (i.e., $\alpha^k = \alpha^k_{opt}$). Furthermore, due to the $\mathbf{Q}$-conjugacy condition (6.20), the gradient $\mathbf{g}^k$ is orthogonal to directions $\mathbf{d}^j$, i.e.,

$$\left(\mathbf{g}^k\right)^T \mathbf{d}^j = \left(\mathbf{d}^j\right)^T \mathbf{g}^k = 0 \text{ for } 0 \leq j < k.$$

To use the conjugate direction algorithm, we need to specify the $\mathbf{Q}$-conjugate directions, which can be generated as we perform iterations, as in the conjugate gradient method.

In the *conjugate gradient* method, the direction $\mathbf{d}^k$ is obtained as a linear combination of $\mathbf{d}^{k-1}$ and the gradient $\mathbf{g}^k$ – all the directions are mutually $\mathbf{Q}$-conjugate. We select our first search direction $\mathbf{d}^0$ to be the steepest descent direction, i.e., for an initial point $\mathbf{x}^0$,

$$\mathbf{d}^0 = -\mathbf{g}^0 = \mathbf{Q}\mathbf{x}^0 + \mathbf{r}. \tag{6.43}$$

Then, we have the conjugate gradient algorithm, composed of the following recursive sequences of $\mathbf{x}^k$ and the *conjugate direction* $\mathbf{d}^k$:

$$\mathbf{x}^{k+1} = \mathbf{x}^k + \alpha^k \mathbf{d}^k, \tag{6.44a}$$
$$\mathbf{d}^{k+1} = -\mathbf{g}^{k+1} + \beta^k \mathbf{d}^k, \tag{6.44b}$$

with

$$\mathbf{g}^k = \mathbf{Q}\mathbf{x}^k + \mathbf{r}, \tag{6.45a}$$

$$\alpha^k = -\frac{(\mathbf{g}^k)^T \mathbf{d}^k}{(\mathbf{d}^k)^T \mathbf{Q}\mathbf{d}^k}, \text{ and } \beta^k = \frac{(\mathbf{g}^{k+1})^T \mathbf{Q}\mathbf{d}^k}{(\mathbf{d}^k)^T \mathbf{Q}\mathbf{d}^k}, \tag{6.45b}$$

which converges to the unique minimizer $\mathbf{x}^\dagger$. In addition, $\mathbf{g}^k$ is orthogonal to a set of previous gradients $\{\mathbf{g}^0, \ldots, \mathbf{g}^{k-1}\}$:

$$\left(\mathbf{g}^k\right)^T \mathbf{g}^j = 0 \text{ for } 0 \le j < k. \tag{6.46}$$

Here, $\alpha^k = \alpha^k_{opt}$ as in Colloquy 6.1, and $\beta^k$ is chosen such that $\mathbf{d}^{k+1}$ is $\mathbf{Q}$-conjugate to a set of previous directions $\{\mathbf{d}^0, \ldots, \mathbf{d}^k\}$. Detailed derivations of $\alpha^k$ and $\beta^k$ are referred to in Chong and Żak (2013). See Algorithm 6.3 for the minimization process using the conjugate gradient method, simple examples in Example 6.5, and comparison with the steepest method in Example 6.6.

The conjugate gradient method is efficient because it is based on the conjugate direction algorithm, thus minimizing a quadratic function of $n$ variables with the positive definite Hessian in $n$ steps (see Example 6.5). In this method, the convergence rate is quadratic (i.e., the order of convergence is $p = 2$), and line searches are not required. For the first iteration, it adopts the steepest descent direction, thus giving a large decrease in the function (see Example 6.6). Although the Hessian should be supplied, calculating the inverse of the Hessian is unnecessary.

**Algorithm 6.3** Minimization process of the conjugate gradient method

```
/* k: iteration number; K: maximum iteration; ε_a: minimization tolerance           */
/* F(x) = ½ x^T Q x + x^T r + c: a quadratic function with the Hessian Q             */
/* x†: minimizer of F(x); F† = F(x†): minimum value of F(x)                          */
/* g^k: gradient (= ∇F(x^k)); S: Hessian (= ∇²F(x))                                  */
1  Input: x^0, F^0 = F(x^0), S = Q, ε_a              ! Initial guess x^0 at k = 0
2  Compute g^0 and assign d^0 = −g^0                 ! Set d^0 to the steepest gradient direction
3  while (k ≤ K or |ΔF^{k+1}| ≥ ε_a) do              ! Loop for termination condition
4      α^k = −((g^k)^T d^k) / ((d^k)^T Q d^k)        ! α^k that minimize F(x^k + α^k d^k)
5      Increment:                                    ! Calculate an increment Δx^k
6          Δx^k = α^k d^k                            ! Δx^k in Newton direction
7      Update:                                       ! Move one step forward to the minimum
8          x^{k+1} = x^k + Δx^k                      ! New x at k + 1
9          F^{k+1} = F(x^{k+1})                      ! Evaluate F(x) at k + 1
10         g^{k+1} = Q x^{k+1} + r                   ! New g at k + 1
11         Compute β^k = ((g^{k+1})^T Q d^k) / ((d^k)^T Q d^k)   ! β^k to calculate d^{k+1}
12         d^{k+1} = −g^{k+1} + β^k d^k              ! New d at k + 1
13     Convergence:                                  ! Compare F(x) between k and k + 1
14         ΔF^{k+1} = F^{k+1} − F^k                  ! Difference between F^{k+1} and F^k
15         |ΔF^{k+1}| < ε_a?                         ! Convergence check using (6.25a)
16 endwhile
17 Output: x† ← x^{k+1}, F† ← F^{k+1}                ! Assign x^{k+1} to x† and F^{k+1} to F†
```

**Example 6.5 Minimization using the conjugate gradient method**

The conjugate gradient method is applied to the minimization of quadratic functions

$$F(\mathbf{x}) = 20x_1^2 + x_2^2 \text{ and } F(\mathbf{x}) = 4x_1^2 - 4x_1x_2 + 8x_2^2,$$

both having two variables and positive definite Hessians (i.e., the eigenvalues are all positive). Therefore, starting from any initial point, the conjugate gradient method converges to the minimum in two iterations, as shown in Figure 6.6.

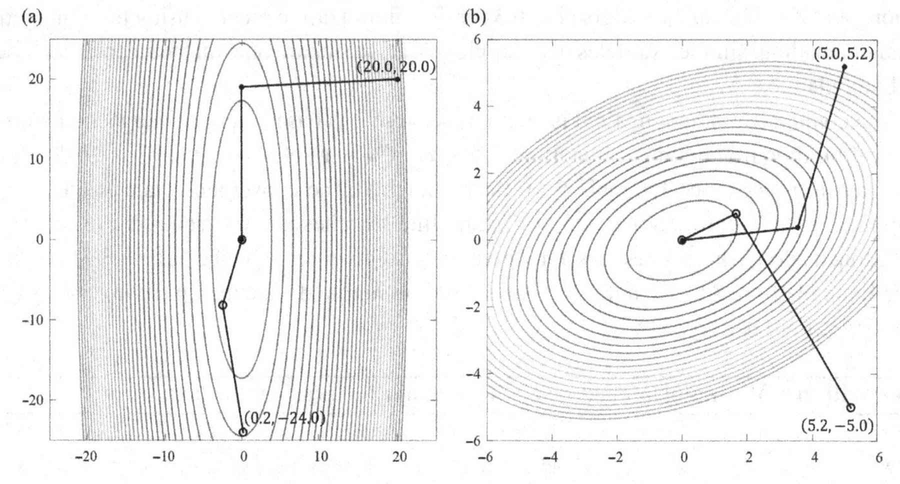


Figure 6.6 Minimization using the conjugate gradient method for two quadratic functions: (a) $F(\mathbf{x}) = 20x_1^2+x_2^2$ and (b) $F(\mathbf{x}) = 4x_1^2-4x_1x_2+8x_2^2$, both starting with two initial points $\mathbf{x}^0 = (x_1^0, x_2^0)^T$. Contours are plotted (a) up to 9000 at an interval of 300 and (b) up to 200 at an interval of 10.

**Example 6.6 Conjugate gradient vs steepest descent method**

Figure 6.7 depicts a comparison between the conjugate gradient method and the steepest descent method in minimizing the quadratic function $F(\mathbf{x}) = 20x_1^2 + x_2^2$: The step length $\alpha^k$ used for the former is in (6.44b) and that for the latter is in (6.32), both minimizing $F(\mathbf{x}^k + \alpha\mathbf{d}^k)$. The initial point, $\mathbf{x}^0 = (0.8, 80)^T$, was intentionally set to be quite far from the minimum. As this is a quadratic function, the conjugate gradient method converges in just 2 steps while the steepest descent method reaches the minimum in 21 steps with successive zigzags for the same tolerance of absolute improvement, $\epsilon_a = 1 \times 10^{-4}$ (see (6.25a)). Both methods made the same large reductions for the first iterations, from $\mathbf{x}^0 = (0.8, 80)^T$ to $\mathbf{x}^1 = (-8.44, -33.78)^T$, because they use the steepest descent direction, i.e., $\mathbf{d}^0 = -\mathbf{g}^0$, for this step.

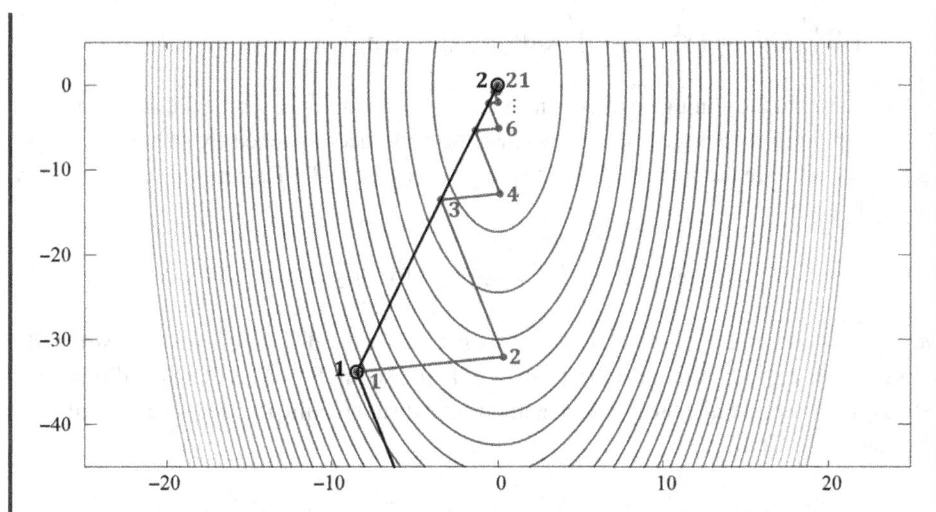


Figure 6.7 Comparison of the conjugate gradient method (darker line with empty circles) and the steepest descent method (lighter line with filled circles) in minimizing $F(\mathbf{x}) = 20x_1^2 + x_2^2$ with an initial point $\mathbf{x}^0 = (0.8, 80)^T$. Contours are plotted up to 9000 at an interval of 300. Minimization paths and iteration numbers are shown from the first iterations: The paths from $\mathbf{x}^0$ to $\mathbf{x}^1$ are identical for both methods.

**Practice 6.12 Minimization using the conjugate gradient method**

Show the step-by-step minimization process using the conjugate gradient method for the following quadratic function:

$$F(\mathbf{x}) = \frac{1}{2}\left(3x_1^2 + 2x_2^2 + 3x_3^2\right) + x_1x_2 + 2x_1x_3 - x_2 - 2x_3$$

to find the minimum $\mathbf{x}^\dagger$, starting from $\mathbf{x}^0 = (0, 0, 0)^T$. How many iterations are required to reach $\mathbf{x}^\dagger$?

As the Hessian is required only for computing $\alpha^k$ and $\beta^k$, the conjugate gradient algorithm can be constructed in a more efficient way, even to be extended to nonquadratic problems. First, the $\alpha^k$ in (6.44b) that minimizes $F(\mathbf{x}^k + \alpha\mathbf{d}^k)$ can be obtained by 1D line-search methods (see Antoniou and Lu, 2007; Chong and Żak, 2013; Arora, 2015). The Hessian in $\beta^k$ (see (6.44b)) can be also eliminated with some manipulation. From (6.24) and (6.45a), we have

$$\mathbf{g}^{k+1} - \mathbf{g}^k = \mathbf{Q}\left(\mathbf{x}^{k+1} - \mathbf{x}^k\right) = \alpha^k\mathbf{Q}\mathbf{d}^k, \tag{6.47}$$

and using the $\mathbf{Q}$-conjugacy condition (6.20) for a set of nonzero conjugate directions $\{\mathbf{d}^0, \ldots, \mathbf{d}^{n-1}\}$,

$$\alpha^k \left(\mathbf{d}^k\right)^T \mathbf{Q}\mathbf{d}^j = \left(\mathbf{x}^{k+1} - \mathbf{x}^k\right)^T \mathbf{Q}\mathbf{d}^j = \left(\mathbf{g}^{k+1} - \mathbf{g}^k\right)^T \mathbf{d}^j = 0 \text{ for } k \neq j. \tag{6.48}$$

This implies that the conjugate directions $\mathbf{d}^j$ are orthogonal to the gradient changes $\mathbf{g}^{k+1} - \mathbf{g}^k = \Delta\mathbf{g}^k$ in successive iterations and that the subspaces spanned by $\{\mathbf{d}^0, \ldots, \mathbf{d}^n\}$ and $\{\Delta\mathbf{g}^0, \ldots, \Delta\mathbf{g}^n\}$ are the same (see also Navon and Legler, 1987; Hagan et al., 2014), which brings about

$$\left(\mathbf{g}^k\right)^T \mathbf{d}^j = \left(\mathbf{g}^k\right)^T \mathbf{g}^j = 0 \text{ for } 0 \leq j < k. \tag{6.49}$$

We again start with the steepest descent direction, i.e., $\mathbf{d}^0 = -\mathbf{g}^0$, and construct $\mathbf{d}^1$ orthogonal to $\Delta\mathbf{g}^0$, $\mathbf{d}^2$ orthogonal to $\{\Delta\mathbf{g}^0, \Delta\mathbf{g}^1\}$, and so on. Based on these conditions, we can develop simple conjugate gradient methods that do not require the Hessian in calculating $\beta^k$ as shown in Colloquy 6.2.

**COLLOQUY 6.2**

**Formulas for $\beta^k$ without the Hessian information**

Using (6.47), we can rewrite $\beta^k$ from (6.45b) as

$$\beta^k = \frac{\left(\mathbf{g}^{k+1}\right)^T \left(\mathbf{g}^{k+1} - \mathbf{g}^k\right)}{\left(\mathbf{d}^k\right)^T \left(\mathbf{g}^{k+1} - \mathbf{g}^k\right)} = \frac{\left(\mathbf{g}^{k+1}\right)^T \Delta\mathbf{g}^k}{\left(\mathbf{d}^k\right)^T \Delta\mathbf{g}^k}. \tag{6.50}$$

This is known as the *Hestenes–Stiefel* formula (Hestenes and Stiefel, 1952).

From (6.49), we have

$$\left(\mathbf{d}^k\right)^T \mathbf{g}^{k+1} = \left(\mathbf{d}^{k-1}\right)^T \mathbf{g}^k = 0, \tag{6.51}$$

and from (6.44b),

$$\left(\mathbf{d}^k\right)^T \mathbf{g}^k = -\left(\mathbf{g}^k\right)^T \mathbf{g}^k + \beta^{k-1} \left(\mathbf{d}^{k-1}\right)^T \mathbf{g}^k = -\left(\mathbf{g}^k\right)^T \mathbf{g}^k. \tag{6.52}$$

Then, $\beta^k$ in (6.50) becomes

$$\beta^k = \frac{\left(\mathbf{g}^{k+1}\right)^T \Delta\mathbf{g}^k}{\left(\mathbf{g}^k\right)^T \mathbf{g}^k}, \tag{6.53}$$

known as the *Polak–Ribière* formula (Polak and Ribiére, 1969).

Another common expression for $\beta^k$ is the *Fletcher–Reeves* formula (Fletcher and Reeves, 1964):

$$\beta^k = \frac{\left(\mathbf{g}^{k+1}\right)^T \mathbf{g}^{k+1}}{\left(\mathbf{g}^k\right)^T \mathbf{g}^k}. \tag{6.54}$$

**Practice 6.13 Conjugate gradient method: Fletcher–Reeves formula**

Derive (6.54) from (6.50).

### *6.3.4 Quasi-Newton's Method*

We have discussed the property of Newton's method, which has a quadratic convergence order; however, the convergence strongly depends on the location of initial point $\mathbf{x}^0$ and the characteristics of the given objective function. For a general nonquadratic function, the algorithm may not have the descent property, especially when $\mathbf{x}^0$ is quite far from the solution $\mathbf{x}^\dagger$. Computationally, Newton's method is very expensive for large $n$ since it requires to evaluate the $n \times n$ Hessian and its inverse at every iterative step. The quasi-Newton methods deals with these defects to make Newton's method practically feasible in large-scale minimization problems by enforcing the algorithm to have the descent property and by employing an approximation to the inverse Hessian or the Hessian itself.

The update of Newton's method (6.40) for a quadratic equation with the Hessian $\mathbf{Q}$ (6.42) is rewritten as

$$\mathbf{x}^{k+1} = \mathbf{x}^k + \Delta\mathbf{x}^k = \mathbf{x}^k - \left(\mathbf{Q}^k\right)^{-1} \mathbf{g}^k. \tag{6.55}$$

In the quasi-Newton's method, this update is modified to include $\alpha^k$ and to approximate the Hessian ($\mathbf{B} \approx \mathbf{Q}$) as

$$\mathbf{x}^{k+1} = \mathbf{x}^k - \alpha^k \left(\mathbf{B}^k\right)^{-1} \mathbf{g}^k. \tag{6.56}$$

Here, a positive line-search parameter $\alpha^k$ is chosen to minimize $F\left(\mathbf{x}^k - \alpha \left(\mathbf{B}^k\right)^{-1} \mathbf{g}^k\right)$ or to satisfy the Wolfe conditions (see (6.33) and (6.34a)) to keep the algorithm's descent property, that is, to ensure $F\left(\mathbf{x}^{k+1}\right) < F\left(\mathbf{x}^k\right)$.

A positive definite $n \times n$ matrix $\mathbf{B}^k$ is an approximation to the exact Hessian $\mathbf{Q}^k$, which satisfies the *quasi-Newton* (or *secant*) condition:

$$\mathbf{B}^{k+1}\Delta\mathbf{x}^k = \Delta\mathbf{g}^k, \tag{6.57}$$

where $\Delta\mathbf{x}^k = \mathbf{x}^{k+1} - \mathbf{x}^k$ and $\Delta\mathbf{g}^k = \mathbf{g}^{k+1} - \mathbf{g}^k$ (see Antoniou and Lu, 2007; Chong and Żak, 2013). In terms of the inverse Hessian, $\left(\mathbf{Q}^k\right)^{-1}$, an approximation can be made by defining $\mathbf{C}^k = \left(\mathbf{B}^k\right)^{-1}$; then the quasi-Newton condition for the inverse Hessian becomes

$$\mathbf{C}^{k+1}\Delta\mathbf{g}^k = \Delta\mathbf{x}^k. \tag{6.58}$$

Assume that $\mathbf{B}^{k+1}$ can be obtained by updating $\mathbf{B}^k$ with a rank 1 matrix $a\mathbf{u}\mathbf{u}^T$ as

$$\mathbf{B}^{k+1} = \mathbf{B}^k + a\mathbf{u}\mathbf{u}^T; \tag{6.59}$$

then, using the quasi-Newton condition (6.57) and putting $\mathbf{u} = \Delta\mathbf{g}^k - \mathbf{B}^k\Delta\mathbf{x}^k$, we have

$$\mathbf{B}^{k+1} = \mathbf{B}^k + \frac{\left(\Delta\mathbf{g}^k - \mathbf{B}^k\Delta\mathbf{x}^k\right)\left(\Delta\mathbf{g}^k - \mathbf{B}^k\Delta\mathbf{x}^k\right)^T}{\left(\Delta\mathbf{g}^k - \mathbf{B}^k\Delta\mathbf{x}^k\right)^T \Delta\mathbf{x}^k}, \tag{6.60}$$

called the *symmetric rank-1 (SR1)* update. The SR1 update for the inverse Hessian approximation is given by

$$\mathbf{C}^{k+1} = \mathbf{C}^k + \frac{\left(\Delta\mathbf{x}^k - \mathbf{C}^k\Delta\mathbf{g}^k\right)\left(\Delta\mathbf{x}^k - \mathbf{C}^k\Delta\mathbf{g}^k\right)^T}{\left(\Delta\mathbf{x}^k - \mathbf{C}^k\Delta\mathbf{g}^k\right)^T\Delta\mathbf{g}^k} \tag{6.61}$$

(do Practice 6.14).

By taking a rank-2 update, we have

$$\mathbf{B}^{k+1} = \mathbf{B}^k + a\mathbf{u}\mathbf{u}^T + b\mathbf{v}\mathbf{v}^T, \tag{6.62}$$

and again using (6.57) and putting $\mathbf{u} = \Delta\mathbf{g}^k$ and $\mathbf{v} = \mathbf{B}^k\Delta\mathbf{x}^k$, we solve for $a$ and $b$ to get

$$\mathbf{B}^{k+1}_{BFGS} = \mathbf{B}^k - \frac{\mathbf{B}^k\Delta\mathbf{x}^k\left(\Delta\mathbf{x}^k\right)^T\mathbf{B}}{\left(\Delta\mathbf{x}^k\right)^T\mathbf{B}\Delta\mathbf{x}^k} + \frac{\Delta\mathbf{g}^k\left(\Delta\mathbf{g}^k\right)^T}{\left(\Delta\mathbf{g}^k\right)^T\Delta\mathbf{x}^k}, \tag{6.63}$$

called the *Broyden–Fletcher–Goldfarb–Shanno (BFGS)* update. This most popular quasi-Newton method was proposed by Broyden (1970), Fletcher (1970), Goldfarb (1970), and Shanno (1970) around the same time. The BFGS update for the Hessian approximation ($\mathbf{C}^{k+1}_{BFGS}$) is expressed, by establishing $\mathbf{B}^{k+1}_{BFGS}\mathbf{C}^{k+1}_{BFGS} = \mathbf{I}$, as

$$\mathbf{C}^{k+1}_{BFGS} = \left(\mathbf{I} - \frac{\Delta\mathbf{x}^k\left(\Delta\mathbf{g}^k\right)^T}{\left(\Delta\mathbf{x}^k\right)^T\Delta\mathbf{g}^k}\right)\mathbf{C}^k\left(\mathbf{I} - \frac{\Delta\mathbf{g}^k\left(\Delta\mathbf{x}^k\right)^T}{\left(\Delta\mathbf{x}^k\right)^T\Delta\mathbf{g}^k}\right) + \frac{\Delta\mathbf{x}^k\left(\Delta\mathbf{x}^k\right)^T}{\left(\Delta\mathbf{x}^k\right)^T\Delta\mathbf{g}^k} \tag{6.64}$$

(do Practice 6.14). The quasi-Newton method using the BFGS update is shown in Algorithm 6.4.

**Practice 6.14 Quasi-Newton updates**

Answer the following:

1. Derive (6.60) from (6.59).
2. Derive (6.61) from (6.60) using the Sherman–Morrison formula:

$$\left(\mathbf{A} + \mathbf{u}\mathbf{v}^T\right)^{-1} = \mathbf{A}^{-1} - \frac{\mathbf{A}^{-1}\mathbf{u}\mathbf{v}^T\mathbf{A}^{-1}}{1 + \mathbf{v}^T\mathbf{A}^{-1}\mathbf{u}}.$$

3. Derive (6.63) from (6.62).
4. Derive (6.64) from (6.63) using the Woodbury formula:

$$(\mathbf{A} + \mathbf{U}\mathbf{D}\mathbf{V})^{-1} = \mathbf{A}^{-1} - \mathbf{A}^{-1}\mathbf{U}\left(\mathbf{D}^{-1} + \mathbf{V}\mathbf{A}^{-1}\mathbf{U}\right)^{-1}\mathbf{V}\mathbf{A}^{-1}.$$

You may assume

$$\mathbf{U} = \mathbf{V}^T = \begin{pmatrix}\mathbf{B}\Delta\mathbf{x} & \Delta\mathbf{g}\end{pmatrix}, \quad \mathbf{D} = \begin{pmatrix} -\frac{1}{\Delta\mathbf{x}^T\mathbf{B}\Delta\mathbf{x}} & 0 \\ 0 & \frac{1}{\Delta\mathbf{g}^T\Delta\mathbf{x}} \end{pmatrix}.$$

Another quasi-Newton method, called the Davidon–Fletcher–Powell update, is described in Colloquy 6.3. The limited memory BFGS (L-BFGS), shown in Colloquy 6.4, is a version

**Algorithm 6.4** Quasi-Newton method using the BFGS update

/* $k$: iteration number; $K$: maximum iteration; $\varepsilon_a$: minimization tolerance */
/* $F(\mathbf{x}) = \frac{1}{2}\mathbf{x}^T\mathbf{Q}\mathbf{x} + \mathbf{x}^T\mathbf{r} + c$: a quadratic function with the Hessian $\mathbf{Q}$ */
/* $\mathbf{x}^\dagger$: minimizer of $F(\mathbf{x})$; $F^\dagger = F(\mathbf{x}^\dagger)$: minimum value of $F(\mathbf{x})$ */
/* $\mathbf{g}^k$: gradient ($= \nabla F(\mathbf{x}^k)$); $\mathbf{C} \approx (\mathbf{Q})^{-1}$: approximation to the inverse Hessian */

1 ***Input***: $\mathbf{x}^0$, $F^0 = F(\mathbf{x}^0)$, $\mathbf{C}^0 = \mathbf{I}$, $\varepsilon_a$ ! Initial guess $\mathbf{x}^0$ at $k = 0$
2 Compute $\mathbf{g}^0 = \mathbf{C}^0\mathbf{x}^0 + \mathbf{r}$ ! Compute initial gradient $\mathbf{g}^0$
3 **while** $\left(k \le K \text{ or } \left|\Delta F^{k+1}\right| \ge \varepsilon_a\right)$ **do** ! Loop for termination condition
4 $\mathbf{d}^k = -\mathbf{C}^k\mathbf{g}^k$ ! Find the search direction $\mathbf{d}^k$
5 $\alpha^k = -\frac{(\mathbf{g}^k)^T\mathbf{d}^k}{(\mathbf{d}^k)^T\mathbf{Q}\mathbf{d}^k}$ ! $\alpha^k$ that minimize $F\left(\mathbf{x}^k + \alpha^k\mathbf{d}^k\right)$
6 ***Increment:*** ! Calculate an increment $\Delta\mathbf{x}^k$
7 $\Delta\mathbf{x}^k = \alpha^k\mathbf{d}^k$ ! $\Delta\mathbf{x}^k$ in quasi-Newton direction
8 ***Update:*** ! Move one step forward to the minimum
9 $\mathbf{x}^{k+1} = \mathbf{x}^k + \Delta\mathbf{x}^k$ ! New $\mathbf{x}$ at $k+1$
10 $F^{k+1} = F(\mathbf{x}^{k+1})$ ! Evaluate $F(\mathbf{x})$ at $k+1$
11 $\mathbf{g}^{k+1} = \mathbf{Q}\mathbf{x}^{k+1} + \mathbf{r}$ ! New $\mathbf{g}$ at $k+1$
12 $\Delta\mathbf{g}^k = \mathbf{g}^{k+1} - \mathbf{g}^k$ ! New $\Delta\mathbf{g}$ at $k+1$
13 $\mathbf{C}^{k+1} = \left(\mathbf{I} - \frac{\Delta\mathbf{x}^k(\Delta\mathbf{g}^k)^T}{(\Delta\mathbf{x}^k)^T\Delta\mathbf{g}^k}\right)\mathbf{C}^k\left(\mathbf{I} - \frac{\Delta\mathbf{g}^k(\Delta\mathbf{x}^k)^T}{(\Delta\mathbf{x}^k)^T\Delta\mathbf{g}^k}\right) + \frac{\Delta\mathbf{x}^k(\Delta\mathbf{x}^k)^T}{(\Delta\mathbf{x}^k)^T\Delta\mathbf{g}^k}$ ! New $\mathbf{C}$ at $k+1$
14 ***Convergence:*** ! Compare $F(\mathbf{x})$ between $k$ and $k+1$
15 $\Delta F^{k+1} = F^{k+1} - F^k$ ! Difference between $F^{k+1}$ and $F^k$
16 $\left|\Delta F^{k+1}\right| < \varepsilon_a$? ! Convergence check using (6.25a)
17 **endwhile**
18 ***Output***: $\mathbf{x}^\dagger \leftarrow \mathbf{x}^{k+1}$, $F^\dagger \leftarrow F^{k+1}$ ! Assign $\mathbf{x}^{k+1}$ to $\mathbf{x}^\dagger$ and $F^{k+1}$ to $F^\dagger$

that just uses a limited number of recent updates $\left\{\left(\mathbf{x}^i, \mathbf{g}^i\right)\right\}_{i=\max(0,k-m)}^{k-1}$ rather than stores $\mathbf{C}^k$ explicitly or the whole previous data $\left\{\left(\mathbf{x}^i, \mathbf{g}^i\right)\right\}_{i=0}^{k-1}$, thus computing $\mathbf{d}^k = -\mathbf{C}^k\mathbf{g}^k$ directly from the recent data (e.g., Liu and Nocedal, 1989).

**COLLOQUY 6.3**

### Davidon–Fletcher–Powell (DFP) update

We apply a rank-2 update directly on the inverse Hessian approximation, $\mathbf{C}$, i.e.,

$$\mathbf{C}^{k+1} = \mathbf{C}^k + a\mathbf{u}\mathbf{u}^T + b\mathbf{v}\mathbf{v}^T.$$

Using the quasi-Newton condition (6.58) and setting $\mathbf{u} = \Delta\mathbf{x}$ and $\mathbf{v} = \mathbf{C}\Delta\mathbf{g}$, we can solve for $a$ and $b$ to get

$$\mathbf{C}_{DFP}^{k+1} = \mathbf{C}^k - \frac{\mathbf{C}^k\Delta\mathbf{g}^k\left(\Delta\mathbf{g}^k\right)^T\mathbf{C}^k}{\left(\Delta\mathbf{g}^k\right)^T\mathbf{C}^k\Delta\mathbf{g}^k} + \frac{\Delta\mathbf{x}^k\left(\Delta\mathbf{x}^k\right)^T}{\left(\Delta\mathbf{x}^k\right)^T\Delta\mathbf{g}^k}.$$

This is called the *Davidon–Fletcher–Powell (DFP)* update, following Davidon (1959) and Fletcher and Powell (1963). Note that this is related to $\mathbf{B}^{k+1}_{BFGS}$ (see (6.63)) by the transformation

$$\mathbf{B}^k \longleftrightarrow \mathbf{C}^k, \;\; \mathbf{B}^{k+1} \longleftrightarrow \mathbf{C}^{k+1}, \;\; \Delta\mathbf{x}^k \longleftrightarrow \Delta\mathbf{g}^k$$

(see, e.g., Dennis and Moré, 1977; Navon and Legler, 1987). This relation between the direct update (e.g., $\mathbf{B}^{k+1}_{BFGS}$) and the inverse update (e.g., $\mathbf{C}^{k+1}_{DFP}$) is often called *dual* or *complementary* updates; thus, the BFGS is called the complementary DFP formula (Dennis and Moré, 1977). Using the transformation and from (6.64), we can easily obtain the DFP update for $\mathbf{B}^{k+1}_{DFP}$ as

$$\mathbf{B}^{k+1}_{DFP} = \left(\mathbf{I} - \frac{\Delta\mathbf{g}^k \left(\Delta\mathbf{x}^k\right)^T}{\left(\Delta\mathbf{g}^k\right)^T \Delta\mathbf{x}^k}\right) \mathbf{B}^k \left(\mathbf{I} - \frac{\Delta\mathbf{x}^k \left(\Delta\mathbf{g}^k\right)^T}{\left(\Delta\mathbf{g}^k\right)^T \Delta\mathbf{x}^k}\right) + \frac{\Delta\mathbf{g}^k \left(\Delta\mathbf{g}^k\right)^T}{\left(\Delta\mathbf{g}^k\right)^T \Delta\mathbf{x}^k}.$$

COLLOQUY 6.4

### Limited Memory BFGS (L-BFGS)

The basic idea behind the limited memory quasi-Newton method is to avoid storing the fully dense $n \times n$ matrix $\mathbf{C}^k$ by storing just a few vectors of length $n$. Assume that the inverse BFGS update (6.64) can be expressed as

$$\mathbf{C}^{k+1} = \mathbf{C}^0 + \mathbf{U}^0 + \cdots + \mathbf{U}^k,$$

where $\left\{\mathbf{U}^0, \ldots, \mathbf{U}^k\right\}$ are the correction matrices as a function of $(\Delta\mathbf{x}, \Delta\mathbf{g}, \mathbf{C})$ at the corresponding iteration. Letting $m$ be the maximum number of $\mathbf{U}$ that can be stored, one can discard $\mathbf{U}^0$ when $\mathbf{C}^m$ is generated by storing $\Delta\mathbf{x}$, $\Delta\mathbf{g}^T\mathbf{C}$, and $\Delta\mathbf{x}^T\Delta\mathbf{g}$ at every iteration (Nocedal, 1980). We can also have the following relation from (6.64):

$$\mathbf{C}^{k+1}\mathbf{g}^k = \mathbf{v} + \left(\eta_1^k - \eta_2^k\right)\Delta\mathbf{x}^k,$$

where

$$\eta_1^k = \frac{\left(\Delta\mathbf{x}^k\right)^T \mathbf{g}^k}{\left(\Delta\mathbf{x}^k\right)^T \Delta\mathbf{g}^k}, \;\; \mathbf{u} = \mathbf{g}^k - \eta_1^k \Delta\mathbf{g}^k, \;\; \mathbf{v} = \mathbf{C}^k\mathbf{u}, \;\text{and}\; \eta_2^k = \frac{\left(\Delta\mathbf{g}^k\right)^T \mathbf{v}^k}{\left(\Delta\mathbf{x}^k\right)^T \Delta\mathbf{g}^k}.$$

Then, the L-BFGS updates are performed through the following steps:

**Step 1:** Start with $\mathbf{u} = \mathbf{g}^k$

**Step 2:** $\eta_1^j = \frac{\left(\Delta\mathbf{x}^j\right)^T \mathbf{u}}{\left(\Delta\mathbf{x}^j\right)^T \Delta\mathbf{g}^j}$; $\mathbf{u} = \mathbf{u} - \eta_1^j \Delta\mathbf{g}^j$ for $j = k-1, \ldots, k-m$

**Step 3:** Put $\mathbf{v} = \tilde{\mathbf{C}}^k \mathbf{u}$

**Step 4:** $\eta_2^j = \frac{(\Delta \mathbf{g}^j)^T \mathbf{v}}{(\Delta \mathbf{x}^j)^T \Delta \mathbf{g}^j}$; $\mathbf{v} = \mathbf{v} + \left(\eta_1^j - \eta_2^j\right) \Delta \mathbf{x}^j$ for $j = k-m, \ldots, k-1$

**Step 5:** Assign $\mathbf{d}^k \longleftarrow -\mathbf{v}$,

where $\tilde{\mathbf{C}}^k$ is a guess of $\mathbf{C}^k$, which is not stored. More details on L-BFGS are referred to in Nocedal (1980) and Liu and Nocedal (1989).

In summary, the quasi-Newton method approximates the inverse Hessian $\left(\mathbf{Q}^k\right)^{-1}$ by a symmetric positive definite matrix $\mathbf{C}^k$, which has a descent property and is updated based on a set of data $\{(\mathbf{x}, \mathbf{g})\}$ from previous iterations; thus, it requires just first-order derivatives. Starting with any symmetric and positive definite matrix (e.g., $\mathbf{C}^0 = \mathbf{I}$), the $n \times n$ matrix $\mathbf{C}^k$ becomes identical to $\left(\mathbf{Q}^k\right)^{-1}$ in $n+1$ iterations, and the search directions $\mathbf{d}^0, \ldots, \mathbf{d}^{k+1}$ are $\mathbf{Q}$-conjugate. It has the superlinear convergence order and shows global convergence for strong convex functions. For more detailed derivations and discussion of the quasi-Newton methods, see books by Antoniou and Lu (2007) and Chong and Żak (2013) as well as reviews by Dennis and Moré (1977) and Navon and Legler (1987).

**Practice 6.15 Comparison of minimization algorithms**

Perform minimization on a function in the Rosenbrock's function family:

$$F(\mathbf{x}) = \tfrac{1}{2}(1 - x_1)^2 + 5(x_2 - x_1^2)^2,$$

as depicted in Figure 6.8, using 1) the steepest descent method with the optimum step length ($\alpha_{opt}^k$ in (6.32)), 2) Newton's method (Algorithm 6.2), 3) the conjugate gradient method (Algorithm 6.3), and 4) the quasi-Newton method (Algorithm 6.4). You can choose any initial point $\mathbf{x}^0$ inside the domain in Figure 6.8. Draw the minimization paths from each method over the contour plot and discuss your results.

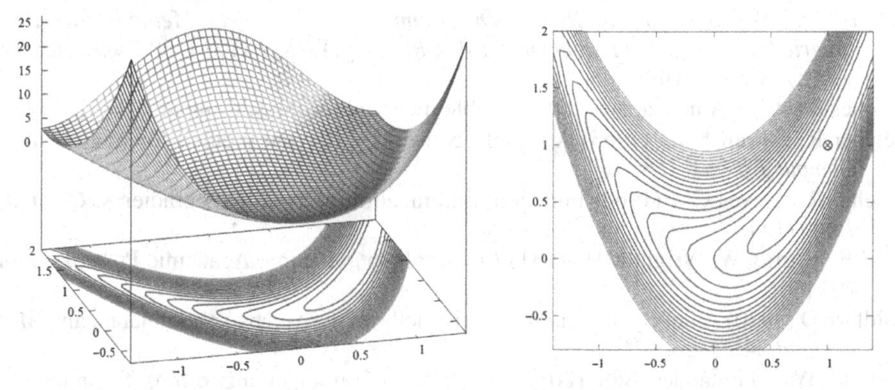


Figure 6.8 Plottings for $F(\mathbf{x}) = \frac{1}{2}(1 - x_1)^2 + 5(x_2 - x_1^2)^2$ in the $x_1$–$x_2$ domain with $x_1 = [-1.4, 1.4]$ and $x_2 = [-0.8, 2.0]$. Contours are plotted up to 5 with an interval of 0.25. The symbol ⊗ indicates the minimum $F(\mathbf{x}^\dagger) = 0$ at $\mathbf{x}^\dagger = (1, 1)^T$.

## References

Ahmadianfar I, Bozorg-Haddad O, Chu X (2020) Gradient-based optimizer: A new metaheuristic optimization algorithm. *Inform Sci* 540:131–159.

Albani RAS, Albani VVL, Silva Neto AJ (2020) Source characterization of airborne pollutant emissions by hybrid metaheuristic/gradient-based optimization techniques. *Environ Pollut* 267:115618, doi:10.1016/j.envpol.2020.115618

Antoniou A, Lu W-S (2007) *Practical Optimization: Algorithms and Engineering Applications*. Springer, New York, 690 pp.

Arora RK (2015) *Optimization: Algorithms and Applications*. CRC Press, Boca Raton, FL, 466 pp.

Arsham H (2008) Gradient-based optimization techniques for discrete event systems simulation. In *Wiley Encyclopedia of Computer Science and Engineering*, (ed.) Wah BW, American Cancer Society, 1–17, doi:10.1002/9780470050118.ecse371

Axelsson O, Barker VA (1984) *Finite Element Solution of Boundary Value Problems. Theory and Computation*. Academic Press, Orlando, FL, 432 pp.

Broyden CG (1970) The convergence of a class of double-rank minimization algorithms 1. General considerations. *IMA J Appl Math* 6:76–90.

Chaudhuri S, Goswami S, Das D, Middey A (2014) Meta-heuristic ant colony optimization technique to forecast the amount of summer monsoon rainfall: Skill comparison with Markov chain model. *Theor Appl Climatol* 116:585–595.

Chong EKP, Żak SH (2013) *An Introduction to Optimization*. 4th ed., John Wiley & Sons, Inc., Hoboken, NJ, 640 pp.

Davidon WC (1959) *Variable Metric Method for Minimization*. Tech. Rep. ANL-5990, Argonne National Laboratory, Lemont, IL, 27 pp.

Dennis JE Jr, Moré JJ (1977) Quasi-newton methods, motivation and theory. *SIAM Rev* 19:46–89.

Esentürk E, Abraham NL, Archer-Nicholls S, et al. (2018) Quasi-Newton methods for atmospheric chemistry simulations: Implementation in UKCA UM vn10. 8. *Geosci Model Dev* 11:3089–3108.

Feng Z-K, Niu W-J, Liu S (2021) Cooperation search algorithm: A novel metaheuristic evolutionary intelligence algorithm for numerical optimization and engineering optimization problems. *Appl Soft Comput J* 98:106734, doi:10.1016/j.asoc.2020.106734

Fisher M (1998) Minimization algorithms for variational data assimilation. In *Proc. of ECMWF Seminar on Recent Development in Numerical Methods for Atmospheric Modelling, 7–11 September 1998, Reading*, ECMWF, 364–385, www.ecmwf.int/node/9400

Fletcher R (1970) A new approach to variable metric algorithms. *Comput J* 13:317–322.

Fletcher R, Powell MJD (1963) A rapidly convergent descent method for minimization. *Comput J* 6:163–168.

Fletcher R, Reeves CM (1964) Function minimization by conjugate gradients. *Comput J* 7:149–154.

Gill PE, Murray W, Wright MH (1981) *Practical Optimization*. Academic Press, London, 401 pp.

Goldfarb D (1970) A family of variable-metric methods derived by variational means. *Math Comput* 24:23–26.

Golfetto WA, Fernandes SdS (2012) A review of gradient algorithms for numerical computation of optimal trajectories. *J Aerosp Technol Manag* 4:131–143.

Hagan MT, Demuth HB, Beal MH, De Jesús O (2014) *Neural Network Design*. 2nd ed., Martin T. Hagan and Howard B. Demuth, 1012 pp. https://hagan.okstate.edu/nnd.html

Haji SH, Abdulazeez AM (2021) Comparison of optimization techniques based on gradient descent algorithm: A review. *PalArch's J Archaeol Egypt/Egyptol* 18:2715–2743.

Hestenes MR, Stiefel E (1952) Methods of conjugate gradients for solving linear systems. *J Res Natl Bur Stand* 49:409–436.

Hong S, Yu X, Park S, Choi Y-S, Myoung B (2014) Assessing optimal set of implemented physical parameterization schemes in a multi-physics land surface model using genetic algorithm. *Geosci Model Dev* 7:2517–2529.

Kalnay E, Park SK, Pu Z-X, Gao J (2000) Application of the quasi-inverse method to data assimilation. *Mon Wea Rev* 128:864–875.

Kim S (2006) Gradient-based simulation optimization. In *Proceedings of the 38th Conference on Winter Simulation*, WSC '06, Winter Simulation Conference, 159–167.

Kruger J (1993) Simulated annealing: A tool for data assimilation into an almost steady model state. *J Phys Oceanogr* 23:679–688.

Le Dimet F-X, Navon IM, Daescu DN (2002) Second-order information in data assimilation. *Mon Wea Rev* 130:629–648.

Lee YH, Park SK, Chang DE (2006) Parameter estimation using the genetic algorithm and its impact on quantitative precipitation forecast. *Ann Geophys* 24:3185–3189.

Liu DC, Nocedal J (1989) On the limited memory BFGS method for large scale optimization. *Math Program* 45:503–528.

Maier HR, Razavi S, Kapelan Z, et al. (2019) Introductory overview: Optimization using evolutionary algorithms and other metaheuristics. *Environ Modell Softw* 114: 195–213.

Navon IM, Legler DM (1987) Conjugate-gradient methods for large-scale minimization in meteorology. *Mon Wea Rev* 115:1479–1502.

Nocedal J (1980) Updating quasi-newton matrices with limited storage. *Math Comput* 35:773–782.

Nocedal J, Wright S (2006) *Numerical Optimization*. 2nd ed., Springer, New York, 664 pp.

Park SK, Kalnay E (2004) Inverse three-dimensional variational data assimilation for an advection-diffusion problem: Impact of diffusion and hybrid application. *Geophys Res Lett* 31:L04102, doi:10.1029/2003GL018830

Polak E, Ribiere G (1969) Note sur la convergence de méthodes de directions conjuguées. *Rev Fr Inf Rech Oper* 3:35–43.

Ruder S (2017) An overview of gradient descent optimization algorithms. *arXiv:160904747v2* `https://arxiv.org/abs/1609.04747`

Salajegheh F, Salajegheh E (2019) PSOG: Enhanced particle swarm optimization by a unit vector of first and second order gradient directions. *Swarm Evol Comput* 46:28–51.

Shanno DF (1970) Conditioning of quasi-Newton methods for function minimization. *Math Comput* 24:647–656.

Stodola P, Michenka K, Nohel J, Rybanský M (2020) Hybrid algorithm based on ant colony optimization and simulated annealing applied to the dynamic traveling salesman problem. *Entropy* 22:884, doi:10.3390/e22080884

Stork J, Eiben AE, Bartz-Beielstein T (2020) A new taxonomy of global optimization algorithms. *Nat Comput.* doi:10.1007/s11047-020-09820-4

Thacker WC (1989) The role of the Hessian matrix in fitting models to measurements. *J Geophys Res Oceans* 94:6177–6196.

Upadhyaya A, Upadhyaya A (2021) Optimization of water productivity in Bhagwanpur distributary command of India employing TLBO and cuckoo search algorithms. *Water Policy* 23:274–290.

Wang J, Zhou B, Zhou S (2016) An improved cuckoo search optimization algorithm for the problem of chaotic systems parameter estimation. *Comput Intell Neurosci* 2016:2959370, doi:10.1155/2016/2959370

Wang Z, Droegemeier KK, White L, Navon IM (1997) Application of a new adjoint Newton algorithm to the 3D ARPS storm-scale model using simulated data. *Mon Wea Rev* 125:2460–2478.

Wu J, Jin L, Liu M (2006) Modeling meteorological prediction using particle swarm optimization and neural network ensemble. In *Advances in Neural Networks–ISNN 2006, Lecture Notes in Computer Science* (vol. 3973) (eds.) Wang J, Yi Z, Zurada JM, Lu BL, Yin H, Springer, Berlin, Heidelberg, 1202–1209.

Yang X-S, He X (2016) Nature-inspired optimization algorithms in engineering: Overview and applications. In *Nature-Inspired Computation in Engineering*, (ed.) Yang X-S, Springer, Cham, 1–20.

Yu X, Lu Y, Cai M (2018) Evaluating agro-meteorological disaster of China based on differential evolution algorithm and VIKOR. *Nat Hazards* 94:671–687.

Zhao W, Wang Y, Zhang Z, Wang H (2021) Multicriteria ship route planning method based on improved particle swarm optimization-genetic algorithm. *J Mar Sci Eng* 9:357, doi:10.3390/jmse9040357

Zou X, Navon IM, Berger M, et al. (1993) Numerical experience with limited-memory quasi-newton and truncated newton methods. *SIAM J Optim* 3:582–608.

Županski M (1993) A preconditioning algorithm for large-scale minimization problems. *Tellus A* 45:478–492.

Županski M (1996) A preconditioning algorithm for four-dimensional variational data assimilation. *Mon Wea Rev* 124:2562–2573.

Županski M (2005) Maximum likelihood ensemble filter: Theoretical aspects. *Mon Wea Rev* 133:1710–1726.

Županski M, Navon IM, Županski D (2008) The Maximum Likelihood Ensemble Filter as a non-differentiable minimization algorithm. *Quart J Roy Meteor Soc* 134:1039–1050.

# Part III

## Methods and Issues

# 7
# Variational Data Assimilation

## 7.1 Introduction

Traditionally data assimilation is largely divided into two groups in terms of methodology – *sequential* (e.g., Evensen, 1994; Bertino et al., 2003; Piazzi et al., 2021) and *variational* (e.g., Navon et al., 1992; Courtier, 1997; Park and Županski, 2003; Le Dimet et al., 2009; Cummings and Smedstad, 2013): Both are suited to the framework of estimation theory (Gelb, 1974; Cohn, 1997) though the latter is closer to optimal control theory (Lions, 1971; Bruneau et al., 1997; Le Dimet et al., 2017) which is extended from the calculus of variations (Gelfand and Fomin, 1963; Lanczos, 1970). Currently, data assimilation further includes *ensemble* and *hybrid* approaches (see the details in Chapter 8).

We start with a brief introduction to the calculus of variations in Colloquy 7.1. We also provide a historical overview of the calculus of variations and optimal control theory and their difference in mathematical formulations in Colloquys 7.2 and 7.3, respectively. As both the calculus of variations and optimal control theory solve constrained minimization problems (see Colloquy 7.3), we introduce the Lagrange multiplier method in Colloquy 7.4 (see also Section 4.4.1.2) to solve a minimization problem with an equality constraint.

Inspired by the pioneering works of Yoshi Sasaki in variational analysis (Sasaki, 1955, 1958, 1969, 1970a, 1970b, 1970c), scientists extensively applied variational method to meteorological analysis and initialization in the 1970s and 1980s (e.g., Sasaki and Lewis, 1970; Stephens, 1970; Lewis, 1971; Sasaki, 1971a, 1971b; Wilkins, 1971; Lewis, 1972; Lewis and Grayson, 1972; Achtemeier, 1975; Wahba and Wendelberger, 1980; Ikawa, 1984; McGinley, 1984; Temperton, 1984; McGinley, 1987). Through these studies, the term *variational objective analysis* had emerged, which is defined by the *Glossary of Meteorology* (AMS, 2021) as

> An objective analysis technique used to create an estimate of the atmospheric state that maximizes (or minimizes) a mathematical measure of desirable (or undesirable) characteristics. The analysis characteristics usually include measures of the fit to data, background field, and dynamical constraints.

Applications of variational objective analysis have been extended to recent studies (e.g., Waliser et al., 2002; Bonavita and Torrisi, 2005; Narkhedkar and Sinha, 2008). Sasaki's pathway and contribution to data assimilation were described in essays by Lewis and Lakshmivarahan (2008) and Lewis (2009).

Rapid progress had been made in variational analysis and assimilation in the 1980s through the seminal works by Lewis and Derber (1985), Le Dimet and Talagrand (1986),

Talagrand and Courtier (1987), Courtier and Talagrand (1987), and Thacker and Long (1988). Early applications of variational methods in geosciences were presented in Sasaki (1986). Since then, variational data assimilation (VAR) has been extensively used in various disciplines, including oceanography (e.g., Weaver et al., 2003; Moore et al., 2011; Cummings and Smedstad, 2013; Ngodock and Carrier, 2014; Smith et al., 2017), hydrology (e.g., Reichle et al., 2001; Le Dimet et al., 2009), meteorology (e.g., Županski and Mesinger, 1995; Schlatter, 2000; Park and Županski, 2003; Gao, 2017), atmospheric chemistry (e.g., Elbern et al., 1997; Park et al., 2016), air quality (e.g., Elbern et al., 2000; Baker et al., 2006), water quality (e.g., Shao et al., 2016), seismology (e.g., Kano et al., 2020), space weather (e.g., Lang and Owens, 2019), magnetohydrodynamics (e.g., Li et al., 2014), etc.

With the availability of the adjoint technique (see Chapter 4) and numerical minimization algorithms (see Chapter 6), VAR had been implemented in most operational centers in 1990s. Three-dimensional VAR (3DVAR) had been in operational use in NCEP[1] (Derber et al., 1991; Parrish and Derber, 1992), ECMWF[2] (Courtier et al., 1998; Rabier et al., 1998), CMC[3] (Gauthier et al., 1999; Laroche et al., 1999), Météo France (Thépaut et al., 1998), UKMO[4] (Andrews, 1998; Lorenc et al., 2000), JMA[5] (Takeuchi and Tsuyuki, 2002), and the High Resolution Limited Area Model (HIRLAM) group (Gustafsson, 1998; Gustafsson et al., 2001; Lindskog et al., 2001). Encouraged by development of the efficient *incremental* formulation in VAR (Courtier et al., 1994; Laroche and Gauthier, 1998; Janisková et al., 1999), operational centers moved on to four-dimensional VAR (4DVAR) – mostly adopting the incremental 4DVAR scheme – including ECMWF (Rabier et al., 2000), Météo France (Gauthier and Thépaut, 2001), JMA (Kadowaki, 2005; Koizumi et al., 2005), UKMO (Rawlins et al., 2007), MSC[6] (Gauthier et al., 2007), the HIRLAM group (Gustafsson et al., 2012), the US Navy (Ngodock and Carrier, 2014), and MSS[7] (Heng et al., 2020).

In this chapter, we introduce the essential elements and mathematical formulations of VAR, including 3DVAR and 4DVAR, and their applications.

**COLLOQUY 7.1**

**The calculus of variations and the Euler–Lagrange equation**

In the calculus of variations (or variational calculus), one seeks the minimum of a *functional* – a function involving the unknown function and its derivatives in a definite integral form – by finding a function that minimizes the functional.

Consider a functional $J = J\left[\mathbf{x}(t)\right]$, for continuous functions $\mathbf{x}(t)$ and $\mathbf{x}' = \frac{d\mathbf{x}}{dt}$, of the form

$$J = \int_{t_0}^{t_1} F\left(t, \mathbf{x}(t), \mathbf{x}'(t)\right) dt, \tag{7.1}$$

[1] National Center for Environmental Prediction – formerly National Meteorological Center.
[2] European Centre for Medium-Range Weather Forecasts.
[3] Canadian Meteorological Center.
[4] Met Office of the United Kingdom.
[5] Japan Meteorological Agency.
[6] Meteorological Service of Canada – formerly Canadian Meteorological Center.
[7] Meteorological Service Singapore.

which maps $\mathbf{x}(t)$ to real numbers and satisfies the following conditions at the terminal points:

$$\mathbf{x}(t_0) = \mathbf{x}_I, \quad \mathbf{x}(t_1) = \mathbf{x}_F, \tag{7.2}$$

with the initial condition $\mathbf{x}_I$ and the final (terminal) condition $\mathbf{x}_F$, assuming that $t_0, t_1, \mathbf{x}(t_0)$, and $\mathbf{x}(t_1)$ are fixed. We seek the function $\mathbf{x}^\dagger(t)$, which is a *stationary function* of $J[\mathbf{x}(t)]$.

Let $\delta\mathbf{x}(t)$ be a sufficiently small deviation around the desired solution $\mathbf{x}^\dagger(t)$, with $\delta\mathbf{x}(t_0) = \delta\mathbf{x}(t_1) = \mathbf{0}$. The total variation of $J$ is given by

$$\Delta J = J\left(t, \mathbf{x}^\dagger + \delta\mathbf{x}, \left.\mathbf{x}'\right|_{\mathbf{x}^\dagger} + \delta\mathbf{x}'\right) - J\left(t, \mathbf{x}^\dagger, \left.\mathbf{x}'\right|_{\mathbf{x}^\dagger}\right).$$

By taking the linear term of the Taylor series expansion, we have the *first variation* (or *Gâteaux derivative*) of $J$:

$$\delta J = \frac{\partial J}{\partial \mathbf{x}}\delta\mathbf{x} + \frac{\partial J}{\partial \mathbf{x}'}\delta\mathbf{x}'. \tag{7.3}$$

The necessary condition for $\mathbf{x}^\dagger(t)$ to be a minimum of $J$ is that $\delta J = 0$ for all $\delta\mathbf{x}$ and $\delta\mathbf{x}'$. For a given continuous function $g(t)$ on $[t_0, t_1]$, if

$$\int_{t_0}^{t_1} g(t)h(t)dt = 0 \tag{7.4}$$

for all $h(t)$, then $g(t) = 0$ for $t_0 \le t \le t_1$: This is called the *fundamental lemma of the calculus of variations*.

By applying the first variation to (7.1) and setting it to 0, we have

$$\delta J = \int_{t_0}^{t_1} \left(\frac{\partial F}{\partial \mathbf{x}}\delta\mathbf{x} + \frac{\partial F}{\partial \mathbf{x}'}\delta\mathbf{x}'\right) dt = 0. \tag{7.5}$$

By integrating the second term of (7.5) by parts and using the lemma (7.4) and the condition $\delta\mathbf{x}(t_0) = \delta\mathbf{x}(t_1) = \mathbf{0}$, we have

$$\frac{\partial F}{\partial \mathbf{x}} - \frac{d}{dt}\frac{\partial F}{\partial \mathbf{x}'} = 0, \tag{7.6}$$

which is called the *Euler–Lagrange* (or *Euler*) *equation*. The solution of (7.6), $\mathbf{x} = \mathbf{x}^\dagger$, is a stationary function of $J$ that requires $\delta J = 0$.

**Practice 7.1 Euler–Lagrange Equation**

Answer the following.

1. Derive the Euler–Lagrange equation (7.6) from (7.5).
2. Find the Euler–Lagrange equation for the functional

$$J[y(x)] = \int_{x_0}^{x_1} \sqrt{1 + (y')^2} dx$$

with the boundary conditions $y(x_0) = y_0$ and $y(x_1) = y_1$.

COLLOQUY 7.2

**The calculus of variations and optimal control: Brief historical overview**

Joseph-Louis Lagrange (1736–1813) is regarded as the founder of the *calculus of variations*, in terms of analytic viewpoint, by inventing the methods of *variations* and *multipliers*. He had corresponded with Leonard Euler (1707–1783), who adopted his ideas to develop the *Euler–Lagrange equation* – the first-order necessary condition for a stationary solution – and named the subject "the calculus of variations."

An earlier concept of the calculus of variations may be traced back to the brachistochrone ("shortest time") problem – the first variational problem to be formulated mathematically (Sagan, 1969) – to find a curve of fastest descent by gravity between two end points in a vertical plane, which was proposed in 1696 by Johann Bernoulli (1667–1748) and solved by five mathematicians, including himself and his brother Jacob Bernoulli (1655–1705), Isaac Newton (1643–1727), Gottfried Leibniz (1646–1716), and Guillaume de L'Hôpital (1661–1704). The methods developed by the Bernoulli brothers were generalized by Euler in 1744. Geometric approaches had essentially been employed to solve the problems until Lagrange described an analytic approach using the variations of optimal curves and undetermined multiplies (Sargent, 2000).

In 1786, Adrien-Marie Legendre (1752–1833) formulated the second variation and the second-order necessary condition for optimality for scalars that requires the second-order matrix to be positive definite, which was extended to vectors by Alfred Clebsch (1833–1872) and named the *Legendre–Clebsch condition*. The *Hamilton–Jacobi equation* had been formulated based on the principle of least action, which was posed by William Rowan Hamilton (1805–1865) and further simplified by Karl Gustav Jacob Jacobi (1804–1851) in 1838. By introducing the excess function, Karl Wilhelm Theodor Weierstrass (1815–1897) discovered the so-called *Weierstrass condition* for a strong local minimum in 1879.

Optimal control theory is rooted in the calculus of variations with a general extension. It is generally accepted that optimal control theory began when the maximum principle had been developed by Lev Semyonovich Pontryagin (1908–1988) and his group between 1956 and 1962; however, Sussmann and Willems (1997) traced its origin back to 1696 when Johann Bernoulli proposed the brachistochrone problem. In fact, the maximum principle had emerged from previous important works.

Oskar Bolza (1857–1942) stated the most general single-integral problem of the calculus of variations in 1913, called the *problem of Bolza*: considered to be the predecessor of the modern control problem. Lawrence Murray Graves (1896–1973) distinguished between *state* and *control* variables by treating the derivative as an independent function in 1924: He also provided a control theory formulation of the Weierstrass condition for a Bolza type problem in

1933. Around 1950, Magnus Rudolph Hestenes (1906–1991) formulated the first optimal control problems and introduced an early formulation of the maximum principle. Richard Ernest Bellman (1920–1984) worked on multistage decision problems and devised the *principle of optimality* (also called the *Bellman equations*), which became the basis of *dynamic programming* (Bellman, 1957).

In response to the military need for a time optimal control problem to determine the least time of aircraft trajectory between two fixed points in the range-altitude space, Pontryagin's group developed the early form of the maximum principle in 1956: It proved to be a necessary and sufficient condition for linear problems but just a necessary condition in the general case. Through further research efforts, they established the *Pontryagin maximum* (or *minimum*) *principle* (PMP) (Pontryagin et al., 1961) for more general optimal control problems, which became the basis of optimal control theory: 4DVAR is shown to be a special case of PMP (Lakshmivarahan et al., 2013). Jacques-Louis Lions (1928–2001) developed the modern framework of the optimal control of partial differential equations (Lions, 1971).

COLLOQUY 7.3

**The calculus of variations vs optimal control**

We formulate a *calculus of variations problem* as a constrained minimization problem for a functional $J$ as follows:

$$\min\ J = \int_{t_0}^{t_1} F\left(t, \mathbf{x}(t), \mathbf{x}'(t)\right) dt$$

$$\text{subject to } \mathbf{x}(t) \geq 0,\ \ \mathbf{x}(t_0) = \mathbf{x}_I,\ \ \mathbf{x}(t_1) = \mathbf{x}_F, \tag{7.7}$$

for a continuously differentiable function $\mathbf{x}(t)$ in a time interval $t_0 \leq t \leq t_1$ and $\mathbf{x}'(t) = \frac{d\mathbf{x}(t)}{dt}$. Here, $t_0$ and $t_1$ represent the *initial time* and the *final* (or *terminal*) *time*, respectively: $\mathbf{x}_I$ and $\mathbf{x}_F$ represent the *initial state* and the *final* (or *terminal*) *state*, respectively. By putting $\mathbf{u}(t) = \mathbf{x}'(t)$, the *optimal control problem* equivalent to (7.7) is formulated as

$$\min\ J = \int_{t_0}^{t_1} F\left(t, \mathbf{x}(t), \mathbf{u}(t)\right) dt$$

$$\text{subject to } \mathbf{x}(t) \geq 0,\ \ \mathbf{x}'(t) = \mathbf{u}(t),\ \ \mathbf{x}(t_0) = \mathbf{x}_I,\ \ \mathbf{x}(t_1) = \mathbf{x}_F. \tag{7.8}$$

In general, an optimal control problem is formulated by defining two variables – *states* or *trajectories* ($\mathbf{x}$) and *controls* ($\mathbf{u}$) – as follows:

$$\min\ J = \int_{t_0}^{t_1} f(t, \mathbf{x}(t), \mathbf{u}(t)) dt \tag{7.9a}$$

$$\text{subject to } \mathbf{x}'(t) = g(t, \mathbf{x}(t), \mathbf{u}(t)), \tag{7.9b}$$

$$\mathbf{x}(t_0) = \mathbf{x}_I, \ \ \mathbf{x}(t_1) = \mathbf{x}_F, \tag{7.9c}$$

for a given time, $t_0 \leq t \leq t_1$ and continuously differential functions $f$ and $g$ of $t$, $\mathbf{x}(t)$, and $\mathbf{u}(t)$. Here, $t_1, \mathbf{x}_I$, and $\mathbf{x}_F$ can be either *fixed* or *free*. The pair $(\mathbf{x}(t), \mathbf{u}(t))$ is called a *controlled trajectory*. The dynamics of the state variables, i.e., changes of $\mathbf{x}(t)$ over time, is represented by (7.9b), which is called a *control system*. The control $\mathbf{u}$ is piecewise continuous in the given time interval and influences the functional $J$ – directly by itself and indirectly through $\mathbf{x}'(t)$. An optimal control problem involves finding an *optimal trajectory* $(\mathbf{x}^\dagger(t), \mathbf{u}^\dagger(t))$ by solving (7.9b). Optimal control theory is considered to be an extension of the calculus of variations with a first-order necessary condition based on *Pontryagin's minimum* (or *maximum*) *principle* (e.g., Pontryagin et al., 1961; Kopp, 1962; Onori et al., 2016). See more details in textbooks such as Kamien and Schwartz (1991) and Lebedev and Cloud (2003).

**Practice 7.2 The calculus of variations vs optimal control**

Transform the following calculus of variations problem into its equivalent optimal control problem:

$$\min \ J = \int_0^T \left(\alpha \mathbf{x}'(t) + \beta \mathbf{x}(t)\right) dt$$

$$\text{subject to } \ \mathbf{x}'(t) \geq 0, \ \ \mathbf{x}(0) = \mathbf{0}, \ \ \mathbf{x}(T) = \mathbf{C},$$

for a continuously differential function $\mathbf{x}$ in a time interval $0 \leq t \leq T$ and constants $\alpha$ and $\beta$.

COLLOQUY 7.4

**Constrained minimization and Lagrange multipliers**

Consider a minimization problem with an equality constraint

$$\begin{aligned} &\min \ f(\mathbf{x}) \\ &\text{such that } \ g_j(\mathbf{x}) = 0, \ \ j = 1, \ldots, k < n \end{aligned} \tag{7.10}$$

for $\mathbf{x} = (x_1, \ldots, x_n)$. By introducing the *Lagrange multipliers* $\lambda_j$, $j = 1, \ldots, k$, we can transform the constrained problem to minimize $f(\mathbf{x})$ to an unconstrained problem to minimize the *Lagrange function*

$$\mathcal{L}(\mathbf{x},\boldsymbol{\lambda}) = f(\mathbf{x}) + \sum_{j=1}^{k} \lambda_j g_j(\mathbf{x}) = f(\mathbf{x}) + \boldsymbol{\lambda}^T \mathbf{g}(\mathbf{x}). \tag{7.11}$$

Let $(\mathbf{x}^\dagger, \boldsymbol{\lambda}^\dagger)$ be the stationary point of $\mathcal{L}$. Then, the necessary condition for $\mathbf{x}^\dagger$ to be a local minimum of (7.10) is that $(\mathbf{x}^\dagger, \boldsymbol{\lambda}^\dagger)$ and $\mathbf{x}^\dagger$ should coincide with each other; thus, there exists $\boldsymbol{\lambda}^\dagger$ that satisfies

$$\frac{\partial \mathcal{L}\left(\mathbf{x}^\dagger, \boldsymbol{\lambda}^\dagger\right)}{\partial \mathbf{x}} = 0; \quad \frac{\partial \mathcal{L}\left(\mathbf{x}^\dagger, \boldsymbol{\lambda}^\dagger\right)}{\partial \boldsymbol{\lambda}} = 0. \tag{7.12}$$

For an integral equality constrained problem,

$$\min\ J = \int_{t_0}^{t_1} F\left(t, \mathbf{x}(t), \mathbf{x}'(t)\right) dt$$

$$\text{subject to } \mathbf{x}(t) \geq 0, \ \ \mathbf{x}(t_0) = \mathbf{x}_I, \ \ \mathbf{x}(t_1) = \mathbf{x}_F, \text{ and}$$

$$\int_{t_0}^{t_1} G_j\left(t, \mathbf{x}(t), \mathbf{x}'\right) dt = C_j, \ \ j = 1, \ldots, k, \tag{7.13}$$

the Lagrange function is given by

$$\mathcal{L}(\mathbf{x},\boldsymbol{\lambda}) = \int_{t_0}^{t_1} \left( F\left(t, \mathbf{x}(t), \mathbf{x}'(t)\right) + \sum_{j=1}^{k} \lambda_j G_j\left(t, \mathbf{x}(t), \mathbf{x}'(t)\right) \right) dt; \tag{7.14}$$

thus, the necessary condition for a minimum is

$$\frac{\partial}{\partial \mathbf{x}} \left( F + \sum_{j=1}^{k} \lambda_j G_j \right) - \frac{d}{dt} \left( \frac{\partial}{\partial \mathbf{x}'} \left( F + \sum_{j=1}^{k} \lambda_j G_j \right) \right) = 0. \tag{7.15}$$

**Practice 7.3 Lagrange multipliers**

Solve the following constrained minimization problems using the Lagrange multipliers method:

1. Minimize $f(\mathbf{x})$ given by

$$f(\mathbf{x}) = (x_1 - 1)^2 + (x_2 - 2)^2$$
$$\text{subject to } g(\mathbf{x}) = x_1 + x_2 - 4 = 0.$$

2. Find the local extrema for a function $f(\mathbf{x})$ given by

$$f(\mathbf{x}) = 0.5x_2^2 - x_1 x_3$$
$$\text{suject to } g_1(\mathbf{x}) = x_1^2 + x_3 - 1 = 0; \ \ g_2(\mathbf{x}) = x_1 + x_2 - 1 = 0.$$

3. Find the extremals for a functional $J[v(u)]$ given by

$$J[v(u)] = \int_0^1 (v')^2 \, du$$

subject to $G_1(v) = \int_0^1 uv \, du = 0.5; \; G_2(v) = \int_0^1 v \, du = 2,$

and $v(0) = v(1) = 0$.

## 7.2 Variational Data Assimilation as an Optimal Control Problem

Optimal control theory was introduced by Pontryagin et al. (1961) through the PMP and generalized for partial differential equations by Lions (1971) (see Colloquy 7.1). In VAR, a functional is defined as measuring the misfit between the model and observations, which is minimized to achieve the best fit: This is based on optimal control theory where the function (or *control variable*) is sought from a set of admissible functions that minimizes the functional under some constraints. The control variable can vary with time and affect the system's outputs: The *optimal control* is a specific control, among all the admissible controls, that minimizes the functional.

In order to utilize optimal control theory, we expect our assimilation system to consist of the following: 1) a model of partial differential equations whose state $\mathbf{x}$ can be obtained in a certain functional space; 2) a model control $\mathbf{u} \in \mathcal{R}^c$ in a set of admissible controls, where $\mathcal{R}^c$ is the control space; and 3) observations that concern both $\mathbf{x}$ and $\mathbf{u}$ (Devenon, 1990). By assimilating the observations and by minimizing the functional, $J$, measuring the distance between the observations and the modeled quantities, we seek the optimal control $\hat{\mathbf{u}}$ and the corresponding optimal model state $\hat{\mathbf{x}}$ as a function of $\hat{\mathbf{u}}$, i.e., $\hat{\mathbf{x}} = \mathbf{x}\left(\hat{\mathbf{u}}\right)$. As $J = J\left((\mathbf{x}(\mathbf{u}), \mathbf{u}\right) = J(\mathbf{u})$, the minimization of $J$ occurs in the control space via a descent algorithm (see Chapter 6) to obtain $\hat{\mathbf{u}}$ and consequently $\hat{\mathbf{x}}$.

In Colloquy 7.1, we have shown that a constraint minimization can be converted to an unconstrained problem using the Lagrange function (or Lagrangian), $\mathcal{L}$. Lakshmivarahan et al. (2013) provided an excellent interpretation on optimal control theory, based on the PMP that exploited the relation between the Lagrange function and the Hamiltonian function, in application to 4DVAR. Let us express a discrete nonlinear model (NLM) dynamics in the model space $\mathcal{R}^m$ as

$$\overline{\mathbf{x}}_{n+1} = \overline{M}\left(\overline{\mathbf{x}}_n, \boldsymbol{\eta}_n\right), \tag{7.16}$$

where $\overline{M} \in \mathcal{R}^{m \times m}$ maps dynamics to $\mathcal{R}^m$, $\overline{\mathbf{x}}_n \in \mathcal{R}^m$ is the state of the time-invariant dynamics, and $\boldsymbol{\eta}_n \in \mathcal{R}^m$ is the given forcing in the model.

We add the control $\mathbf{u}_n$ as an external forcing term by defining the forced dynamics as

$$\mathbf{x}_{n+1} = M\left(\mathbf{x}_n, \boldsymbol{\eta}_n, \mathbf{u}_n\right) = \overline{M}\left(\mathbf{x}_n, \boldsymbol{\eta}_n\right) + \mathbf{B}\mathbf{u}_n, \tag{7.17}$$

where $\mathbf{u}_n \in \mathcal{R}^c$ is the control vector, with $1 \leq c \leq m$ the number of admissible controls, and $\mathbf{B} \in \mathcal{R}^{m \times c}$, whose structure is determined by $c$, maps from $\mathcal{R}^m$ to $\mathcal{R}^c$. Let the observation vector, $\mathbf{y}_n^o \in \mathcal{R}^n$, be given in the observation space $\mathcal{R}^n$ by

$$\mathbf{y}_n^o = H_n \mathbf{x}_n + \boldsymbol{\varepsilon}_n^o, \tag{7.18}$$

where $H_n \in \mathbf{R}^m \to \mathbf{R}^n$ denotes the observation operator that relates the model state $\mathbf{x}_n$ to the observation $\mathbf{y}_n^o$, and $\boldsymbol{\varepsilon}_n^o \sim N(0, \mathbf{R})$ is the observation error with $\mathbf{R} \in \mathcal{R}^{n\times n}$ a known positive definite matrix.

We define the cost function as

$$\bar{J} = \sum_{n=0}^{N-1} J_n\left(\mathbf{x}_n, \mathbf{y}_n^o, \mathbf{u}_n\right) = \frac{1}{2}\left(\mathbf{y}_n^o - H_n\mathbf{x}_n\right)^T \mathbf{R}^{-1}\left(\mathbf{y}_n^o - H_n\mathbf{x}_n\right) + \frac{1}{2}\mathbf{u}_n^T \mathbf{C}\mathbf{u}_n, \tag{7.19}$$

where $N$ is the number of observations and $\mathbf{C} \in \mathcal{R}^{p\times p}$ is a given symmetric and positive definite matrix. The Lagrangian $\mathcal{L}$ is defined by augmenting the dynamical constraint (7.17) using the Lagrangian multipliers $\lambda_n$:

$$\mathcal{L} = \sum_{n=0}^{N-1}\left[J_n + \boldsymbol{\lambda}_{n+1}^T\left(M\left(\mathbf{x}_n, \boldsymbol{\eta}_n, \mathbf{u}_n\right) - \mathbf{x}_{n+1}\right)\right], \tag{7.20}$$

where $\boldsymbol{\lambda}_n \in \mathcal{R}^n$ for $1 \le k \le N$ denotes the set of $N$ undetermined Lagrangian multipliers or the adjoint variables.

For optimal control theory, we further define the associated Hamiltonian function

$$\mathcal{H}_n = \mathcal{H}_n\left(\mathbf{x}_n, \mathbf{u}_n, \boldsymbol{\eta}_n, \boldsymbol{\lambda}_{n+1}\right) = J_n + \boldsymbol{\lambda}_{n+1}^T M\left(\mathbf{x}_n, \boldsymbol{\eta}_n, \mathbf{u}_n\right), \tag{7.21}$$

making (7.20)

$$\mathcal{L} = \mathcal{H}_0 - \boldsymbol{\lambda}_N^T \mathbf{x}_N + \sum_{n=1}^{N-1}\left[\mathcal{H}_n - \boldsymbol{\lambda}_n^T \mathbf{x}_n\right]. \tag{7.22}$$

Let $\delta\mathcal{L}$ be the induced variation resulting from the variations $\delta\mathbf{x}_n$ and $\delta\mathbf{u}_n$ for $0 \le n \le N-1$ and $\delta\boldsymbol{\lambda}_n$ for $0 \le n \le N$; $\delta\boldsymbol{\eta}_n = 0$ as $\boldsymbol{\eta}_n$ is specified. Since $\mathcal{H}_n$ is a scalar-valued function, we have $\delta\mathcal{L}$ as

$$\begin{aligned}\delta\mathcal{L} &= \left(\nabla_{\mathbf{x}_0}\mathcal{H}_0\right)^T \delta\mathbf{x}_0 + \left(\nabla_{\mathbf{u}_0}\mathcal{H}_0\right)^T \delta\mathbf{u}_0 - \boldsymbol{\lambda}_N^T \delta\mathbf{x}_N \\ &\quad + \underbrace{\sum_{n=1}^{N}\left(\nabla_{\boldsymbol{\lambda}}\mathcal{H}_{n-1} - \mathbf{x}_n\right)^T \delta\boldsymbol{\lambda}_n}_{A} + \underbrace{\sum_{n=1}^{N-1}\left(\nabla_{\mathbf{x}}\mathcal{H}_n - \boldsymbol{\lambda}_k\right)^T \delta\mathbf{x}_n}_{B} + \underbrace{\sum_{n=1}^{N-1}\left(\nabla_{\mathbf{u}}\mathcal{H}_n\right)^T \delta\mathbf{u}_k}_{C},\end{aligned} \tag{7.23}$$

where $\nabla_{\mathbf{x}}\mathcal{H}_n \in \mathcal{R}^m, \nabla_{\mathbf{u}}\mathcal{H}_n \in \mathcal{R}^p$, and $\nabla_{\boldsymbol{\lambda}}\mathcal{H}_n \in \mathcal{R}^m$ are the gradients of $\mathcal{H}_n$ with respect to $\mathbf{x}_n, \mathbf{u}_n$, and $\boldsymbol{\lambda}_{n+1}$, respectively.

By setting $\delta\mathcal{L} = 0$, we obtain a set of necessary conditions for the minimum as follows, all for $0 \le n \le N-1$:

1. *State equation*: Term $A$ becomes 0 when

$$\mathbf{x}_n = \nabla_{\boldsymbol{\lambda}}\mathcal{H}_{n-1} = M\left(\mathbf{x}_{n-1}, \boldsymbol{\eta}_{n-1}, \mathbf{u}_{n-1}\right) \tag{7.24}$$

from (7.21), which is the state equation (7.17).

2. *Adjoint equation*: Term $B$ is 0 when

$$\boldsymbol{\lambda}_n = \nabla_{\mathbf{x}}\mathcal{H}_n = \mathbf{M}_n^T\boldsymbol{\lambda}_{n+1} - \mathbf{H}_n^T\mathbf{R}^{-1}\left(\mathbf{y}_n^o - H_n\mathbf{x}_n\right) \tag{7.25}$$

   from (7.21), where $\mathbf{M}_n = \frac{\partial M}{\partial \mathbf{x}}\big|_{\mathbf{x}=\mathbf{x}_n}$ and $\mathbf{H}_n = \frac{\partial H}{\partial \mathbf{x}}\big|_{\mathbf{x}=\mathbf{x}_n}$ are the Jacobians of $M_n$ and $H_n$, respectively. This is called the adjoint equation.
3. *Optimality* (*stationary*) *condition*: Term $C$, combined with the second term in (7.23), vanishes when

$$0 = \nabla_{\mathbf{u}}H_n = \mathbf{C}\mathbf{u}_n + \left(\frac{\partial M}{\partial \mathbf{u}}\right)^T\boldsymbol{\lambda}_{n+1} = \mathbf{C}\mathbf{u}_n + \mathbf{B}^T\boldsymbol{\lambda}_{n+1} \tag{7.26}$$

   from (7.21); thus, the *optimal control* is given by

$$\mathbf{u}_n = -\mathbf{C}^{-1}\mathbf{B}^T\boldsymbol{\lambda}_{n+1}. \tag{7.27}$$

The remaining terms in (7.23) provide the required boundary conditions. Note that $\mathbf{x}_0$ is given, thus $\delta\mathbf{x}_0 = 0$, making the first term in (7.23) 0; $\mathbf{x}_N$ is free (not specified), thus $\delta\mathbf{x}_N$ is arbitrary, making the third term in (7.23) 0 by forcing

$$\boldsymbol{\lambda}_N = 0. \tag{7.28}$$

Equations (7.24)–(7.28) constitute the framework for optimal control and are called the *optimality system*, i.e., the set of equations characterizing the optimal control solution: the state equation, the adjoint equation, and the first-order necessary optimality condition, including the boundary conditions.

## 7.3 Elements of Variational Data Assimilation

In VAR, one seeks to find an optimal initial state that minimize a functional, so-called the *cost function* – measuring a square distance (i.e., quadratic) between observations and the corresponding model states. As it involves the model states in minimizing the cost function, it corresponds to a nonlinear least squares problem under a constraint of dynamical model. Therefore, the solutions of VAR can be obtained through the Euler–Lagrange equations (i.e., the calculus of variations problem) or Pontryagin's minimum principle (i.e., the optimal control problem), depending on the complexity of the problem. Since the governing equations of most atmospheric models consist of partial differential equations, atmospheric VAR can be considered an optimal control problem generalized by Lions (1971).

### *7.3.1 Cost Function*

#### *7.3.1.1 Least Squares Estimation Revisited*

The cost function is one of the main ingredients in VAR. Recall that the cost function in a least squares estimation problem (see Section 1.4.1) is defined as in (1.13) as

$$J = (\boldsymbol{\varepsilon}^r)^T\boldsymbol{\varepsilon}^r = \|\boldsymbol{\varepsilon}^r\|_2^2 = \left(\mathbf{y}^o - \mathbf{H}\hat{\mathbf{x}}\right)^T\left(\mathbf{y}^o - \mathbf{H}\hat{\mathbf{x}}\right), \tag{7.29}$$

where $\boldsymbol{\varepsilon}^r$ is the residual error between the observation $\mathbf{y}^o$ and the estimated observation $\hat{\mathbf{y}} = \mathbf{H}\hat{\mathbf{x}}$ with a linear observation operator $\mathbf{H}$ (see (1.12)), and $\|\boldsymbol{\varepsilon}^r\|_2 = \left((\varepsilon_1^r)^2 + \cdots + (\varepsilon_n^r)^2\right)^{1/2}$ represents the Euclidean or $L^2$ norm; then, the desired estimate of model state, $\hat{\mathbf{x}}$, which minimizes $J$ in a least squares sense is given by

$$\hat{\mathbf{x}} = \left(\mathbf{H}^T\mathbf{H}\right)^{-1} \mathbf{H}^T\mathbf{y}^o, \tag{7.30}$$

as in (1.17). Here, $\mathbf{H}$: $\mathcal{R}^m \rightarrow \mathcal{R}^n$ is a forward mapping from model state space ($\mathcal{R}^m$) to observation space ($\mathcal{R}^n$) and assumed to be of full rank (see Section 1.3.3), whereas its adjoint operator $\mathbf{H}^* \left(\equiv \mathbf{H}^T\right) : \mathcal{R}^n \rightarrow \mathcal{R}^m$ represents the corresponding inverse mapping; $\hat{\mathbf{x}}$ in (7.30) is the solution of linear least squares estimation problem for an overdetermined case, i.e., $n > m$ (Lewis et al., 2006).

Most geoscientific least squares estimation problems are underdetermined, i.e., $n < m$. In this case, one can determine the estimate through a *constrained minimization* problem to find $\hat{\mathbf{x}}$ that minimizes $J = \mathbf{x}^T\mathbf{x}$ under a linear constraint $\mathbf{y}^o = \mathbf{H}\mathbf{x}$: This can be solved as an *unconstrained minimization* problem in $\mathcal{R}^{n+m}$ using the Lagrangian multipliers method (see Colloquy 7.4) by defining a Lagrange function as

$$\mathcal{L}(\mathbf{x}, \boldsymbol{\lambda}) = J + \boldsymbol{\lambda}^T \left(\mathbf{y}^o - \mathbf{H}\mathbf{x}\right), \tag{7.31}$$

where $\boldsymbol{\lambda} \in \mathcal{R}^n$ is a vector of $n$ Lagrangian multipliers. By minimizing $\mathcal{L}(\mathbf{x}, \boldsymbol{\lambda})$, we have the solution of the original constrained minimization problem as

$$\hat{\mathbf{x}} = \mathbf{H}^T \left(\mathbf{H}\mathbf{H}^T\right)^{-1} \mathbf{y}^o. \tag{7.32}$$

Both the overdetermined and underdetermined estimation problems are generally *ill-posed*; thus, a regularization is essentially required to obtain the least squares estimate (see Colloquy 7.5).

**COLLOQUY 7.5**

### Ill-posed problem and Tikhonov regularization

Suppose that we aim to find a solution vector $\mathbf{x}$ of the following system:

$$\mathbf{y} = \mathbf{H}\mathbf{x}. \tag{7.33}$$

We may consider $\mathbf{y}$ as a measured response (say, observation) to a given condition $\mathbf{x}$ (say, model state) through an operator $\mathbf{H}$ that relates the model states to the measured data: $\mathbf{H}$ can be nonlinear in general, but it is considered to be linear here.

Equation (7.33) is *ill-posed* if it counters any of the following well-posedness conditions: 1) it has at least one solution (i.e., $\mathbf{H}$ is surjective); 2) its solution is unique (i.e., $\mathbf{H}$ is injective); and 3) its solution depends continuously on the data, thus being stable with respect to perturbations in the data (i.e., a small perturbation in the data cannot cause a large perturbation in the solution). Ill-posed problems often occur in the inverse problems to determine unknown inputs or the internal structure of a dynamical system from the measured behavior of the system, which generally bring about nonunique solutions. Furthermore, the inverse mapping of $\mathbf{H}$ can cause amplification of the measurement errors.

In a least squares estimation, one seeks to minimize a functional represented as the sum of squared residuals as in (7.29), i.e.,

$$J = \|\mathbf{y} - \mathbf{H}\mathbf{x}\|_2^2, \tag{7.34}$$

where $\|\cdot\|_2$ is the $L^2$ norm. However, in an underdetermined problem, observations are insufficient; thus, the least squares estimation does not provide a unique optimal solution and needs additional information. One can impose desirable properties on a particular solution, e.g., to constrain the initial state estimate to keep close to the prior estimate, by including a regularization term in $J$:

$$J_\alpha = \|\mathbf{y} - \mathbf{H}\mathbf{x}\|_2^2 + \|\mathbf{\Gamma}\mathbf{x}\|_2^2, \tag{7.35}$$

where $\mathbf{\Gamma}$ is called the *Tikhonov regularization matrix*. This is $L_2$ regularization and improves the conditioning of the problem. Then, the least squares estimate is the solution of the following normal equation

$$\left(\mathbf{H}^T\mathbf{H} + \mathbf{\Gamma}^T\mathbf{\Gamma}\right)\mathbf{x} = \left(\mathbf{H}^T\mathbf{H} + \alpha\mathbf{I}\right)\mathbf{x} = \mathbf{H}^T\mathbf{y}, \tag{7.36}$$

where we put $\mathbf{\Gamma} = \sqrt{\alpha}\mathbf{I}$ with the *regularization parameter* $\alpha > 0$. If the matrix $\left(\mathbf{H}^T \ \mathbf{\Gamma}^T\right)^T$ has full rank, then there exists a unique solution. When $\mathbf{H}$ is rank deficient, both $\mathbf{H}\mathbf{H}^T$ and $\mathbf{H}^T\mathbf{H}$ are rank deficient (i.e., singular); however, the matrices $(\alpha\mathbf{I} + \mathbf{H}\mathbf{H}^T)$ and $(\mathbf{H}^T\mathbf{H} + \alpha\mathbf{I})$ become nonsingular, making the least squares problem be well-posed (see Lewis et al., 2006).

#### 7.3.1.2 Tikhonov Regularization on Cost Function

An ill-posed problem can be converted to a well-posed problem by adopting the so-called *Tikhonov regularization* method (Tikhonov, 1963a, 1963b), which establishes a unified framework that deals with both the overdetermined and underdetermined problems (Lewis et al., 2006). In this framework, we define the cost function following (7.35) as

$$J_\alpha = \underbrace{\frac{1}{2}\left\|\mathbf{y}^o - \mathbf{H}\mathbf{x}\right\|_2^2}_{J^o} + \underbrace{\frac{1}{2}\|\mathbf{\Gamma}\mathbf{x}\|_2^2}_{J^r} = \frac{1}{2}\left(\mathbf{y}^o - \mathbf{H}\mathbf{x}\right)^T\left(\mathbf{y}^o - \mathbf{H}\mathbf{x}\right) + \frac{\alpha}{2}\mathbf{x}^T\mathbf{x}, \tag{7.37}$$

where $\mathbf{\Gamma} = \sqrt{\alpha}\mathbf{I}$, and $J^o$ and $J^r$ reflect the effects of observation and regularization, respectively. Here, $J_\alpha$ is called the *Tikhonov functional*; $J^r$ is also called the *penalty term* in VAR and varies with the value of $\alpha$ or the scale of $\mathbf{\Gamma}$, exerting a damping (or smoothing) effect on the cost function and the solution. When $\alpha = 0$ (or $\mathbf{\Gamma} = \mathbf{0}$), (7.37) represents the cost function for an unregularized least squares estimation problem, provided that $\left(\mathbf{\Gamma}^T\mathbf{\Gamma}\right)^{-1}$ exists.

The regularized least squares estimate, which is the solution of the normal equation (7.36) and minimizes $J_\alpha$ in (7.37), is given by

$$\hat{\mathbf{x}}_\alpha = \left(\mathbf{H}^T\mathbf{H} + \alpha\mathbf{I}\right)^{-1}\mathbf{H}^T\mathbf{y}^o; \tag{7.38}$$

for $\alpha = 0$, this corresponds to the overdetermined least squares estimate (7.30), i.e., minimizing the residual error (7.29). With some matrix manipulation on (7.38), we can obtain

$$\hat{\mathbf{x}}_\alpha = \mathbf{H}^T\left(\alpha\mathbf{I} + \mathbf{H}\mathbf{H}^T\right)^{-1}\mathbf{y}^o, \tag{7.39}$$

which, for $\alpha = 0$, reduces to the underdetermined least squares estimate (7.32), i.e., minimizing the state norm $\|\mathbf{x}\|_2^2$ with a constraint $\mathbf{y}^o = \mathbf{H}\mathbf{x}$.

**Practice 7.4 Regularized least squares estimation**

Answer the following:

1. Derive (7.38) by minimizing $J_\alpha$ in (7.37).
   (*Hint*: Find $\mathbf{x}$ that satisfies $\nabla J_\alpha = 0$.)
2. Derive (7.39) from (7.38) using the following vector identity:

$$\left(\mathbf{C}^T\mathbf{B}^{-1}\mathbf{C} + \mathbf{D}^{-1}\right)\mathbf{C}^T\mathbf{B}^{-1} = \mathbf{D}\mathbf{C}^T\left(\mathbf{B} + \mathbf{C}\mathbf{D}\mathbf{C}^T\right)^{-1}.$$

   (*Hint*: Put $\mathbf{C} = \mathbf{H}$, $\mathbf{B} = \mathbf{I}$, and $\mathbf{D}^{-1} = \alpha\mathbf{I}$.)
3. Show that the Hessian of $J$ in (7.37) is positive definite when $\alpha > 0$.
   (*Hint*: An $n \times n$ matrix $\mathbf{A}$ is positive definite when $\mathbf{b}^T\mathbf{A}\mathbf{b} > 0$ for an arbitrary vector $\mathbf{b} \in \mathcal{R}^n$ and its transpose $\mathbf{b}^T$.)

In the framework of data assimilation with *cycling* (see Algorithm 1.1 in Section 1.3.5), the background or first guess $\mathbf{x}^b$ is introduced. For a weighted least squares estimation problem (see Section 1.4.2) with $\mathbf{x}^b$, one can define the inhomogeneous Tikhonov functional as

$$J_\alpha = \frac{1}{2}\left\|\mathbf{y}^o - \mathbf{H}\mathbf{x}\right\|_{\mathbf{W}_R}^2 + \frac{1}{2}\left\|\mathbf{x} - \mathbf{x}^b\right\|_{\mathbf{W}_B}^2, \tag{7.40}$$

where, for any vector $\mathbf{z}$, $\|\mathbf{z}\|_{\mathbf{W}}^2 = \mathbf{z}^T\mathbf{W}\mathbf{z}$ denotes the weighted norm square, $\mathbf{W}_R$ and $\mathbf{W}_B$ are symmetric and positive definite weight matrices, and $\mathbf{x}^b$ is the background or first guess. Here, $\mathbf{W}_B$ can be factored as $\mathbf{W}_B = \mathbf{\Gamma}^T\mathbf{\Gamma}$ (see Colloquy 3.1). Then, the weighted least squares estimate is given by

$$\hat{\mathbf{x}}_\alpha = \mathbf{x}^b + \left(\mathbf{H}^T\mathbf{W}_R\mathbf{H} + \mathbf{W}_B\right)^{-1}\mathbf{H}^T\mathbf{W}_R\left(\mathbf{y}^o - \mathbf{H}\mathbf{x}^b\right). \tag{7.41}$$

**Practice 7.5 Regularized weighted least squares estimation**

Derive (7.41) by minimizing $J_\alpha$ in (7.40), and show that $\mathbf{x}_\alpha$ in (7.41) can be represented as

$$\hat{\mathbf{x}}_\alpha = \mathbf{x}^b + \left(\mathbf{H}^T\mathbf{W}_R\mathbf{H} + \alpha\mathbf{I}\right)^{-1}\mathbf{H}^T\mathbf{W}_R\left(\mathbf{y}^o - \mathbf{H}\mathbf{x}^b\right).$$

(*Hint*: Set $\mathbf{z} = \mathbf{x} - \mathbf{x}^b$, then transform $J_\alpha$ in (7.40) into the form in (7.37) in terms of $\mathbf{z}$, say, $J'_\alpha = J'_\alpha(\mathbf{z})$, to find $\mathbf{z}$ that satisfies $\nabla J'_\alpha = 0$. Assume that the regularization matrix $\mathbf{\Gamma} = \sqrt{\alpha}\mathbf{I}$.)

The weight matrices in (7.40) are further assumed to be invertible and are usually chosen to be the inverses of corresponding error covariance matrices; that is, $\mathbf{W}_R = \mathbf{R}^{-1}$ and $\mathbf{W}_B = \mathbf{B}^{-1}$, where $\mathbf{R}$ and $\mathbf{B}$ are the observation error covariance and the background error covariance, respectively. Then, the Tikhonov functional is given by

$$\begin{aligned} J_\alpha &= \frac{1}{2}\left\|\mathbf{y}^o - \mathbf{H}\mathbf{x}\right\|^2_{\mathbf{R}^{-1}} + \frac{1}{2}\left\|\mathbf{x} - \mathbf{x}^b\right\|^2_{\mathbf{B}^{-1}} \\ &= \left(\mathbf{y}^o - \mathbf{H}\mathbf{x}\right)^T\mathbf{R}^{-1}\left(\mathbf{y}^o - \mathbf{H}\mathbf{x}\right) + \left(\mathbf{x} - \mathbf{x}^b\right)^T\mathbf{B}^{-1}\left(\mathbf{x} - \mathbf{x}^b\right). \end{aligned} \tag{7.42}$$

An alternative regularization can be formulated using the $L^1$ norm and by defining the cost function as

$$J_\alpha = \frac{1}{2}\left\|\mathbf{y}^o - \mathbf{H}\mathbf{x}\right\|^2_2 + \frac{\alpha}{2}\left\|\mathbf{x}\right\|_1, \tag{7.43}$$

where $\|\mathbf{x}\|_1 \equiv \sum_i |x_i|$ is the $L^1$ norm of $\mathbf{x} \in \mathcal{R}^m$. This $L_1$ regularization is proved to produce more accurate and stable solutions than using the conventional $L_2$ regularization in VAR of fluid flow with sharp fronts and shocks in the advection (Budd et al., 2011) and advection–diffusion (Ebtehaj et al., 2014) processes.

### *7.3.1.3 Strong-versus Weak-constraint Variational Formulations*

Incorporation of model dynamics, represented by prognostic or diagnostic equations, into data assimilation can be achieved by taking dynamical constraints – either *strong* or *weak* – as suggested by (Sasaki, 1958, 1969, 1970a, 1970b, 1970c). Assume that the model dynamics is expressed as the following constraint condition:

$$\mathbf{G} = \mathbf{0}, \tag{7.44}$$

quasi-geostrophic equation, hydrostatic equation, etc., where $\mathbf{G}$ is the function of the model state and its gradient, i.e., $\mathbf{G} = \mathbf{G}(\mathbf{x}, \nabla\mathbf{x})$.

Sasaki (1970c) devised variational formulations of the strong-constraint problem through the Lagrangian multiplier method and the weak-constraint problem by adding a weighted penalty term as follows, for the variational operator $\delta$ and the functionals $J_\lambda$ and $J_\gamma$:

***Strong constraint***: The dynamical constraint (7.44) is assumed to be true and is strictly applied by defining the functional to be the Langrange function, i.e., $J_\lambda = \mathcal{L}$ (see Colloquy 7.4), as

$$\delta J_\lambda = \delta \left[ \|\mathbf{y}^o - \mathbf{Hx}\|_2^2 + \boldsymbol{\lambda}^T \mathbf{G} \right] = 0, \tag{7.45}$$

where $\lambda$ is the Lagrangian multiplier and $\mathbf{G} = \mathbf{0}$.

***Weak constraint***: The dynamical constraint (7.44) is regarded as an approximation to reality (i.e., the model is not perfect) and is included in the functional as a weighted penalty term as

$$\delta J_\gamma = \delta \left[ \|\mathbf{y}^o - \mathbf{Hx}\|_2^2 + \boldsymbol{\gamma}^T \|\mathbf{G}\|_2^2 \right] = 0, \tag{7.46}$$

where $\boldsymbol{\gamma}$ is the weight and $\mathbf{G} \approx \mathbf{0}$.

The above variational constraint problems are formulated when no prior information but just dynamical constraint (e.g., prognostic/diagnostic equation) is available. When neither prior information nor dynamical constraint is provided, one can define the functional by adding the regularization term as in (7.37): The Tikhonov regularization method is quite similar to the weak-constraint formulation in (7.46) – see also Lewis et al. (2006).

#### *7.3.1.4 Bayesian Formulation of Cost function*

In this section, we derive the cost function of VAR in terms of the Bayesian framework. As discussed in Section 3.5, the Bayesian perspective of data assimilation includes seeking the model state that maximizes the probability of the posterior distribution, which corresponds to the optimal state in VAR. The combination of the observational data $\mathbf{y}^o \in \mathcal{R}^n$ and the model state $\mathbf{x} \in \mathcal{R}^m$ and its corresponding probability distribution can be described by Bayes' rule:

$$p\left(\mathbf{x} \mid \mathbf{y}^o\right) = \frac{p\left(\mathbf{y}^o \mid \mathbf{x}\right) p(\mathbf{x})}{p(\mathbf{y}^o)}, \tag{7.47}$$

where $p(\mathbf{x} \mid \mathbf{y}^o)$ represents the posterior probability distribution, i.e., the probability of $\mathbf{x}$ given $\mathbf{y}^o$, which is expected to be obtained through data assimilation; $p(\mathbf{y}^o \mid \mathbf{x})$ depicts the distribution of data for any given unobservable (i.e., the likelihood distribution); $p(\mathbf{x})$ is the prior distribution of states with no observational information (i.e., a priori knowledge such as the 6-h cycle backgrounds); and $p(\mathbf{y}^o)$ is the marginal distribution of the data, serving as a normalizing constant for any given observation (see also (3.50) and description of different probability distributions).

We define the errors of the prior or background ($\boldsymbol{\varepsilon}^b$) and the observation ($\boldsymbol{\varepsilon}^o$) as follows:

$$\boldsymbol{\varepsilon}^b = \mathbf{x} - \mathbf{x}^b; \;\; E\left(\boldsymbol{\varepsilon}^b\right) = 0; \;\; \mathbf{B} = E\left(\boldsymbol{\varepsilon}^b (\boldsymbol{\varepsilon}^b)^T\right) \tag{7.48a}$$

$$\boldsymbol{\varepsilon}^o = \mathbf{y}^o - H\mathbf{x}; \;\; E\left(\boldsymbol{\varepsilon}^o\right) = 0; \;\; \mathbf{R} = E\left(\boldsymbol{\varepsilon}^o (\boldsymbol{\varepsilon}^o)^T\right), \tag{7.48b}$$

where $\mathbf{B} \in \mathcal{R}^{m \times m}$ and $\mathbf{R} \in \mathcal{R}^{n \times n}$ are symmetric and positive definite matrices of the background error covariance and the observation error covariance, respectively, and

$H: \mathcal{R}^m \to \mathcal{R}^n$ is the general (nonlinear) observation operator. The probability distributions $p(\mathbf{x})$ and $p(\mathbf{y}^o \mid \mathbf{x})$, for $\mathbf{x} \sim N(\mathbf{x}^b, \mathbf{B})$ and $\mathbf{y}^o \sim N(H\mathbf{x}, \mathbf{R})$, are given by

$$p(\mathbf{x}) = \frac{1}{\sqrt{2\pi^m |\mathbf{B}|}} \exp\left(-\frac{1}{2}(\boldsymbol{\varepsilon}^b)^T \mathbf{B}^{-1} \boldsymbol{\varepsilon}^b\right) \tag{7.49a}$$

$$p(\mathbf{y}^o \mid \mathbf{x}) = \frac{1}{\sqrt{2\pi^n |\mathbf{R}|}} \exp\left(-\frac{1}{2}(\boldsymbol{\varepsilon}^o)^T \mathbf{R}^{-1} \boldsymbol{\varepsilon}^o\right), \tag{7.49b}$$

where $|\mathbf{B}|$ and $|\mathbf{R}|$ are the determinants of $\mathbf{B}$ and $\mathbf{R}$, respectively. From Bayes' rule (7.47), we have

$$\begin{aligned} p\left(\mathbf{x} \mid \mathbf{y}^o\right) &\propto \exp\left(-\frac{1}{2}\left(\mathbf{y}^o - H\mathbf{x}\right)^T \mathbf{R}^{-1}\left(\mathbf{y}^o - H\mathbf{x}\right) - \frac{1}{2}\left(\mathbf{x} - \mathbf{x}^b\right)^T \mathbf{B}^{-1}\left(\mathbf{x} - \mathbf{x}^b\right)\right) \\ &= e^{-(J^o + J^b)} = e^{-J}, \end{aligned} \tag{7.50}$$

where $J^o$ and $J^b$ represent the observation and background terms of the cost function $J$:

$$J(\mathbf{x}) = \underbrace{\frac{1}{2}\left(\mathbf{y}^o - H\mathbf{x}\right)^T \mathbf{R}^{-1}\left(\mathbf{y}^o - H\mathbf{x}\right)}_{J^o} + \underbrace{\frac{1}{2}\left(\mathbf{x} - \mathbf{x}^b\right)^T \mathbf{B}^{-1}\left(\mathbf{x} - \mathbf{x}^b\right)}_{J^b}. \tag{7.51}$$

Equation (7.50) indicates that the posterior distribution $p\left(\mathbf{x} \mid \mathbf{y}^o\right)$ is maximized when the cost function $J$ is minimized.

**Practice 7.6 Posterior distribution and cost function**

Using (7.47), (7.48), and (7.49), show that

$$\ln p\left(\mathbf{x} \mid \mathbf{y}^o\right) = \ln C - \left(J^o + J^b\right),$$

where $C$ is a constant.

### *7.3.1.5 Incremental Cost Function*

The 3DVAR cost function (7.51) can be further manipulated by linearizing the nonlinear observation operator $H$ around the background $\mathbf{x}^b$ for sufficiently small $\mathbf{x} - \mathbf{x}^b$:

$$H\mathbf{x} \equiv H\left(\mathbf{x}\right) = H\left(\mathbf{x}^b + \mathbf{x} - \mathbf{x}^b\right) \approx H\left(\mathbf{x}^b\right) + \mathbf{H}\left(\mathbf{x} - \mathbf{x}^b\right) \equiv H\mathbf{x}^b + \mathbf{H}\left(\mathbf{x} - \mathbf{x}^b\right), \tag{7.52}$$

where $\mathbf{H} = \left.\frac{\partial H(\mathbf{x})}{\partial \mathbf{x}}\right|_{\mathbf{x}=\mathbf{x}^b}$, i.e., the linearized (tangent linear) operator of $H$; then, (7.51) can be written as

$$\begin{aligned} J(\mathbf{x}) = &\frac{1}{2}\left(\mathbf{y}^o - H\mathbf{x}^b - \mathbf{H}\left(\mathbf{x} - \mathbf{x}^b\right)\right)^T \mathbf{R}^{-1}\left(\mathbf{y}^o - H\mathbf{x}^b - \mathbf{H}\left(\mathbf{x} - \mathbf{x}^b\right)\right) \\ &+ \frac{1}{2}\left(\mathbf{x} - \mathbf{x}^b\right)^T \mathbf{B}^{-1}\left(\mathbf{x} - \mathbf{x}^b\right). \end{aligned} \tag{7.53}$$

By letting the increment $\mathbf{x} - \mathbf{x}^b = \delta\mathbf{x}$ and the innovation $\mathbf{y}^o - H\mathbf{x}^b = \mathbf{d}_b^o$, we can express (7.53) in the incremental form as

$$J(\delta\mathbf{x}) = \frac{1}{2}\left(\mathbf{d}_b^o - \mathbf{H}\delta\mathbf{x}\right)^T \mathbf{R}^{-1}\left(\mathbf{d}_b^o - \mathbf{H}\delta\mathbf{x}\right) + \frac{1}{2}\delta\mathbf{x}^T\mathbf{B}^{-1}\delta\mathbf{x}. \tag{7.54}$$

### *7.3.2 Error Covariances*

The cost function in (7.51) represents a summed measure of squared distances between the state and observations ($J_o$) and between the state and the background ($J_b$). The squared distances are scaled by their corresponding error covariance matrices, i.e., the observation error covariance **R** and the background error covariance **B**, respectively. Furthermore, the analysis increment occurs within the subspace spanned by **B** (Kalnay, 2003; Bannister, 2008a), and the model error covariance **Q** appears in the weak-constraint formulation of VAR. Therefore, the error covariance matrices play important roles in practical VAR, and we explore their characteristics and estimation methods here.

#### *7.3.2.1 Background Error Covariance*

The background field $\mathbf{x}^b$ is a main component of VAR as represented in the cost function (7.51) and is assumed to have unbiased Gaussian errors with the error covariance matrix **B**: $\mathbf{x}^b$ contains observational information assimilated at previous analysis cycle and has shown larger contribution (82%) to analysis than new observations (18%) in the ECMWF 4DVAR system (Cardinali, 2013). Matrix **B** depends on the analysis errors at the previous assimilation and on the model error that evolved from the previous analysis. Therefore, the analysis accuracy is strongly dependent on the background error covariance.

The background error covariance **B** is the static approximation to the forecast error covariance $\mathbf{P}^f = E\left(\boldsymbol{\varepsilon}^f(\boldsymbol{\varepsilon}^f)^T\right)$, thus having no flow dependency. We can assume that $\mathbf{B} = \mathbf{DCD}$ where **D** is the diagonal matrix of background-error standard deviation and **C** is the background error correlation matrix. Therefore, **B** plays an important role in VAR essentially through **C** as follows (see also Bouttier and Courtier, 2002; Bannister, 2008a):

- to spread the information from the observation points to a finite model domain, especially into data-sparse areas;
- to smooth the analysis increments, thus providing statistically consistent increments at the nearby model grid points;
- to impose the balance properties in the atmosphere and to spread information to other variables, thus assuring dynamical consistency in the increments among different variables;
- to incorporate the flow-dependent structure function in an area with specific events and wave patterns; and
- to promote synergy among different observations when assimilated together to improve the analysis more than the combined effects of separate assimilation of each observation.

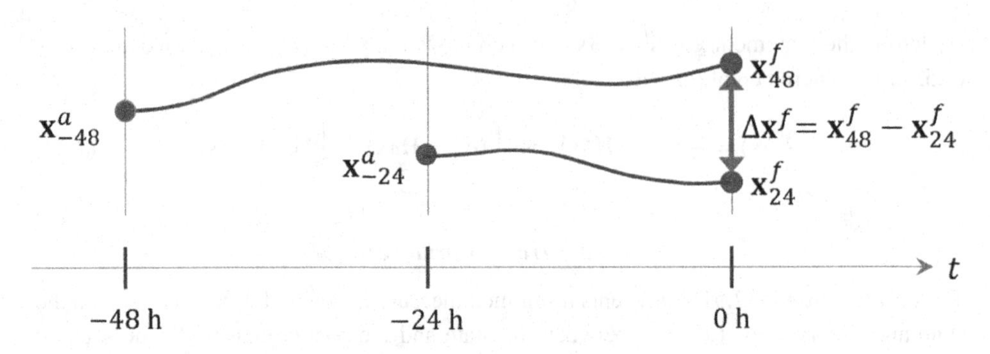


Figure 7.1 The NMC method to estimate the background error covariance **B**. The vectors $\mathbf{x}^a_{-48}$ and $\mathbf{x}^a_{-24}$ represent the analyses at $t = -48$ h and $t = -24$ h, respectively; $\mathbf{x}^f_{48}$ and $\mathbf{x}^f_{24}$ represent the 48-h and 24-h forecasts, valid at $t = 0$ h, using $\mathbf{x}^a_{-48}$ and $\mathbf{x}^a_{-24}$, respectively.

We can obtain the background error covariance as a proxy of the forecast error covariance. A common method to estimate **B** is the so-called *NMC*[8] *method*, which was suggested by Parrish and Derber (1992) and has been widely employed by operational centers – see Colloquy 7.6 and Figure 7.1.

**Colloquy 7.6**

**Estimation of B The NMC method**

In this approach, the background error $\boldsymbol{\varepsilon}^b$ is approximated by the difference between pairs of forecasts at the same verification time (say, $t = 0$ h) but started from different analysis time (say, $t = -48$ h and $t = -24$ h, respectively) as follows:

$$\boldsymbol{\varepsilon}^b = \mathbf{x}^b - \mathbf{x}^t \approx \frac{\mathbf{x}^f_{48} - \mathbf{x}^f_{24}}{\sqrt{2}}, \tag{7.55}$$

where $\mathbf{x}^f_{48}$ is a 48-h forecast and $\mathbf{x}^f_{24}$ is a 24-h forecast, both valid at $t = 0$ h (see Figure 7.1). Assuming that the bias is 0 or remains constant, we have only the random error for each forecast; thus, the forecast error $\Delta\mathbf{x}^f$ is expressed in terms of the random errors as

$$\Delta\mathbf{x}^f = \mathbf{x}^f_{48} - \mathbf{x}^f_{24} = \boldsymbol{\epsilon}_{48} - \boldsymbol{\epsilon}_{24}, \tag{7.56}$$

where $\boldsymbol{\epsilon}_{48}$ and $\boldsymbol{\epsilon}_{24}$ are the random errors of the corresponding forecasts. Then, the background error covariance is given by

[8] National Meteorological Center – now National Center for Environmental Prediction.

$$\mathbf{B} = E\left(\boldsymbol{\varepsilon}^b\left(\boldsymbol{\varepsilon}^b\right)^T\right) \approx \frac{1}{2}E\left(\left(\mathbf{x}_{48}^f - \mathbf{x}_{24}^f\right)\left(\mathbf{x}_{48}^f - \mathbf{x}_{24}^f\right)^T\right)$$
$$= \frac{1}{2}E\left(\left(\boldsymbol{\epsilon}_{48} - \boldsymbol{\epsilon}_{24}\right)\left(\boldsymbol{\epsilon}_{48} - \boldsymbol{\epsilon}_{24}\right)^T\right)$$
$$= \frac{1}{2}\left[E\left(\boldsymbol{\epsilon}_{48}\boldsymbol{\epsilon}_{48}^T\right) + E\left(\boldsymbol{\epsilon}_{24}\boldsymbol{\epsilon}_{24}^T\right) - E\left(\boldsymbol{\epsilon}_{48}\boldsymbol{\epsilon}_{24}^T\right) - E\left(\boldsymbol{\epsilon}_{24}\boldsymbol{\epsilon}_{48}^T\right)\right]. \quad (7.57)$$

By further assuming that $\boldsymbol{\epsilon}_{48}$ and $\boldsymbol{\epsilon}_{24}$ are uncorrelated and that each of them has approximately the same error covariance as $\boldsymbol{\varepsilon}^b$, we have

$$\mathbf{B} \approx \frac{1}{2}\left[E\left(\boldsymbol{\epsilon}_{48}\boldsymbol{\epsilon}_{48}^T\right) + E\left(\boldsymbol{\epsilon}_{24}\boldsymbol{\epsilon}_{24}^T\right)\right] \approx E\left(\boldsymbol{\varepsilon}^b\left(\boldsymbol{\varepsilon}^b\right)^T\right). \quad (7.58)$$

This indicates that the covariance obtained from the forecast differences $\left(\Delta\mathbf{x}^f\right)$, calculated for 30 days (Parrish and Derber, 1992) or more, can be a proper estimate of **B**. See also Bannister (2008a).

The **B** matrix obtained from the NMC method has the following characteristics: 1) **B** is likely to be overestimated in a region of dense observations and at large scales but underestimated in a region of sparse observations. This can be improved through posterior diagnosis and tuning (e.g., Desroziers et al., 2005; Cheng et al., 2021); 2) the forecast differences are computed for longer time length (e.g., 48 h and 24 h) than the backgrounds are usually produced (i.e., 6 h), making the forecast difference-based **B** be broader, both horizontally and vertically, than the actual **B**. The forecast differences can be also calculated from different pairs of forecasts, e.g., the 36-h and 12-h forecasts (e.g., Bello Pereira and Berre, 2006); 3) the forecast errors are produced in model space, describing the global statistics of model variables at all spatial scales represented by the model; 4) both the analysis error and the forecast error are included in the statistics; and 5) **B** is more like a climatoligical covariance as it is based on the forecast differences computed over 30 days or more.

Hollingsworth and Lönnberg (1986) introduced the "*analysis of innovations method*," which had been widely used in the 1980s and 1990s. By assuming that the background error correlations **C** are homogeneous, isotropic and separable into horizontal and vertical components, and that $H$ is linearized about $\mathbf{x}^b$, this method approximates the innovation in terms of the observation and background errors as $\mathbf{y}^o - H\mathbf{x}^b \approx \boldsymbol{\varepsilon}^o - \mathbf{H}\boldsymbol{\varepsilon}^b$; then, by further assuming that $\boldsymbol{\varepsilon}^o$ and $\boldsymbol{\varepsilon}^b$ are uncorrelated, the covariance of the innovations is related to the the background and observation error covariance matrices as

$$E\left(\left(\mathbf{y}^o - H\mathbf{x}^b\right)\left(\mathbf{y}^o - H\mathbf{x}^b\right)^T\right) \approx E\left(\left(\boldsymbol{\varepsilon}^o - \mathbf{H}\boldsymbol{\varepsilon}^b\right)\left(\boldsymbol{\varepsilon}^o - \mathbf{H}\boldsymbol{\varepsilon}^b\right)^T\right)$$
$$\approx \mathbf{R} + \mathbf{H}\mathbf{B}\mathbf{H}^T. \quad (7.59)$$

This method calculates the errors in observation space, and hence the covariances are strongly dependent on the observation network density (or horizontal scale); thus, it is good for those regions with dense radiosonde observations but not appropriate for numerical prediction using a model grid size much smaller (e.g., 10 km) than the average distance of a radiosonde network (e.g., 100 km). Although the error statistics are quite successful, the assumptions are not realistic, and it is hard to separate the background and observation error variances. The innovation covariance has also been used to tune the forecast difference-based **B** in an operational 3DVAR system (Buehner et al., 2005).

Fisher (2003) introduced the "*analysis-ensemble method*" of ECMWF, which replaced the NMC method in October 1999. The basic idea is to generate a perturbed analysis first by putting randomly perturbed inputs into the analysis system, then making a short forecast from the perturbed analysis to generate a perturbed forecast, which can be used as a perturbed background field for the next cycle: The differences between pairs of the background fields will represent statistical characteristics of the background error, which are independent of the initial background perturbation after a few days of analysis cycles. Similar to the NMC method, it obtains the background errors as forecast differences, thus accounting for all the spatial scales.

The mathematical formulation to estimate **B** from the analysis-ensemble method is given by

$$\mathbf{B} = E\left(\boldsymbol{\varepsilon}^b\left(\boldsymbol{\varepsilon}^b\right)^T\right) \approx E\left(\left(\mathbf{x}^b - E\left(\mathbf{x}_k^b\right)\right)\left(\mathbf{x}^b - E\left(\mathbf{x}_k^b\right)\right)^T\right), \tag{7.60}$$

where $E\left(\mathbf{x}_k^b\right)$ is evaluated for the given ensemble members (i.e., ensemble mean for $k = 1, \ldots, K$) at a particular forecast time. To figure out a typical number of ensemble runs to estimate **B** with statistical significance, we refer to Lee and Lee (2011) who used 12-h forecasts in a 30-day period with 8 ensemble members, resulting in 8 members $\times$ 2 times/day $\times$ 30 days $=$ 480 ensemble forecasts, in the MM5[9] 3DVAR system (Barker et al., 2004). Bello Pereira and Berre (2006) had 6-h forecasts with 5 ensemble members in a 49-day period, giving 5 members $\times$ 4 times/day $\times$ 49 days $=$ 980 ensemble forecasts, in the ARPEGE[10] global 4DVAR system.

The analysis-ensemble method has shown improvement in estimating **B** over the NMC method in various studies. For example, Lee and Lee (2011) reported that, in a domain of East Asia where more than half the area is covered by the ocean, the broadness problem of **B** from the NMC method was reduced in wind, temperature, and humidity when the analysis-ensemble method was employed. In Figure 7.2, Bello Pereira and Berre (2006) compared the two methods in terms of the vertical correlations of temperature background error. In both methods, the tropics had narrower vertical correlations and the background errors in the mid-troposphere were negatively correlated with those near the tropopause. The vertically extended (i.e., broader) correlation functions, both positive and negative, and larger negative correlations near the tropopause, which appeared in the NMC method, have significantly improved in the analysis-ensemble method.

[9] Fifth-generation NCAR/Penn State Mesoscale Model.
[10] Action de Recherche Petite Echelle Grande Echelle.

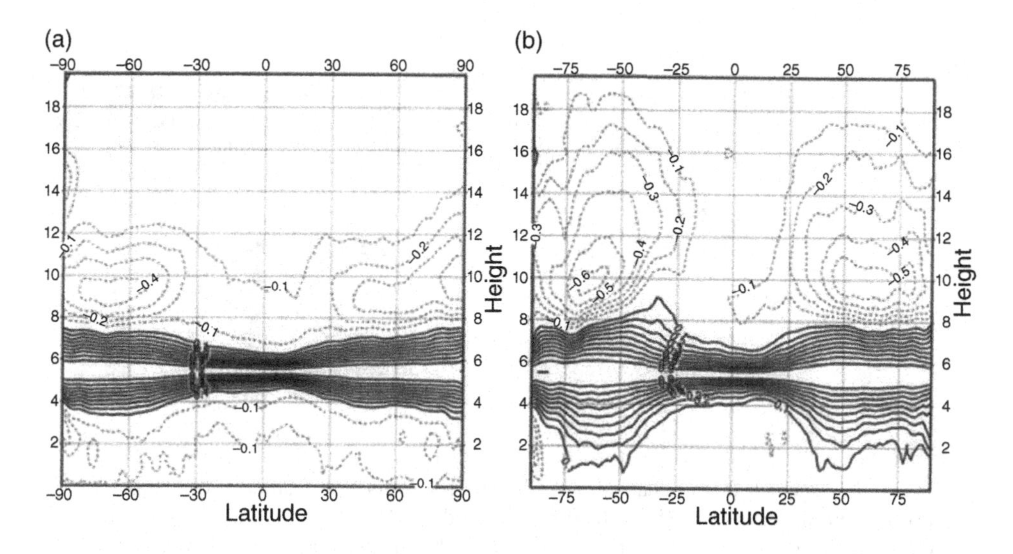


Figure 7.2 Meridional variation of the vertical correlations of temperature background error (around 500 hPa), estimated by (a) the ensemble method and (b) the NMC method, with a contour interval of 0.1. From Bello Pereira and Berre (2006). ©American Meteorological Society. Used with permission.

Some other methods to estimate **B**, including the lagged NMC method and the so-called Canadian Quick method, are referred to Bannister (2008a) and references therein. The background error covariance modeling to construct an approximate but feasible **B** matrix, based on the control variable transforms technique, is extensively discussed in Bannister (2008b).

### *7.3.2.2 Observation Error Covariance*

Observation errors ($\boldsymbol{\varepsilon}^o$) generally come from four main sources: 1) errors in the observing system (e.g., instrument noise); 2) errors of representativeness; 3) errors in the observation operator or forward model; and 4) errors in quality control or preprocessing. Among them, the instrument noise is considered to be an uncorrelated error after calibration while the other sources can create correlations. The observation error covariance **R** is usually treated as a diagonal matrix by assuming that the observations made at different locations have uncorrelated errors or the observations are independent; otherwise, **R** may have nonzero off-diagonal elements. Therefore, it is essential to account for observation error correlations in data assimilation.

Although both **R** and **B** are important components in VAR, more attention has been payed to the estimating, modeling, and preconditioning of **B** (e.g., Fisher, 2003; Bannister, 2008a, 2008b, Arcucci et al., 2017). Recently, as more remotely sensed observations are available, various studies have shown that improved specifications of **R** led to a reduction of the analysis and forecast errors (e.g., Stewart et al., 2013; Weston et al., 2014; Bormann et al., 2016; Wang et al., 2018; Tabeart et al., 2020), especially when the correlated observation

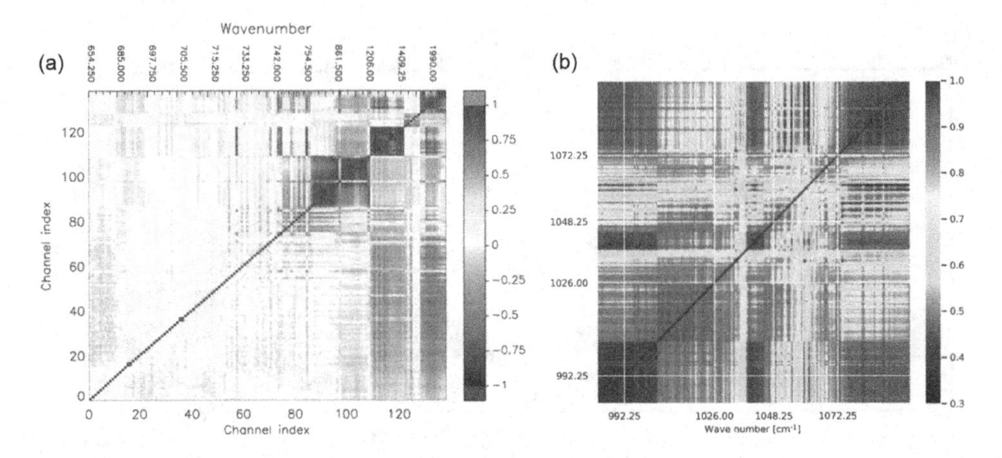


Figure 7.3 Diagnosed inter-channel observation error correlation matrices from IASI for (a) the 139 channels (648.75–1996.00 cm$^{-1}$) used in 4DVAR and (b) a subset of 280 channels (980.00–1100.00 cm$^{-1}$) in 3DVAR. The labeled bars indicate correlation values. (a) is modified from Stewart et al. (2014) ©2013 The Authors and Crown Copyright, used with permission, and (b) is modified from El Aabaribaoune et al. (2021) ©Authors 2021, distributed under CC BY 4.0 License.

errors were considered. In fact, observations can have notable error correlations, and such error statistics can be diagnosed using certain diagnostics, e.g., the one developed by Desroziers et al. (2005) (see Section 7.3.2.4). Observation error correlations are mostly due to errors of representativeness rather than instrument noise; thus, they are predominated by observing systems with complex observation operators, scales different to those of the model, and high-level preprocessing, e.g., radiances from satellites, radial winds from Doppler radars, satellite-derived atmospheric motion vectors, etc. (Fowler et al., 2018).

Figure 7.3 illustrates nonnegligible off-diagonal error statistics in the inter-channel observation error correlations from IASI.[11] In Figure 7.3a, from Stewart et al. (2014), four significant correlation blocks are found around the diagonal for the window (i.e., surface-sensitive) channels (773.50–1206.00 cm$^{-1}$) and three water vapor-sensitive channels (1149.50, 1174.50, 1330.00–1996.00 cm$^{-1}$). Figure 7.3b, from El Aabaribaoune et al. (2021), shows high correlations between most channels, especially among the surface-sensitive channels, but relatively lower correlations among the ozone-sensitive channels (1014.50–1062.50 cm$^{-1}$) and between the ozone- and surface-sensitive channels. Both show the block error structures, which suggests strong error correlations within the corresponding channels.

For high-resolution numerical prediction, assimilation of denser observations is desired, and hence a proper accounting for observation error correlations is essentially required. Fowler et al. (2018) further investigated the interaction of the prior and observation error correlations and its impact on the analysis. They found that, in an optimal system, the

[11] Infrared Atmospheric Sounding Instrument, onboard the EUMETSAT MetOp-A satellite.

analysis error variance and spread of information significantly improved when the statistics of observation and prior errors were complementary or in opposite signs and when dense observations provided a more accurate estimate of the small-scale state than the prior estimate did.

#### *7.3.2.3 Model Error Covariance*

The model error covariance $\mathbf{Q}$ is usually considered in a weak-constraint formulation of VAR. The model error ($\boldsymbol{\varepsilon}_i^m$) at $t = t_i$ is defined as

$$\boldsymbol{\varepsilon}_i^m = \mathbf{x}_i - M_i\left(\mathbf{x}_{i-1}\right), \tag{7.61}$$

where $M_i$ is a nonlinear forward propagator (i.e., model), evolving the state $\mathbf{x}$ from $t_{i-1}$ to $t_i$. Then, the error covariance is given by $\mathbf{Q}_i = E\left(\boldsymbol{\varepsilon}_i^m\left(\boldsymbol{\varepsilon}_i^m\right)^T\right)$.

Model errors generally stem from unknown physical processes, numerical errors in discretizing continuous equations, uncertainties in parameterizing subgrid scale processes, etc. It is recognized that neglecting model errors has detrimental effects in data assimilation (e.g., Županski and Županski, 2006). Many efforts have been made to estimate model error statistics. Daley (1992) estimated $\mathbf{Q}$ using the covariance of the difference between model forecasts and observations (i.e., innovation) along with the evolved analysis error and observation error. By minimizing the difference between the actual innovation covariance and its a priori estimate, Dee (1995) estimated parameters in a model calculating $\mathbf{Q}$.

An ensemble approach was also employed to simulate model errors: In a so-called system simulation experiment, Houtekamer et al. (1996) simulated model error by perturbing some surface fields (roughness length, albedo, and sea surface temperature) and choosing different parameterization options in horizontal diffusion, deep convection, radiation, orography, and gravity wave drag; Houtekamer et al. (2009) introduced various ways to simulate model errors – using additive isotropic perturbations, different ensemble members from different versions of the model, stochastic perturbations to physical tendencies, and stochastic kinetic energy backscatter.

Furthermore, model errors were estimated within data assimilation systems: Županski and Županski (2006) estimated model errors, both bias and parameter errors, by employing the ensemble data assimilation and state augmentation approaches; based on the Newtonian relaxation scheme, Danforth et al. (2007) developed a low-dimensional scheme to separate model errors into three components – the bias (errors averaged over several years), periodic errors including the seasonal and diurnal cycles, and state-dependent (nonperiodic) errors – from a large sample of 6-h analysis corrections with the NCEP reanalysis approximating the true atmosphere; Trémolet (2007) calculated $\mathbf{Q}$ using the time tendencies derived from an ensemble of 4DVAR, using 10 members and 4 tendencies; Todling (2015) diagnosed $\mathbf{Q}$ using residual statistics related to a pair of analyses produced by a filter and a fixed-lag smoother (see Sections 3.3 and 3.4.2, respectively), in the context of a data assimilation system where all the observations were treated in batches. Bowler (2017) extended the Desroziers diagnostics (see Section 7.3.2.4) to estimate $\mathbf{Q}$ in a weak-constraint VAR, by dealing with a set of observations over a time window and using the temporal distribution to separate model errors from errors in the background forecast.

#### 7.3.2.4 Diagnosis of Error Covariances

As shown in the previous subsections of Section 7.3.2, estimation of the error covariances inherently includes uncertainties. In order to adjust or tune the error covariance matrices in VAR, an objective tool to diagnose the errors statistics is essentially required. Desroziers et al. (2005) developed error covariance diagnostics based on linear estimation theory (see Section 1.4.4):

$$\mathbf{x}^a = \mathbf{x}^b + \mathbf{K}\left(\mathbf{y}^o - H\mathbf{x}^b\right) \tag{7.62a}$$

$$\mathbf{K} = \mathbf{B}\mathbf{H}^T\left(\mathbf{H}\mathbf{B}\mathbf{H}^T + \mathbf{R}\right)^{-1}, \tag{7.62b}$$

where $\mathbf{x}^a$ is the analysis, $\mathbf{x}^b$ is the background, $\mathbf{y}^o$ is the observations, and $H$ and $\mathbf{H}$ are the nonlinear and linearized observation operators, respectively. The matrices $\mathbf{B}$ and $\mathbf{R}$ are the background and observation error covariances, respectively, and $\mathbf{K}$ is the Kalman gain.

The following properties are then defined in observation space:

$$\mathbf{d}_b^o = \mathbf{y}^o - H\mathbf{x}^b \approx \boldsymbol{\varepsilon}^o - \mathbf{H}\boldsymbol{\varepsilon}^b \tag{7.63a}$$

$$\mathbf{d}_b^a = H\mathbf{x}^a - H\mathbf{x}^b \approx \mathbf{H}\mathbf{K}\mathbf{d}_b^o \tag{7.63b}$$

$$\mathbf{d}_a^o = \mathbf{y}^o - H\mathbf{x}^a \approx \mathbf{R}\left(\mathbf{H}\mathbf{B}\mathbf{H}^T + \mathbf{R}\right)^{-1}\mathbf{d}_b^o, \tag{7.63c}$$

where $\mathbf{d}_b^o$ is the innovation vector or *observation-minus-background* (O−B) difference, $\mathbf{d}_b^a$ is the *analysis-minus-background* (A−B) difference, and $\mathbf{d}_a^o$ is the *observation-minus-analysis* (O−A) difference.

$$E\left(\mathbf{d}_b^o\left(\mathbf{d}_b^o\right)^T\right) \approx E\left(\left(\boldsymbol{\varepsilon}^o - \mathbf{H}\boldsymbol{\varepsilon}^b\right)\left(\boldsymbol{\varepsilon}^o - \mathbf{H}\boldsymbol{\varepsilon}^b\right)^T\right) = \mathbf{H}\mathbf{B}\mathbf{H}^T + \mathbf{R}, \tag{7.64}$$

when $\mathbf{R}$ and $\mathbf{H}\mathbf{B}\mathbf{H}^T$ (i.e., the background error covariance in observation space) are correctly specified; thus, it can check the specification of those covariances and serves as a consistency check on the innovations.

Additional diagnostics on various errors statistics are derived using (7.63) and (7.64):

$$E\left(\mathbf{d}_b^a\left(\mathbf{d}_b^o\right)^T\right) \approx E\left(\mathbf{H}\mathbf{K}\mathbf{d}_b^a\left(\mathbf{d}_b^o\right)^T\right) = \mathbf{H}\mathbf{B}\mathbf{H}^T \tag{7.65a}$$

$$E\left(\mathbf{d}_a^o\left(\mathbf{d}_b^o\right)^T\right) \approx \mathbf{R}\left(\mathbf{H}\mathbf{B}\mathbf{H}^T + \mathbf{R}\right)^{-1} E\left(\mathbf{d}_b^o\left(\mathbf{d}_b^o\right)^T\right) = \mathbf{R} \tag{7.65b}$$

$$E\left(\mathbf{d}_b^a\left(\mathbf{d}_a^o\right)^T\right) \approx \mathbf{H}\mathbf{B}\mathbf{H}^T\left(\mathbf{H}\mathbf{B}\mathbf{H}^T + \mathbf{R}\right)^{-1}\mathbf{R}, \tag{7.65c}$$

when $\mathbf{H}\mathbf{B}\mathbf{H}^T\left(\mathbf{H}\mathbf{B}\mathbf{H}^T + \mathbf{R}\right)^{-1}$ agrees with the true background and observation error covariances. Equations (7.65a) and (7.65b) are the consistency diagnostics on the background and observation errors, respectively, and (7.65c) is a diagnostic on the analysis errors in observation space. Detailed derivations on these diagnostics are provided in Desroziers et al. (2005).

Bathmann (2018) justified that this approach is best suited when the background and observation errors are uncorrelated with each other, and when the assumed background error covariance matrix is confident. Despite its known limitations, it has been successfully applied to diagnose and tune the observation error covariances of various observing systems (e.g., Weston et al., 2014; Waller et al., 2016; Wang et al., 2018; Tabeart et al., 2020; Mirza et al., 2021; Yin et al., 2021).

Based on the diagnostics of Desroziers et al. (2005), a simple framework for a fixed-point iteration was devised by Mattern et al. (2018) using the diagonal elements of the matrices in (7.65a) and (7.65b):

$$\hat{\sigma}_b^i = \left(\frac{1}{|O_i|}\sum_{j\in O_i} (\mathbf{d}_b^a)_j (\mathbf{d}_b^o)_j\right)^{\frac{1}{2}} \approx \left(\frac{1}{|O_i|}\sum_{j\in O_i} \left(\mathbf{HBT}^T\right)_{jj}\right)^{\frac{1}{2}} = \check{\sigma}_b^i \tag{7.66a}$$

$$\hat{\sigma}_o^i = \left(\frac{1}{|O_i|}\sum_{j\in O_i} (\mathbf{d}_a^o)_j (\mathbf{d}_b^o)_j\right)^{\frac{1}{2}} \approx \left(\frac{1}{|O_i|}\sum_{j\in O_i} \mathbf{R}_{jj}\right)^{\frac{1}{2}} = \check{\sigma}_o^i, \tag{7.66b}$$

where $O_i$ denotes the $i$-th subset of observations (e.g., different observation types). The error covariance diagnostics $\check{\sigma}_b^i$ and $\check{\sigma}_o^i$ are based on predefined $\mathbf{B}$ and $\mathbf{R}$, whereas $\hat{\sigma}_b^i$ and $\hat{\sigma}_o^i$ are determined after data assimilation.

From the results of Weston et al. (2014), we show in Figure 7.4 an example of applying the Desroziers diagnostic to the observation error covariance of the high-resolution sounder IASI and its corresponding tuning (reconditioning). For reconditioning, they reduced the condition number ($\kappa$) (see Section 6.2.6) of a matrix by modifying the eigenvalues ($\lambda$) to increase the matrix diagonal by a quantity that brings about the required condition number ($\kappa_{req}$):

$$\lambda_{inc} = \frac{\lambda_{max} - \lambda_{min} \cdot \kappa_{req}}{\kappa_{req} - 1}, \tag{7.67}$$

where $\lambda_{inc}$ is the eigenvalue increment, and $\lambda_{max}$ and $\lambda_{min}$ are the existing maximum and minimum eigenvalues, respectively; $\kappa_{req}$ is empirically set to 67. Figure 7.4a, from the diagnostic based on (7.65b), indicates that the water vapour and surface-sensitive channels have the strongest positive error correlations (the off-diagonal red blocks toward the bottom right corner), and the temperature-sounding channels have weaker error correlations among themselves (the paler-colored blocks in the top left corner). Figure 7.4b, after reconditioning based on (7.67), shows that the error variances of all channels increase by $\sim$0.11 K$^2$: The pattern of the off-diagonal elements of $\mathbf{R}$ are not much affected while their error correlation values are remarkably modified with a reduction by as much as 0.25 between the channels having the strongest correlations (see Weston et al., 2014).

### *7.3.3 Issues on Minimization*

#### *7.3.3.1 Gradient*

The gradient information is one of the important elements in VAR as it is essential in the performance of numerical minimization (see Chapter 6) to find the minimizer of the cost function. Some algorithms require the first-order derivative (i.e., gradient) information

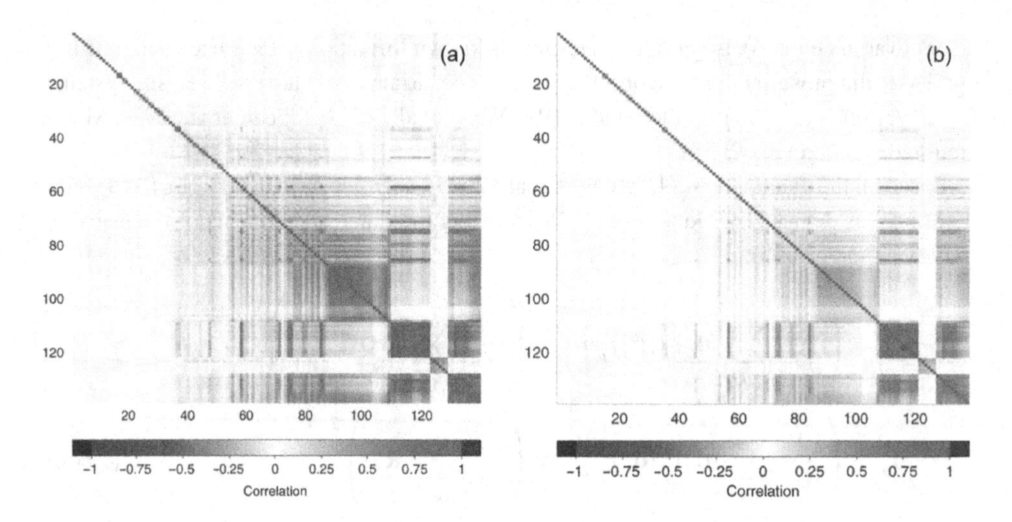


Figure 7.4 Matrices of inter-channel observation error correlations of IASI from 4DVAR output: (a) diagnostic and (b) reconditioned. Shown on the axis are the channel indices: 1–86 represent the temperature-sounding channels (648.75–756.00 cm$^{-1}$), 87–102 the surface-sensitive channels (759.00–962.50 cm$^{-1}$), and 103–138 the water vapor-sensitive channels (1096.00–1996.00 cm$^{-1}$). From Weston et al. (2014). 

(e.g., steepest descent method – see Section 6.3.1) while some others further require the second-order derivative (i.e., Hessian) information (e.g., Newton method – see Section 6.3.2). The line search algorithms have the following iterative process as in Algorithm 6.1:

$$\mathbf{x}^{k+1} = \mathbf{x}^k + \Delta\mathbf{x}^k = \mathbf{x}^k + \alpha^k \mathbf{d}^k, \tag{7.68}$$

where $k$ is the iteration index, $\Delta\mathbf{x}^k$ is the increment to adjust $\mathbf{x}^k$ toward the next step, $\alpha^k$ is the step length, and $\mathbf{d}^k$ is the search direction. The steepest descent method, for instance, is solely dependent on the gradient information because the steepest descent direction is defined as

$$\mathbf{d}^k = -\mathbf{g}^k, \tag{7.69}$$

where $\mathbf{g}^k = \nabla J^k$. The quasi-Newton method, one of the most widely used minimization algorithms in large-scale minimization of meteorological problems, using the BFGS update (see Algorithm 6.4) defines the search direction as

$$\mathbf{d}^k = -\mathbf{C}^k \mathbf{g}^k, \tag{7.70}$$

where $\mathbf{C}^k$ is the approximation to the inverse Hessian and $\alpha^k$ is chosen to minimize $J\left(\mathbf{x}^k + \alpha^k \mathbf{d}^k\right)$.

In VAR, the gradients of the cost function with respect to the initial states (i.e., $\nabla_{\mathbf{x}_0} J$) are obtained by running the ADJM backward in time (see Chapter 4), whose code can be produced using an automatic differentiation tool (see Chapter 5). Various issues in evaluating the gradients using the TLM and ADJM, which include discontinuous/nondifferentiable physical processes, have been discussed in Chapter 4. To treat the problems related to physical processes, Janisková et al. (1999) proposed to regularize, i.e., to smooth the parameterized discontinuities, so as to make the scheme differentiable (see also Verlinde and Cotton, 1993; Županski, 1993a). Janisková and Lopez (2013) introduced the linearized physics package of ECMWF, including radiation, vertical diffusion, orographic gravity wave drag, moist convection, large-scale condensation/precipitation, and nonorographic gravity wave activity.

#### 7.3.3.2 Preconditioning

The $\mathbf{B}$ matrix often can be badly conditioned, and this problem can be alleviated by representing $\mathbf{B}$ through a control variable transformation as $\mathbf{B} = \mathbf{U}\mathbf{U}^T$, as suggested by Parrish and Derber (1992) and Lorenc (1997). For the incremental cost function (7.54), which is rewritten as

$$J(\delta\mathbf{x}) = \frac{1}{2}\left(\mathbf{d}_b^o - \mathbf{H}\delta\mathbf{x}\right)^T \mathbf{R}^{-1}\left(\mathbf{d}_b^o - \mathbf{H}\delta\mathbf{x}\right) + \frac{1}{2}\delta\mathbf{x}^T\mathbf{B}^{-1}\delta\mathbf{x}, \tag{7.71}$$

we can define $\delta\mathbf{x} = \mathbf{U}\delta\mathbf{v}$, which satisfies the following properties:

$$E\left(\boldsymbol{\varepsilon}^b\left(\boldsymbol{\varepsilon}^b\right)^T\right) = \mathbf{B},\ \ E\left(\delta\mathbf{v}\delta\mathbf{v}^T\right) = \mathbf{I},\ \text{and}\ E\left(\left(\mathbf{U}^{-1}\boldsymbol{\varepsilon}^b\right)\left(\mathbf{U}^{-1}\boldsymbol{\varepsilon}^b\right)^T\right) = \mathbf{I}. \tag{7.72}$$

Then, (7.71) is formulated in terms of $\delta\mathbf{v}$ as

$$J(\delta\mathbf{v}) = \frac{1}{2}\left(\mathbf{d}_b^o - \mathbf{H}\mathbf{U}\delta\mathbf{v}\right)^T \mathbf{R}^{-1}\left(\mathbf{d}_b^o - \mathbf{H}\mathbf{U}\delta\mathbf{v}\right) + \frac{1}{2}\delta\mathbf{v}^T\delta\mathbf{v}; \tag{7.73}$$

thus, its gradient and Hessian become

$$\nabla_{\delta\mathbf{v}}J = \mathbf{U}^T\mathbf{H}^T\mathbf{R}^{-1}\left(\mathbf{d}_b^o - \mathbf{H}\mathbf{U}\delta\mathbf{v}\right) + \delta\mathbf{v}, \tag{7.74a}$$

$$\nabla_{\delta\mathbf{v}}^2 J = \mathbf{U}^T\mathbf{H}^T\mathbf{R}^{-1}\mathbf{H}\mathbf{U} + \mathbf{I}, \tag{7.74b}$$

respectively. As (7.74b) includes the identity matrix, the minimum eigenvalue of the Hessian satisfies $\lambda_{min} \geq 1$, which makes the minimization better conditioned.

Numerical minimization can be accelerated through preconditioning that relaxes the ill-conditioned Hessian matrix by adjusting the condition number (Axelsson and Barker, 1984; Županski, 1993c), (see also Section 6.2.6). For an ill-conditioned Hessian, the cost function varies much more rapidly along some specific directions than along others: This implies that the cost function shape is ellipsoidal, and hence similar changes in $\mathbf{x}$ of different directions do not bring about similar changes in the cost function $J(\mathbf{x})$ (i.e., badly scaled). As the condition number increases, the cost function becomes more elongated, which causes a slow convergence in minimization. Through preconditioning, the cost function shape can be significantly less elongated or more spherical.

Assume that we are minimizing a cost function in the following quadratic form:

$$J(\mathbf{x}) = \frac{1}{2}\mathbf{x}^T\mathbf{S}\mathbf{x} - \mathbf{r}^T\mathbf{x} + c, \tag{7.75}$$

where $\mathbf{S}$ is a symmetric positive definite Hessian matrix, $\mathbf{r}$ is an arbitrary vector, and $c$ is a constant. Through a variable transformation $\mathbf{z} = \mathbf{E}^T\mathbf{x}$, following Axelsson and Barker (1984), the problem alters to minimizing the transformed cost function $\bar{J}(\mathbf{z})$ instead of $J(\mathbf{x})$:

$$\bar{J}(\mathbf{z}) = J\left(\mathbf{E}^{-T}\mathbf{z}\right) = \frac{1}{2}\mathbf{z}^T\bar{\mathbf{S}}\mathbf{z} - \bar{\mathbf{r}}^T\mathbf{z} + \bar{c}, \tag{7.76}$$

where $\bar{\mathbf{S}} = \mathbf{E}^{-1}\mathbf{S}\mathbf{E}^{-T}$ is called the *preconditioned matrix*, $\bar{\mathbf{r}} = \mathbf{E}^{-1}\mathbf{r}$, and $\bar{c} = c$. The convergence rate increases if the condition number of $\bar{\mathbf{S}}$ is less than that of $\mathbf{S}$, i.e.,

$$\kappa(\bar{\mathbf{S}}) < \kappa(\mathbf{S}). \tag{7.77}$$

By defining the *preconditioning matrix* $\mathcal{C} = \mathbf{E}\mathbf{E}^T$, one notices that the eigenvalues of $\bar{\mathbf{S}}$ and $\mathcal{C}^{-1}\mathbf{S}$ are the same (see Axelsson and Barker, 1984). Equation (7.77) is one of the properties that makes $\mathcal{C}$ a good preconditioning matrix.

Preconditioning can be directly applied to the conjugate gradient minimization algorithm (see Algorithm 6.3) by defining the search direction as

$$\mathbf{d}^{k+1} = -\mathbf{s}^{k+1} + \beta^k\mathbf{d}^k, \tag{7.78}$$

where

$$\mathbf{s}^{k+1} = \mathcal{C}^{-1}\mathbf{g}^{k+1} \text{ and } \beta^k = \frac{\left(\mathbf{g}^{k+1}\right)^T\mathbf{h}^{k+1}}{\left(\mathbf{g}^k\right)^T\mathbf{h}^k}. \tag{7.79}$$

To calculate the update equation (7.68), this *preconditioned conjugate algorithm* defines $\alpha^k$ and updates $\mathbf{g}^{k+1}$ as

$$\alpha^k = \frac{\left(\mathbf{g}^k\right)^T\mathbf{h}^k}{\left(\mathbf{d}^k\right)^T\mathbf{S}\mathbf{d}^k} \text{ and } \mathbf{g}^{k+1} = \mathbf{g}^k + \alpha^k\mathbf{S}\mathbf{d}^k, \tag{7.80}$$

by initially putting $\mathbf{g}^0 = \mathbf{S}\mathbf{x}^0$, $\mathbf{s}^0 = \mathcal{C}^{-1}\mathbf{g}^0$, and $\mathbf{d}^0 = -\mathbf{s}^0$ for an initial guess $\mathbf{x}^0$.

Županski (1993c) developed a preconditioning algorithm that can be applied to any minimization algorithm when the first iteration is conducted using the preconditioned steepest descent. It is based on the relation between the control variable adjustment and the calculated gradient norm and assumes that a second-order Taylor series expansion is valid in the vicinity of the first guess. The preconditioning matrix was defined to be a product of two positive definite diagonal matrices, each playing its own role to scale the control variable: one is the weight related to the Hessian for the identity model operator and performs a rough scaling while the other is regarded as a refined scaling. This preconditioning is further generalized by Županski (1996) to include direct observations as well as different forms of the cost function, which is suitable to 4DVAR. Županski (2013) discussed the effect of the Hessian preconditioning on all-sky satellite radiance assimilation related to nonlinearity of cloud microphysical processes and the radiative transfer operator for all-sky radiances (see also Section 12.2.3 and Figure 12.3).

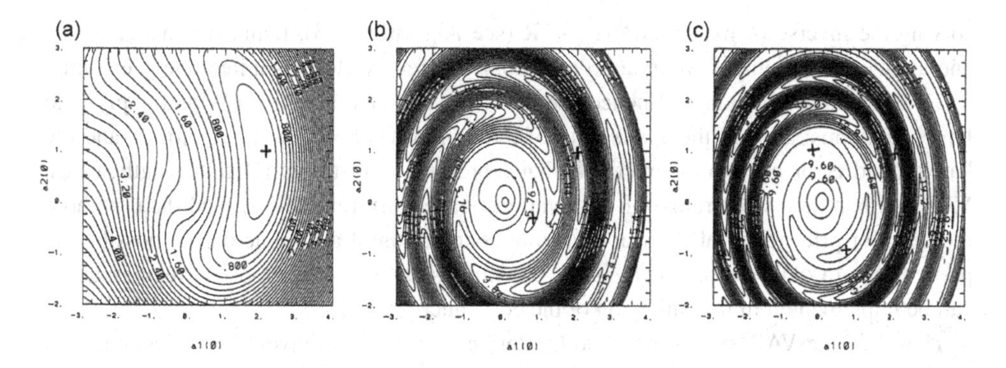

Figure 7.5 Cost functions for the assimilation periods of (a) $T = 0.2$, (b) $T = 0.8$, and (c) $T = 1.2$, respectively, where minima are marked by +. From Li (1991). ©1991 Meteorological Society of Japan (MSJ). The original version was published in Journal of the MSJ.

#### *7.3.3.3 Length of Assimilation Window*

In 4DVAR, minimization is performed within a given assimilation window (i.e., period). A proper choice of the assimilation period is important as nonlinearity may increase during the forward integration of the NLM used in data assimilation: When the length of the assimilation window exceeds a certain threshold time when nonlinearity significantly increases, the number of local minima in the cost function may also increase, thus making the corresponding 4DVAR minimization a multiple-minima problem.

As an example, Figure 7.5 compares the cost functions in terms of non-dimensional assimilation periods of $T = 0.2, 0.8$, and 1.2 in a 4DVAR problem using a simple model for nonlinear resonant Rossby waves (see the details in Li, 1991). When $T = 0.2$ (Figure 7.5a), the cost function has only one (i.e., global) minimum at $(a_1(0), a_2(0)) = (2.0, 1.0)$, and this property continues up to $T = 0.7$. But more than one minimum appears when $T = 0.8$ (Figure 7.5b): The global minimum still remains there but a local minimum exists approximately at $(0.9, -0.2)$. The cost function becomes more complex as the assimilation period increases, e.g., at $T = 1.2$ (Figure 7.5c). This study indicates that the cost function may have multiple local minima when the assimilation period is long enough; thus, one should choose an appropriate assimilation period to avoid undesired complexity in the cost function due to nonlinearity. Using weak constraint 4DVAR that accounts for model errors (see Section 7.4.3.2), longer assimilation windows may be allowable (e.g., Trémolet, 2006, 2007).

## 7.4 From 3DVAR to 4DVAR

Theoretical aspect of minimizing the cost function $J(\mathbf{x})$ in 3DVAR corresponds to the OI (and BLUE) problem in the linear case (see Sections 1.4.4 and 1.5, and Practice 1.7 #5); however, operational centers that replaced the previous OI scheme with the 3DVAR system have demonstrated improvement in forecast skills. The major advantages of the VAR system over OI include the following: 1) OI obtains the Kalman gain by directly solving the inverse of matrix $\mathbf{HP}^b\mathbf{H}^T + \mathbf{R}$ (see Algorithm 1.3), requiring data selection and allowing only simple observation operators, whereas VAR does it iteratively, enabling us to avoid data selection and hence use all observations simultaneously and apply it to larger problems; 2) with the availability of the nonlinear observation operator $H$ and the background error covariance matrix $\mathbf{B}$ that represents realistic correlation length-scales, VAR can handle a wide range of observations, especially remotely sensed data, without the need for prior retrieval; 3) near-real-time assimilation of asynoptic data is possible, in particular with 4DVAR; and 4) dynamic constraints as well as appropriate regularization can be implemented in minimization of the cost function.

However, the VAR system has also limitations: 1) VAR iteratively minimizes the cost function, which essentially requires the adjoint of the direct model and/or observation operator, and hence linearized physical processes, potentially giving problems with highly nonlinear processes and 2) the quality of VAR analysis is strongly dependent on the property and accuracy of prescribed errors, e.g., the background and observation error statistics.

In this section, we introduce 3DVAR and its time-dimension expansion – 4DVAR – including mathematical formalism and computational aspects, and some alternative approaches to the standard 3/4DVAR.

### *7.4.1 3DVAR*

In 3DVAR, one seeks the optimal estimate (i.e., analysis) that minimizes the cost function – see (7.51) and (7.53) – which is rewritten here as

$$J(\mathbf{x}) = \frac{1}{2}\left(\mathbf{x}-\mathbf{x}^b\right)^T \mathbf{B}^{-1}\left(\mathbf{x}-\mathbf{x}^b\right) + \frac{1}{2}\left(\mathbf{y}^o - H\mathbf{x}\right)^T \mathbf{R}^{-1}\left(\mathbf{y}^o - H\mathbf{x}\right) \tag{7.81a}$$

$$= \frac{1}{2}\left(\mathbf{x}-\mathbf{x}^b\right)^T \mathbf{B}^{-1}\left(\mathbf{x}-\mathbf{x}^b\right) + \frac{1}{2}\left(\mathbf{y}^o - H\mathbf{x}^b - \mathbf{H}\left(\mathbf{x}-\mathbf{x}^b\right)\right)^T \mathbf{R}^{-1}\left(\mathbf{y}^o - H\mathbf{x}^b - \mathbf{H}\left(\mathbf{x}-\mathbf{x}^b\right)\right). \tag{7.81b}$$

A penalty term, which is omitted here, can be added to (7.81a and 7.81b) in practical implementation for the purpose of noise control. The analysis $\mathbf{x}^a$ is obtained as a minimization problem:

$$\mathbf{x}^a = \arg\min J(\mathbf{x}). \tag{7.82}$$

The quadratic functional $J$ reaches its minimum when

$$\nabla_{\mathbf{x}} J(\mathbf{x}^a) = \mathbf{0}, \tag{7.83}$$

where $\nabla_{\mathbf{x}} J$ is the gradient of $J(\mathbf{x})$ with respect to $\mathbf{x}$:

$$\nabla_{\mathbf{x}} J(\mathbf{x}) = \mathbf{B}^{-1}\left(\mathbf{x}-\mathbf{x}^b\right) - \mathbf{H}^T\mathbf{R}^{-1}\left(\mathbf{y}^o - H\mathbf{x}\right) \tag{7.84a}$$

$$= \left(\mathbf{B}^{-1} + \mathbf{H}^T\mathbf{R}^{-1}\mathbf{H}\right)\left(\mathbf{x}-\mathbf{x}^b\right) - \mathbf{H}^T\mathbf{R}^{-1}\left(\mathbf{y}^o - H\mathbf{x}^b\right). \tag{7.84b}$$

By setting $\nabla_{\mathbf{x}} J(\mathbf{x}^a) = \mathbf{0}$, we obtain an equation for the analysis increment:

$$\mathbf{x}^a - \mathbf{x}^b = \mathbf{K}\left(\mathbf{y}^o - H\mathbf{x}^b\right), \tag{7.85}$$

where

$$\mathbf{K} = \left(\mathbf{B}^{-1} + \mathbf{H}^T\mathbf{R}^{-1}\mathbf{H}\right)^{-1}\mathbf{H}^T\mathbf{R}^{-1}, \tag{7.86}$$

which is equivalent to the Kalman gain (1.57) in the BLUE with a background (see Section 1.4.4), i.e.,

$$\mathbf{K}^{\mathrm{BLUE}} = \mathbf{B}\mathbf{H}^T\left(\mathbf{H}\mathbf{B}\mathbf{H}^T + \mathbf{R}\right)^{-1}. \tag{7.87}$$

Equation (7.85) indicates that the analysis increment relies upon the credibility of the error covariance matrices, **B** and **R**, through the gain matrix **K**.

**Practice 7.7 Equivalence of 3DVAR and BLUE**

Show that the Kalman gain from 3DVAR (7.86) and that from BLUE (7.87) are equivalent.

In an incremental form, the cost function $J(\delta\mathbf{x})$ (rewritten from (7.54)) and its gradient with respect to $\delta\mathbf{x}$ are given by

$$J(\delta\mathbf{x}) = \frac{1}{2}\delta\mathbf{x}^T\mathbf{B}^{-1}\delta\mathbf{x} + \frac{1}{2}\left(\mathbf{d}_b^o - \mathbf{H}\delta\mathbf{x}\right)^T\mathbf{R}^{-1}\left(\mathbf{d}_b^o - \mathbf{H}\delta\mathbf{x}\right) \tag{7.88a}$$

$$\nabla_{\delta\mathbf{x}} J = \left(\mathbf{B}^{-1} + \mathbf{H}^T\mathbf{R}^{-1}\mathbf{H}\right)\delta\mathbf{x} - \mathbf{H}^T\mathbf{R}^{-1}\mathbf{d}_b^o, \tag{7.88b}$$

where $\delta\mathbf{x} = \mathbf{x}^a - \mathbf{x}^b$ and $\mathbf{d}_b^o = \mathbf{y}^o - H\mathbf{x}^b$.

Figure 7.6 illustrates how 3DVAR is performed. At each analysis time, observations in a $\pm 3$ h window are gathered and used to compute the cost function (7.81) using the

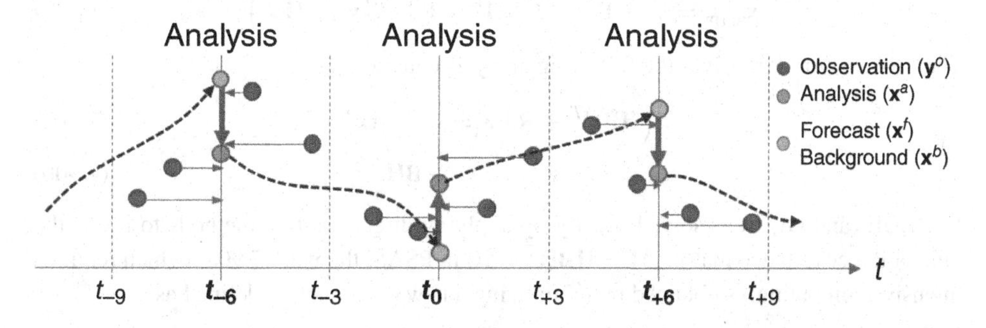


Figure 7.6 3DVAR: At the analysis times ($t = t_{-6}, t_0$, and $t_{+6}$), observations in a $\pm 3$ h window are used for computing the cost function. Dashed lines with arrows depict forecast trajectories from the analysis fields. Thick solid arrows indicate the analysis increment.

**Algorithm 7.1** Computational procedure of 3DVAR

```
/* k: iteration number; K: maximum iteration; ε: tolerance                                  */
/* x^0: initial state; x^b: background; x^a: analysis; y^o: observation                   */
/* H: observation operator; H^T: adjoint of H                                              */
/* B: background error covariance; R: observation error covariance                         */
/* x̂: minimizer of J(x); J_min = J(x̂): minimum value of J(x)                               */
/* d^k: search direction; α^k: step length (minimization algorithm)                        */
1  Input: x^0 = x^b, y^o, B, R, k = 0, ε                ! Initial state x^0 = x^b at k = 0
2  while (k ≤ K or ‖∇J^{k+1}‖ ≥ ε) do                   ! Loop for termination condition
3     Cost function:                                     ! Calculate the cost function J^k = J(x)^k
4        Calculate J^k using (7.54) via H
5     Gradient of cost function:                         ! Calculate the gradient of J^k (∇J^k)
6        Calculate ∇J^k using (7.54) via H^T
7     Update (minimization):                             ! Use any descent algorithm (see Chapter 6)
8        x^{k+1} = x^k + α^k d^k                         ! Update x at k + 1
9     Next iteration:                                    ! Move to the next iteration k + 1
10       Set k = k + 1
11 endwhile
12 Output: x̂ ← x^{k+1}, J_min ← J^{k+1}, x^a ← x̂       ! Assign x^{k+1} to x^a, and J^{k+1} to J_min
```

nonlinear operator $H$ and its gradient (7.84) using the adjoint operator $\mathbf{H}^T$; then a descent algorithm is applied to seek the minimum of the cost function and to obtain the analysis. The computational procedure is provided in Algorithm 7.1.

### *7.4.2 Physical-space Statistical Analysis System*

An alternative VAR scheme that avoids computing the inverse of $\mathbf{B}$, called the Physical-space Statistical Analysis System (PSAS), was introduced by Cohn et al. (1998). The BLUE solution $\mathbf{x}^a_{\text{BLUE}}$ in (1.50) is rewritten, using the Kalman gain in (7.87) as

$$\mathbf{x}^a_{\text{BLUE}} = \mathbf{x}^b + \mathbf{B}\mathbf{H}^T\left(\mathbf{H}\mathbf{B}\mathbf{H}^T + \mathbf{R}\right)^{-1}\left(\mathbf{y}^o - \mathbf{H}\mathbf{x}^b\right). \tag{7.89}$$

The PSAS algorithm solves the following analysis equations:

$$\left(\mathbf{H}\mathbf{B}\mathbf{H}^T + \mathbf{R}\right)\mathbf{w} = \mathbf{y}^o - \mathbf{H}\mathbf{x}^b \tag{7.90a}$$

$$\mathbf{x}^a_{\text{PSAS}} = \mathbf{x}^b + \mathbf{B}\mathbf{H}^T\mathbf{w}. \tag{7.90b}$$

In BLUE (and OI), $\mathbf{w}$ is solved exactly by applying direct solution methods to invert the innovation covariance matrix, $\mathbf{M} = \mathbf{H}\mathbf{B}\mathbf{H}^T + \mathbf{R}$; in PSAS, the $\mathbf{w}$ in (7.90a), which requires intensive computing, is obtained by minimizing a new cost function defined as

$$J_{\text{PSAS}}(\mathbf{w}) = \frac{1}{2}\mathbf{w}^{\mathrm{T}}\left(\mathbf{H}\mathbf{B}\mathbf{H}^{\mathrm{T}} + \mathbf{R}\right)\mathbf{w} - \mathbf{w}^{\mathrm{T}}\left(\mathbf{y}^o - \mathbf{H}\mathbf{x}^b\right), \tag{7.91}$$

which brings about the PSAS solution, $\mathbf{x}^a_{\text{PSAS}}$, as in (7.90b).

Note that (7.90a) is regarded as a *dual* formalism of the *primal* approach, i.e., the analysis equation (7.85) using the Kalman gain (7.86) whose quadratic minimization is performed directly in $\mathcal{R}^m$ (model space). The PSAS algorithm solves $\mathbf{w}$ with a length of $n$ in the observation space where $\mathbf{y}^o \in \mathcal{R}^n$ and $\mathbf{M} \in \mathcal{R}^n \times \mathcal{R}^n$: this requires $O\left(N_{descent}n^2\right)$ operations, where $N_{descent}$ is the number of iterations by a descent algorithm, e.g., the conjugate gradient method. Additional computation is done by (7.90b) to transfer the solution $\mathbf{w}$ to the state space, i.e., $\mathbf{BH}^T\mathbf{w} \in \mathcal{R}^m$, which requires $O(mn)$ operations for the state vector length $m$. Therefore, PSAS is an efficient method when $n \ll m$ in producing similar results as 3DVAR. The 4D-PSAS algorithm is also available; however, PSAS is equivalent to 3/4DVAR only when the observation operator is linear, though an incremental formalism of PSAS can include nonlinearity through incremental updates (Bouttier and Courtier, 2002).

### *7.4.3 4DVAR*

The 4DVAR is an extension of 3DVAR in the time dimension; that is, all the observations in a given assimilation window are incorporated and used to compute the cost function at the correct time. In other words, 4DVAR seeks an optimal initial condition that produces a forecast best fitting the observations and consistent with the system dynamics, within a certain assimilation period. Therefore, the dynamical model can be included in the function of observation operator, which enables us to obtain more information from the observations.

In 4DVAR, the cost function is computed over space and time (assimilation window), and its minimum is sought using a descent algorithm (see Chapter 6). Figure 7.7 shows the iterative process in standard 4DVAR starting from the initial guess. The NLM ($M$) propagates the model state to $t = t_N$, i.e., the end of assimilation window to produce the

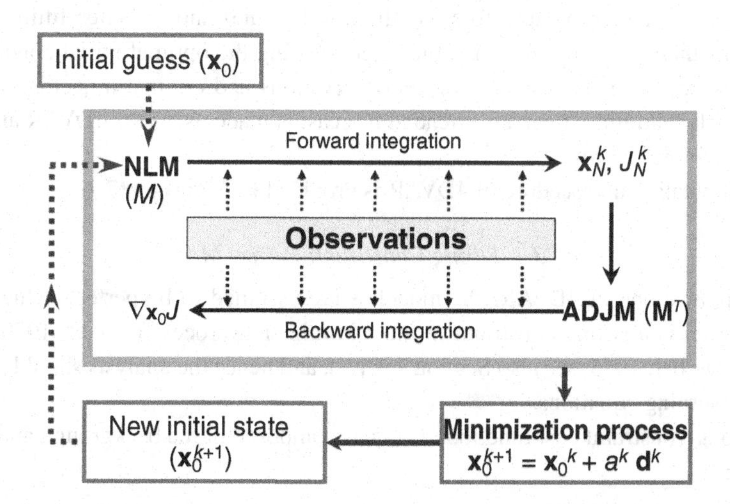


Figure 7.7 Iterative process of standard 4DVAR: The model state is denoted $\mathbf{x}$ where subscripts mean the time step ($t_N$ is the end of assimilation window) while superscripts indicate the iteration index. NLM ($M$) represents the nonlinear model and ADJM ($\mathbf{M}^T$) is the linearized adjoint model. Modified from Park and Županski (2003). ©Springer-Verlag 2002. Used with permission.

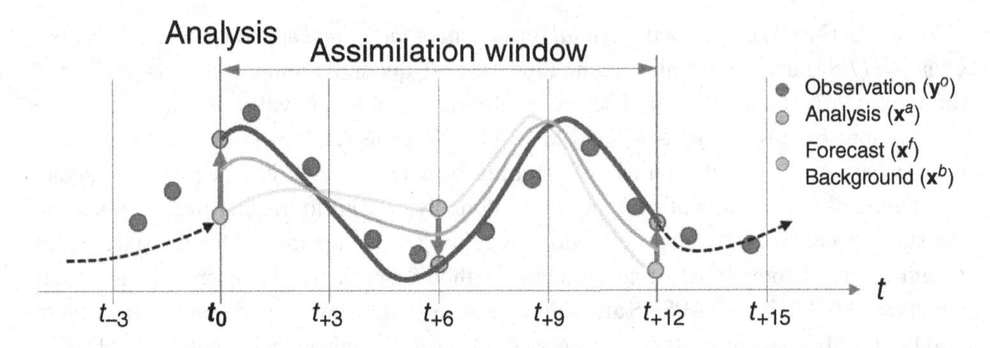


Figure 7.8 4DVAR: The analysis at $t = t_0$ is obtained via iterative minimization process using all the observations in a 12-h assimilation window. Solid lines show forecast trajectories within the assimilation window during the iterative 4DVAR, starting from the initial guess (background; lightest solid line), from the updated initial condition in an intermediate iterative step (light solid line), and from the optimal initial condition (analysis; dark solid line) after the minimization of cost function is completed. Dashed lines with arrows depict the forecast from the previous analysis ($t < t_0$) and the extended forecast from the current analysis ($t > t_{+12}$). Thick solid arrows indicate the analysis increment.

cost function. The ADJM ($\mathbf{M}^T$) is integrated backward to produce the gradient of the cost function with respect to the initial state ($\nabla_{\mathbf{x}_0} J$). Then, the minimization process is performed using a descent algorithm to update the initial state, which will be the input to the NLM for the next iteration.

In Figure 7.8, the initial forecast is made (lightest solid line) and its corresponding cost function is computed within a 12-h assimilation window, starting from an initial guess (i.e., $\mathbf{x}_0 = \mathbf{x}^b$); an intermediate forecast (light solid line) shows better fitting with the observations than the initial forecast; the forecast using the optimal initial condition, i.e., the analysis ($\mathbf{x}^a$), depicts best fit with the observations and optimal trajectory within the window (dark solid line); then, an extended forecast is made using the 4DVAR analysis up to the desired forecast time.

The computational procedure of 4DVAR is provided in Algorithm 7.2.

#### *7.4.3.1 Strong Constraint: Perfect Model*

In early applications of 4DVAR, the models were assumed to be perfect, which implies that the model is a strong constraint in the minimization process (Sasaki, 1970a, 1970b, Lewis and Derber, 1985, see also Section 7.3.1.3), and hence the analysis should satisfy the model's governing equations.

The strong-constraint cost function, $J_{\text{SC}}(\mathbf{x})$, is composed of the background and observation terms:

$$\begin{aligned} J_{\text{SC}}(\mathbf{x}_0) &= J^b(\mathbf{x}_0) + J^o(\mathbf{x}_0) \\ &= \frac{1}{2}\left(\mathbf{x}_0 - \mathbf{x}_0^b\right)^T \mathbf{B}^{-1}\left(\mathbf{x}_0 - \mathbf{x}_0^b\right) + \frac{1}{2}\sum_{n=0}^{N}\left(\mathbf{y}_n^o - H_n\mathbf{x}_n\right)^T \mathbf{R}_n^{-1}\left(\mathbf{y}_n^o - H_n\mathbf{x}_n\right). \end{aligned} \tag{7.92}$$

**Algorithm 7.2** Computational procedure of 4DVAR

/* $k$: iteration number (superscript); $K$: maximum iteration; $\varepsilon$: tolerance */
/* $n$: time step (subscript); $N$: maximum time step in assimilation window */
/* $\mathbf{x}_0^0$: initial state; $\mathbf{x}^b$: background; $\mathbf{x}^a$: analysis; $\mathbf{y}_n^o$: observation */
/* $M_n$: direct model; $H_n$: observation operator */
/* $\mathbf{M}_n^T$: adjoint of $M_n$; $\mathbf{H}_n^T$: adjoint of $H_n$ */
/* $\mathbf{B}$: background error covariance; $\mathbf{R}_n$: observation error covariance */
/* $\hat{\mathbf{x}}$: minimizer of $J(\mathbf{x}_0)$; $J_{min} = J(\hat{\mathbf{x}})$: minimum value of $J(\mathbf{x}_0)$ */
/* $\mathbf{d}^k$: search direction; $\alpha^k$: step length (minimization algorithm) */

1 ***Input***: $\mathbf{x}_0^0 = \mathbf{x}^b$, $\mathbf{y}_0^o$, $\mathbf{B}$, $\mathbf{R}_0$, $k = 0$, $\varepsilon$ ! Initial state $\mathbf{x}_0^0 = \mathbf{x}^b$ at $k = 0$
2 **while** $\left(k \leq K \text{ or } \left\|\nabla J^{k+1}\right\| \geq \varepsilon\right)$ **do** ! Loop for termination condition
3 ***Cost function:*** ! Compute the cost function $J^k = J(\mathbf{x}_0^k)$
4 Compute $J^k$ using (7.54) via $M_n$ and $H_n$
5 ***Gradient of cost function:*** ! Compute the gradient of $J^k$ $\left(\nabla J^k = \nabla J(\mathbf{x}_0^k)\right)$
6 Compute $\nabla J^k$ using (7.54) via $\mathbf{M}_n^T$ and $\mathbf{H}_n^T$ (reverse mode)
7 ***Update (minimization):*** ! Use any descent algorithm (see Chapter 6)
8 $\mathbf{x}_0^{k+1} = \mathbf{x}_0^k + \alpha^k \mathbf{d}^k$ ! Update $\mathbf{x}_0$ at $k + 1$
9 ***Next iteration:*** ! Move to the next iteration $k + 1$
10 Set $k = k + 1$
11 **endwhile**
12 ***Output***: $\hat{\mathbf{x}} \leftarrow \mathbf{x}_0^{k+1}$, $J_{min} \leftarrow J^{k+1}$, $\mathbf{x}^a \leftarrow \hat{\mathbf{x}}$ ! Assign $\mathbf{x}_0^{k+1}$ to $\mathbf{x}^a$, and $J^{k+1}$ to $J_{min}$

The model state $\mathbf{x}$ evolves via the nonlinear propagator $M$ as

$$\mathbf{x}_n = M_n \mathbf{x}_{n-1} = M_n M_{n-1} \cdots M_2 M_1 \mathbf{x}_0 \equiv M_{n\leftarrow 1}\mathbf{x}_0; \tag{7.93}$$

then, $J^o(\mathbf{x}_0)$ can be expressed as

$$J^o(\mathbf{x}_0) = \sum_{n=0}^{N} \left(\mathbf{y}_n^o - H_n M_{n\leftarrow 1}\mathbf{x}_0\right)^T \mathbf{R}_n^{-1} \left(\mathbf{y}_n^o - H_n M_{n\leftarrow 1}\mathbf{x}_0\right). \tag{7.94}$$

The gradient of $J_{\text{sc}}(\mathbf{x})$ is then given by

$$\begin{aligned}\nabla J_{\text{sc}}(\mathbf{x}_0) &= \nabla J^b(\mathbf{x}_0) + \nabla J^o(\mathbf{x}_0)\\ &= \mathbf{B}^{-1}\left(\mathbf{x}_0 - \mathbf{x}_0^b\right) - \sum_{n=0}^{N} \underbrace{\mathbf{M}_1^T \mathbf{M}_2^T \cdots \mathbf{M}_{n-1}^T \mathbf{M}_n^T}_{\mathbf{M}_{1\leftarrow n}^T} \mathbf{H}_n^T \mathbf{R}_n^{-1} \left(\mathbf{y}_n^o - H_n \mathbf{x}_n\right),\end{aligned} \tag{7.95}$$

where $\mathbf{H}$ and $\mathbf{M}$ are the tangent linear operators of $H$ and $M$, respectively. Equation (7.94) clearly shows that the cost function's observation term calculates the innovation, $\mathbf{y}_n^o - H_n\mathbf{x}_n$, at the observation times $t_n$ as the model propagates forward from $\mathbf{x}_0$; Eq. (7.95) indicates that $\nabla J^o(\mathbf{x}_0)$ integrates the weighted innovation, $\mathbf{H}_n^T \mathbf{R}_n^{-1}\left(\mathbf{y}_n^o - H_n\mathbf{x}_n\right)$, backward through the ADJM run, i.e., $\mathbf{M}_{1\leftarrow n}^T$.

Note that the $\mathbf{B}$ matrix spreads information among different model states and parameter variables through the off-diagonal elements when correlations are included (Kalnay, 2003).

In 4DVAR, the forecast model propagates the **B** matrix implicitly; this enables the propagated diagonal **B** matrix to develop correlations throughout an assimilation window, and hence improve data assimilation results (Bannister, 2008a).

#### *7.4.3.2 Weak Constraint: Model with Errors*

Sasaki (1970a) introduced weak-constraint VAR, which approximately satisfies the model equations as they are not exact (see Section 7.46 and Eq. (7.46)). Operational implementation of full weak-constraint 4DVAR has not been feasible due to computational cost difficulty in defining the model error covariance. Even with approximate estimation of model error statistics, however, several studies showed promising results (e.g., Županski, 1993b, 1997; Lee and Lee, 2003; Vidard et al., 2004). Trémolet (2006) formulated three weak-constraint 4DVAR problems using model bias, model-error forcing and model state for the control variable: The results indicated that using the former two as the control variable corresponded to an initial-condition problem with parameter estimation regarding model error as an additional parameter, whereas weak-constraint 4DVAR using the model state as the control variable was interpreted as a coupling between successive strong-constraint assimilation cycles.

As the model is assumed to be imperfect, we include the model error in the forecast model as in (7.61) where the model error $\boldsymbol{\varepsilon}^m$ can be represented in three control variables, including the bias ($\boldsymbol{\beta}$), model-error forcing ($\boldsymbol{\eta}$), and model state ($\mathbf{x}$). Here, we introduce the weak-constraint 4DVAR formulation by choosing $\eta$ and $\mathbf{x}_0$ as the control variables, following Laloyaux et al. (2020). The model trajectory $\mathbf{x}_n$ and the weak-constraint cost function $J_{\mathrm{WC}}(\mathbf{x}_0, \boldsymbol{\eta})$ are formulated, assuming that $\boldsymbol{\eta}$ is additive and constant within the assimilation window with a Gaussian distribution having no cross-correlation with the background error, as

$$\mathbf{x}_n = M_n \mathbf{x}_{n-1} + \boldsymbol{\eta} \tag{7.96a}$$

$$\begin{aligned} J_{\mathrm{WC}}(\mathbf{x}_0, \boldsymbol{\eta}) &= J^b(\mathbf{x}_0, \boldsymbol{\eta}) + J^o(\mathbf{x}_0, \boldsymbol{\eta}) + J^q(\mathbf{x}_0, \boldsymbol{\eta}) \\ &= \frac{1}{2}\left(\mathbf{x}_0 - \mathbf{x}_0^b\right)^T \mathbf{B}^{-1}\left(\mathbf{x}_0 - \mathbf{x}_0^b\right) + \frac{1}{2}\sum_{n=0}^{N}\left(\mathbf{y}_n^o - H_n\mathbf{x}_n\right)^T \mathbf{R}_n^{-1}\left(\mathbf{y}_n^o - H_n\mathbf{x}_n\right) \\ &\quad + \frac{1}{2}\left(\boldsymbol{\eta} - \boldsymbol{\eta}^b\right)^T \mathbf{Q}^{-1}\left(\boldsymbol{\eta} - \boldsymbol{\eta}^b\right)^T, \end{aligned} \tag{7.96b}$$

where $\boldsymbol{\eta}^b$ is the prior model forcing estimate and $\mathbf{Q}$ is the model error covariance (see Section 7.3.2.3).

For simplicity in derivation, we consider just one time-step integration ($t_0 \rightarrow t_1$) with observations available at both $t_0$ and $t_1$ and assume that the model and observation operators are linear, as in Laloyaux et al. (2020), i.e., $\mathbf{x} = (\mathbf{x}_0\ \ \mathbf{x}_1)^T$, $\mathbf{y} = (\mathbf{y}_0\ \ \mathbf{y}_1)^T$, and $\mathbf{R}$ and $\mathbf{H}$ are diagonal matrices whose diagonal elements are $(\mathbf{R}_0, \mathbf{R}_1)$ and $(\mathbf{H}_0, \mathbf{H}_1)$, respectively. By introducing another control vector $\mathbf{p} = (\mathbf{x}_0\ \ \eta)^T$, which satisfies

$$\mathbf{p} = \mathbf{L}\mathbf{x} = \begin{pmatrix} I & 0 \\ -\mathbf{M}_1 & I \end{pmatrix}\mathbf{x} \;\Rightarrow\; \mathbf{x} = \mathbf{L}^{-1}\mathbf{p} = \begin{pmatrix} I & 0 \\ \mathbf{M}_1 & I \end{pmatrix}\mathbf{p}, \tag{7.97}$$

where $\mathbf{L}$ connects the model state $\mathbf{x}$ and the vector $\mathbf{p}$ containing the initial state and the model error. Then, the weak-constraint cost function and its gradient are expressed as a function of $\mathbf{p}$ as

$$J_{\mathrm{WC}}(\mathbf{p}) = \frac{1}{2}\left(\mathbf{p}-\mathbf{p}^b\right)^T \mathbf{D}^{-1}\left(\mathbf{p}-\mathbf{p}^b\right) + \frac{1}{2}\left(\mathbf{y}^o - \mathbf{H}\mathbf{L}^{-1}\mathbf{p}\right)^T \mathbf{R}^{-1}\left(\mathbf{y}^o - \mathbf{H}\mathbf{L}^{-1}\mathbf{p}\right), \tag{7.98a}$$

$$\nabla J_{\mathrm{WC}}(\mathbf{p}) = \mathbf{D}^{-1}\left(\mathbf{p}-\mathbf{p}^b\right) + \left(\mathbf{H}\mathbf{L}^{-1}\right)^T \mathbf{R}^{-1}\left(\mathbf{y}^o - \mathbf{H}\mathbf{L}^{-1}\mathbf{p}\right), \tag{7.98b}$$

where $\mathbf{p}^b$ is the background of the new control space, and $\mathbf{D}$ is a diagonal matrix whose elements are $\mathbf{B}$ and $\mathbf{Q}$.

Laloyaux et al. (2020) reported that the model errors and initial state can be accurately estimated through weak-constraint 4DVAR only with different spatial scales between the background and model errors and with unbiased and spatially homogeneous observations. Longer assimilation windows may be allowable using weak-constraint 4DVAR by accounting for model errors; however, its formulation reveals that the window length is still affected by the validity of tangent linear approximation (e.g., Trémolet, 2007).

#### *7.4.3.3 Incremental 4DVAR*

In an incremental form, the cost function of a strong-constraint 4DVAR is defined by

$$J_{\mathrm{SC}}(\delta\mathbf{x}_0) = \frac{1}{2}(\delta\mathbf{x}_0)^T \mathbf{B}^{-1}(\delta\mathbf{x}_0) + \frac{1}{2}\sum_{n=0}^{N}\left(\mathbf{d}_n^o - \mathbf{H}_n\mathbf{M}_{n\leftarrow 1}\delta\mathbf{x}_0\right)^T \mathbf{R}_n^{-1}\left(\mathbf{d}_n^o - \mathbf{H}_n\mathbf{M}_{n\leftarrow 1}\delta\mathbf{x}_0\right), \tag{7.99}$$

where $\mathbf{d}_n^o = \mathbf{y}_n^o - H_n\mathbf{x}_n$. The gradient of the incremental cost function is given by

$$\nabla J_{\mathrm{SC}}(\delta\mathbf{x}_0) = \mathbf{B}^{-1}\delta\mathbf{x}_0 - \sum_{n=0}^{N}\mathbf{M}_{1\leftarrow n}^T\mathbf{H}_n^T\mathbf{R}_n^{-1}\left(\mathbf{d}_n^o - \mathbf{H}_n\mathbf{M}_{n\leftarrow 1}\delta\mathbf{x}_0\right). \tag{7.100}$$

Note that the incremental cost function (7.99) is linear quadratic, and hence has a unique minimum. Therefore, incremental 4DVAR can be solved as a series of linearized minimization problems and is often performed in low resolution and with simplified (linearized) physical parameterization schemes (e.g., Courtier et al., 1994; Janisková et al., 1999; Janisková and Lopez, 2013). Lawless et al. (2005) showed that incremental 4DVAR algorithm is equivalent to solving the full nonlinear 4DVAR problem using a Gauss–Newton iteration.

Trémolet (2006) formulated the incremental weak-constraint 4DVAR by defining the departure $(\delta\mathbf{x}_0,\ \delta\boldsymbol{\eta})$ from a first guess $(\mathbf{x}_0^g,\ \boldsymbol{\eta}^g)$ as the control variables. For sufficiently small perturbations, the evolution of perturbations can be described as

$$\delta\mathbf{x}_n = \mathbf{M}_n\delta\mathbf{x}_{n-1} + \delta\boldsymbol{\eta}_n = \mathbf{M}_{n\leftarrow 1}\delta\mathbf{x}_0 + \sum_{j=1}^{n}\mathbf{M}_{n\leftarrow j+1}\delta\boldsymbol{\eta}_j. \tag{7.101}$$

The incremental cost function of weak-constraint 4DVAR is given by

$$J_{\mathrm{WC}}\left(\delta\mathbf{x}_0,\delta\boldsymbol{\eta}_n\right)=\frac{1}{2}\left(\delta\mathbf{x}_0+\mathbf{b}\right)^T\mathbf{B}^{-1}\left(\delta\mathbf{x}_0+\mathbf{b}\right)$$
$$+\frac{1}{2}\sum_{n=0}^{N}\left(\mathbf{d}_n^o-\mathbf{H}_n\mathbf{M}_{n\leftarrow 1}\delta\mathbf{x}_0\right)^T\mathbf{R}_n^{-1}\left(\mathbf{d}_n^o-\mathbf{H}_n\mathbf{M}_{n\leftarrow 1}\delta\mathbf{x}_0\right)$$
$$+\frac{1}{2}\sum_{n=1}^{N}\left(\delta\boldsymbol{\eta}_n+\boldsymbol{\eta}_n^g\right)^T\mathbf{Q}_n^{-1}\left(\delta\boldsymbol{\eta}_n+\boldsymbol{\eta}_n^g\right), \tag{7.102}$$

where $\mathbf{b}=\mathbf{x}_0^g-\mathbf{x}^b$; then the gradient $\nabla J_{\mathrm{WC}}^o(\delta\mathbf{x}_0)$ and $\nabla J_{\mathrm{WC}}^q(\delta\boldsymbol{\eta}_n)$ are given by

$$\nabla J_{\mathrm{WC}}^o(\delta\mathbf{x}_0)=\mathbf{B}^{-1}\left(\delta\mathbf{x}_0+\mathbf{b}\right)-\sum_{n=0}^{N}\mathbf{M}_{1\leftarrow n}^T\mathbf{H}_n^T\mathbf{R}_n^{-1}\left(\mathbf{d}_n^o-\mathbf{H}_n\delta\mathbf{x}_n\right), \tag{7.103a}$$

$$\nabla J_{\mathrm{WC}}^q(\delta\boldsymbol{\eta}_n))=\mathbf{Q}^{-1}\left(\delta\boldsymbol{\eta}_n+\boldsymbol{\eta}_n^g\right)-\sum_{n=i}^{N}\mathbf{M}_{i+1\leftarrow n}^T\mathbf{H}_n^T\mathbf{R}_n^{-1}\left(\mathbf{d}_n^o-\mathbf{H}_n\delta\mathbf{x}_n\right). \tag{7.103b}$$

Note that $\nabla J_{\mathrm{WC}}^q(\delta\boldsymbol{\eta}_n)$ has the same form as $\nabla J_{\mathrm{WC}}^o(\delta\mathbf{x}_0)$ except that the observations contribute to the sum only in the steps $n \geq i$. This is due to the backward integration of the adjoint from the end of window to $t = t_i$. This weak-constraint 4DVAR gives new estimates of both the initial condition and the model-error forcing. As the model-error forcing directly modifies the model state in the forward integration, the corresponding adjoint integration appears in the gradient computation (see Trémolet, 2006). Further formulations and discussions on using the model bias and the model state as the control variables in weak-constraint 4DVAR are referred to in Trémolet (2006) and Laloyaux et al. (2020). The computational procedure of incremental 4DVAR is shown in Algorithm 7.3.

### 7.4.4 First Guess at Appropriate Time

One of the approximate approaches to 4DVAR for practical use is the first guess at appropriate time approach (FGAT) (Lawless, 2010), which is considered a midway process between 3DVAR and 4DVAR in incremental form. The 3D-FGAT has the same formalism as the incremental 4DVAR in (7.99) except that $\delta\mathbf{x}_0$ is stationary within the assimilation window, i.e., $\delta\mathbf{x}_n=\delta\mathbf{x}_0$ for all $t_n$. This implies that the tangent linear propagator is assumed to be the identity operator, i.e., $\mathbf{M}_{n\leftarrow 1}=\mathbf{I}$, and hence $\mathbf{M}_{1\leftarrow n}^T=\mathbf{I}$. Then, the cost function for the FGAT is given by

$$J_{\mathrm{FGAT}}\left(\delta\mathbf{x}_0\right)=\frac{1}{2}\left(\delta\mathbf{x}_0\right)^T\mathbf{B}^{-1}\left(\delta\mathbf{x}_0\right)+\frac{1}{2}\sum_{n=-N/2}^{N/2}\left(\mathbf{d}_n^o-\mathbf{H}_n\delta\mathbf{x}_0\right)^T\mathbf{R}_n^{-1}\left(\mathbf{d}_n^o-\mathbf{H}_n\delta\mathbf{x}_0\right), \tag{7.104}$$

where the increment is defined at the center of an observation window rather than at the beginning (Lawless, 2010).

In 3D-FGAT, the innovations are exact as the observations over the assimilation window are compared with their model counterparts at an *appropriate* time during the time

**Algorithm 7.3** Computational procedure of incremental 4DVAR

```
/* k: iteration number (superscript); K: maximum iteration; ε: tolerance */
/* n: time step (subscript); N: maximum time step in assimilation window */
/* x_0^0: initial state; x^b: background; x^a: analysis; y_n^o: observation */
/* M_n: direct model; H_n: observation operator */
/* M_n^T: adjoint of M_n; H_n^T: adjoint of H_n */
/* B: background error covariance; R_n: observation error covariance */
/* x̂: minimizer of J(x_0); J_min = J(x̂): minimum value of J(x_0) */
/* d^k: search direction; α^k: step length (minimization algorithm) */
1  Input: x_0^0 = x^b, y_0^o, B, R_0, k = 0, ε         ! Initial state x_0^0 = x^b at k = 0

   /* OUTER LOOP: START */
2  repeat                                              ! Loop for termination condition
3  |  Nonlinear propagation:                           ! NLM integration
4  |      x_n = M_{n←1}(x_0) using (7.93) via NLM M_n
5  |  Innovation vector:                               ! Compute innovations
6  |      d_n^o = y_n^o − H_n x_n via nonlinear operator H_n

   |  /* INNER LOOP: START */
7  |  while (k ≤ K or ‖∇J^{k+1}‖ ≥ ε) do              ! Loop for termination condition
8  |  |  Cost function:                                ! Compute the cost function J^k = J(δx_0^k)
9  |  |      Compute J^k using (7.99) via M_n and H_n
10 |  |  Gradient of cost function:                    ! Compute the gradient (∇J^k = ∇J(δx_0^k))
11 |  |      Compute ∇J^k using (7.100) via M_n^T and H_n^T (reverse mode)
12 |  |  Update (minimization):                        ! Use any descent algorithm (see Chapter 6)
13 |  |      δx_0^{k+1} = δx_0^k + α^k d^k             ! Update δx_0 at k + 1
14 |  |  Next iteration:                               ! Move to the next iteration k + 1
15 |  |      Set k = k + 1
16 |  endwhile
   |  /* INNER LOOP: END */

17 |  Analysis increment:                              ! Update the analysis increment
18 |      δx_0^a ← δx_0^{k+1}
19 |  Reference state:                                 ! Update the reference state
20 |      x_0^{k+1} = x_0^{k+1} + δx_0^a
21 until optimal solution is obtained
   /* OUTER LOOP: END */

22 Output: x̂ ← x_0^{k+1}, J_min ← J^{k+1}, x_0^a ← x̂   ! Assign x_0^{k+1} to x_0^a, and J^{k+1} to J_min
```

integration (i.e., four-dimensional), whereas no explicit dynamics via model integration is included in the cost function minimization (i.e., three-dimensional). This approach has shown improvement over 3DVAR where all observations in a time window are compared at the same time with the first guess. Furthermore, as it does not require the ADJM, it can be used in the development stage of an operational 4DVAR.

## References


Achtemeier GL (1975) On the initialization problem: A variational adjustment method. *Mon Wea Rev* 103:1089–1103.

AMS (2021) Glossary of Meteorology: Variational objective analysis. American Meteorological Society, https://glossary.ametsoc.org/wiki/Variational_objective_analysis

Andrews P (1998) Details of the mesoscale version of the UKMO's variational analysis scheme. In *Workshop Report, HIRLAM 4 Workshop on Variational Analysis in Limited Area Models*, HIRLAM 4 Project, Met Éireann, 23–25 February 1998, Toulouse, 25–31.

Arcucci R, D'Amore L, Toumi R (2017) Preconditioning of the background error covariance matrix in data assimilation for the Caspian Sea. In *AIP Conference Proceedings*, (vol. 1836), AIP Publishing LLC, 020002, doi:10.1063/1.4981942

Axelsson O, Barker VA (1984) *Finite Element Solution of Boundary Value Problems. Theory and Computation*. Academic Press, Orlando, FL, 432 pp.

Baker DF, Doney SC, Schimel DS (2006) Variational data assimilation for atmospheric $CO_2$. *Tellus B* 58:359–365.

Bannister RN (2008a) A review of forecast error covariance statistics in atmospheric variational data assimilation. I: Characteristics and measurements of forecast error covariances. *Quart J Roy Meteor Soc* 134:1951–1970.

Bannister RN (2008b) A review of forecast error covariance statistics in atmospheric variational data assimilation. II: Modelling the forecast error covariance statistics. *Quart J Roy Meteor Soc* 134:1971–1996.

Barker DM, Huang W, Guo Y-R, Bourgeois AJ, Xiao QN (2004) A three-dimensional variational data assimilation system for MM5: Implementation and initial results. *Mon Wea Rev* 132:897–914.

Bathmann K (2018) Justification for estimating observation-error covariances with the Desroziers diagnostic. *Quart J Roy Meteor Soc* 144:1965–1974.

Bellman R (1957) *Dynamic Programming*. Princeton University Press, Princeton, NJ, 342 pp.

Bello Pereira M, Berre L (2006) The use of an ensemble approach to study the background error covariances in a global NWP model. *Mon Wea Rev* 134:2466–2489.

Bertino L, Evensen G, Wackernagel H (2003) Sequential data assimilation techniques in oceanography. *Int Stat Rev* 71:223–241.

Bonavita M, Torrisi L (2005) Impact of a variational objective analysis scheme on a regional area numerical model: The Italian Air Force Weather Service experience. *Meteor Atmos Phys* 88:39–52.

Bormann N, Bonavita M, Dragani R, et al. (2016) Enhancing the impact of iasi observations through an updated observation-error covariance matrix. *Quart J Roy Meteor Soc* 142:1767–1780.

Bouttier F, Courtier P (2002) *Data Assimilation Concepts and Methods*. Meteorological Training Course Lecture Series, ECMWF, Reading, 59 pp., www.ecmwf.int/node/16928

Bowler NE (2017) On the diagnosis of model error statistics using weak-constraint data assimilation. *Quart J Roy Meteor Soc* 143:1916–1928.

Bruneau CH, Fabrie P, Veersé F (1997) Optimal control data assimilation with an atmospheric model. *Numer Funct Anal Optim* 18:691–722.

Budd CJ, Freitag MA, Nichols NK (2011) Regularization techniques for ill-posed inverse problems in data assimilation. *Comput Fluids* 46:168–173.

Buehner M, Gauthier P, Liu Z (2005) Evaluation of new estimates of background-and observation-error covariances for variational assimilation. *Quart J Roy Meteor Soc* 131:3373–3383.

Cardinali C (2013) Observation influence diagnostic of a data assimilation system. In *Data Assimilation for Atmospheric, Oceanic and Hydrologic Applications* (vol. II), (eds.) Park SK, Xu L, Springer-Verlag, Berlin, Heidelberg, 89–110.

Cheng S, Argaud J-P, Iooss B, Lucor D, Ponçot A (2021) Error covariance tuning in variational data assimilation: Application to an operating hydrological model. *Stochastic Environ Res Risk Assess* 35:1019–1038.

Cohn SE (1997) An introduction to estimation theory. *J Meteor Soc Japan* 75:257–288.

Cohn SE, Da Silva A, Guo J, Sienkiewicz M, Lamich D (1998) Assessing the effects of data selection with the DAO physical-space statistical analysis system. *Mon Wea Rev* 126:2913–2926.

Courtier P (1997) Variational methods. *J Meteor Soc Japan* 75:211–218.

Courtier P, Talagrand O (1987) Variational assimilation of meteorological observations with the adjoint vorticity equation. II: Numerical results. *Quart J Roy Meteor Soc* 113:1329–1347, doi:10.1002/qj.49711347813

Courtier P, Thépaut JN, Hollingsworth A (1994) A strategy for operational implementation of 4D-Var, using an incremental approach. *Quart J Roy Meteor Soc* 120:1367–1387.

Courtier P, Andersson E, Heckley W, et al. (1998) The ECMWF implementation of three-dimensional variational assimilation (3D-Var). I: Formulation. *Quart J Roy Meteor Soc* 124:1783–1807.

Cummings JA, Smedstad OM (2013) Variational data assimilation for the global ocean. In *Data Assimilation for Atmospheric, Oceanic and Hydrologic Applications* (vol. II), (eds.) Park SK, Xu L, Springer-Verlag, Berlin, Heidelberg, 303–343.

Daley R (1992) Estimating model-error covariances for application to atmospheric data assimilation. *Mon Wea Rev* 120:1735–1746.

Danforth CM, Kalnay E, Miyoshi T (2007) Estimating and correcting global weather model error. *Mon Wea Rev* 135:281–299.

Dee DP (1995) On-line estimation of error covariance parameters for atmospheric data assimilation. *Mon Wea Rev* 123:1128–1145.

Derber JC, Parrish DF, Lord SJ (1991) The new global operational analysis system at the National Meteorological Center. *Wea Forecasting* 6:538–547.

Desroziers G, Berre L, Chapnik B, Poli P (2005) Diagnosis of observation, background and analysis-error statistics in observation space. *Quart J Roy Meteor Soc* 131:3385–3396.

Devenon J-L (1990) Optimal control theory applied to an objective analysis of a tidal current mapping by HF radar. *J Atmos Ocean Technol* 7:269–284.

Ebtehaj AM, Županski M, Lerman G, Foufoula-Georgiou E (2014) Variational data assimilation via sparse regularisation. *Tellus A* 66:21789, doi:10.3402/tellusa.v66.21789

El Aabaribaoune M, Emili E, Guidard V (2021) Estimation of the error covariance matrix for IASI radiances and its impact on the assimilation of ozone in a chemistry transport model. *Atmos Meas Tech* 14:2841–2856.

Elbern H, Schmidt H, Ebel A (1997) Variational data assimilation for tropospheric chemistry modeling. *J Geophys Res Atmos* 102:15967–15985.

Elbern H, Schmidt H, Talagrand O, Ebel A (2000) 4D-variational data assimilation with an adjoint air quality model for emission analysis. *Environ Modell Softw* 15:539–548.

Evensen G (1994) Sequential data assimilation with a nonlinear quasi-geostrophic model using Monte Carlo methods to forecast error statistics. *J Geophys Res Oceans* 99:10143–10162.

Fisher M (2003) Background error covariance modelling. In *Seminar on Recent Developments in Data assimilation for Aatmosphere and Ocean, 8–12 September 2003*, ECMWF, Shinfield Park, Reading, 45–64, www.ecmwf.int/node/9404

Fowler AM, Dance SL, Waller JA (2018) On the interaction of observation and prior error correlations in data assimilation. *Quart J Roy Meteor Soc* 144:48–62.

Gao J (2017) A three-dimensional variational radar data assimilation scheme developed for convective scale NWP. In *Data Assimilation for Atmospheric, Oceanic and Hydrologic Applications* (vol. III), (eds.) Park SK, Xu L, Springer International Publishing, Cham, 285–326.

Gauthier P, Thépaut J-N (2001) Impact of the digital filter as a weak constraint in the preoperational 4DVAR assimilation system of Météo-France. *Mon Wea Rev* 129:2089–2102.

Gauthier P, Charette C, Fillion L, Koclas P, Laroche S (1999) Implementation of a 3D variational data assimilation system at the Canadian Meteorological Centre. Part I: The global analysis. *Atmos Ocean* 37:103–156.

Gauthier P, Tanguay M, Laroche S, Pellerin S, Morneau J (2007) Extension of 3DVAR to 4DVAR: Implementation of 4DVAR at the Meteorological Service of Canada. *Mon Wea Rev* 135:2339–2354.

Gelb A (ed.) (1974) *Applied Optimal Estimation*. MIT Press, Cambridge, MA, and London, 374 pp.

Gelfand IM, Fomin SV (1963) *Calculus of Variations*. Prentice-Hall, Inc., Englewood Cliffs, NJ, 232 pp.

Gustafsson N (1998) Status of the HIRLAM 3D-VAR. In *Workshop Report, HIRLAM 4 Workshop on Variational Analysis in Limited Area Models*, HIRLAM 4 Project, Met Éireann, Ireland, 23–25 February 1998, Toulouse, 25–31.

Gustafsson N, Berre L, Hörnquist S, et al. (2001) Three-dimensional variational data assimilation for a limited area model: Part I: General formulation and the background error constraint. *Tellus A* 53:425–446.

Gustafsson N, Huang X-Y, Yang X, et al. (2012) Four-dimensional variational data assimilation for a limited area model. *Tellus A* 64:14985, doi:10.3402/tellusa.v64i0.14985

Heng BCP, Tubbs R, Huang XY, et al. (2020) SINGV-DA: A data assimilation system for convective-scale numerical weather prediction over Singapore. *Quart J Roy Meteor Soc* 146:1923–1938.

Hollingsworth A, Lönnberg P (1986) The statistical structure of short-range forecast errors as determined from radiosonde data. Part I: The wind field. *Tellus A* 38:111–136.

Houtekamer PL, Mitchell HL, Deng X (2009) Model error representation in an operational ensemble Kalman filter. *Mon Wea Rev* 137:2126–2143.

Houtekamer PL, Lefaivre L, Derome J, Ritchie H, Mitchell HL (1996) A system simulation approach to ensemble prediction. *Mon Wea Rev* 124:1225–1242.

Ikawa M (1984) An alternative method of solving weak constraint problem and a unified expression of weak and strong constraints in variational objective analysis. *Papers Meteor Geophys* 35:71–79.

Janisková M, Lopez P (2013) Linearized physics for data assimilation at ECMWF. In *Data Assimilation for Atmospheric, Oceanic and Hydrologic Applications* (vol. II), (eds.) Park SK, Xu L, Springer-Verlag, Berlin, Heidelberg, 251–286.

Janisková M, Thépaut J-N, Geleyn J-F (1999) Simplified and regular physical parameterizations for incremental four-dimensional variational assimilation. *Mon Wea Rev* 127:26–45.

Kadowaki T (2005) A 4-dimensional variational assimilation system for the JMA Global Spectrum Model. In *Research Activities in Atmosphric and Oceanic Modelling*, WMO/TD-No. 1276, WMO, 1.17–1.18.

Kalnay E (2003) *Atmospheric Modeling, Data Assimilation and Predictability*. Cambridge University Press, New York, 341 pp.

Kamien MI, Schwartz NL (1991) *Dynamic Optimization: The Calculus of Variations and Optimal Control in Economics and Management*. 2nd ed., Elsevier Science B.V., Amsterdam, 377 pp.

Kano M, Miyazaki S, Ishikawa Y, Hirahara K (2020) Adjoint-based direct data assimilation of GNSS time series for optimizing frictional parameters and predicting postseismic deformation following the 2003 Tokachi-oki earthquake. *Earth Planets Space* 72:1–24.

Koizumi K, Ishikawa Y, Tsuyuki T (2005) Assimilation of precipitation data to the JMA mesoscale model with a four-dimensional variational method and its impact on precipitation forecasts. *SOLA* 1:45–48.

Kopp RE (1962) Pontryagin maximum principle. In *Optimization Techniques with Applications to Aerospace Systems*, (ed.) Leitmann G, Academic Press, New York, London, 255–279.

Lakshmivarahan S, Lewis JM, Phan D (2013) Data assimilation as a problem in optimal tracking: Application of Pontryagin's minimum principle to atmospheric science. *J Atmos Sci* 70:1257–1277.

Laloyaux P, Bonavita M, Chrust M, Gürol S (2020) Exploring the potential and limitations of weak-constraint 4D-Var. *Quart J Roy Meteor Soc* 146:4067–4082.

Lanczos C (1970) *The Variational Principles of Mechanics*. University of Toronto Press, Toronto, 464 pp.

Lang M, Owens MJ (2019) A variational approach to data assimilation in the solar wind. *Space Weather* 17:59–83.

Laroche S, Gauthier P (1998) A validation of the incremental formulation of 4D variational data assimilation in a nonlinear barotropic flow. *Tellus A* 50:557–572.

Laroche S, Gauthier P, St-James J, Morneau J (1999) Implementation of a 3D variational data assimilation system at the Canadian Meteorological Centre. Part II: The regional analysis. *Atmos Ocean* 37:281–307.

Lawless AS (2010) A note on the analysis error associated with 3D-FGAT. *Quart J Roy Meteor Soc* 136:1094–1098.

Lawless AS, Gratton S, Nichols NK (2005) An investigation of incremental 4D-Var using non-tangent linear models. *Quart J Roy Meteor Soc* 131:459–476.

Le Dimet F-X, Talagrand O (1986) Variational algorithms for analysis and assimilation of meteorological observations: Theoretical aspects. *Tellus A* 38:97–110.

Le Dimet F-X, Navon IM, Ştefănescu R (2017) Variational data assimilation: Optimization and optimal control. In *Data Assimilation for Atmospheric, Oceanic and Hydrologic Applications* (vol. III), (eds.) Park SK, Xu L, Springer International Publishing, Cham, 1–53.

Le Dimet F-X, Castaings W, Ngnepieba P, Vieux B (2009) Data assimilation in hydrology: Variational approach. In *Data Assimilation for Atmospheric, Oceanic and Hydrologic Applications*, (eds.) Park SK, Xu L, Springer-Verlag, Berlin, Heidelberg, 367–405.

Lebedev LP, Cloud MJ (2003) *The Calculus of Variations and Functional Analysis*. World Scientific, Singapore, 436 pp.

Lee M-S, Lee D-K (2003) An application of a weakly constrained 4DVAR to satellite data assimilation and heavy rainfall simulation. *Mon Wea Rev* 131:2151–2176.

Lee S-W, Lee D-K (2011) Improvement in background error covariances using ensemble forecasts for assimilation of high-resolution satellite data. *Adv Atmos Sci* 28:758–774.

Lewis JM (1971) Variational subsynoptic analysis with applications to severe local storms. *Mon Wea Rev* 99:786–795.

Lewis JM (1972) An operational upper air analysis using the variational method. *Tellus* 24:514–530.

Lewis JM (2009) Sasaki's pathway to deterministic data assimilation. In *Data Assimilation for Atmospheric, Oceanic and Hydrologic Applications*, (eds.) Park SK, Xu L, Springer-Verlag, Berlin, Heidelberg, 1–9.

Lewis JM, Derber JC (1985) The use of adjoint equations to solve a variational adjustment problem with advective constraints. *Tellus A* 37:309–322.

Lewis JM, Grayson TH (1972) The adjustment of surface wind and pressure by Sasaki's variational matching technique. *J Appl Meteor Climatol* 11:586–597.

Lewis JM, Lakshmivarahan S (2008) Sasaki's pivotal contribution: Calculus of variations applied to weather map analysis. *Mon Wea Rev* 136:3553–3567.

Lewis JM, Lakshmivarahan S, Dhall S (2006) *Dynamic Data Assimilation: A Least Squares Approach*. Cambridge University Press, Cambridge, 654 pp.

Li K, Jackson A, Livermore PW (2014) Variational data assimilation for a forced, inertia-free magnetohydrodynamic dynamo model. *Geophys J Int* 199:1662–1676.

Li Y (1991) A note on the uniqueness problem of variational adjustment approach to four-dimensional data assimilation. *J Meteor Soc Japan* 69:581–585.

Lindskog M, Gustafsson N, Navascues B, et al. (2001) Three-dimensional variational data assimilation for a limited area model: Part II: Observation handling and assimilation experiments. *Tellus A* 53:447–468.

Lions JL (1971) *Optimal Control of Systems Governed by Partial Differential Equations*. Springer-Verlag, Berlin, Heidelberg, New York, 400 pp., Translated by Mitter, SK, from the French edition *Contrôle Optimal de Systèmes Gouvernés par des Équations aux Dérivées Partielles*, in the series "Etudes Mathematiques" edited by Lelong P, published by Dunod and Gauther-Villas, Paris, 1968.

Lorenc AC (1997) Development of an operational variational assimilation scheme. *J Meteor Soc Japan* 75:339–346.

Lorenc AC, Ballard SP, Bell RS, et al. (2000) The Met. Office global three-dimensional variational data assimilation scheme. *Quart J Roy Meteor Soc* 126:2991–3012.

Mattern JP, Edwards CA, Moore AM (2018) Improving variational data assimilation through background and observation error adjustments. *Mon Wea Rev* 146:485–501.

McGinley JA (1984) Scaling and theoretical considerations in variational analysis of flow around mountains. *Beitr Phys Atmos* 57:527–535.

McGinley JA (1987) A variational objective analysis scheme for analysis of the ALPEX data set. *Meteor Atmos Phys* 36:5–23.

Mirza AK, Dance SL, Rooney GG, et al. (2021) Comparing diagnosed observation uncertainties with independent estimates: A case study using aircraft-based observations and a convection-permitting data assimilation system. *Atmos Sci Lett* 22:el01029, doi: 10.1002/asl.1029

Moore AM, Arango HG, Broquet G, et al. (2011) The Regional Ocean Modeling System (ROMS) 4-dimensional variational data assimilation systems: Part I–System overview and formulation. *Prog Oceanogr* 91:34–49.

Narkhedkar SG, Sinha SK (2008) Variational method for objective analysis of scalar variable and its derivative. *J Earth Syst Sci* 117:621–635.

Navon IM, Zou X, Derber J, Sela J (1992) Variational data assimilation with an adiabatic version of the NMC spectral model. *Mon Wea Rev* 120:1433–1446.

Ngodock H, Carrier M (2014) A 4DVAR system for the Navy Coastal Ocean Model. Part I: System description and assimilation of synthetic observations in Monterey Bay. *Mon Wea Rev* 142:2085–2107.

Onori S, Serrao L, Rizzoni G (2016) Pontryagin's minimum principle. In *Hybrid Electric Vehicles: Energy Management Strategies*, Springer, London, 51–63.

Park SK, Županski D (2003) Four-dimensional variational data assimilation for mesoscale and storm-scale applications. *Meteor Atmos Phys* 82:173–208.

Park S-Y, Kim D-H, Lee S-H, Lee HW (2016) Variational data assimilation for the optimized ozone initial state and the short-time forecasting. *Atmos Chem Phys* 16:3631–3649.

Parrish DF, Derber JC (1992) The National Meteorological Center's spectral statistical-interpolation analysis system. *Mon Wea Rev* 120:1747–1763.

Piazzi G, Thirel G, Perrin C, Delaigue O (2021) Sequential data assimilation for streamflow forecasting: Assessing the sensitivity to uncertainties and updated variables of a conceptual hydrological model at basin scale. *Water Resour Res* 57:e2020WR028390, doi:https://doi.org/10.1029/2020WR028390

Pontryagin LS, Boltyanskii VG, Gamkrelidze RV, Mischenko EF (1961) *Matematicheskaya Teoriya Optimal'nykh Prozessov*. Fizmatgiz, Moscow, English translation *The Mathematical Theory of Optimal Control Processes* by Trirogoff KN, published in 1962 by Interscience Publishers, John Wiley & Sons, Inc., New York, London, Sydney, 360 pp.

Rabier F, Järvinen H, Klinker E, Mahfouf JF, Simmons A (2000) The ECMWF operational implementation of four-dimensional variational assimilation. I: Experimental results with simplified physics. *Quart J Roy Meteor Soc* 126:1143–1170.

Rabier F, McNally A, Andersson E, et al. (1998) The ECMWF implementation of three-dimensional variational assimilation (3D-Var). II: Structure functions. *Quart J Roy Meteor Soc* 124:1809–1829.

Rawlins F, Ballard SP, Bovis KJ, et al. (2007) The Met Office global four-dimensional variational data assimilation scheme. *Quart J Roy Meteor Soc* 133:347–362.

Reichle RH, McLaughlin DB, Entekhabi D (2001) Variational data assimilation of microwave radiobrightness observations for land surface hydrology applications. *IEEE Geosci Remote Sens* 39:1708–1718.

Sagan H (1969) *Introduction to the Calculus of Variations*. McGraw-Hill, New York, 449 pp.

Sargent RWH (2000) Optimal control. *J Comput Appl Math* 124:361–371.

Sasaki Y (1955) A fundamental study of the numerical prediction based on the variational principle. *J Meteor Soc Japan* 33:262–275.

Sasaki Y (1958) An objective analysis based on the variational method. *J Meteor Soc Japan* 36:77–88.

Sasaki Y (1969) Proposed inclusion of time variation terms, observational and theoretical, in numerical variational objective analysis. *J Meteor Soc Japan* 47:115–124.

Sasaki Y (1970a) Some basic formalisms in numerical variational analysis. *Mon Wea Rev* 98:875–883.

Sasaki Y (1970b) Numerical variational analysis formulated under the constraints as determined by longwave equations and a low-pass filter. *Mon Wea Rev* 98:884–898.

Sasaki Y (1970c) Numerical variational analysis with weak constraint and application to surface analysis of severe storm gust. *Mon Wea Rev* 98:899–910.

Sasaki Y (1971a) Low-pass and band-pass filters in numerical variational optimization. *J Meteor Soc Japan* 49:766–773.

Sasaki Y (1971b) A theoretical interpretation of anisotropically weighted smoothing on the basis of numerical variational analysis. *Mon Wea Rev* 99:698–707.

Sasaki Y (1986) *Variational Methods in Geosciences*. Elsevier Science Publishers B.V., Amsterdam, 309 pp.

Sasaki Y, Lewis JM (1970) Numerical variational objective analysis of the planetary boundary layer in conjunction with squall line formation. *J Meteor Soc Japan* 48:381–399.

Schlatter TW (2000) Variational assimilation of meteorological observations in the lower atmosphere: A tutorial on how it works. *J Atmos Sol Terr Phys* 62:1057–1070.

Shao D, Wang Z, Wang B, Luo W (2016) A water quality model with three dimensional variational data assimilation for contaminant transport. *Water Resour Manage* 30:4501–4512.

Smith S, Ngodock H, Carrier M, et al. (2017) Validation and operational implementation of the Navy Coastal Ocean Model four dimensional variational data assimilation system (NCOM 4DVAR) in the Okinawa Trough. In *Data Assimilation for Atmospheric, Oceanic and Hydrologic Applications* (vol. III), (eds.) Park SK, Xu L, Springer International Publishing, Cham, 405–427.

Stephens JJ (1970) Variational initialization with the balance equation. *J Appl Meteor Climatol* 9:732–739.

Stewart LM, Dance SL, Nichols NK (2013) Data assimilation with correlated observation errors: Experiments with a 1-D shallow water model. *Tellus A* 65:19546, doi:10.3402/tellusa.v65i0.19546

Stewart LM, Dance SL, Nichols NK, Eyre JR, Cameron J (2014) Estimating interchannel observation-error correlations for IASI radiance data in the Met Office system. *Quart J Roy Meteor Soc* 140:1236–1244.

Sussmann HJ, Willems JC (1997) 300 years of optimal control: From the brachystochrone to the maximum principle. *IEEE Control Syst Mag* 17:32–44.

Tabeart JM, Dance SL, Lawless AS, et al. (2020) The impact of using reconditioned correlated observation-error covariance matrices in the Met Office 1D-Var system. *Quart J Roy Meteor Soc* 146:1372–1390.

Takeuchi Y, Tsuyuki T (2002) The operational 3D-Var assimilation system of JMA for the global spectral model and the typhoon model. In *Research Activities in Atmosphric and Oceanic Modelling*, WMO/TD-No. 1105, WMO, 1.59–1.60.

Talagrand O, Courtier P (1987) Variational assimilation of meteorological observations with the adjoint vorticity equation. I: Theory. *Quart J Roy Meteor Soc* 113:1311–1328.

Temperton C (1984) Variational normal mode initialization for a multilevel model. *Mon Wea Rev* 112:2303–2316.

Thacker WC, Long RB (1988) Fitting dynamics to data. *J Geophys Res* 93:1227–1240.

Thépaut J-N, Alary P, Caille P, et al. (1998) The operational global data assimilation system at Météo-France. In *Workshop Report, HIRLAM 4 Workshop on Variational Analysis in Limited Area Models*, HIRLAM 4 Project, Met Éireann, 23–25 February 1998, Toulouse, 25–31.

Tikhonov AN (1963a) Solution of incorrectly formulated problems and the regularization method. *Soviet Math Dokl* 4:1035–1038, Translated from the Russian version published in *Dokl Akad Nauk SSSR* 151:501–504.

Tikhonov AN (1963b) Regularization of incorrectly posed problem. *Soviet Math Dokl* 4:1624–1627, Translated from the Russian version published in *Dokl Akad Nauk SSSR* 153:49–52.

Todling R (2015) A lag-1 smoother approach to system-error estimation: Sequential method. *Quart J Roy Meteor Soc* 141:1502–1513.

Trémolet Y (2006) Accounting for an imperfect model in 4D-Var. *Quart J Roy Meteor Soc* 132:2483–2504.

Trémolet Y (2007) Model-error estimation in 4D-Var. *Quart J Roy Meteor Soc* 133:1267–1280.

Verlinde J, Cotton WR (1993) Fitting microphysical observations of nonsteady convective clouds to a numerical model: An application of the adjoint technique of data assimilation to a kinematic model. *Mon Wea Rev* 121:2776–2793.

Vidard PA, Piacentini A, Le Dimet F-X (2004) Variational data analysis with control of the forecast bias. *Tellus A* 56:177–188.

Wahba G, Wendelberger J (1980) Some new mathematical methods for variational objective analysis using splines and cross validation. *Mon Wea Rev* 108:1122–1143.

Waliser DE, Ridout JA, Xie S, Zhang M (2002) Variational objective analysis for atmospheric field programs: A model assessment. *J Atmos Sci* 59:3436–3456.

Waller JA, Simonin D, Dance SL, Nichols NK, Ballard SP (2016) Diagnosing observation error correlations for Doppler radar radial winds in the Met Office UKV model using observation-minus-background and observation-minus-analysis statistics. *Mon Weather Rev* 144:3533–3551.

Wang T, Fei J, Cheng X, Huang X, Zhong J (2018) Estimating the correlated observation-error characteristics of the Chinese FengYun Microwave Temperature Sounder and Microwave Humidity Sounder. *Adv Atmos Sci* 35:1428–1441.

Weaver AT, Vialard J, Anderson DLT (2003) Three-and four-dimensional variational assimilation with a general circulation model of the tropical Pacific Ocean. Part I: Formulation, internal diagnostics, and consistency checks. *Mon Wea Rev* 131:1360–1378.

Weston PP, Bell W, Eyre JR (2014) Accounting for correlated error in the assimilation of high-resolution sounder data. *Quart J Roy Meteor Soc* 140:2420–2429.

Wilkins EM (1971) Variational principle applied to numerical objective analysis of urban air pollution distributions. *J Appl Meteor Climatol* 10:974–981.

Yin J, Han W, Gao Z, Chen H (2021) Assimilation of Doppler radar radial wind data in the GRAPES mesoscale model with observation error covariances tuning. *Quart J Roy Meteor Soc* 147:2087–2102.

Županski D (1993a) The effects of discontinuities in the Betts-Miller cumulus convection scheme on four-dimensional variational data assimilation. *Tellus A* 45:511–524.

Županski D (1997) A general weak constraint applicable to operational 4DVAR data assimilation systems. *Mon Wea Rev* 125:2274–2292.

Županski D, Mesinger F (1995) Four-dimensional variational assimilation of precipitation data. *Mon Wea Rev* 123:1112–1127.

Županski D, Županski M (2006) Model error estimation employing an ensemble data assimilation approach. *Mon Wea Rev* 134:1337–1354.

Županski M (1993b) Regional four-dimensional variational data assimilation in a quasi-operational forecasting environment. *Mon Wea Rev* 121:2396–2408.

Županski M (1993c) A preconditioning algorithm for large-scale minimization problems. *Tellus A* 45:478–492.

Županski M (1996) A preconditioning algorithm for four-dimensional variational data assimilation. *Mon Wea Rev* 124:2562–2573.

Županski M (2013) All-sky satellite radiance data assimilation: Methodology and challenges. In *Data Assimilation for Atmospheric, Oceanic and Hydrologic Applications* (vol. II), (eds.) Park SK, Xu L, Springer-Verlag, Berlin, Heidelberg, 465–488.

# 8
# Ensemble and Hybrid Data Assimilation

## 8.1 Introduction

In this chapter we describe additional improvement of data assimilation to include flow-dependent forecast error covariance. Most processes in geosciences are time dependent and nonlinear, so it is only natural to desire the same characteristics for the uncertainty. In addition, one would like to use the uncertainty from the past in the present data assimilation cycle, as dictated by Bayes' theorem, eventually producing a system that can improve by learning from past performances.

Unfortunately, variational data assimilation (VAR) does not have the capability to address these tasks. As seen in Chapter 7, a common assumption of VAR is that background (i.e., forecast) error covariance is modeled using covariance functions (e.g., Gaspari and Cohn, 1999) Since this modeled covariance is often kept the same in all data assimilation cycles, it is also referred to as *static* or time-independent error covariance. Such static error covariance is used in the definition of the cost function of VAR methods. One should be aware, however, that in four-dimensional variational data assimilation (4DVAR) the static error covariance begins to evolve during the data assimilation window, effectively introducing some flow dependence in the process. Regarding Bayesian inference in VAR, uncertainty from the previous data assimilation cycle is not used in defining uncertainty in the current data assimilation cycle. Every time a new data assimilation cycle begins, the uncertainty is redefined starting from a prescribed covariance function, as this is the very first cycle. Therefore, in order to include flow-dependent error covariance and update that covariance throughout data assimilation cycles, an improvement in the data assimilation methodology is required. One such methodology is ensemble data assimilation (ENS).

As will be described in Section 8.4, neither VAR nor ENS methods could fulfill all desirable characteristics of data assimilation, especially in practical high-dimensional applications. This motivated the development of hybrid data assimilation (HYB) methods. In principle, HYB methods refer to the combined use of basic data assimilation methods. In this chapter, however, we will focus on hybrid variational-ensemble methods, as they are the most developed of all hybrid methods and used in operational weather centers.

Before proceeding to a detailed description of ENS and HYB methods, we first discuss the role of forecast error covariance in data assimilation, as it is one of the most relevant sources of difference between various data assimilation methods.

## 8.2 Role of Forecast Error Covariance

### *8.2.1 Algebraic View*

Forecast error covariance is typically used as a measure of uncertainty of the forecast, and could be defined as

$$\mathbf{P}_f = \left\langle \left[\mathbf{x}^f - \mathbf{x}^t\right]\left[\mathbf{x}^f - \mathbf{x}^t\right]^T \right\rangle, \tag{8.1}$$

where $\mathbf{x}^f$ and $\mathbf{x}^t$ are the first-guess forecast and the (unknown) truth, respectively, $\langle\cdot\rangle$ denotes mathematical expectation, and the superscript $T$ denotes a transpose.

Forecast error covariance has a fundamental role in data assimilation as it defines the subspace where the analysis correction can be defined.

To see that, let the singular value decomposition (SVD) of a square root forecast error covariance (e.g., Golub and Van Loan, 2013) be

$$\mathbf{P}_f^{1/2} = \sum_i \sigma_i \mathbf{q}_i \mathbf{v}_i^T, \tag{8.2}$$

where $\{\mathbf{q}_i\}$ and $\{\mathbf{v}_i\}$ are singular vectors and $\{\sigma_i\}$ are singular values. Kalman filter (KF) and related ensemble Kalman filtering methods solve the analysis equation

$$\mathbf{x}^a = \mathbf{x}^f + \mathbf{P}_f\mathbf{H}^T\left[\mathbf{H}\mathbf{P}_f\mathbf{H}^T + \mathbf{R}\right]^{-1}\left[\mathbf{y}^o - \mathbf{H}\mathbf{x}^f\right], \tag{8.3}$$

where the superscripts $a$ and $f$ denote analysis and forecast, respectively, $\mathbf{R}$ is the observation error covariance, $H$ is a nonlinear observation operator, and $\mathbf{H}$ is its Jacobian. When observation and/or prediction model operators are nonlinear, the iterated data assimilation methods are used to search for the analysis as a limit of an iterative sequence

$$\mathbf{x}^a = \lim_{m\to M_I} \mathbf{x}_m, \tag{8.4}$$

where $m = 1, \ldots, M_I$ is the iteration index and $M_I$ denotes the maximum number of iterations required to satisfy a convergence criteria. A typical starting point is the background vector $\mathbf{x}_0 = \mathbf{x}^f$. Using the above KF analysis equation as an example, in the case of nonlinear operators

$$\mathbf{x}_m = \mathbf{x}^f + \mathbf{P}_f\mathbf{H}^T\left[\mathbf{H}\mathbf{P}_f\mathbf{H}^T + \mathbf{R}\right]^{-1}\left[\mathbf{y}^o - H(\mathbf{x}_{m-1})\right]. \tag{8.5}$$

Given that $\mathbf{x}^f$ and $\mathbf{P}_f$ are not functions of the current iteration, it is possible to combine (8.4) and (8.5) to get the analysis

$$\mathbf{x}^a = \lim_{m\to M_I} \mathbf{x}_m = \mathbf{x}^f + \mathbf{P}_f \lim_{m\to M_I}\left(\mathbf{H}^T\left[\mathbf{H}\mathbf{P}_f\mathbf{H}^T + \mathbf{R}\right]^{-1}\left[\mathbf{y}^o - H(\mathbf{x}_{m-1})\right]\right). \tag{8.6}$$

In variational methods, using $\mathbf{P}_f$ for preconditioning, one has the iterative formula in the form

$$\mathbf{x}_m = \mathbf{x}^f + \mathbf{P}_f\mathbf{H}^T\mathbf{R}^{-1}\left[\mathbf{y}^o - H(\mathbf{x}_{m-1})\right], \tag{8.7}$$

which in the limit produces the analysis

$$\mathbf{x}^a = \lim_{m \to M_I} \mathbf{x}_m = \mathbf{x}^f + \mathbf{P}_f \lim_{m \to M_I} \left( \mathbf{H}^T \mathbf{R}^{-1} \left[ \mathbf{y}^o - H(\mathbf{x}_{m-1}) \right] \right). \tag{8.8}$$

The above reasoning can be extended in a straightforward manner to the so-called *square root* filters that use the square root instead of the full forecast error covariance matrix. For example, the iterative formula (8.5) can be transformed into the square root form

$$\mathbf{x}_m = \mathbf{x}^f + \mathbf{P}_f^{1/2} \left[ \mathbf{I} + \mathbf{Z}^T \mathbf{Z} \right]^{-1} \mathbf{Z}^T \mathbf{R}^{-1/2} \left[ \mathbf{y}^o - H(\mathbf{x}_{m-1}) \right], \tag{8.9}$$

where $\mathbf{Z} = \mathbf{R}^{-1/2} \mathbf{H} \mathbf{P}_f^{1/2}$. The final analysis is

$$\mathbf{x}^a = \lim_{m \to M_I} \mathbf{x}_m = \mathbf{x}^f + \mathbf{P}_f^{1/2} \lim_{m \to M_I} \left( \left[ \mathbf{I} + \mathbf{Z}^T \mathbf{Z} \right]^{-1} \mathbf{Z}^T \mathbf{R}^{-1/2} \left[ \mathbf{y}^o - H(\mathbf{x}_{m-1}) \right] \right). \tag{8.10}$$

Note that using $\mathbf{P}_f = \mathbf{P}_f^{1/2} \mathbf{P}_f^{T/2}$ the analysis equations (8.3), (8.6), (8.8), and (8.10) can be represented by a comprehensive formula

$$\mathbf{x}^a = \mathbf{x}^f + \mathbf{P}_f^{1/2} \mathbf{w}, \tag{8.11}$$

where $\mathbf{w}$ is a transformed vector of observation increments and can be easily derived from the above formulas for each of the methods. Although here we have demonstrated for a limited number of methods only, the formulation (8.11) can be shown to be valid for most data assimilation methods. After substituting (8.2) into (8.11), the analysis increment $\mathbf{x}^a - \mathbf{x}^f$ is

$$\mathbf{x}^a - \mathbf{x}^f = \left( \sum_i \sigma_i \mathbf{q}_i \mathbf{v}_i^T \right) \mathbf{w} = \sum_i \gamma_i \mathbf{q}_i, \tag{8.12}$$

where $\gamma_i = \sigma_i (\mathbf{v}_i^T \mathbf{w})$. Therefore, the analysis increment is defined as a linear combination of left singular vectors of the forecast error covariance. In other words, the space in which the analysis can change the forecast guess is defined by the range of forecast error covariance. Since analysis increments define the impact of assimilated observations, one can say that forecast error covariance defines the space where observations can impact the analysis.

The structure of forecast error covariance in data assimilation is often examined using a single-observation framework to assimilate one observation at a chosen location and perform standard data assimilation analysis calculations. The result will give analysis increments at that point, but also reveal the structure of forecast error covariance. Let us assume there is an observation $y$ at point $k$. The corresponding KF analysis solution (e.g., Thépaut et al., 1996; Suzuki et al., 2017) is

$$\mathbf{x}^a - \mathbf{x}^f = \mathbf{P}_f \begin{pmatrix} 0 & \cdots & \left( \frac{\mathbf{y}^o - \mathbf{x}^f}{\sigma_o^2 + \sigma_f^2} \right)_k & \cdots & 0 \end{pmatrix}^T = \left( \frac{\mathbf{y}^o - \mathbf{x}^f}{\sigma_o^2 + \sigma_f^2} \right)_k \mathbf{p}_k^f, \tag{8.13}$$

where $\sigma_o$and $\sigma_f$ are standard deviations of observation and forecast errors, respectively, and $\mathbf{p}_k^f$ is the $k$-th column of the forecast error covariance.

It is also of interest to relate covariance to correlation, as correlation is often used to examine the structure of covariance. Introducing the correlation coefficient $\rho$, the covariance can be written as

$$\mathbf{P}_f = \begin{pmatrix} [\sigma_f]_1^2 & [\sigma_f]_1[\sigma_f]_2\rho_{12} & \cdots & [\sigma_f]_1[\sigma_f]_{N_S}\rho_{1N_S} \\ [\sigma_f]_2[\sigma_f]_1\rho_{21} & [\sigma_f]_2^2 & & [\sigma_f]_2[\sigma_f]_{N_S}\rho_{2N_S} \\ \vdots & & \ddots & \vdots \\ [\sigma_f]_{N_S}[\sigma_f]_1\rho_{N_S1} & [\sigma_f]_{N_S}[\sigma_f]_2\rho_{N_S2} & \cdots & [\sigma_f]_{N_S}^2 \end{pmatrix}, \tag{8.14}$$

where $N_S$ is the state dimension. Noting that $\rho_{ij} = \rho_{ji}$, this can be further decomposed into

$$\mathbf{P}_f = \mathbf{\Sigma}\mathbf{C}_f\mathbf{\Sigma}, \tag{8.15}$$

where

$$\mathbf{\Sigma} = \begin{pmatrix} [\sigma_f]_1 & 0 & \cdots & 0 \\ 0 & [\sigma_f]_2 & & 0 \\ \vdots & & \ddots & \vdots \\ 0 & 0 & \cdots & [\sigma_f]_{N_S} \end{pmatrix} \tag{8.16}$$

is the standard deviation matrix, and the correlation matrix is

$$\mathbf{C}_f = \begin{pmatrix} 1 & \rho_{12} & \cdots & \rho_{1N_S} \\ \rho_{12} & 1 & & \rho_{2N_S} \\ \vdots & & \ddots & \vdots \\ \rho_{1N_S} & \rho_{2N_S} & \cdots & 1 \end{pmatrix}. \tag{8.17}$$

Note that the standard deviation matrix includes the physical dimensions of related variables, while the correlation matrix is nondimensional. Although standard deviation is important for representing true uncertainty, its dependence on physical dimensions can sometimes hide the underlying structure, i.e., the dependence between variables and points. For example, even though variables at points 1 and 2 could be highly correlated ($\rho_{12} \approx 1$), and thus dependent, extremely small values of standard deviation at one of these locations ($(\sigma_f)_1 \approx 0$) can make the covariance very small ($(\sigma_f)_1(\sigma_f)_2\rho_{12} \approx 0$). Therefore, the correlation matrix reveals the structure of covariance by emphasizing dependence relationships.

The analysis increment from the single-observation experiment (8.13) can be further adjusted to include correlations, and thus better reveal the covariance structure. Using (8.13), (8.15), and (8.16), the analysis increment of a single-observation experiment can be expressed in terms of standard deviation and correlation coefficients

$$\begin{aligned} \mathbf{x}^a - \mathbf{x}^f &= [\mathbf{y}^o - \mathbf{x}^f]_k \begin{pmatrix} \frac{[\sigma_f]_1[\sigma_f]_k}{[\sigma_o]_k^2+[\sigma_f]_k^2}\rho_{1k} \cdots \frac{[\sigma_f]_k^2}{[\sigma_o]_k^2+[\sigma_f]_k^2} \cdots \frac{[\sigma_f]_{N_S}[\sigma_f]_k}{[\sigma_o]_k^2+[\sigma_f]_k^2}\rho_{N_Sk} \end{pmatrix}^T \\ &= \frac{[\mathbf{y}^o - \mathbf{x}^f]_k}{1+\left[\frac{\sigma_o}{\sigma_f}\right]_k^2} \begin{pmatrix} \frac{[\sigma_f]_1}{[\sigma_f]_k}\rho_{1k} \cdots 1 \cdots \frac{[\sigma_f]_{N_S}}{[\sigma_f]_k}\rho_{N_Sk} \end{pmatrix}^T. \end{aligned} \tag{8.18}$$

The formulation (8.18) may be more revealing than (8.13), as the analysis adjustment at each point can be traced back to correlation and/or to standard deviation. For example, one can consider the multiplicative term in (8.18) to be a constant for given forecast error covariance and observation characteristics. Then the vector term in (8.18) describes the response of the analysis. If forecast error standard deviations at surrounding points are of similar magnitude, then the response will mostly depend on correlation coefficients and eventually reveal the structure. However, if the forecast error standard deviation is much larger in one of the surrounding points than at the observation point, then the correlation structure will be considerably changed, with likely larger responses away from the central observation point.

Another characteristic of analysis increments is related to the smoothness of the forecast error covariance, which is implied from characteristics of the left singular vectors $\{\mathbf{q}_i\}$. From (8.12), the linear combination of smooth singular vectors will also be smooth, while the linear combination of noisy singular vectors will create noisy increments.

All these considerations are important in order to correctly interpret the results of single-observation experiments.

### 8.2.2 Flow Dynamics View

It is also of interest to understand the role of error covariance in terms of the flow dynamics. Let us consider an isotropic initial uncertainty, meaning that uncertainty change is the same in all directions and therefore can be represented by a circle in a 2D plot. We assume that a nonlinear prediction model is used to evolve this uncertainty in time and follow the uncertainty evolution at one grid point of the model. At some later time $t$ we define two observations, one located near ($obs_1$) and one further away ($obs_2$) from the mentioned grid point. In order to keep the problem tractable, we assume that both observations have the same observation increment $[\mathbf{y}^o - H(\mathbf{x})]_1 = [\mathbf{y}^o - H(\mathbf{x})]_2$ and the same observation error $(\sigma_o)_1 = (\sigma_o)_2$, and that they are not correlated.

Let us first assume that data assimilation is using adequate flow-dependent uncertainty at time $t$. This is represented in Figure 8.1 by ellipses. Due to the elongated shape of the flow-dependent uncertainty in one direction, the distant observation will have a larger impact on the analysis increment than the geographically close observation. This appears counterintuitive, but it can be explained using (8.14). Given that the observation characteristics and the standard deviation at the central grid point $[\sigma_f]_k$ are the same for both observations, the difference between analysis increments at points 1 and 2 comes from $[\sigma_f]_1\rho_{1k}$ and $[\sigma_f]_2\rho_{2k}$. Since forecast uncertainty implies that $[\sigma_f]_2\rho_{2k} \gg [\sigma_f]_1\rho_{1k}$, the analysis increment at point 1 is much smaller than at point 2.

Now, let us assume that the flow-dependent forecast error covariance is not known, and instead a static, isotropic error covariance is used. This error covariance is represented by a circle on a 2D plot (Figure 8.2). From such defined static error covariance $[\sigma_f]_2\rho_{2k} \ll [\sigma_f]_1\rho_{1k}$, implying from (8.14) that the analysis increment at point 1 is much larger than at point 2.

Figures 8.1 and 8.2 illustrate the anticipated impact of using static or flow-dependent error covariance. Flow-dependent error is clearly preferable as it corresponds to actual

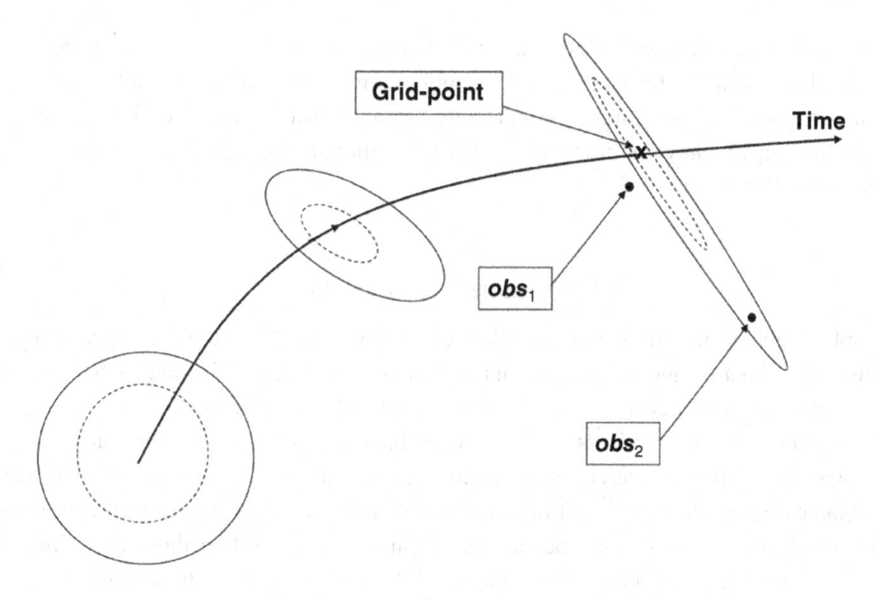


Figure 8.1 Schematic representation of data assimilation with flow-dependent error covariance in a 2D system. The analysis grid point is represented by "x" and the observations by large dots. Circles and ellipses represent lines of equal uncertainty.

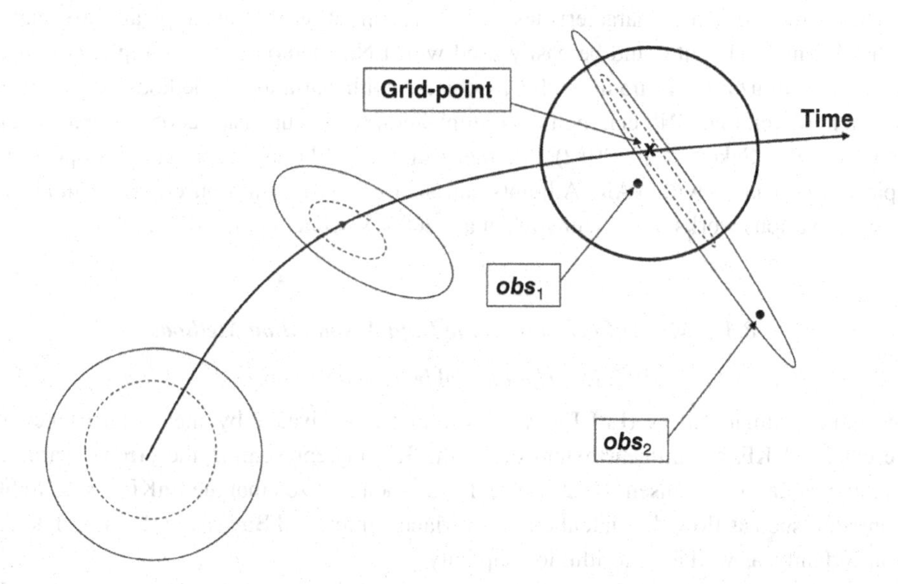


Figure 8.2 Same as in Figure 8.1, except for data assimilation with static error covariance.

dynamics that changes in space and time. However, a common practical problem associated with flow-dependent error covariance is that often it does not have sufficient number of degrees of freedom (DOF), making its estimate noisy and relatively inaccurate. By DOF, we formally define the number of nonzero singular values in (8.2). In commonly used ensemble, hybrid, and related data assimilation methods that incorporate ensemble forecasting, DOF are closely related to the number of ensembles. Since for realistic complex and high-resolution models it is feasible to compute only a small subset of ensembles, it is challenging to use flow-dependent error covariance without additional processing. This will be further discussed within Section 8.3.

## 8.3 Ensemble Data Assimilation

Ensemble data assimilation (ENS) implies estimation and propagation of uncertainty, in addition to estimation and propagation of the optimal state. One can distinguish between two main groups of ENS algorithms: (1) Monte Carlo (MC) methods and (2) direct transform (DT) methods. In MC methods PDF moments have a sample-based representation, and ensembles are generally interpreted as random realizations. In DT methods PDF moments are obtained by direct transformation from a previously available estimate, and ensembles generally represent vectors that span an uncertainty subspace. DT methods are sometimes confused with MC methods, as the same word "ensemble" is interchangeably used with two different meanings. Common MC methodologies include the ensemble KF (EnKF) (Evensen, 1994), while common DT methodologies include the KF (Kalman, 1960; Kalman and Bucy, 1961).

There are also some characteristics of data assimilation that are typically associated with VAR methods but could be easily used with ENS methods. For example, nonlinear iterative minimization is traditionally associated with variational methods. It has been shown, however, that ENS can include minimization (e.g., Županski, 2005; Lorentzen and Nævdal, 2011; Sakov et al., 2012). Another example is the use of the adjoint operators, typically associated with VAR. Adjoints can also be used within ensemble methods, as shown in various works (e.g., Szunyogh et al., 2005; Buehner et al., 2017).

### *8.3.1 Monte Carlo Ensemble Data Assimilation Methods*

#### *8.3.1.1 The Ensemble Kalman Filter*

Ensemble Kalman filters (EnKFs) were originally motivated by the inconsistency of the extended KF, a nonlinear extension of the KF, in representing the error covariance evolution equation (Evensen, 1992, 1994). It was soon realized that the EnKF has desirable properties, such as flow-dependent error covariance, improved Bayesian inference in terms of uncertainty, as well as algorithmic simplicity.

Since the EnKF is developed with Gaussian assumption, it is limited to estimating only the first two PDF moments of analysis/forecast errors. In fact, the system is typically assumed *unbiased*, implying that the mean of the forecast and analysis errors is 0.

Given a set of initial states $\{x_i^a : i = 1, N_E\}$, where $N_E$ denotes the number of ensembles, and a nonlinear dynamical model $M$, one can compute ensemble forecasts

$$\mathbf{x}_i^f = M(\mathbf{x}_i^a) \quad (i = 1, N_E). \tag{8.19}$$

Using sample mean to define the mathematical expectation $E$

$$\mathbf{x}^f = E(\mathbf{x}_i^f) = \bar{\mathbf{x}} = \frac{1}{N_E}\sum_{i=1}^{N_E} x_i^f, \tag{8.20}$$

where $\mathbf{x}^f$ denotes the expectation value, the forecast error sample can be defined as $\varepsilon_i^f = \mathbf{x}_i^f - \mathbf{x}^f$, with $\mathbf{x}^f = \bar{\mathbf{x}}$. It is clear that the first PDF moment of this error is 0, as $E(\varepsilon_i^f) = E(\mathbf{x}_i^f) - \bar{\mathbf{x}} = 0$, where we used $E(\bar{\mathbf{x}}) = \frac{1}{N_E}\sum_{i=1}^{N_E}\bar{\mathbf{x}} = \frac{1}{N_E}N_E\bar{\mathbf{x}} = \bar{\mathbf{x}}$ and that $E$ is a linear operator. The second PDF moment (covariance) is defined as an unbiased sample estimate

$$\mathbf{P}_f = \frac{1}{N_E - 1}\sum_{i=1}^{N_E}(\mathbf{x}_i^f - \bar{\mathbf{x}})(\mathbf{x}_i^f - \bar{\mathbf{x}})^T. \tag{8.21}$$

Therefore, the forecast guess and the forecast error covariance are defined using a sampling approach. An implied assumption of the sampling approach is that $N_E$ is large enough to satisfy the conditions of the central limit theorem, as well as that each sample is independently obtained. Since in practice $N_E$ is relatively small, and the initial conditions perturbations in ensemble forecasting are not independent, formulas (8.20) and (8.21) are only estimates and therefore have some additional error.

For the correct application of the EnKF in the analysis, the observations also need to be random. This is achieved by adding a random perturbation sampled from a prespecified observation error PDF to the actual observation, and creating *perturbed observations*

$$\mathbf{y}_i^o = \mathbf{y}^o + \varepsilon_i \qquad (i = 1, N_E) \qquad \mathbf{R} = \langle \varepsilon\varepsilon^T \rangle, \tag{8.22}$$

so that the mean of the perturbed observations is equal to the actual observations ($\overline{\mathbf{y}^o} = \mathbf{y}^o$).

The analysis of the EnKF follows the original KF analysis equations, however, using the ensemble forecast error covariance (8.21) instead of the full-rank KF covariance, for each perturbed observation (Houtekamer and Mitchell, 1998)

$$\mathbf{x}_i^a = \mathbf{x}_i^f + [\mathbf{P}_f\mathbf{H}^T]_{ENS}\left([\mathbf{H}\mathbf{P}_f\mathbf{H}^T]_{ENS} + \mathbf{R}\right)^{-1}\left[\mathbf{y}_i^o - h(\mathbf{x}_i^f)\right], \tag{8.23}$$

where

$$[\mathbf{P}_f\mathbf{H}^T]_{ENS} = \frac{1}{N_E - 1}\sum_i\left[\mathbf{x}_i^f - \bar{\mathbf{x}}\right]\left[H(\mathbf{x}_i^f) - \overline{H(\mathbf{x}_i^f)}\right]^T \text{ and} \tag{8.24}$$

$$[\mathbf{H}\mathbf{P}_f\mathbf{H}^T]_{ENS} = \frac{1}{N_E - 1}\sum_i\left[H(\mathbf{x}_i^f) - \overline{H(\mathbf{x}_i^f)}\right]\left[H(\mathbf{x}_i^f) - \overline{H(\mathbf{x}_i^f)}\right]^T. \tag{8.25}$$

Therefore, for each perturbed observation the analysis equation is solved, resulting in new analysis: For $N_E$ ensembles there are $N_E$ analysis equations to solve. Each analysis equation is using a forecast guess from the corresponding ensemble. After solving the analysis equation, one can form the analysis as a sample mean

$$\mathbf{x}^a = \frac{1}{N_E}\sum_i \mathbf{x}_i^a \tag{8.26}$$

and analysis error covariance as unbiased sample covariance

$$\mathbf{P}_a = \frac{1}{N_E - 1}\sum_i (\mathbf{x}_i^a - \mathbf{x}^a)(\mathbf{x}_i^a - \mathbf{x}^a)^T. \tag{8.27}$$

These last equations, (8.26) and (8.27), are diagnostic and therefore not need to be computed, as the EnKF only requires the initial conditions for the ensuing ensemble forecast. These ensemble initial conditions are defined by $N_E$ analyses (8.23).

The most common examples of EnKFs are the algorithms by Evensen (1994, 2003) and Houtekamer and Mitchell (1998, 2001).

*8.3.1.2 Square Root Ensemble Kalman Filter*

One of the computational challenges of the EnKF is the need to calculate $N$ analyses. Mathematically, however, a larger number of samples implies a better estimate of the state and its uncertainty. Therefore, there is a motivation to reduce the computation burden in the EnKF by calculating one analysis only, as is done in VAR. This issue is addressed in square root ensemble KFs (SR-EnKFs), by considering an average of the EnKF analysis equations

$$\begin{aligned}
\mathbf{x}^a = \overline{\mathbf{x}_i^a} &= \overline{\left(\mathbf{x}_i^f + [\mathbf{P}_f\mathbf{H}^T]_{ENS}\left([\mathbf{HP}_f\mathbf{H}^T]_{ENS} + \mathbf{R}\right)^{-1}\left[\mathbf{y}_i^o - H(\mathbf{x}_i^f)\right]\right)} \\
&= \mathbf{x}^f + [\mathbf{P}_f\mathbf{H}^T]_{ENS}\left([\mathbf{HP}_f\mathbf{H}^T]_{ENS} + \mathbf{R}\right)^{-1}\overline{\left[\mathbf{y}_i^o - H(\mathbf{x}_i^f)\right]} \\
&= \mathbf{x}^f + [\mathbf{P}_f\mathbf{H}^T]_{ENS}\left([\mathbf{HP}_f\mathbf{H}^T]_{ENS} + \mathbf{R}\right)^{-1}\left[\mathbf{y}^o - \overline{H(\mathbf{x}_i^f)}\right],
\end{aligned} \tag{8.28}$$

where the averaging operator does not affect the matrices since they are not a function of $\mathbf{x}$. Since $\overline{H(\mathbf{x}_i^f)} = \frac{1}{N_E}\sum_{i=1}^{N_E} H(\mathbf{x}_i^f)$ implies a need for $N_E$ calculations of the observation operator, which can be quite costly for realistic observation systems, it is common to assume

$$\overline{H(\mathbf{x}_i^f)} \approx H(\overline{\mathbf{x}_i^f}) = H(\mathbf{x}^f), \tag{8.29}$$

i.e., to compute the observation guess from the ensemble mean, and reduce it to only one observation operator calculation. With the above simplification, the analysis equation is

$$\mathbf{x}^a = \mathbf{x}^f + [\mathbf{P}_f\mathbf{H}^T]_{ENS}\left([\mathbf{HP}_f\mathbf{H}^T]_{ENS} + \mathbf{R}\right)^{-1}\left[\mathbf{y}^o - H(\mathbf{x}^f)\right]. \tag{8.30}$$

The analysis uncertainty of the KF is given by $\mathbf{P}_a = (\mathbf{I} - \mathbf{KH})\mathbf{P}_f(\mathbf{I} - \mathbf{KH})^T + \mathbf{KRK}^T$ where $\mathbf{K} = \mathbf{P}_f\mathbf{H}^T(\mathbf{HP}_f\mathbf{H}^T + \mathbf{R})^{-1}$ (e.g., Jazwinski, 1970). Following EnKF equations, however, the analysis error covariance is (e.g., Burgers et al., 1998)

$$\mathbf{P}_a = (\mathbf{I} - \mathbf{KH})\mathbf{P}_f(\mathbf{I} - \mathbf{KH})^T, \tag{8.31}$$

i.e., the positive definite term $\mathbf{KRK}^T$ is missing, producing an underestimated analysis error covariance estimate in the EnKF. This, however, can be accounted for by adding observation noise (e.g., Burgers et al., 1998).

Therefore, such SR-EnKFs will considerably reduce the analysis calculation of the EnKF, still producing comparable results. There are numerous "flavors" of SR-EnKFs, such as the ensemble square root filter (Whitaker and Hamill, 2002), the ensemble adjustment filter (Anderson, 2001), the ensemble transform KF (Bishop et al., 2001), and many more.

#### *8.3.1.3 Particle Filters*

Particle filters (PFs) are the most typical representation of sequential MC methods, and are also more general than EnKF and/or SR-EnKF methods. *Particle* and *ensemble* have the same meaning. PF methods rely on estimating the complete PDF from a sample, not only the first two moments. This allows straightforward estimation of non-Gaussian PDFs, but also imposes a computational burden as non-Gaussian PDFs generally include higher-order moments and, in principle, PFs require a larger number of ensembles than EnKFs. We will not go through the derivation of these methods here but will refer the reader to the works by Doucet et al. (2001), van Leeuwen (2009), and Chorin and Tu (2009).

### *8.3.2 Direct Transform Ensemble Data Assimilation Methods*

As indicated above, the KF is the best known example of this methodology. Given the initial state and uncertainty of that state, one can form vectors that span the uncertainty subspace.

Let us assume the analysis $\mathbf{x}^a$ is available. Let us also assume that a subspace of the analysis uncertainty is represented by vectors $\{\mathbf{u}_i^a : i = 1, N_E\}$. Then one can define perturbed analyses as initial conditions for ensemble forecasting

$$x_i^a = x^a + u_i^a. \tag{8.32}$$

Given a nonlinear model (NLM) $M$ and perturbed analyses one can compute ensemble forecasts $\mathbf{x}_i^f = M(\mathbf{x}_i^a)$ as in MC methods (Eq. (5.15)). The forecast guess, however, is now

$$\mathbf{x}^f = M(\mathbf{x}^a), \tag{8.33}$$

i.e., it is a deterministic forecast. The forecast guess is obtained by direct transformation from the analysis. One can still define the forecast error as $\varepsilon_i^f = \mathbf{x}_i^f - \mathbf{x}^f$. Note that this error will in general have a nonzero expectation, i.e., it will be biased. On the positive side, the use of a deterministic forecast for the guess is more appealing from the dynamical point of view, as it reflects a dynamically consistent estimate of the state. The use of the sample mean can be problematic for a small sample, however, as it represents a "smeared" forecast without dynamical characteristics of an actual state.

The forecast error can also be interpreted as forecast uncertainty

$$\mathbf{x}_i^f = \mathbf{x}^f + \mathbf{u}_i^f, \tag{8.34}$$

i.e., $\varepsilon_i^f = \mathbf{u}_i^f$. Using (8.32) and (8.33) the forecast uncertainty is

$$\mathbf{u}_i^f = M(\mathbf{x}^a + \mathbf{u}_i^a) - M(\mathbf{x}^a). \tag{8.35}$$

Therefore, forecast uncertainty is obtained by direct transformation from analysis uncertainty.

For a Gaussian PDF, the uncertainty is defined by square root error covariances

$$\mathbf{P}_a^{1/2} = \left( \begin{array}{cccc} \mathbf{p}_1^a & \mathbf{p}_2^a & \cdots & \mathbf{p}_{N_E}^a \end{array} \right), \tag{8.36}$$

$$\mathbf{P}_f^{1/2} = \left( \begin{array}{cccc} \mathbf{p}_1^f & \mathbf{p}_2^f & \cdots & \mathbf{p}_N^f \end{array} \right) \tag{8.37}$$

with $\mathbf{p}_i^f = \mathbf{u}_i^f$ and $\mathbf{p}_i^a = \mathbf{u}_i^a$ representing span vectors of forecast and analysis uncertainty subspaces, respectively. Therefore, uncertainty evolution (8.35) can be represented as evolution of the column vectors of the square root error covariance

$$\mathbf{p}_i^f = M(\mathbf{x}^a + \mathbf{p}_i^a) - M(\mathbf{x}^a). \tag{8.38}$$

With additional linear model assumption $\mathbf{p}_i^f = M(\mathbf{x}^a + \mathbf{p}_i^a) - M(\mathbf{x}^a) \approx \mathbf{M}\mathbf{p}_i^a$, where $\mathbf{M} = \frac{\partial M}{\partial \mathbf{x}}$ is a linearized model. Therefore, the square root forecast error covariance is

$$\mathbf{P}_f^{1/2} = \left( \begin{array}{cccc} \mathbf{M}\mathbf{p}_1^a & \mathbf{M}\mathbf{p}_2^a & \cdots & \mathbf{M}\mathbf{p}_{N_E}^a \end{array} \right) = \mathbf{M} \left( \begin{array}{cccc} \mathbf{p}_1^a & \mathbf{p}_2^a & \cdots & \mathbf{p}_{N_E}^a \end{array} \right) = \mathbf{M}\mathbf{P}_a^{1/2} \tag{8.39}$$

implying that there is a direct linear transformation of covariances. After forming the full forecast error covariance

$$\mathbf{P}_f = \mathbf{P}_f^{1/2}(\mathbf{P}_f^{1/2})^T = \mathbf{M}\mathbf{P}_a^{1/2}(\mathbf{M}\mathbf{P}_a^{1/2})^T = \mathbf{M}\mathbf{P}_a\mathbf{M}^T, \tag{8.40}$$

which is a well-known KF formula assuming no model errors.

The analysis step of DT methods can be the KF analysis equation or using a minimization of a cost function. The analysis uncertainty can be obtained by exact algebraic transformation: in the case of KF

$$\mathbf{P}_a = \mathbf{P}_f - \mathbf{P}_f\mathbf{H}^T(\mathbf{R} + \mathbf{H}\mathbf{P}_f\mathbf{H}^T)^{-1}\mathbf{H}\mathbf{P}_f \tag{8.41}$$

or in case of cost function minimization using the inverse Hessian

$$\mathbf{P}_a = (\mathbf{P}_f^{-1} + \mathbf{H}^T\mathbf{R}^{-1}\mathbf{H})^{-1}. \tag{8.42}$$

Advanced formulation of DT methods would normally use nonlinear transformations (e.g., (8.38)) and also include minimization of an arbitrary nonlinear cost function in the analysis. The analysis uncertainty formulation (8.38) is still applicable, however, if calculated at the analysis point.

The DT methods also have their square root formulations (e.g., Bierman, 1977; Bernstein and Hyland, 1985; Verlaan and Heemink, 1997; Farrell and Ioannou, 2001; Županski, 2005).

### *8.3.3 Error Covariance Localization*

In considering MC and DT ensemble filters, neither method has a clear general advantage in realistic high-dimensional problems. The number of ensembles will very likely be too small to fully describe uncertainty, requiring us to find other ways to address the insufficient number of DOF. A common remedy is to increase the DOF by localizing error covariance based on geographical distance (e.g., Houtekamer and Mitchell, 1998; Hamill et al., 2001).

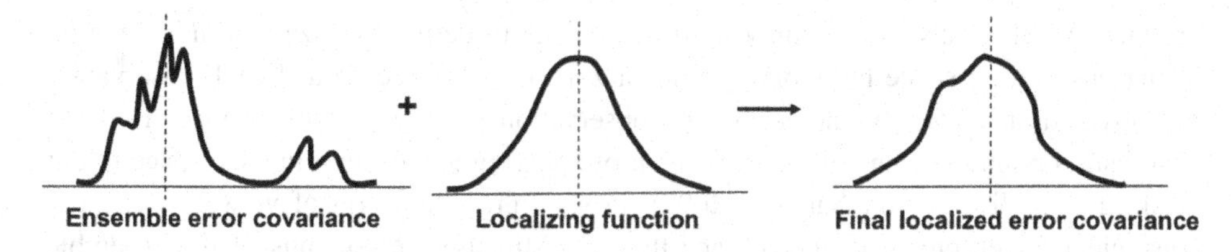


Figure 8.3 Error covariance localization. Convolution of ensemble covariance with localized error covariance produces a smooth and localized covariance.

An error covariance with fewer DOF would be noisy and with possibly large spurious correlations at distant points. The localization procedure typically implies a convolution of noisy ensemble error covariance with a pre-defined smooth and localized error covariance (e.g., Figure 8.3).

Formally, given an ensemble covariance $\mathbf{P}_{ENS}$ and a localizing error covariance $\mathbf{L}$, their pointwise (e.g., Hadamard or Schur) product produces a new covariance

$$\mathbf{P}_L = \mathbf{L} \circ \mathbf{P}_{ENS}. \tag{8.43}$$

The localizing covariance $L$ can be chosen to be the same as static error covariance used in VAR, for example, or simply as a banded Toeplitz matrix (e.g., Županski, 2016). A property of the Schur product implies the rank (e.g., the number of nonzero singular values) of final covariance

$$rank(\mathbf{P}_L) \leqq rank(\mathbf{L}) \cdot rank(\mathbf{P}_{ENS}), \tag{8.44}$$

which can increase by several orders of magnitude. Since $rank(\mathbf{P}_{ENS}) \leq N_E$, one has $N_E \ll rank(\mathbf{L}) \leq N_S$. For a diagonal matrix $\mathbf{L}$ the upper limit is reached, i.e., $rank(\mathbf{L}) = N_S$. Due to correlations (i.e., banded matrix), the rank of $\mathbf{L}$ is reduced, but overall it is large enough to ensure a considerable increase of DOF of the final error covariance.

However, if the ensemble error covariance is negligible in some areas, then localization cannot help, as the convolution with 0 will still be 0. This means that, although the number of DOF will be mathematically increased to allow successful data assimilation, the available DOF may still be insufficient to correctly represent uncertainty at all analysis points. This issue will be addressed by hybrid methods.

An additional issue with localization is related to the dynamical consistency of the final covariance. While the ensemble error covariance is obtained consistently with model dynamics, the localization covariance is not. Therefore, the resulting covariance matrix may be dynamically inconsistent. Since the error covariance defines the subspace of analysis increments, the produced ensemble initial conditions may not be dynamically balanced.

One can distinguish between error covariance localization in state and in observation space (e.g., Houtekamer and Mitchell, 2001). State-space localization is essentially defined by (8.43), as both matrices, $P_{ENS}$ and $L$, are defined in state space and so is their Schur product. An alternative, commonly used in ENS, is to define localization and changing in the observation space by applying a local zone of influence around each observation the observation error. By increasing the observation error with the distance from the observation point one can get the equivalent of applying localization in state space (Hunt et al., 2007; Miyoshi and Yamane, 2007). From a practical point of view this is more convenient for ensemble methods since they can still use the columns of the ensemble square root forecast error covariance (e.g., Hunt et al., 2007). However, vertically integrated observations, such as satellite radiances or aerosol optical depth, do not have a precisely defined vertical location, which can adversely impact the assimilation of such observations in ENS.

**Example 8.1 1D localizing matrix using the Gaussian function**

Let

$$\mathbf{L} = \begin{pmatrix} r_{1,1} & r_{1,2} & \cdots & r_{1,N_S} \\ r_{2,1} & r_{2,2} & \cdots & r_{2,N_S} \\ \vdots & \vdots & \ddots & \vdots \\ r_{N_S,1} & r_{N_{state},2} & \cdots & r_{N_S,N_S} \end{pmatrix} \tag{8.45}$$

be a symmetric square matrix in state space. The elements $r_{i,j}$ denote the correlation between points with indexes $i$ and $j$ and can be defined using a Gaussian function (e.g., Miyoshi and Yamane, 2007)

$$r_{i,j} = \exp\left(-\frac{d_{ij}^2}{2F^2}\right), \tag{8.46}$$

where $d_{ij}$ is the geographic distance between points $i$ and $j$ and F is the characteristic length associated with the Gaussian function. Since a Gaussian function theoretically extends to infinity, in practical applications it is helpful to limit its impact to the neighboring points only. One typically defines first a *decorrelation* length $d_{max}$ which defines the maximum distance from central point at which the correlation becomes negligible. The "negligible" value of the correlation can be defined using a symbol $\varepsilon$, which should be a small number between 0 and 1. After defining $d_{max}$ and $\varepsilon$ and substituting in (8.46) it is possible to calculate the characteristic length $F$ from

$$\varepsilon = \exp\left(-\frac{d_{max}^2}{2F^2}\right) \tag{8.47}$$

to obtain

$$F = \frac{d_{max}}{\sqrt{2ln\frac{1}{\varepsilon}}}. \tag{8.48}$$

**Example 8.2 2D localizing matrix using the Gaussian function**

We still use the localizing matrix formulation given by (8.45). However, the elements $r_{ij}$ can be defined using (i) straightforward extension to denote the correlation between points in 2D state space or (ii) a product of correlations along each coordinate direction. Following approach (i) one applies (8.46) with distance in 2D space defined as

$$d_{ij} = \sqrt{dx_{ij}^2 + dy_{ij}^2}, \tag{8.49}$$

where $dx_{ij}$ and $dy_{ij}$ are geographic distances between points $i$ and $j$ along the $x$ and $y$ coordinate directions, respectively. Following approach (ii) one redefines (8.46) to

$$r_{i,j} = \exp\left(-\frac{dx_{ij}^2}{2F_x^2}\right)\exp\left(-\frac{dy_{ij}^2}{2F_y^2}\right), \tag{8.50}$$

where $F_x$ and $F_y$ are characteristic lengths of Gaussian functions in the $x$ and $y$ coordinate directions, respectively. The approach in (i) is more correct in the sense that the 2D distances between points $i$ and $j$ are correctly calculated. The approach in (ii), however, has some practical advantages when the directions $x$ and $y$ are of different orders of magnitude, such as the horizontal and vertical direction in atmospheric applications.

**Practice 8.1 Characteristic lengths**

Find characteristic lengths of the Gaussian function for a decorrelation length of 500 km and correlation thresholds (a) $\varepsilon = 0.1$ and (b) $\varepsilon = 0.0001$.

**Practice 8.2 Correlations**

Find correlations calculated using the two approaches described in Example 8.2 at distances from the central point $dx_{ij} = dy_{ij} = 100$ km, for correlation threshold $\varepsilon = 0.01$ and characteristic lengths $F = F_x = F_y = 200$ km.

## 8.4 Hybrid Data Assimilation

Hybrid data assimilation (HYB) is data assimilation developed as a combination between two or more data assimilation methods. However, *hybrid* is often used to denote a combination of static and ensemble forecast error covariances, and ensemble-variational

(EnVar) to denote a combination of ensemble and variational methods (e.g., Lorenc, 2003; Bannister, 2017). Since most often EnVar is used to refer to a practical two-system data assimilation, with separate ENS and VAR components, we will use *EnVar* to denote such a two-component system. Furthermore, if such a system includes a combination of variational and ensemble forecast error covariance, we will refer to it as a *hybrid EnVar* system. Other possibilities of combining ensemble and variational data assimilation within a single system, such as the *maximum likelihood ensemble filter* (MLEF), we will refer to by their commonly used name.

Our discussion of HYB is motivated by the limited capability of basic VAR and ENS methods to simultaneously address two preferences for forecast error covariance: (1) all DOF and (2) flow dependence. The first preference is typically associated with variational methods, and the second preference is associated with ensemble methods. In a wider sense, HYB can also include ENS methods with nonlinear minimization (e.g., Županski, 2005; Županski et al., 2008; Sakov et al., 2012), a common property of VAR, and ENS that employs tangent linear and adjoint models (e.g., Szunyogh et al., 2008; Buehner et al., 2017), also a common property of VAR.

### *8.4.1 Hybrid Ensemble Variational*

The basic idea in developing hybrid error covariance is to form a linear combination of VAR and ENS covariances, as described in Hamill and Snyder (2000)

$$\mathbf{P}_f = \gamma \mathbf{P}_{ENS} + (1-\gamma)\,\mathbf{P}_{VAR}, \tag{8.51}$$

where $\gamma$ is the relative weight of flow-dependent ensemble covariance. A more practical scheme was proposed by Lorenc (2003) and further adjusted by Wang et al. (2008) and Zhang et al. (2009). Such a hybrid approach essentially includes two data assimilation systems: The ENS algorithm is used for uncertainty estimation (e.g., second PDF moment), while the VAR algorithm is used for state estimation (e.g., first PDF moment). These two systems interact: The ensemble forecast error covariance from the ENS is combined with the VAR covariance to form a hybrid error covariance, eventually producing hybrid analysis. This hybrid analysis is then used in ENS by overwriting the ENS-produced analysis and for recentering of ENS analysis uncertainties.

During the hybrid analysis calculation, the forecast error covariance from ENS is combined with the VAR static covariance so that an augmented cost function is minimized

$$\begin{aligned} f(\mathbf{x},\alpha) = {} & \beta_f \left(\delta\mathbf{x}^f\right)^T \mathbf{P}_{VAR}^{-1} \left(\delta\mathbf{x}^f\right) + \beta_e(\alpha)^T \left(\mathbf{P}_{ENS} \circ \mathbf{L}\right)^{-1} (\alpha) \\ & + \frac{1}{2}\left[\mathbf{y}^o - \mathbf{H}\delta\mathbf{x}^{tot}\right]^T \mathbf{R}^{-1} \left[\mathbf{y}^o - \mathbf{H}\delta\mathbf{x}^{tot}\right], \end{aligned} \tag{8.52}$$

where $\alpha$ is the additional control variable and

$$\delta\mathbf{x}^{tot} = \delta\mathbf{x}^f + \sum_k \alpha_k \circ \left[\mathbf{P}_{ENS}^{1/2}\right]_k \tag{8.53}$$

with $\beta_e$ and $\frac{1}{\beta_f} + \frac{1}{\beta_e} = 1$ and defining the relative weights. Since this is essentially an augmented VAR method, there is a nonlinear iterative minimization included as well. A consequence of using an augmented VAR system for the analysis, however, is that the minimization space is still high dimensional, and any estimation of the analysis error uncertainty is not reliable or practical. One can also note that there are two additive forecast error covariance terms in (8.52), implying that the corresponding (Gaussian) PDFs multiply with each other. From the theory of probability this automatically implies that the errors of $\delta\mathbf{x}^f$ and $\alpha$ are independent, and therefore that covariances $\mathbf{P}_{VAR}$ and $\mathbf{P}_{ENS} \circ \mathbf{L}$ are uncorrelated. However, this assumption may not be always valid since both covariances attempt to represent uncertainties of the same variables.

The main theoretical limitation of operational hybrid EnVar methods is that one component, VAR, produces the analysis (e.g., the first PDF moment), while the other component, ENS, estimates the posterior error covariance (e.g., the second PDF moment). Since the ENS posterior error covariance is calculated for the ENS analysis, it is generally not representative of the VAR analysis uncertainty. This inconsistency in estimating the moments of posterior PDF effectively prevents Bayesian inference (e.g., Jaynes, 2003) and ultimately make the method suboptimal. A possible numerical remedy for suboptimal Bayesian inference could be to apply ENS and VAR in an iterative manner, i.e., repeat the ENS $\rightarrow$ VAR sequence several times. There is no assurance that such system would converge, though, so this would need to be established first. The computational cost of such system would clearly increase, but even conducting the minimal sequence ENS $\rightarrow$ VAR $\rightarrow$ ENS may produce a more consistent estimate of the posterior error covariance.

From the practical point of view, the main difficulty of the current hybrid systems is that there are essentially two separate systems that need maintenance and updates. It seems natural to define a single hybrid system that has all positive qualities of VAR and ENS but is also more efficient and theoretically more consistent. Current practice suggests that error covariance localization in state space, as in (8.43) and (8.52), is more advantageous for vertically integrated observations. Given that the majority of observations currently assimilated in operational numerical weather prediction (NWP) is satellite and from other space-borne remote sensing instruments, this issue becomes important. The ENS methods have a natural way of accounting for uncertainties, but a straightforward use of hybrid error covariance is difficult because static error covariance with a large number of DOF cannot be used efficiently within ENS methods. New DOF would effectively increase $N_E$ by several orders of magnitude, making calculations by (8.24) and (8.25) impractical.

Other practical HYB procedures also include ensemble of VAR methods, typically 4DVAR (e.g., Isaksen et al., 2010; Bonavita et al., 2016). This type of hybrid is essentially based on the EnKF, however defined with 4DVAR as the analysis method, not the KF. This method includes assimilation of perturbed observations, as with EnKF, but often has additional perturbation for model error. The benefit of using 4DVAR versus KF is that (1) 4D variational analysis is nonlinear while KF analysis is linear, (2) 4DVAR is smoother, which has some advantages for dynamical stability and assimilation of time-integrated observations such as precipitation, and (3) 4DVAR is stable as has full-rank covariance, whereas EnKF employs a reduced-rank ensemble covariance. Recent developments of hybrid methods introduce numerous alternatives to existing hybrid systems, and this is a very active area of research. An equivalent alternative would be to use 3DVAR instead of 4DVAR, which would reduce the cost. However, the application of ensemble of 4DVAR appears to produce better results in practice.

In summary, by requiring two relatively independent systems, ENS and VAR, current hybrid methods do not produce consistent and reliable estimates of forecast and analysis uncertainties. Current research is focusing on those issues, and it is likely that an acceptable and reliable estimate of the optimal state and its uncertainty will be found (Auligné et al., 2016).

### 8.4.2 Maximum Likelihood Ensemble Filter

The MLEF was originally developed without covariance localization (Županski, 2005; Županski et al., 2008), although its practical extensions include covariance localization in the observation space (Zhang et al., 2013a, 2013b; Zhang et al., 2017; Suzuki et al., 2017; Suzuki and Županski, 2018). Efforts to define MLEF with state-space localization are also underway. Here, we present the original MLEF algorithm. The novel idea of MLEF was to combine VAR and ENS into a single system, by selectively including the beneficial characteristics of each method. This was essentially translated into calculating the analysis using an iterative nonlinear minimization and estimating the analysis error covariance from the inverse Hessian matrix that is a by-product of minimization. The basic elements of this idea have already been mentioned in Section 8.3.2. Here we include a few additional details related to the MLEF algorithm.

The forecast step of the MLEF is a time evolution of the analysis state and the analysis uncertainty by a nonlinear prediction model. The analysis step is the nonlinear numerical optimization of the Gaussian-like cost function

$$f(\mathbf{x}) = \frac{1}{2}\left[\mathbf{x} - \mathbf{x}^f\right]^T \mathbf{P}_f^{-1}\left[\mathbf{x} - \mathbf{x}^f\right] + \frac{1}{2}\left[\mathbf{y}^o - H(\mathbf{x})\right]^T \mathbf{R}^{-1}\left[\mathbf{y}^o - H(\mathbf{x})\right], \tag{8.54}$$

where $\mathbf{P}_f$ is defined over the ensemble subspace. Since calculating its inverse in (8.54) is not practical, a change of variable common in VAR is introduced similar to (8.11), resulting in the transformed cost function

$$f(\mathbf{w}) = \frac{1}{2}\mathbf{w}^T\mathbf{w} + \frac{1}{2}\left[\mathbf{y}^o - H\left(\mathbf{x}^f + \mathbf{P}_f^{1/2}\mathbf{w}\right)\right]^T \mathbf{R}^{-1}\left[\mathbf{y}^o - H\left(\mathbf{x}^f + \mathbf{P}_f^{1/2}\mathbf{w}\right)\right]. \tag{8.55}$$

The above cost function is solved using an iterative numerical optimization algorithm

$$\mathbf{w}_m = \mathbf{w}_{m-1} + \alpha_m \mathbf{d}_m, \tag{8.56}$$

where $m = 1, \ldots, M_I$ is the iteration index, $\mathbf{d}$ is the descent direction vector, and $\alpha$ is the step length calculated from the line search. The initial value $\mathbf{w}_0 = 0$ corresponds to the choice of the initial guess $\mathbf{x}_0 = \mathbf{x}^f$. The numerical optimization method that calculates $\mathbf{d}$ could be any minimization algorithm discussed in Section 6.3. When solving the minimization problem, the MLEF incorporates the optimal Hessian preconditioning by using an additional change of variable

$$\mathbf{w} = \left[\mathbf{I} + \mathbf{Z}^T\mathbf{Z}\right]^{-\frac{1}{2}} \zeta, \tag{8.57}$$

where the columns of matrix $\mathbf{Z}$ are defined as

$$\mathbf{z}_i = \mathbf{R}^{-\frac{1}{2}} \left[H(\mathbf{x}^f + \mathbf{p}_i^f) - H(\mathbf{x}^f)\right]. \tag{8.58}$$

The index $i$ denotes the ensemble member. The inversion and square root calculation of the matrix in (8.53) is accomplished using eigenvalue decomposition (EVD), which is computationally efficient given that this matrix is defined in the low-dimensional ensemble subspace. After the analysis solution $\mathbf{x}^a = \mathbf{P}_f^{\frac{1}{2}}\mathbf{w}^a$ is found, the initial conditions for ensemble forecasting are updated using the columns of the square root analysis covariance as perturbations

$$\mathbf{x}_i = \mathbf{x}^a + \left(\mathbf{P}_f^{\frac{1}{2}} \left[\mathbf{I} + \mathbf{Z}^T\mathbf{Z}\right]^{-\frac{1}{2}}\right)_i. \tag{8.59}$$

The matrix $\mathbf{Z}$ used for the ensemble update differs from the one calculated in (8.54) in that it employs the analysis $\mathbf{x}^a$ instead of the initial guess $\mathbf{x}^f$.

There are several notable distinctions between the MLEF and the related EnKF algorithms. One obvious difference is that the analysis is obtained as the minimizing solution of an iterative minimization, rather than from the EnKF direct linear solution (8.23). Other distinctions include the calculation of uncertainties by directly estimating the analysis error covariance rather than by employing a sample-based approach (e.g., (8.21) and (8.27)). Also, the first guess in the MLEF is obtained as a result of deterministic prediction from the analysis (8.19) rather than using the ensemble mean (8.20).

**Colloquy 8.1**

**Error of applying observation-space covariance localization in the EnKF**

We investigate the implication of using *observation-space* or *state-space* covariance localization in the EnKF, both widely used in ensemble and hybrid variational-ensemble algorithms. Following Houtekamer and Mitchell (2001), the state-space localization (also referred to as model-space localization) is defined in the KF analysis equation

$$\mathbf{K}_S = \left[(\mathbf{L} \circ \mathbf{P}_f)\mathbf{H}^T\right]\left[\mathbf{H}(\mathbf{L} \circ \mathbf{P}_f)\mathbf{H}^T + \mathbf{R}\right]^{-1}, \tag{8.60}$$

where $\mathbf{K}_S$ is the Kalman gain matrix obtained with localization in the state space. This formulation can be traced back to the prior PDF that employs a localized forecast error covariance. However, a straightforward application of the above equation in realistic problems requires localization of a prohibitively large matrix, so it does not have a practical value. Consequently, an alternative formulation with observation-space localization is introduced

$$\mathbf{K}_O = \left[\mathbf{L} \circ (\mathbf{P}_f \mathbf{H}^T)\right] \left[\mathbf{L} \circ (\mathbf{H}\mathbf{P}_f \mathbf{H}^T) + \mathbf{R}\right]^{-1}, \tag{8.61}$$

where $\mathbf{K}_O$ is the Kalman gain matrix obtained with localization in the observation space, which is eventually used in the actual EnKF algorithm.

First to note is that the equality $\mathbf{K}_S = \mathbf{K}_O$ implies

$$(\mathbf{L} \circ \mathbf{P}_f)\mathbf{H}^T = \mathbf{L} \circ (\mathbf{P}_f \mathbf{H}^T) \tag{8.62}$$

and

$$\mathbf{H}(\mathbf{L} \circ \mathbf{P}_f)\mathbf{H}^T = \mathbf{L} \circ (\mathbf{H}\mathbf{P}_f \mathbf{H}^T). \tag{8.63}$$

Therefore, it is assumed in (8.61) that the localizing matrix $\mathbf{L}$ can exchange order with other matrices $\mathbf{P}_f$, $\mathbf{H}$, and $\mathbf{H}^T$ in the Hadamard product. An equality for matrix multiplication using the Hadamard product states that

$$\mathbf{D}\,(\mathbf{A} \circ \mathbf{B})\,\mathbf{E} = \mathbf{A} \circ (\mathbf{DBE}), \tag{8.64}$$

where $\mathbf{A}$ and $\mathbf{B}$ are square matrices while $\mathbf{D}$ and $\mathbf{E}$ are *diagonal* matrices. By comparing (8.64) and (8.63) one can identify $\mathbf{D} = \mathbf{H}$, $\mathbf{A} = \mathbf{L}$, $\mathbf{B} = \mathbf{P}_f$, and $\mathbf{E} = \mathbf{H}^T$. Consequently, (8.63) can be true *only if* the observation operator Jacobian $\mathbf{H}$ is a diagonal matrix, also implying that $\mathbf{H}^T$ is diagonal. Similarly, (8.62) can be obtained from (8.64) by identifying $\mathbf{D} = \mathbf{I}$, $\mathbf{A} = \mathbf{L}$, $\mathbf{B} = \mathbf{P}_f$, and $\mathbf{E} = \mathbf{H}^T$, also implying the equality only if $\mathbf{H}^T$ is a diagonal matrix. Both examples confirm that the introduced observation-space error covariance localization in the EnKF is valid only for diagonal observation operators. This implies that (i) observations have to be colocated with model grid points and (ii) model variables are observed (i.e., there is no transformation of variable involved). Since this is generally not the case and since the typical $\mathbf{H}$ is not a square matrix, the use of (8.61) instead of (8.60) is an approximation.

We are now interested in quantifying the approximation errors that are introduced by applying the observation-space localization instead of the state-space localization in the EnKF. In particular, we focus on assessing errors of localization when the observation operator $\mathbf{H}$ is nondiagonal, and not a square matrix. To keep the arguments simple, we examine errors in assuming (8.63) to be true. Although the same notation is used for the localizing matrix $\mathbf{L}$ on both sides of (8.63), different dimensions of state and observation spaces require a state-space localizing matrix $\mathbf{L}_S$ and an observation-space localizing matrix $\mathbf{L}_O$ to be introduced. Note that both of these matrices include the same correlation scales, implying that the impact of localization as a function of distance they produce is the same. Therefore, the approximation error matrix can be defined as

$$\mathbf{E} = \mathbf{L}_O \circ (\mathbf{H}\mathbf{P}_f \mathbf{H}^T) - \mathbf{H}(\mathbf{L}_S \circ \mathbf{P}_f)\mathbf{H}^T. \tag{8.65}$$

In order to quantify the error, we choose the commonly used Frobenius norm

$$\|\mathbf{E}\|_F = \sqrt{\sum_i \sum_j e_{ij}^2}, \tag{8.66}$$

where $e_{ij}$ denotes an element of this matrix, and calculate the normalized error

$$\epsilon = \frac{\|\mathbf{L}_O \circ (\mathbf{H}\mathbf{P}_f\mathbf{H}^T) - \mathbf{H}(\mathbf{L}_S \circ \mathbf{P}_f)\mathbf{H}^T\|_F}{\|\mathbf{H}(\mathbf{L}_S \circ \mathbf{P}_f)\mathbf{H}^T\|_F}. \tag{8.67}$$

We now consider a low-dimensional and simplified example to demonstrate a possible adverse impact of observation-space localization in the EnKF. The chosen state dimension is $N_S = 3$ and the observation-space dimension is $N_O = 2$. The forecast error covariance is assigned to be

$$\mathbf{P}_f = \begin{pmatrix} 1 & 0.7 & 0.2 \\ 0.7 & 1 & 0.7 \\ 0.2 & 0.7 & 1 \end{pmatrix} \tag{8.68}$$

and the observation operator is

$$\mathbf{H} = \begin{pmatrix} 0.5 & 0.5 & 0 \\ 0 & 0.5 & 0.5 \end{pmatrix}, \tag{8.69}$$

which implies that observations are located at the center between two state point. The corresponding localizing matrices are defined to be

$$\mathbf{L}_S = \begin{pmatrix} 1 & 0.5 & 0 \\ 0.5 & 1 & 0.5 \\ 0 & 0.5 & 1 \end{pmatrix} \qquad \mathbf{L}_O = \begin{pmatrix} 1 & 0.5 \\ 0.5 & 1 \end{pmatrix}. \tag{8.70}$$

After substituting (8.68) and (8.69) in (8.67) the normalized error is

$$\epsilon = \frac{0.285}{1.128} \approx 0.25. \tag{8.71}$$

Therefore, the error of using observation-space localization in the EnKF could be 25%, which is not negligible. This example shows that using observation-space localization in the EnKF is an approximation. It is important to note that the above consideration is applicable only to EnKF algorithms that use the formulation (8.61), not necessarily to all algorithms that use observation-space localization.

## References

Anderson JL (2001) An ensemble adjustment Kalman filter for data assimilation. *Mon Wea Rev* 129:2884–2903.

Auligné T, Ménétrier B, Lorenc AC, Buehner M (2016) Ensemble-variational integrated localized data assimilation. *Mon Wea Rev* 144:3677–3696.

Bannister RN (2017) A review of operational methods of variational and ensemble-variational data assimilation. *Quart J Roy Meteor Soc* 143:607–633.

Bernstein DS, Hyland DC (1985) The optimal projection equations for reduced-order state estimation. *IEEE Trans Autom Control* 30:583–585.
Bierman GJ (1977) *Factorization Methods for Discrete Sequential Estimation*. Academic Press, New York, 241 pp.
Bishop CH, Etherton BJ, Majumdar SJ (2001) Adaptive sampling with the ensemble transform Kalman filter. Part I: Theoretical aspects. *Mon Wea Rev* 129:420–436.
Bonavita M, Hólm E, Isaksen L, Fisher M (2016) The evolution of the ECMWF hybrid data assimilation system. *Quart J Roy Meteor Soc* 142:287–303.
Buehner M, McTaggart-Cowan R, Heilliette S (2017) An ensemble Kalman filter for numerical weather prediction based on variational data assimilation: VarEnKF. *Mon Wea Rev* 145:617–635.
Burgers G, Jan van Leeuwen P, Evensen G (1998) Analysis scheme in the ensemble Kalman filter. *Mon Wea Rev* 126:1719–1724.
Chorin AJ, Tu X (2009) Implicit sampling for particle filters. *Proc Natl Acad Sci USA* 106:17249–17254.
Doucet A, de Freitas N, Gordon N (eds.) (2001) *Sequential Monte Carlo Methods in Practice*. Springer, New York, 616 pp.
Evensen G (1992) Using the extended Kalman filter with a multilayer quasi-geostrophic ocean model. *J Geophys Res Oceans* 97:17905–17924.
Evensen G (1994) Sequential data assimilation with a nonlinear quasi-geostrophic model using Monte Carlo methods to forecast error statistics. *J Geophys Res Oceans* 99:10143–10162.
Evensen G (2003) The ensemble Kalman filter: Theoretical formulation and practical implementation. *Ocean Dyn* 53:343–367.
Farrell BF, Ioannou PJ (2001) State estimation using a reduced-order Kalman filter. *J Atmos Sci* 58:3666–3680.
Gaspari G, Cohn SE (1999) Construction of correlation functions in two and three dimensions. *Quart J Roy Meteor Soc* 125:723–757.
Golub GH, Van Loan CF (2013) *Matrix Computations*. 4th ed., Johns Hopkins University Press, Baltimore, MD, 756 pp.
Hamill TM, Snyder C (2000) A hybrid ensemble Kalman filter–3D variational analysis scheme. *Mon Wea Rev* 128:2905–2919.
Hamill TM, Whitaker JS, Snyder C (2001) Distance-dependent filtering of background error covariance estimates in an ensemble Kalman filter. *Mon Wea Rev* 129:2776–2790.
Houtekamer PL, Mitchell HL (1998) Data assimilation using an ensemble Kalman filter technique. *Mon Wea Rev* 126:796–811.
Houtekamer PL, Mitchell HL (2001) A sequential ensemble Kalman filter for atmospheric data assimilation. *Mon Wea Rev* 129:123–137.
Hunt BR, Kostelich EJ, Szunyogh I (2007) Efficient data assimilation for spatiotemporal chaos: A local ensemble transform Kalman filter. *Physica D* 230:112–126.
Isaksen L, Bonavita M, Buizza R, et al. (2010) *Ensemble of Data Assimilations at ECMWF*. Tech. Memo. 636, ECMWF, Reading, 45 pp.
Jaynes ET (2003) *Probability Theory: The Logic of Science*. Cambridge University Press, New York, 753 pp.
Jazwinski AH (1970) *Stochastic Processes and Filtering Theory*. Academic Press, New York, 376 pp.
Kalman RE (1960) A new approach to linear filtering and prediction problems. *J Basic Eng* 82:35–45.
Kalman RE, Bucy RS (1961) New results in linear filtering and prediction theory. *J Basic Eng* 83:95–108.

Lorenc AC (2003) The potential of the ensemble Kalman filter for NWP – a comparison with 4D-Var. *Quart J Roy Meteor Soc* 129:3183–3203.
Lorentzen RJ, Nævdal G (2011) An iterative ensemble Kalman filter. *IEEE Trans Autom Control* 56:1990–1995.
Miyoshi T, Yamane S (2007) Local ensemble transform kalman filtering with an AGCM at a T159/L48 resolution. *Mon Wea Rev* 135:3841–3861.
Sakov P, Oliver DS, Bertino L (2012) An iterative enkf for strongly nonlinear systems. *Mon Wea Rev* 140:1988–2004.
Suzuki K, Županski M (2018) Uncertainty in solid precipitation and snow depth prediction for siberia using the Noah and Noah-MP land surface models. *Front Earth Sci* 12:672–682.
Suzuki K, Županski M, Županski D (2017) A case study involving single observation experiments performed over snowy siberia using a coupled atmosphere-land modelling system. *Atmos Sci Lett* 18:106–111.
Szunyogh I, Kostelich EJ, Gyarmati G, et al. (2005) Assessing a local ensemble kalman filter: Perfect model experiments with the national centers for environmental prediction global model. *Tellus A* 57:528–545.
Szunyogh I, Kostelich EJ, Gyarmati G, et al. (2008) A local ensemble transform Kalman filter data assimilation system for the NCEP global model. *Tellus A* 60:113–130.
Thépaut J-N, Courtier P, Belaud G, Lemaître G (1996) Dynamical structure functions in a four-dimensional variational assimilation: A case study. *Quart J Roy Meteor Soc* 122:535–561.
van Leeuwen PJ (2009) Particle filtering in geophysical systems. *Mon Wea Rev* 137:4089–4114.
Verlaan M, Heemink AW (1997) Tidal flow forecasting using reduced rank square root filters. *Stoch Hydrol Hydraul* 11:349–368.
Wang X, Barker DM, Snyder C, Hamill TM (2008) A hybrid ETKF–3DVAR data assimilation scheme for the WRF model. Part I: Observing system simulation experiment. *Mon Wea Rev* 136:5116–5131.
Whitaker JS, Hamill TM (2002) Ensemble data assimilation without perturbed observations. *Mon Wea Rev* 130:1913–1924.
Zhang F, Weng Y, Sippel JA, Meng Z, Bishop CH (2009) Cloud-resolving hurricane initialization and prediction through assimilation of doppler radar observations with an ensemble Kalman filter. *Mon Wea Rev* 137:2105–2125.
Zhang M, Županski M, Kim M-J, Knaff JA (2013a) Assimilating AMSU-A radiances in the TC core area with NOAA operational HWRF (2011) and a hybrid data assimilation system: Danielle (2010). *Mon Wea Rev* 141:3889–3907.
Zhang SQ, Matsui T, Cheung S, Županski M, Peters-Lidard C (2017) Impact of assimilated precipitation-sensitive radiances on the NU-WRF simulation of the west African monsoon. *Mon Wea Rev* 145:3881–3900.
Zhang SQ, Županski M, Hou AY, Lin X, Cheung SH (2013b) Assimilation of precipitation-affected radiances in a cloud-resolving WRF ensemble data assimilation system. *Mon Wea Rev* 141:754–772.
Županski M (2005) Maximum likelihood ensemble filter: Theoretical aspects. *Mon Wea Rev* 133:1710–1726.
Županski M (2016) Reduced rank static error covariance for high-dimensional applications. *Int J Numer Methods Fluids* 83:245–262.
Županski M, Navon IM, Županski D (2008) The Maximum Likelihood Ensemble Filter as a non-differentiable minimization algorithm. *Quart J Roy Meteor Soc* 134:1039–1050.

# 9
# Coupled Data Assimilation

Coupled data assimilation refers to data assimilation that employs coupled models. In general, coupled models are numerical prediction models that include interactions between and an explicit account of various physical processes such as atmosphere, land surface, ocean, chemistry, and aerosol, etc. In this chapter we present the fundamentals of coupled data assimilation and associated challenges when applied within variational, ensemble, and/or hybrid methods.

## 9.1 Introduction

The development of coupled modeling systems is critical for improving the predictability of geophysical systems. Through absorbing and emitting the energy from the Sun our planet changes its temperature and energy. The relationship between temperature and energy change is typically expressed by heat capacity, which defines how much energy is needed to change the temperature of a given mass. Various media/materials have different heat capacities. For example, air has a heat capacity of 700 $\mathrm{Jkg^{-1}K^{-1}}$, while water has a heat capacity of 4184 $\mathrm{Jkg^{-1}K^{-1}}$. This means that the ocean requires much greater energy to change its temperature compared to the atmosphere. One can also interpret heat capacity as the amount of information that is stored in material. From that point of view the ocean can store much more information than the atmosphere. When coupling an ocean model with an atmospheric model, the interaction between them implies that the ocean can help in providing additional information to the atmosphere and consequently extend atmospheric prediction through this exchange. The role of coupled data assimilation is to simultaneously assimilate all available observations of the coupled system components. Such assimilation brings additional information to each component, compared to a standalone system, achieved through correlations between the coupled system components. This aspect of coupled data assimilation, formally expressed through coupled forecast error covariance, will be examined in further detail.

However, due to increased complexity, coupled modeling systems are potentially more sensitive to the initial conditions than the individual components, and thus require special attention in data assimilation (e.g., Sakaguchi et al., 2012). In general, one can distinguish three main regimes of coupled data assimilation systems: (a) uncoupled, (b) weakly coupled, and (c) strongly coupled. Most interesting and powerful is strongly coupled data assimilation, owing to its direct way of combining the information from models and observations.

Coupled data assimilation has been an active area of research and several coupled data assimilation systems have already been developed (Sugiura et al., 2008; Rasmy et al., 2012; Han et al., 2013). Most of these systems are based on variational data assimilation (VAR), although a few ensemble-based coupled data assimilation systems have also been developed (Zhang et al., 2007; Tardif et al., 2014).

The chapter is organized as follows. We begin by schematically describing coupled modeling assimilation in part 1, followed by classifying coupled data assimilation in part 2, and the challenges of coupled data assimilation in part 3. An example of the two-component coupled data assimilation is presented in part 4, followed by realistic applications of coupled data assimilation to investigate the coupled forecast error covariance and uncertainty interaction in part 6. A summary and future developments are then presented in part 6.

## 9.2 Coupled Modeling System

The development of coupled modeling systems is motivated by a desire to have a holistic prediction system which includes all interactions between relevant physical processes, such as the atmosphere, ocean, land surface, hydrology, cryosphere (e.g., ice, snow), and carbon cycle. The anticipated benefit of having such a system is that an overall "knowledge" of the whole Earth system is improved, and that predictability of individual components can be extended through interactions. It is clear that such a system is extremely complex and potentially very sensitive to its initial state and empirical parameters. By utilizing all available observations to adjust the initial conditions and empirical parameters, coupled data assimilation can improve the prediction of coupled modeling systems. Because of correlations originating from complex interactions between the model components, it is generally assumed that analysis adjustment in coupled data assimilation is in better physical agreement with the real world than analysis adjustment in data assimilation of individual components.

A coupled modeling system is schematically represented in Figure 9.1. While the interaction between the driver and the model components is always present, the interaction between model components is optional and related to the degree of dependence between relevant physical processes. Although practical implementation of such a system can differ, the shown schematic illustrates a need for the coupled systems' driver to combine all components.

## 9.3 Classification Based on Coupling Strength

There are several possibilities for characterizing coupled data assimilation based on the strength of coupling.

*Uncoupled* data assimilation is certainly the most basic since it implies that each forecast and assimilation system is completely independent. There are no interactions between

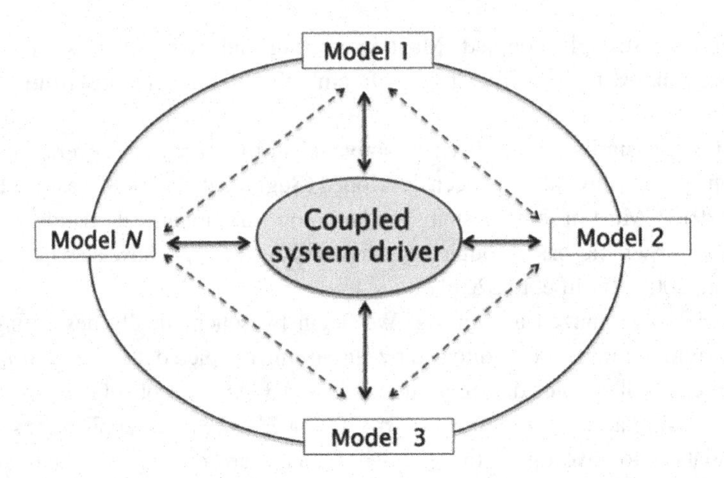


Figure 9.1 Schematic representation of an $N$-component coupled modeling system. The full lines represent the mandatory interaction between the model component and the driver, while the dashed lines represent the optional interaction between components. From Županski (2017). ©Springer International Publishing Switzerland 2017. Used with permission.

components, in both the prediction and the analysis. Since the calculation of prediction and analysis is done separately for each component, a standalone algorithm likely existed before and therefore does not require any additional development. For example, considering the atmosphere and hydrology, each system applies its own system for prediction and analysis, also implying the assimilation of its own observations only. The hydrology model may still use precipitation forcing from the atmospheric model, and the atmospheric model may still use the updated surface conditions from hydrology; however, there is no two-way interaction between hydrology and the atmosphere.

*Weakly coupled* data assimilation is defined as a system that is coupled through the use of a coupled forecast system, but not through data assimilation. There is a two-way interaction between components in the prediction model, but data assimilation is performed independently for each component. The initial conditions and parameters calculated in the standalone analysis of each component are eventually fed back into the coupled model. From the assimilation perspective, weakly coupled data assimilation implies coupling in the forecast step, but no coupling in the analysis step. More importantly, information from observations is still not shared between the components, also implying the forecast error covariances are standalone covariances, i.e., there is no use of the cross-component part of the forecast error covariance in data assimilation.

*Strongly coupled* data assimilation means that both the coupled forecast and coupled data assimilation are conducted in a single system that combines the information from all components. There is a full two-way interaction between the components. Since there is a cross-component covariance used in such data assimilation, each component has the potential to influence all the other components.

It is useful to quantify the coupling strength in terms of Shannon information theory (Shannon and Weaver, 1949). Already discussed in Chapter 2, the entropy is defined as

$$H_E(X) = -\int_x \varphi(x) \ln \varphi(x) dx, \tag{9.1}$$

where $\varphi$ denotes PDF and $x$ is a random variable. In coupled systems we are interested in the probabilistic interaction between at least two processes, formally expressed as joint entropy of processes, $X_1$ and $X_2$,

$$H_E(X_1, X_2) = -\int_{x_1}\int_{x_2} \varphi(x_1, x_2) \ln \varphi(x_1, x_2) dx_1 dx_2, \tag{9.2}$$

where $\varphi(x_1, x_2)$ is their joint PDF.

For random variables, $X_1$ and $X_2$, the information content of variable $X_2$ given $X_1$ can be quantified by conditional entropy $H_E(X_2 \mid X_1)$

$$H_E(X_2 \mid X_1) = H_E(X_1, X_2) - H_E(X_1). \tag{9.3}$$

The information content gives the value-added by one of the variables to the coupled system. The exchange of information in coupled systems can be quantified by mutual information (see also (2.44)), which measures the information shared by the multicomponent coupled system

$$I_E(X;Y) = H_E(X_1) + H_E(X_2) - H_E(X_1, X_2). \tag{9.4}$$

Since $H_E(X_1, X_2) \leq H_E(X_1) + H_E(X_2)$, with the equality sign true for independent processes $X_1$ and $X_2$, the mutual information is nonnegative with a zero value corresponding to independent processes. Given that the motivation of developing coupled systems modeling is to improve the interaction between related physical processes, the component processes of the coupled system have a well-developed dependence and are intrinsically characterized by positive mutual information. This implies that the information from one component enhances the information about other components. This is quite important for the data assimilation of coupled systems since it effectively reduces DOF of the forecast error covariance and allows for cross-component impacts during data assimilation, as will be shown in detail in the next sections.

**Practice 9.1 Information content and mutual information**

Consider a two-component strongly coupled system with Gaussian variables $X_1 \sim N(0, 0.5)$ and $X_2 \sim N(0, 2.0)$ describing the errors of coupled system components, the index "1" referring to the first coupled component and the index "2" referring to the second coupled component. Calculate the information content of each variable and their mutual information.

One can also introduce a classification of coupling in terms of its strength: Low intensity coupling referring to having only marginal interaction between the components and high intensity coupling implying a profound interaction between the components. Such classification of coupling strength can be quantified in principle by calculating the mutual information of the coupled modeling system, i.e., using their prior and joint PDFs. Using the Gaussian PDF as an example, one can define the mutual information of multivariate Gaussian probability distribution (Silva and Quiroz, 2003) as

$$I_E(X_1; X_2) = -\frac{1}{2} \ln \frac{\det(\Sigma)}{\det(\Sigma_{11}) \det(\Sigma_{22})}, \tag{9.5}$$

where $\Sigma$, $\Sigma_{11}$, and $\Sigma_{22}$ are joint, and marginal covariances of variables $X_1$ and $X_2$, respectively, and *det* denotes the determinant of a matrix. Since the joint covariance is

$$\Sigma = \begin{pmatrix} \Sigma_{11} & \Sigma_{12} \\ \Sigma_{12}^T & \Sigma_{22} \end{pmatrix} \tag{9.6}$$

and

$$\det(\Sigma) = \det(\Sigma_{11}) \det(\Sigma_{22}) \det\left(\mathbf{I} - \Sigma_{11}^{-1} \Sigma_{12} \Sigma_{22}^{-1} \Sigma_{12}^T\right) \tag{9.7}$$

(see Arellano-Valle et al., 2013), one can define the mutual information as

$$I_E(X_1; X_2) = -\frac{1}{2} \ln \left[\det\left(\mathbf{I} - \Sigma_{11}^{-1} \Sigma_{12} \Sigma_{22}^{-1} \Sigma_{12}^T\right)\right]. \tag{9.8}$$

It is implied from the above equation that mutual information can be calculated from the auto- and cross-components of the coupled system error covariance. If the cross-component covariance does not exist in Eq. (9.7) ($\Sigma_{12} = 0$), then the mutual information is 0.

In some applications it may be possible to calculate the exact value of the mutual information according to (9.8), or an approximate value using only the diagonal elements of the matrices. Once a mutual information threshold for low/high intensity of the coupling is defined, it may be possible to obtain a quantitative measure of the coupling strength.

**Example 9.1 Determinant of a square matrix**

Let $\mathbf{A} = \begin{pmatrix} a & b \\ c & d \end{pmatrix}$. The determinant is $\det(\mathbf{A}) = ad - bc$. Let $\lambda_1, \lambda_2, \ldots, \lambda_n$ denote the nonzero eigenvalues of the matrix $\mathbf{A}$. Then $\det(\mathbf{A}) = \prod_{i=1}^{n} \lambda_i = \lambda_1 \lambda_2 \ldots \lambda_n$.

**Example 9.2 Mutual information for a $2 \times 2$ covariance matrix**

Let $\mathbf{A} = \begin{pmatrix} a & b \\ b & c \end{pmatrix}$ be a $2 \times 2$ covariance matrix. From (9.5) one can identify $\Sigma_{11} = a$, $\Sigma_{22} = c$, and $\Sigma_{12} = \Sigma_{12}^T = b$. After calculating the determinants required in (9.5) the mutual information is

$$I_E = -\frac{1}{2} \ln \left[1 - \frac{b^2}{a \cdot c}\right]. \tag{9.9}$$

**Practice 9.2 Mutual information**

Calculate mutual information for a $2 \times 2$ covariance matrix in terms of correlation between two Gaussian random variables.

**Example 9.3 Quantifying coupling strength using mutual information**

One has to first define coupling strength in terms of correlations between variables. Let the critical correlation be denoted $\rho_C$. If the correlation between variables is larger than the predefined critical correlation ($\rho > \rho_C$), we refer to it as *high* coupling strength, whereas if the correlation is smaller than critical ($\rho < \rho_C$), we refer to it as *low* coupling strength. For example, let the critical correlation be $\rho_C = 0.5$. Then the critical mutual information is $I_E(X_1; X_2) = 0.1438$. All values smaller than critical will define low coupling strength and higher values will define high coupling strength. Multiple thresholds of coupling strength could be defined as well.

**Practice 9.3 Coupling strength**

Quantify the coupling strength of the $2 \times 2$ covariance matrix

$$\mathbf{A} = \begin{pmatrix} 1.0 & 0.7 \\ 0.7 & 0.25 \end{pmatrix}.$$

## 9.4 Challenges

Coupled data assimilation has several challenges, some of which will be discussed here. For example, control variable, spatiotemporal scales, forecast error covariance, high-dimensional state, and non-Gaussian errors.

### *9.4.1 Control Variable*

We refer to the control variable as the modeling system parameters that impact its prediction. In particular, the prediction has to be sensitive to the choice of these parameters within the spatiotemporal limits of a data assimilation window. For practical reasons one can define the control variable differently from the state variable. For example, one can assume that the control variable includes a stream function and a velocity potential, while the state (model) variable includes the east–west and north–south wind components (e.g., $u$ and $v$, respectively). For simplicity of presentation, however, we will assume here that the control variable is a simple subset of the state variable, without requiring any additional transformation.

In the Earth system modeling based on using partial differential equations, the control variable typically includes (i) initial conditions, (ii) lateral boundary conditions (for regional systems), (iii) model empirical parameters, and (iv) model and observation operator biases (systematic errors). Therefore, for the $k$-th modeling component the control variable can be written as

$$\mathbf{x}_k = \begin{pmatrix} x_k^{ic} & x_k^{bias} & x_k^{par} \end{pmatrix}^T \quad k = 1, \ldots, N_C, \tag{9.10}$$

where the superscripts *ic*, *bias*, and *par* refer to the initial conditions, bias, and empirical parameters, respectively, and $N_C$ is the number of coupled system components. This formulation includes both single and coupled systems, $N_C = 1$ corresponding to the single system and $N_C \geq 2$ corresponding to the coupled system. A general form of control variable in coupled data assimilation is

$$\mathbf{x} = \begin{pmatrix} \mathbf{x}_1 & \mathbf{x}_2 & \cdots & \mathbf{x}_{N_C} \end{pmatrix}^T. \tag{9.11}$$

The challenge of coupled data assimilation originates from the fact that the practical data assimilation for individual components does not include all possible control parameters as in (9.11). Most often the control variable is defined as the initial conditions. However, some data assimilation systems are focused on adjusting not only the initial conditions, but model empirical parameters (e.g., climate), empirical parameters and the initial conditions (hydrology), and the model bias (carbon cycle). This creates a need to develop a coupled data assimilation system that is general enough so that it can assimilate any type of control variable. In principle, the control variable in coupled data assimilation includes relevant control variables for each component. For example, if the coupled system includes the following components: atmospheric – denoted $a$, carbon transport – denoted $c$, land surface – denoted $l$, and hydrological – denoted $h$, one could define the control variable as

$$\mathbf{x} = \begin{pmatrix} \mathbf{a}^{ic} & \mathbf{c}^{bias} & \mathbf{l}^{ic} & \mathbf{l}^{par} & \mathbf{h}^{par} \end{pmatrix}^T. \tag{9.12}$$

Defining the control variable this way may be desirable and efficient, but it does require a mathematical apparatus that can handle it in practice. In the case where the available algorithmic structure of the coupled data assimilation cannot include all these variables, possibly because it was developed as an extension of a standalone algorithm for one component, one may reduce the list to include only a workable subset of parameters. In the long term, however, it is still preferable to include all relevant control variables even if this requires additional development.

### *9.4.2 Forecast Error Covariance*

Forecast error covariance is intrinsically related to the choice of control variable. Since it represents the uncertainty of the control variable, it includes all control variable components. Forecast error covariance has a fundamental role in data assimilation (e.g., Bannister, 2008a, 2008b) since all analysis adjustments are projected onto the subspace of this matrix. There are multiple papers addressing the role of error covariance in variational and ensemble data assimilation (Hollingsworth and Lönnberg, 1986; Derber and Bouttier, 1999; Buehner, 2005; Bello Pereira and Berre, 2006; Berre and Desroziers, 2010).

For a coupled system control variable (9.11), the forecast error covariance is a matrix

$$\mathbf{P}_f = \begin{pmatrix} \mathbf{P}_{11} & \mathbf{P}_{12} & \cdots & \mathbf{P}_{1N_C} \\ \mathbf{P}_{12}^T & \mathbf{P}_{22} & & \mathbf{P}_{2N_C} \\ \vdots & & \ddots & \vdots \\ \mathbf{P}_{1N_C}^T & \mathbf{P}_{2N_C}^T & \cdots & P_{N_C N_C} \end{pmatrix}, \tag{9.13}$$

where the subscripts refer to the coupled system component. Note that each of the inputs in (9.13) is also a matrix. For example

$$\mathbf{P}_{ij} = \begin{pmatrix} P_{ij}^{ic,ic} & P_{ij}^{ic,bias} & P_{ij}^{ic,par} \\ (P_{ij}^{ic,bias})^T & P_{ij}^{bias,bias} & P_{ij}^{bias,par} \\ (P_{ij}^{ic,par})^T & (P_{ij}^{bias,par})^T & P_{ij}^{par,par} \end{pmatrix}, \tag{9.14}$$

where $i$ and $j$ define the system components. It is clear that defining the elements of such a complex matrix becomes challenging.

In VAR, forecast error covariance is modeled using previous knowledge about correlations and variances, typically based on statistics and climatology. However, knowing a priori the correlations between initial conditions, empirical model parameters, and model biases, that may be required in strongly couple data assimilation, is very difficult. On the other hand, in ensemble data assimilation the correlations are obtained directly from the ensemble forecast, without the need for modeling the correlation functions. Unfortunately, ensemble data assimilation also requires at least some knowledge about the true correlations in order to assess if a low-dimensional approximation of the ensemble error covariance is acceptable.

One should be aware that similar challenges exist even in assimilation of individual components; in coupled systems, however, these issues are magnified, being more complex, and thus are more difficult. The most challenging aspect of defining error covariance in coupled data assimilation is to define the cross-component correlations, since the cross-variable correlations, and in particular the cross-component correlations, are least known.

### 9.4.3 High-Dimensional State Vector

The state vector dimension can considerably impact the design and performance of the coupled data assimilation. Realistic atmospheric data assimilation systems have a control variable dimension of the order of tens to hundreds of millions ($O(10^7) - O(10^8)$). Adding an aerosol to the atmospheric control variable can considerably increase the dimension of the augmented control variable. The same is true with adding a chemistry control variable that may include several chemical species, depending on complexity of the chemistry program. Most of these additional control variables are defined on the same 3D grid as the atmospheric variables. Therefore, by augmenting the atmospheric control variable with chemicals and aerosols only, the control variable dimension can increase by at least one order of magnitude. Some other components, such as land surface, may add only a smaller dimension to the augmented state since typically there are fewer soil layers than atmospheric layers, but this can also considerably increase the dimension of the control variable for more complex land surface models.

With all these possibilities for augmenting the control variable, the coupled data assimilation system has to be able to deal with the resulting high dimensions. This may imply that new development may be necessary when extending the single-component assimilation to the coupled data assimilation. Practical atmospheric data assimilation algorithms often include approximations such as dual resolution (i.e., coarse resolution adjoint or ensemble, and high-resolution control), or the incremental minimization approach adequate for weakly nonlinear problems. This may cause serious algorithmic problems when an additional coupled component is more nonlinear than others, or if there is a highly nonlinear observation operator of one of the components. In such cases it may be necessary to revalidate the original assumptions in the context of the coupled system.

### 9.4.4 Non-Gaussian Errors

Some control variables can have non-Gaussian errors, although a typical data assimilation is based on Gaussian error assumption. In the case where there is only a small number of such variables, their impact on the cost function and other measures of the fit to observations may also be small. For example, specific humidity or cloud hydrometeor variables can be non-Gaussian since they are strictly positive definite (i.e., Gaussian distribution formally requires a range from minus to plus infinity). However, atmospheric observations and relatively well-known correlations between atmospheric variables may ameliorate the problem of non-Gaussianity. In the case of coupled systems this issue can become more relevant, especially in the situation where one of the non-Gaussian variables may be the only variable describing one coupled component. This may apply to atmospheric chemistry and aerosol variables, which are also strictly positive definite. There may be fewer observations of these variables compared to observations of atmospheric variables, and thus less direct information that could improve their estimate. In addition, the cross-component correlations are generally less known or accurate, making their update from observations of other coupled components less reliable. Consequently, in the case of coupled atmosphere-chemistry-aerosol data assimilation, there may be a need to revisit the Gaussian error assumptions. In more extreme cases, it may be necessary to design a mixed Gaussian and non-Gaussian data assimilation system to accommodate all coupling components.

### 9.4.5 Spatiotemporal Scales

Characteristic spatiotemporal scales of control variables may present another potential difficulty of coupled data assimilation. For example, atmospheric variables generally have shorter temporal scales than ocean variables, and possibly the stratospheric ozone too. In the coupled system model this may be resolved by using different time steps for different components of the system. For data assimilation, however, this could become more complicated as it involves the coupled forecast error covariance with generally less reliable estimates due to the insufficient knowledge of cross-component correlations (e.g., variational methods), or due to insufficient rank (e.g., ensemble methods).

Therefore, in VAR the modeled error covariance has to include some knowledge of the spatiotemporal scales in order to produce dynamically balanced increments of both components. Since in addition to limited knowledge of these correlations they can also be situation-dependent, the use of complete cross-component correlations in practical VAR is not very likely. The use of ensemble error covariance may be algorithmically simpler since the information about the spatiotemporal scales is already included in the coupled forecast models used in ensemble forecasting. However, since both variational and ensemble approaches can produce only an approximate error covariance, there is still a need to verify the validity of the coupled error covariance in terms of the spatiotemporal scales. The means resolving this issue may be more intuitive in data assimilation with 4D error covariance, since then one can impose the temporal aspect of error covariance more directly.

## 9.5 One-Point, Two-Component Coupled Data Assimilation System

An idealized two-component coupled modeling system can be used to explain the role of coupled forecast error covariance in coupled data assimilation. Such an approach simplifies the mathematical representation while still maintaining the basic interactions within a data assimilation system. Let us define the two state components as $x_1$ and $x_2$, thus forming a 2D state vector of the coupled system

$$\mathbf{x} = \begin{pmatrix} x_1 & x_2 \end{pmatrix}^T. \tag{9.15}$$

Each coupled component is represented by a value at a single grid point. The location of the single grid point does not have to be the same for each coupled component. For example, in an atmosphere-land-atmosphere coupled system the gridpoint defining the location of the atmospheric variable may be different from the gridpoint defining the location of the soil variable. The forecast step of such systems includes a coupled prediction model (denoted $M$)

$$\mathbf{x}^n = M\left(\mathbf{x}^{n-1}\right), \tag{9.16}$$

where the superscript $n$ defines time. We now proceed with the analysis step at time $n$. The cost function formally appears the same in coupled data assimilation as in any other data assimilation. Assuming for simplicity a sequential data assimilation, the cost function for a two-component coupled system can be formally defined as

$$\begin{aligned} f(x_1,x_2) = {} & \frac{1}{2}\begin{pmatrix} x_1 - x_1^f \\ x_2 - x_2^f \end{pmatrix}^T \begin{pmatrix} P_{11} & P_{12} \\ P_{12}^T & P_{22} \end{pmatrix}^{-1} \begin{pmatrix} x_1 - x_1^f \\ x_2 - x_2^f \end{pmatrix} \\ & + \frac{1}{2}\begin{pmatrix} y_1^o - H_1 x_1^f \\ y_2^o - H_2 x_2^f \end{pmatrix}^T \begin{pmatrix} R_1 & 0 \\ 0 & R_2 \end{pmatrix}^{-1} \begin{pmatrix} y_1^o - H_1 x_1^f \\ y_2^o - H_2 x_2^f \end{pmatrix}. \end{aligned} \tag{9.17}$$

The subscripts 1 and 2 refer to the state vector of the first and second components, respectively. Implied in (9.16) is that the forecast error covariance can have cross-component correlations, but observation errors are independent. A general linear solution to minimization of (9.17) can be obtained after setting $\nabla f = 0$ (e.g., Lorenc, 1986)

$$\mathbf{x}^a = \left(\mathbf{I} + \mathbf{P}_f \mathbf{H}^T \mathbf{R}^{-1} \mathbf{H}\right)^{-1} \left(\mathbf{x}^f + \mathbf{P}_f \mathbf{H}^T \mathbf{R}^{-1} \mathbf{y}^o\right) \tag{9.18}$$

with

$$\mathbf{x}^a = \begin{pmatrix} x_1^a \\ x_2^a \end{pmatrix} \quad \mathbf{x}^f = \begin{pmatrix} x_1^f \\ x_2^f \end{pmatrix} \quad \mathbf{y}^o = \begin{pmatrix} y_1^o \\ y_2^o \end{pmatrix}. \tag{9.19}$$

Let us assume that there is only a single observation of one of the components, say the component 1, defined at that grid point. Then the cost function (9.17) becomes

$$f(x_1, x_2) = \frac{1}{2} \begin{pmatrix} x_1 - x_1^f \\ x_2 - x_2^f \end{pmatrix}^T \begin{pmatrix} (\sigma_f)_1^2 & \rho(\sigma_f)_1(\sigma_f)_2 \\ \rho(\sigma_f)_1(\sigma_f)_2 & (\sigma_f)_2^2 \end{pmatrix}^{-1} \begin{pmatrix} x_1 - x_1^f \\ x_2 - x_2^f \end{pmatrix}$$
$$+ \frac{1}{2}[y_1^o - x_1^f]^T \left(\sigma_o^2\right)_1^{-1} [y_1^o - x_1^f], \tag{9.20}$$

where $\sigma$ denotes the standard deviation, the cross-component error covariance (directly related to the correlation between the coupling components) is denoted $\rho$, and indexes $f$ and $o$ refer to the forecast and the observation, respectively. The analysis solution (9.18) and (9.19) for the cost function (9.20) in terms of the components is

$$x_1^a = \frac{1}{1+\varepsilon^2} x_1^f + \frac{\varepsilon^2}{1+\varepsilon^2} y_1^o, \tag{9.21}$$

$$x_2^a = x_2^f + \rho \left[\frac{(\sigma_f)_2}{(\sigma_f)_1}\right] \frac{\varepsilon^2}{1+\varepsilon^2} \left[y_1^o - x_1^f\right], \tag{9.22}$$

where $\varepsilon$ represents the ratio between forecast and observation standard deviations

$$\varepsilon = \frac{(\sigma_f)_1}{(\sigma_o)_1}. \tag{9.23}$$

The solutions (9.21) and (9.22) illustrate several important aspects of coupled data assimilation. When observing only the first coupled component, the analysis is identical to the standalone assimilation of the first component (e.g., (9.21)). Although the coupled covariance has cross-component correlations, the analysis of the first component does not benefit from the coupling. However, the analysis of the second component is impacted by the observation of the first component, even though the second component is not observed. This is important since in principle it allows a change of the initial conditions of all coupled components by observing any component. A closer inspection reveals that this was possible only because of the nonzero cross-correlation between the components, i.e., $\rho \neq 0$, assuming $y_1^o \neq x_1^f$. For example, consider a coupled atmosphere-land-surface system. There are often instances when land surface is not well observed, while atmosphere is well observed. In that situation the above results imply that land surface initial conditions could still be adjusted, something that would not be possible in the standalone land surface analysis system.

## 9.6 Structure of Coupled Forecast Error Covariance

Following an improved understanding of coupled forecast error covariance revealed by the one-point, two-component coupled system, we now focus on illustrating the impact of coupled error covariance in realistic meteorological systems. As suggested above, the true power of coupled data assimilation comes from the cross-component correlations. They allow a more efficient use of observations by impacting all control variables, and thus produce a more balanced change of control variables. The structure of the forecast error covariance can be assessed using a single-observation experiment setup, where only one observation at a single point is assimilated (e.g., Thépaut et al., 1996; Whitaker et al., 2009). We apply the maximum likelihood ensemble filter (MLEF) (Županski, 2005; Županski et al., 2008) as a coupled data assimilation system. In all experiments we use 32 ensembles. Since this system employs the ensemble coupled error covariance, there is no need to model the cross-components as they are created automatically by the ensemble forecasting.

We first consider a land-atmosphere coupling by using the NASA Unified Weather Research and Forecasting (NU-WRF) model (Peters-Lidard et al., 2015) that has an implicit coupling between the Noah land surface model and the WRF atmospheric component. The horizontal model resolution of the parent domain is 27 km with the nest at 9 km. There are 31 vertical layers in the atmosphere, and 4 vertical layers in the soil. The augmented control variable includes (i) atmospheric variables, such as perturbation surface pressure, perturbation potential temperature, perturbation geopotential, horizontal winds, specific humidity, (ii) cloud variables, such as cloud ice, cloud snow, cloud rain, cloud water, graupel, and (iii) land surface variables that include soil moisture and soil temperature. A pseudo-observation of cloud rain at 700 hPa is assimilated and its impact is evaluated in the analysis, in particular on land surface variables.

In Figure 9.2. the analysis increments ($\mathbf{x}^a - \mathbf{x}^f$) are shown for cloud rain and soil moisture. The auto-correlation of cloud rain appears to be adequately represented by the ensemble, indicated by the maximum near the observed location and the response in the lower troposphere spreading down to the surface (Figure 9.2a). From the coupled data assimilation point of view, the analysis increments of soil variables are more interesting (e.g., Figure 9.2b). One can notice that the soil moisture analysis increments are nonzero, implying the existence of a cross-component correlation between cloud rain and soil moisture. The impact of cloud rain observation has spread into all soil layers, with a stronger impact on the near-surface layers. A possible physical interpretation of these results is that an increase of cloud rain at 700 hPa induces an increase in moisture at the surface, which eventually spreads into the soil causing an increase in soil moisture. Although one could have anticipated the impact of cloud rain on the top layer of soil model, it may be unlikely that it has such an extensive spread throughout the soil layers. The MLEF system used in these experiments did not employ the covariance localization in the vertical, so

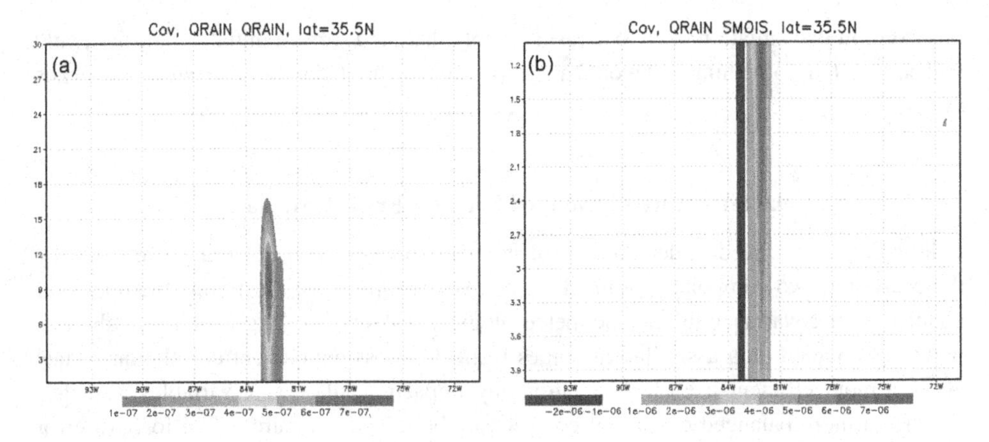


Figure 9.2 Vertical cross section of analysis increments $\mathbf{x}^a - \mathbf{x}^f$ for (a) cloud rain (kg kg$^{-1}$) and (b) soil moisture (kg kg$^{-1}$), as a response to a single pseudo-observation of cloud rain at 700 hPa. Note that the vertical index increases downward for soil moisture, thus defining the depth. The vertical location of single observation is near level 11. From Županski (2017). ©Springer International Publishing Switzerland 2017. Used with permission.

this response should be taken with some caution. An additional result worth noticing is related to the spatiotemporal character of the interaction between the atmospheric and soil variables. A careful investigation of the location of these impacts shows that the analysis increments of the soil moisture lag the analysis increments of the atmospheric variable by about 60 km, with a shift toward the east. This may suggest that in this situation the forecast error does not represent an instantaneous response, which would be characterized by a response at the same location. Rather, it appears to reflect the interactions between the atmosphere and the soil as represented by the coupled modeling system, in this case characterized by a delay of land surface response. This may have some relevance to the issue of spatiotemporal scale interaction in the coupled system (Section 3.5), but certainly requires further evaluation.

Another example is a coupled atmosphere-chemistry model WRF-Chem (Grell et al., 2005). In this case, the control variables include the standard atmospheric variables and five species of chemical constituents ($O_3$, $SO_2$, $SO_3$, $NO_2$, and $NO_3$). The structure of the coupled atmosphere-chemistry error covariance has been investigated in more detail in Park et al. (2015). Several additional cross-component correlations that confirm the inherent complexity of the coupled forecast error covariance are presented here. As in Park et al. (2015), the ozone monitoring instrument (OMI) total column ozone at a single point is assimilated. In Figure 9.3 we present the additional cross-component correlation that such an atmosphere-chemistry coupled system describes, in terms of the specific humidity (Figure 9.3a) and the east–west wind component (Figure 9.3b). It is noticeable that atmospheric variables have well-defined analysis increments in both the vertical and horizontal directions. Specific humidity has somewhat smaller magnitudes than the wind. It is also interesting to note that analysis responses for atmospheric variables are located in the middle and lower troposphere, while the maximum ozone response is at 250 hPa

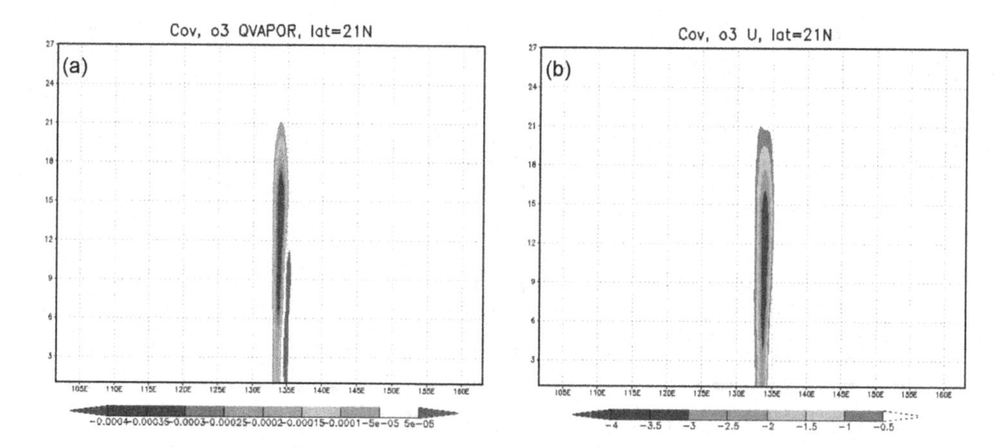


Figure 9.3 Vertical cross section of analysis increments $x^a - x^f$ for (a) specific humidity ($\mathrm{kg\,kg^{-}1}$) and (b) east–west wind component ($\mathrm{m\,s^{-1}}$), as a response to a single observation of an OMI total column ozone. The single observation location is near level 24. From Županski (2017). ©Springer International Publishing Switzerland 2017. Used with permission.

(e.g., Park et al., 2015). These results confirm that the cross-component covariance captures important information that could potentially benefit the coupled atmosphere-chemistry data assimilation.

We also consider data assimilation with a coupled model of atmosphere-aerosol-chemistry, in this case the WRF-Chem model with the Goddard chemistry aerosol radiation and transport (GOCART) aerosol model (Chin et al., 2000) that includes the prediction of sulfates, black carbon, organic carbon, sea salt, and dust. We pay special attention to investigating the correlations between ozone and dust. Although such correlations are generally unknown, the results suggest that they contain information that could be made useful in coupled data assimilation. The augmented control variable includes atmospheric variables, chemistry variables, and the aerosol variables (dust) at 0.5 μm, 1.4 μm, 2.4 μm, 4.5 μm, and 8.0 μm. As in the earlier example, a single-point OMI total column ozone observation is assimilated. In Figure 9.4 we show the analysis increments for selected dust variables. The cross-component correlation between ozone and dust indicates that the maximum analysis response of dust to ozone observation is at 300–400 hPa. One can notice well-defined analysis increments, which suggest that cross-component correlations have at least some physical credibility, with potentially important information about the coupled system uncertainties. A closer inspection of Figures 9.4a and 9.4b shows that, although the responses are similar, larger dust particles have the maximum response at lower levels than small particles, which could be expected due to their weight. The magnitude of the increments is larger for smaller dust particles as well.

The above results indicate that the structure of coupled ensemble forecast error covariance is complex and that these correlations contain important information about cross-component interactions, with potential benefit for coupled data assimilation.

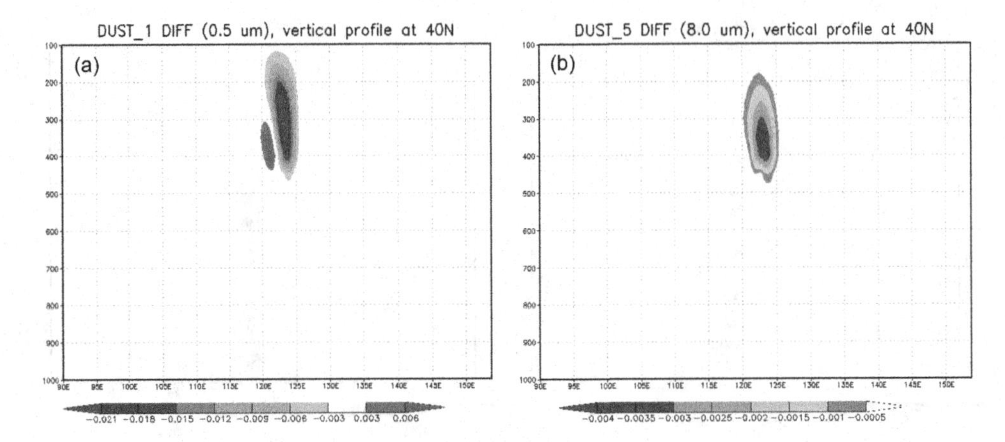


Figure 9.4 Vertical cross section of analysis increments $x^a - x^f$ in response to a single OMI total column ozone observation for (a) dust-1 (0.5 μm) and (b) dust-5 (8.0 μm). From Županski (2017). ©Springer International Publishing Switzerland 2017. Used with permission.

## 9.7 Summary and Future

Coupled data assimilation is an important component of coupled system modeling. It can provide a simultaneous change to the initial conditions and empirical parameters of all coupled components. The simultaneous adjustment also acts as a constraint in data assimilation, effectively preventing the changes in one component that are contradictory to the changes in the other component of a coupled system. In particular, the strongly coupled data assimilation has the potential to improve the forecast of various Earth science components, and also contributes to better understanding of the coupled systems' state and uncertainty.

Formally, coupled data assimilation is a system similar to a single-component data assimilation, however with an increased difficulty in development and implementation. The main challenges that make coupled data assimilation difficult are associated with increased state dimensions, the representation of the cross-component correlations in the coupled forecast error covariance, the increased complexity of the augmented control variable, possible non-Gaussian errors, and the interactions between the coupled components characterized by different spatiotemporal scales.

The results of a one-point, two-component coupled system analysis indicate the relevance of cross-component correlations. These cross-component correlations have the potential to increase the utility of observations in data assimilation by spreading the information throughout the components. The conducted single-observation assimilation experiments confirm that the structure of cross-component correlations is complex and clearly related to the dynamical links between the coupled components and their control variables.

There still remain numerous challenges of coupled data assimilation. A particularly urgent issue is related to the use of hybrid variational-ensemble systems, in which the cross-component correlations of variational and ensemble components of the hybrid error covariance need to be reconciled. However, given the relevance and the increased interest in performing the forecasts with coupled modeling systems, the role of coupled data assimilation and its benefits will likely steadily increase.

COLLOQUY 9.1

**Covariance localization in coupled systems**

Covariance localization is a necessity in realistic ensemble and hybrid variational-ensemble data assimilation systems, as already discussed in Section 8.3.3. In coupled data assimilation systems, there is an additional challenge of covariance localization that originates from different spatiotemporal scales of coupled system components. For example, consider a land surface – atmosphere coupled 3D system. The forecast error covariance in block form can be derived from (9.12)

$$\mathbf{P}_f = \begin{pmatrix} \mathbf{P}_{aa} & \mathbf{P}_{as} \\ \mathbf{P}_{as}^T & \mathbf{P}_{ss} \end{pmatrix}, \tag{9.24}$$

where subscript $a$ denotes atmosphere and $s$ land surface. The localizing matrix $\mathbf{L}$ will have the same components and matching dimensions. The localized forecast error covariance is

$$\mathbf{P}_L = \mathbf{L} \circ \mathbf{P}_f = \begin{pmatrix} \mathbf{L}_{aa} \circ \mathbf{P}_{aa} & \mathbf{L}_{as} \circ \mathbf{P}_{as} \\ [\mathbf{L}_{as} \circ \mathbf{P}_{as}]^T & \mathbf{L}_{ss} \circ \mathbf{P}_{ss} \end{pmatrix}. \tag{9.25}$$

The block diagonal matrices $\mathbf{L}_{aa} \circ \mathbf{P}_{aa}$ and $\mathbf{L}_{ss} \circ \mathbf{P}_{ss}$ are relatively straightforward to localize, having their own prescribed localization scales. However, implementing the cross-component localized covaraince matrix $\mathbf{L}_{as} \circ \mathbf{P}_{as}$ and its transpose is more challenging.

The difficulty can be best understood using the localization in the vertical as an example. In this case, an atmospheric variable such as a 2-meter (i.e., surface) temperature can generally impact the soil moisture in the top soil layers, but not in the bottom soil layers. Using a prescribed atmospheric localization length, which can be of the order of thousands of meters, a straightforward application to the land surface grid would create unrealistically high correlations with soil moisture in all soil layers since the soil depth is typically only a few meters. Therefore, there is a need to adjust localization length when linking the variables from two components of a coupled system.

We now introduce a method for accounting for different spatiotemporal scales in coupled system localization using an analog with optics, in particular the refraction of light when passing through different media. Recall that *refractive index* of light is defined as a ratio between the speed of light in vacuum $c$ and the phase velocity of light in the medium $v$

$$n = \frac{c}{v}. \tag{9.26}$$

The refractive index is used to define the *optical path length* ($d^{opt}$)

$$d^{opt} = n \cdot \ell, \tag{9.27}$$

where $\ell$ is the geometric length. We use an equivalent to optical path length, denoted *localization path length*, and an equivalent to refractive index, denoted *localization refractive index*, when comparing and calculating localization distances in different components of a coupled system.

To proceed, one first needs to define the characteristic spatial ($D$) and temporal ($T$) scales of the particular coupled component, and then form an equivalent to $v$

$$v_i = \frac{D_i}{T_i}, \tag{9.28}$$

where the index $i$ denotes the coupled component. Assuming that localization scales are given for the atmosphere, instead of $c$ one defines the referent speed $c_{ref}$ to be equivalent to the characteristic speed in the atmosphere

$$c = c_{ref} = v_a, \tag{9.29}$$

which makes the localization refractive index in the atmosphere equal to 1 ($n_a = \frac{c_{ref}}{v_a} = 1$) and also implies that the geometric distance and localization optical path length in the atmosphere are the same (i.e., $d_a^{opt} = \ell_a$). The localization refractive index in soil is then

$$n_s = \frac{v_a}{v_s}. \tag{9.30}$$

Finally, the localization length between two points in the atmosphere and in the soil is

$$d^{opt} = d_a^{opt} + d_s^{opt} = n_a \ell_a + n_s \ell_s, \tag{9.31}$$

where $\ell_a$ and $\ell_s$ are geometric distances in the atmosphere and soil, respectively.

As a numerical example, one could define for atmosphere $D_a = 1000$ m, $T_a = 3600$ s, so that $v_a = \frac{D_a}{T_a} \approx 2.8 \times 10^{-1}$ m/s. For soil, one could define $D_s = 1$ m, $T_s = 12 \times 3600$ s, implying $v_s = \frac{D_s}{T_s} \approx 2.3 \times 10^{-5}$ m/s. Consequently, the localization refractive indices are $n_a = 1$ and $n_s = 1.2 \times 10^4$. For $\ell_a = 1500$ m and $\ell_s = 0.2$ m the localization distance is

$$d^{opt} = 1 \cdot 1500 \text{ m} + 1.2 \times 10^4 \cdot 0.2 \text{ m} = 1500 \text{ m} + 2400 \text{ m} = 3900 \text{ m}. \tag{9.32}$$

Similar reasoning can be applied to horizontal length scales. Note that, even though the horizontal grid points of atmospheric and land surface models can be overlapping, the localization optical path lengths can be different. Although our discussion above was applied to state-space localization, the same arguments could be used for observation-space localization as well.

**References**

Arellano-Valle RB, Contreras-Reyes JE, Genton MG (2013) Shannon entropy and mutual information for multivariate skew-elliptical distributions. *Scand J Stat* 40:42–62.

Bannister RN (2008a) A review of forecast error covariance statistics in atmospheric variational data assimilation. I: Characteristics and measurements of forecast error covariances. *Quart J Roy Meteor Soc* 134:1951–1970.

Bannister RN (2008b) A review of forecast error covariance statistics in atmospheric variational data assimilation. II: Modelling the forecast error covariance statistics. *Quart J Roy Meteor Soc* 134:1971–1996.

Bello Pereira M, Berre L (2006) The use of an ensemble approach to study the background error covariances in a global NWP model. *Mon Wea Rev* 134:2466–2489.

Berre L, Desroziers G (2010) Filtering of background error variances and correlations by local spatial averaging: A review. *Mon Wea Rev* 138:3693–3720.

Buehner M (2005) Ensemble-derived stationary and flow-dependent background-error covariances: Evaluation in a quasi-operational NWP setting. *Quart J Roy Meteor Soc* 131:1013–1043.

Chin M, Rood RB, Lin SJ, Müller JF, Thompson AM (2000) Atmospheric sulfur cycle simulated in the global model GOCART: Model description and global properties. *J Geophys Res Atmos* 105:24671–24687.

Derber J, Bouttier F (1999) A reformulation of the background error covariance in the ECMWF global data assimilation system. *Tellus A* 51:195–221.

Grell GA, Peckham SE, Schmitz R, et al. (2005) Fully coupled "online" chemistry within the WRF model. *Atmos Environ* 39:6957–6975.

Han G, Wu X, Zhang S, Liu Z, Li W (2013) Error covariance estimation for coupled data assimilation using a lorenz atmosphere and a simple pycnocline ocean model. *J Clim* 26:10218–10231.

Hollingsworth A, Lönnberg P (1986) The statistical structure of short-range forecast errors as determined from radiosonde data. Part I: The wind field. *Tellus A* 38:111–136.

Lorenc AC (1986) Analysis methods for numerical weather prediction. *Quart J Roy Meteor Soc* 112:1177–1194.

Park SK, Lim S, Županski M (2015) Structure of forecast error covariance in coupled atmosphere-chemistry data assimilation. *Geosci Model Dev* 8:1315.

Peters-Lidard CD, Kemp EM, Matsui T, et al. (2015) Integrated modeling of aerosol, cloud, precipitation and land processes at satellite-resolved scales. *Environ Modell Softw* 67:149–159.

Rasmy M, Koike T, Kuria D, et al. (2012) Development of the coupled atmosphere and land data assimilation system (caldas) and its application over the tibetan plateau. *IEEE Trans Geosci Remote Sens* 50:4227–4242.

Sakaguchi K, Zeng X, Brunke MA (2012) The hindcast skill of the CMIP ensembles for the surface air temperature trend. *J Geophys Res Atmos* 117:D16113, doi:10.1029/2012JD017765

Shannon CE, Weaver W (1949) *The Mathematical Theory of Communication*. University of Illinois Press, Chicago, IL, 117 pp.

Silva C, Quiroz A (2003) Optimization of the atmospheric pollution monitoring network at Santiago de Chile. *Atmos Environ* 37:2337–2345.

Sugiura N, Awaji T, Masuda S, et al. (2008) Development of a four-dimensional variational coupled data assimilation system for enhanced analysis and prediction of seasonal to interannual climate variations. *J Geophys Res Oceans* 113:C10017, doi:10.1029/2008JC004741

Tardif R, Hakim GJ, Snyder C (2014) Coupled atmosphere–ocean data assimilation experiments with a low-order climate model. *Clim Dyn* 43:1631–1643.
Thépaut J-N, Courtier P, Belaud G, Lemaître G (1996) Dynamical structure functions in a four-dimensional variational assimilation: A case study. *Quart J Roy Meteor Soc* 122:535–561.
Whitaker JS, Compo GP, Thépaut JN (2009) A comparison of variational and ensemble-based data assimilation systems for reanalysis of sparse observations. *Mon Wea Rev* 137:1991–1999.
Zhang S, Harrison M, Rosati A, Wittenberg A (2007) System design and evaluation of coupled ensemble data assimilation for global oceanic climate studies. *Mon Wea Rev* 135:3541–3564.
Županski M (2005) Maximum likelihood ensemble filter: Theoretical aspects. *Mon Wea Rev* 133:1710–1726.
Županski M (2017) Data assimilation for coupled modeling systems. In *Data Assimilation for Atmospheric, Oceanic and Hydrologic Applications* (vol. III), (eds.) Park SK, Xu L, Springer International Publishing, Cham, 55–70.
Županski M, Navon IM, Županski D (2008) The Maximum Likelihood Ensemble Filter as a non-differentiable minimization algorithm. *Quart J Roy Meteor Soc* 134:1039–1050.

# 10
# Dynamics and Data Assimilation

## 10.1 Introduction

The goal of this chapter is to show that data assimilation has an important link with prediction and to describe a few basic characteristics of this link. The number of topics related to the analysis step of data assimilation can give the impression that data assimilation is all about an optimal fit to observations, potentially neglecting the impact of the prediction step of data assimilation.

We have already seen in Chatepr 3 that prediction is an essential component of data assimilation. Pointwise estimates of the prediction of the PDF, such as the mean or the mode are used to define the guess field. Uncertainty of prediction is often used to form the forecast error covariance, which defines the subspace of analysis increments in data assimilation. From the definition of cost function in Chatepr 2 the background component refers to prediction and it balances the fit to observations given by the observation cost function. We have seen that optimal analysis is essentially an interpolation between the background and the observations. One important application of data assimilation is to provide optimal initial conditions for the prediction model. It is therefore clear that prediction has to be an intrinsic component of data assimilation. If its impact is artificially altered the analysis would overfit or underfit observations, likely causing instabilities and unnecessary adjustments in the prediction. In other words, incorrect use of prediction in data assimilation will adversely impact dynamical balances in the system and result in suboptimal prediction after data assimilation. This is particularly detrimental in recursive data assimilation since negative effects will accumulate over time. In this chapter we address these issues in more detail.

## 10.2 Probabilistic Prediction

A common assumption is that prediction processes are Markov, implying that the future is dependent on the present only. In terms of densities

$$\varphi\left(\mathbf{x}_k|\mathbf{x}_1,\ldots,\mathbf{x}_{k-1}\right)=\varphi\left(\mathbf{x}_k|\mathbf{x}_{k-1}\right), \tag{10.1}$$

where $\mathbf{x}$ is the state and the index $k$ is the time index. The conditional density $\varphi\left(\mathbf{x}_k|\mathbf{x}_{k-1}\right)$ is often referred to as the transition density, as it represents the change of PDF over time. The Markov process is a stochastic analog of ordinary differential equations describing a deterministic dynamical process. As such, it is commonly represented by a stochastic differential equation. Of particular relevance for data assimilation is the stochastic differential equation with additive Gaussian forcing

$$d\mathbf{x}_t = M(\mathbf{x}_t, t)\, dt + g(\mathbf{x}_t, t)\, d\omega_t, \tag{10.2}$$

where the index $t$ defines time, $M$ denotes the nonlinear dynamical prediction model, and $g$ is a forcing function applied to Gaussian random variable $\omega_t$. The last equation is also referred to as the Itô stochastic differential equation. The Kolmogorov (or Fokker–Planck) equation can be derived from (10.2), assuming that all derivatives exist (e.g., Jazwinski, 1970),

$$\frac{\partial \varphi(\mathbf{x}, t)}{\partial t} = -\frac{\partial \left[M(\mathbf{x}, t)\varphi(\mathbf{x}, t)\right]}{\partial x} + \frac{1}{2}\frac{\partial^2 \left[g^2(\mathbf{x}, t)\varphi(\mathbf{x}, t)\right]}{\partial x^2}. \tag{10.3}$$

All of the above equations imply that prediction is probabilistic. Unfortunately, solving Kolmogorov or similar probabilistic equations in terms of PDF is not feasible in realistic applications, even with moderate dimensions. Given the general availability of deterministic prediction models, a practical solution to probabilistic prediction is to use multiple deterministic predictions, referred to as ensemble prediction.

## 10.3 Error Growth and the Lyapunov Exponent

The repetitive use of prediction in data assimilation over time makes understanding of the characteristics of long-term prediction relevant for data assimilation. Most geosciences prediction models are also chaotic, which introduces additional complexity. We are mostly interested in growth of the error

$$\mathbf{u}_n(\mathbf{x}_0, \epsilon) = M_n(\mathbf{x}_0 + \epsilon) - M_n(\mathbf{x}_0), \tag{10.4}$$

where the index $n$ refers to time, $M$ is a nonlinear dynamical model used for prediction, $\mathbf{x}_0$ is the initial state at $t = 0$, and $\epsilon$ is the initial state error. The last expression indicates that the final error depends not only on the initial error, but also on the initial state.

In general, prediction errors can grow, stay the same (e.g., neutral growth), or decay over time. A detailed investigation of the link between the prediction error subspaces and data assimilation can be found in Gurumoorthy et al. (2017), Bocquet et al. (2017), and Carrassi et al. (2018). Error growth can be assessed in different ways, but a relatively simple way is by looking at the ratio between prediction error and initial error in terms of a norm

$$r = \frac{\| M_n(\mathbf{x}_0 + \epsilon) - M_n(\mathbf{x}_0) \|}{\| \epsilon \|}, \tag{10.5}$$

where $r > 1$ defines growing errors, $r = 1$ defines neutral errors, and $r < 1$ defines decaying errors.

**Example 10.1 Conditions for error growth**

We want to apply Eq. (10.5) for linear and quadratic prediction models. We also assume that $\epsilon \neq 0$. (a) A one-point linear model is given by $Mx_0 = bx_0$ where $x_0$ and $b$ are real numbers for the one-point model. Then $M[x_0 + \epsilon] - M[x_0] = b(x_0 + \epsilon) - bx_0 = b\epsilon$, with $\epsilon$ also a real number. Substitution in (10.5) gives $r = \frac{|b\epsilon|}{|\epsilon|} = |b|$, where the norm is substituted by the absolute value. Therefore, the error growth estimate based on Eq. (10.5) implies that the linear prediction model has neutral error growth for $|b| = 1$, unstable growth for $|b| > 1$, and decaying errors for $|b| < 1$. (b) A quadratic model is given by $M(x_0) = cx_0^2$ where $c$ is a constant. Then $M(x_0 + \epsilon) - M(x_0) = c(x_0 + \epsilon)^2 - cx_0^2 = c\epsilon(2x_0 + \epsilon)$. The critical value for neutral error growth is achieved for $|c| \cdot |2x_0 + \epsilon| = 1$. Therefore, the estimate based on Eq. (10.5) implies that for $|2x_0 + \epsilon| = \frac{1}{|c|}$ there will be neutral error growth. Additionally, for $|2x_0 + \epsilon| > \frac{1}{|c|}$ there is unstable error growth and for $|2x_0 + \epsilon| < \frac{1}{|c|}$ the error is decaying.

**Practice 10.1 Growth/decay of error**

Determine whether the error is growing, decaying, or neutral for a quadratic model $M(x_0) = x_0^2$ with $x_0 = 0$ and $\epsilon = 0.5$.

It is clear that data assimilation would like to capture growing errors, since then the analysis would include only decaying and neutral errors, therefore making the ensuing prediction more accurate, ideally without growing errors. This would make the prediction very good. To see this, consider a simplified linear prediction system

$$\mathbf{x}_n = \mathbf{M}\mathbf{x}_{n-1}. \tag{10.6}$$

Assuming the model is perfect,

$$\mathbf{u}_n = \mathbf{x}_n - \mathbf{x}_n^t = \mathbf{M}\mathbf{x}_{n-1} - \mathbf{M}\mathbf{x}_{n-1}^t = \mathbf{M}\left[\mathbf{x}_{n-1} - \mathbf{x}_{n-1}^t\right] = \mathbf{M}\mathbf{u}_{n-1}, \tag{10.7}$$

where $\mathbf{x}^t$ denotes the truth. Define the initial error that includes growth and decay errors

$$\mathbf{u}_0 = \mathbf{u}_{grow} + \mathbf{u}_{decay}, \tag{10.8}$$

implying that the initial state is

$$\mathbf{x}_0 = \mathbf{x}_0^t + \mathbf{u}_0. \tag{10.9}$$

This gives the guess at next time as

$$\mathbf{x}_1 = \mathbf{M}\mathbf{x}_0 = \mathbf{M}\left[\mathbf{x}_0^t + \mathbf{u}_0\right] = \mathbf{M}\mathbf{x}_0^t + \mathbf{M}\mathbf{u}_{grow} + \mathbf{M}\mathbf{u}_{decay}. \tag{10.10}$$

By definition

$$\frac{\|\mathbf{Mu}_{grow}\|}{\|\mathbf{u}_{grow}\|} > 1, \quad \frac{\|\mathbf{Mu}_{decay}\|}{\|\mathbf{u}_{decay}\|} < 1. \tag{10.11}$$

The analysis is a correction to the guess $\mathbf{x}_1^a = \mathbf{x}_1 + \delta\mathbf{x}$. Ideally, one would like to remove all errors from the initial conditions. However, note that if we can choose $\delta\mathbf{x} = -\mathbf{Mu}_{grow}$ the analysis is

$$\mathbf{x}_1^a = \mathbf{x}_1 - \mathbf{Mu}_{grow} = \mathbf{Mx}_0^t + \mathbf{Mu}_{decay} = \mathbf{x}_1^t + \mathbf{Mu}_{decay}. \tag{10.12}$$

Applying the prediction model operator one more time gives the guess at the next time as

$$\mathbf{x}_2 = \mathbf{Mx}_1 = \mathbf{M}\left[\mathbf{x}_1^t + \mathbf{Mu}_{decay}\right] = \mathbf{x}_2^t + \mathbf{M}^2 u_{decay}. \tag{10.13}$$

After $n$ repetitions

$$\mathbf{x}_n = \mathbf{x}_n^t + \mathbf{M}^n \mathbf{u}_{decay}. \tag{10.14}$$

Therefore,

$$\begin{aligned}\frac{\|\mathbf{M}^n\mathbf{u}_{decay}\|}{\|\mathbf{u}_{decay}\|} &= \frac{\|\mathbf{M}^n\mathbf{u}_{decay}\|}{\|\mathbf{M}^{n-1}\mathbf{u}_{decay}\|} \cdot \frac{\|\mathbf{M}^{n-1}\mathbf{u}_{decay}\|}{\|\mathbf{M}^{n-2}\mathbf{u}_{decay}\|} \cdots \frac{\|\mathbf{M}^2\mathbf{u}_{decay}\|}{\|\mathbf{Mu}_{decay}\|} \cdot \frac{\|\mathbf{Mu}_{decay}\|}{\|\mathbf{u}_{decay}\|} \\ &= r_n r_{n-1} \cdots r_2 r_1,\end{aligned} \tag{10.15}$$

where $r_k = \frac{\|\mathbf{M}^k\mathbf{u}_{decay}\|}{\|\mathbf{M}^{k-1}\mathbf{u}_{decay}\|} < 1$ for all $k$. After denoting $r_{max} = \max_k r_k$, we obtain from (10.15)

$$\frac{\|\mathbf{M}^n\mathbf{u}_{decay}\|}{\|\mathbf{u}_{decay}\|} < r_{max}^n \xrightarrow[n\to\infty]{} 0. \tag{10.16}$$

Because of that, from (10.14) and since $\|\mathbf{u}_{decay}\|$ is finite it follows that

$$\|\mathbf{x}_n - \mathbf{x}^t\| \xrightarrow[n\to\infty]{} 0. \tag{10.17}$$

In other words, if data assimilation correction can be defined in the direction of growing errors, then it has a chance of canceling the growth of errors in the prediction eventually, thus producing an error-free prediction. This is an important new aspect of data assimilation which follows from the prediction component. In order for data assimilation to be successful, it is not only important to improve the fit to observations, but also to make the correction in the subspace of growing errors. This is quite general and can be shown to be valid under much stronger conditions (e.g., Trevisan and Palatella, 2011; Palatella et al., 2013; Bocquet et al., 2017).

There are various ways an error of two trajectories, described by (10.5), can be utilized. One such way is the notion of the Lyapunov exponent, defined as the rate of separation of infinitesimally close trajectories. The maximum Lyapunov exponent, denoted $\lambda$, is of special importance since it describes the maximum separation of trajectories and therefore the largest possible error growth (e.g., Nicolis, 1995)

$$\lambda = \lim_{n\to\infty}\lim_{\epsilon\to 0}\frac{1}{n}\ln\left\|\frac{\mathbf{u}_n(\mathbf{x}_0,\epsilon)}{\mathbf{u}_0(\mathbf{x}_0,\epsilon)}\right\| = \lim_{n\to\infty}\lim_{\epsilon\to 0}\frac{1}{n}\ln\left\|\frac{M_n(\mathbf{x}_0+\epsilon) - M_n(\mathbf{x}_0)}{\epsilon}\right\|. \tag{10.18}$$

One should take Eq. (10.18) with caution, however, since it implies that error is defined in the tangent (linear) space of the system's attractor, which will eventually be exited due to nonlinearity. Therefore, the Lyapunov exponent is more relevant for characterization of the attractor than for the actual error growth and predictability. However, the formulation (10.18) is generally valid for weakly nonlinear systems.

## 10.4 Recursive Data Assimilation

Recursive data assimilation implies repetitive use of Bayes' formula over time. This is important since Bayes' formula defines a framework for data assimilation as a self-learning method. Since Bayes' formula is defined in terms of probability, the learning process can be viewed in terms of the PDF moments. For example, typical variational and hybrid ensemble-variational methods used in weather operations define recursive data assimilation in terms of the first PDF moment only, since the analysis at time $t_{k-1}$ is used to produce the optimal prediction at time $t_k$. The second moment (e.g., covariance), however, is not recursive in these systems: the analysis covariance from time $t_{k-1}$ is not used to produce forecast error covariance at time $t_k$. The covariance is prescribed (e.g., variational), or the feedback between the analysis and forecast error covariance is lost (e.g., hybrid). With Gaussian assumption, the ensemble data assimilation provides complete feedback between the first and second PDF moments and is therefore truly recursive.

### *10.4.1 Recursive Uncertainty Processing in Data Assimilation*

We now proceed by describing the analysis uncertainty processing in recursive ensemble data assimilation, focusing on the impact of potential errors of uncertainty estimation. Let $\mathbf{F} = \mathbf{P}_f^{1/2}$ and $\mathbf{A} = \mathbf{P}_a^{1/2}$ denote the forecast and analysis uncertainties, respectively. For simplicity of presentation, we assume a linear forecast model with no model errors, denoted **M**, and a linear observation operator, denoted **H**. This derivation will closely resemble the maximum likelihood ensemble filter (MLEF) algorithm under the linearity assumption. The uncertainty evolution can be described by

$$\mathbf{F}_k = \mathbf{M}_k \mathbf{A}_{k-1}, \tag{10.19}$$

where the index $k \geq 1$ defines the data assimilation cycle. The uncertainty update in the analysis step is

$$\mathbf{A}_k = \mathbf{F}_k \mathbf{S}_k, \tag{10.20}$$

where **S** can be interpreted as a scaling matrix or observation information matrix (e.g., Županski et al., 2007). In applications with a Gaussian cost function (e.g., Županski, 2005)

$$\mathbf{S} = \left(\mathbf{I} + \mathbf{P}_f^{T/2}\mathbf{H}^T\mathbf{R}^{-1}\mathbf{H}\mathbf{P}_f^{1/2}\right)^{-1/2}. \tag{10.21}$$

By including the time evolution of uncertainty (10.19) and the analysis update of uncertainty (10.20), one can relate the analysis uncertainty in consecutive data assimilation cycles

$$\mathbf{A}_k = \mathbf{M}_k \mathbf{A}_{k-1} \mathbf{S}_k. \tag{10.22}$$

Similarly, the forecast uncertainty in consecutive data assimilation cycles is

$$\mathbf{F}_k = \mathbf{M}_k \mathbf{F}_{k-1} \mathbf{S}_{k-1}. \tag{10.23}$$

Given the initial analysis uncertainty $A_0$ the recursive applications of data assimilation will produce in the first cycle, using (10.22)

$$\mathbf{A}_1 = \mathbf{M}_1 \mathbf{A}_0 \mathbf{S}_1. \tag{10.24}$$

From (10.22) and (10.24) in the second cycle

$$\mathbf{A}_2 = \mathbf{M}_2 \mathbf{A}_1 \mathbf{S}_2 = \mathbf{M}_2 \mathbf{M}_1 \mathbf{A}_0 \mathbf{S}_1 \mathbf{S}_2. \tag{10.25}$$

Continuing through consecutive cycles one obtains at cycle $k$

$$\mathbf{A}_k = \mathbf{M}_k \cdots \mathbf{M}_2 \mathbf{M}_1 \mathbf{A}_0 \mathbf{S}_1 \mathbf{S}_2 \cdots \mathbf{S}_k. \tag{10.26}$$

Denoting

$$\mathbf{M}^{(k)} = \mathbf{M}_k \cdots \mathbf{M}_2 \mathbf{M}_1 \tag{10.27}$$

the prediction from cycle 1 to cycle $k$, and

$$\mathbf{S}^{(k)} = \mathbf{S}_1 \mathbf{S}_2 \cdots \mathbf{S}_k \tag{10.28}$$

the accumulated scaling matrix, the analysis uncertainty after $k$ cycles is

$$\mathbf{A}_k = \mathbf{M}^{(k)} \mathbf{A}_0 \mathbf{S}^{(k)}. \tag{10.29}$$

Given column vectors $\mathbf{a}_0$ of the initial analysis uncertainty

$$\mathbf{A}_0 = \left( \mathbf{a}_0^1, \ldots, \mathbf{a}_0^n \right). \tag{10.30}$$

The $i$-th column of the analysis uncertainty in the $k$-th cycle is given by

$$\mathbf{a}_k^i = \mathbf{M}^{(k)} \mathbf{a}_0^i. \tag{10.31}$$

Note that one can identify the columns of the uncertainty matrix $A$ as errors

$$\mathbf{u}_0 = \mathbf{a}_0^i, \quad \mathbf{u}_k = \mathbf{M}^{(k)} \mathbf{a}_0^i. \tag{10.32}$$

Then, in the limit of small initial errors and large number of data assimilation cycles, it is possible to see that the maximum Lyapunov exponent can be defined for each initial uncertainty vector in $A_0$

$$\lambda_i = \lim_{k \to \infty} \lim_{\mathbf{a}_0^i \to 0} \frac{1}{k} \ln \left\| \frac{\mathbf{M}^{(k)} \mathbf{a}_0^i}{\mathbf{a}_0^i} \right\|. \tag{10.33}$$

Therefore, $n$ initial uncertainties produce $n$ maximum Lyapunov exponents. Given that the above was derived using linear prediction and observation operators, Eq. (10.33) is also relevant for error growth and predictability in such data assimilation systems. Assuming that limits in (10.33) exist, one can identify the subspace spanned by

$$\left( \mathbf{M}^{(\infty)} \mathbf{a}_0^1, \mathbf{M}^{(\infty)} \mathbf{a}_0^2, \ldots, \mathbf{M}^{(\infty)} \mathbf{a}_0^n \right), \tag{10.34}$$

where $\mathbf{M}^{(\infty)}$ denotes a very long prediction. Since this subspace represents the limit of long model integration, it will produce the directions of the dominant error growth of long prediction. Therefore, in view of (10.6)–(10.17), the span vectors (10.34) form the optimal analysis correction subspace. The uninterrupted model prediction over a long time eventually converges to a steady-state subspace defined by (10.34), which is captured by recursive data assimilation. Recursive data assimilation has the potential of correcting the initial conditions in such a way that growth errors are reduced, or even removed.

A side benefit of having the subspace defined by long prediction is the self-localization property of general dynamical systems (Pikovsky and Politi, 1998). As shown by Županski et al. (2006) using a quasi-geostrophic model, error covariance is automatically localized as a result of using long uninterrupted predictions to create the analysis correction subspace.

When the forecast is interrupted, however, the desired subspace is difficult or impossible to create. There may be many reasons for interrupting the long forecast in recursive data assimilation, such as by imposing covariance localization or covariance inflation. The deteriorating impact of covariance localization on dynamical balances in data assimilation are known (e.g., Mitchell et al., 2002; Kepert, 2009). Unfortunately, these procedures are required due to insufficient degrees of freedom (DOF) typical for realistic ensemble data assimilation.

### 10.4.2 Dynamical Balance Errors in Recursive Data Assimilation

To see the impact of such procedures, consider an additive error of the analysis uncertainty caused by dynamical imbalances. Let us denote such an error in the current cycle as $\eta$. The analysis update with correction is then

$$\mathbf{A}_k = \mathbf{M}_k \mathbf{A}_{k-1} \mathbf{S}_k + \eta_k. \tag{10.35}$$

Starting from $k = 1$ and dynamically balanced $\hat{\mathbf{A}}_0$, where symbol ^ indicates a dynamically balanced analysis uncertainty, and applying the recursive formula (10.35)

$$\mathbf{A}_k = \mathbf{M}_k \cdots \mathbf{M}_2 \mathbf{M}_1 \hat{\mathbf{A}}_0 \mathbf{S}_1 \mathbf{S}_2 \cdots \mathbf{S}_k + \mathbf{M}_k \cdots \mathbf{M}_2 \eta_1 \mathbf{S}_2 \cdots \mathbf{S}_k + \cdots + \mathbf{M}_k \eta_{k-1} \mathbf{S}_k + \eta_k. \tag{10.36}$$

The last expression can be also written as a sum of a balanced and an unbalanced part

$$\mathbf{A}_k = \hat{\mathbf{A}}_k + \mathbf{E}_k, \tag{10.37}$$

where the dynamically balanced part is denoted $\hat{\mathbf{A}}_k$ and the unbalanced part (e.g., total error due to dynamical imbalances) is denoted $\mathbf{E}_k$. One can identify the dynamically balanced part by comparison with previously defined uncertainty errors without dynamical imbalances (10.26)

$$\hat{\mathbf{A}}_k = \mathbf{M}_k \cdots \mathbf{M}_2 \mathbf{M}_1 \hat{\mathbf{A}}_0 \mathbf{S}_1 \mathbf{S}_2 \cdots \mathbf{S}_k \tag{10.38}$$

while the unbalanced part is

$$\mathbf{E}_k = \mathbf{M}_k \cdots \mathbf{M}_2 \eta_1 \mathbf{S}_2 \cdots \mathbf{S}_k + \cdots + \mathbf{M}_k \eta_{k-1} \mathbf{S}_k + \eta_k. \tag{10.39}$$

The last expression for the total error illustrates how errors of the analysis uncertainty estimation from all previous data assimilation cycles contribute to the total error in the current data assimilation cycle. The contribution to total error is mostly through evolution of errors from previous cycles, which implies a need to understand error in terms of model dynamics. For example, if errors from individual data assimilation cycles, $\boldsymbol{\eta}_i$, are in the subspace of the fast-growing perturbations, then the total error $\mathbf{E}_k$ will be large, and not corrected in data assimilation. However, if one can enforce errors from individual cycles to be in the subspace of neutral or decaying perturbations, then the total error of the uncertainty estimation will be small. This has important implications on Bayesian inference as well, since small errors of analysis uncertainty estimation imply a more efficient "learning" from the past. On the other hand, the existence of the long prediction in (10.39) may be able to filter out the errors from distant previous cycles and therefore have negligible impact on the current analysis cycle. Therefore, one should not expect that errors will grow indefinitely, i.e., there will be an error saturation even in the linear assimilation system. The more recent dynamical imbalance errors, however, are not sufficiently filtered by the prediction and therefore continue to corrupt the uncertainty estimation.

**Example 10.2 Controlling the dynamically unbalanced analysis error**

The unbalanced part of the analysis error is given by (10.39). To better understand its impact, it is convenient to make some simplifications, such as to consider the one-point model $Mx = bx$ where $b$ is a real number. We also assume that all operators and vectors in each cycle are the same: $M_1 = M_2 = \cdots = M_k = M$, $\eta_1 = \eta_2 = \cdots = \eta_k = \eta$, and $S_1 = S_2 = \cdots = S_k = S$. We further assign $S = \frac{1}{|c|}$, where $|c| > 1$ is a real number. Inserting these values in (10.39) gives

$$E_k = b^{k-1}\eta\frac{1}{|c|^{k-1}} + b^{k-2}\eta\frac{1}{|c|^{k-2}} + \cdots + b\eta\frac{1}{|c|} + \eta.$$

This can be further transformed to

$$E_k = \eta\left(d^{k-1} + \cdots + d + 1\right),$$

where $d = \frac{b}{|c|}$. The summation in brackets is a geometric series with partial sum $\sum_{i=1}^{k} d_{i-1} = \frac{1-d^k}{1-d}$. Therefore,

$$E_k = \eta\frac{1-d^k}{1-d}. \qquad (10.40)$$

Since the geometric series converges for $|d| < 1$ the unbalanced error growth will be limited in this case to

$$E = \lim_{k\to\infty} E_k = \frac{\eta}{1-d}. \qquad (10.41)$$

Therefore, the unbalanced error can be controlled for $|b| < |c|$, while it will be unbounded for $|b| \geq |c|$. The impact of the unbalanced error can be negligibly small for a subset of values of $\eta$, $b$, and $c$, implying that in some instances unbalanced errors may not produce deterioration of the analysis and/or forecast even without additional control. Although the implications of the above example in realistic data assimilation may be limited, a shorter assimilation window can produce a smaller $b$, which in turn may imply a smaller unbalanced error.

**Practice 10.2 Unbalanced error**

Find the unbalanced error of model $Mx = 2x$ with $c = 4$ and $\eta = 0.4$: (a) after 5 data assimilation cycles and (b) after 100 data assimilation cycles.

As was shown in this subsection, through recursive application of data assimilation the dynamical imbalance errors adversely impact Bayesian inference and eventually lead to an inefficient data assimilation system. Therefore, it is desirable to address dynamical imbalances in the analysis.

## 10.5 Addressing Dynamical Imbalances in the Analysis

There may be several methodologies that could be considered for improving dynamical balances in the analysis: (i) increase the number of DOF in ensemble data assimilation and therefore avoid imposed covariance localization and (ii) design a scheme that automatically controls dynamical imbalances. The first approach is currently not feasible, since it is computationally prohibitive to use a large number of ensembles in realistic high-dimensional applications. However, there are some encouraging attempts (Kondo and Miyoshi, 2016). The second approach offers several sub-methods. For example, Bocquet (2016) described a few possible strategies by considering the Liouville equation with localization in ensemble Kalman smoothers. Another approach, described here in more detail, is to introduce a new term in the cost function penalizing dynamical imbalances. This approach is common in variational data assimilation (VAR), but it could be used in ensemble data assimilation as well. One can think of a time filtering of the current state, denoted $q$, that produces a dynamically balanced state $\mathbf{x}_{bal}$

$$\mathbf{x}_{bal} = q(\mathbf{x}). \tag{10.42}$$

The mapping $q$ can denote a digital filter (e.g., Lynch and Huang, 1992; Huang and Lynch, 1993), or an alternative way of producing a dynamically balanced state, such as pressure tendency (e.g., Thépaut and Courtier, 1991; Zou et al., 1992). An additional penalty term can be added to the original cost function, denoted $f_{pen}$,

$$f_{pen}(\mathbf{x}) = \frac{1}{2}\left[\mathbf{x} - q(\mathbf{x})\right]^T \mathbf{U}^{-1}\left[\mathbf{x} - q(\mathbf{x})\right], \tag{10.43}$$

where $\mathbf{U}$ is a prescribed matrix, typically specified as diagonal, and $\mathbf{x}_{unbal} = \mathbf{x} - q(\mathbf{x}) = \mathbf{x} - \mathbf{x}_{bal}$ denotes the unbalanced component of the state. Matrix $\mathbf{U}$ describes the unbalanced error covariance $\mathbf{U} = E\left(\mathbf{x}_{unbal}\mathbf{x}_{unbal}^T\right)$, where $E$ denotes mathematical expectation. Methodologies that introduce digital filtering penalty terms, similar to the above description, are also discussed in Polavarapu et al. (2004) and Caron and Fillion (2010) in applications to variational and hybrid ensemble-variational data assimilation, respectively.

Consider the use of a dynamical imbalance penalty term in an ensemble system, such as the MLEF (Županski, 2005; Županski et al., 2008). The cost function in an ensemble system is

$$f(\mathbf{x}) = \frac{1}{2}\left[\mathbf{x}-\mathbf{x}^f\right]^T \mathbf{P}_f^{-1}\left[\mathbf{x}-\mathbf{x}^f\right] + \frac{1}{2}\left[\mathbf{y}^o - H(\mathbf{x})\right]^T \mathbf{R}^{-1}\left[\mathbf{y}^o - H(\mathbf{x})\right] + \frac{1}{2}\left[\mathbf{x} - q(\mathbf{x})\right]^T \mathbf{U}^{-1}\left[\mathbf{x} - q(\mathbf{x})\right] \quad (10.44)$$

with forecast error covariance $\mathbf{P}_f$ defined over ensemble subspace. After introducing a standard change of variable

$$\mathbf{x} = \mathbf{x}^f + \mathbf{P}_f^{1/2}\mathbf{w}, \quad (10.45)$$

where $\mathbf{w}$ is a new control variable, the transformed cost function is

$$f(\mathbf{w}) = \frac{1}{2}\mathbf{w}^T\mathbf{w} + \frac{1}{2}\left[\mathbf{y}^o - H\left(\mathbf{x}^f + \mathbf{P}_f^{1/2}\mathbf{w}\right)\right]^T \mathbf{R}^{-1}\left[\mathbf{y}^o - H\left(\mathbf{x}^f + \mathbf{P}_f^{1/2}\mathbf{w}\right)\right] + \frac{1}{2}\left[\mathbf{x}^f + \mathbf{P}_f^{1/2}\mathbf{w} - q\left(\mathbf{x}^f + \mathbf{P}_f^{1/2}\mathbf{w}\right)\right]^T \mathbf{U}^{-1}\left[\mathbf{x}^f + \mathbf{P}_f^{1/2}\mathbf{w} - q\left(\mathbf{x}^f + \mathbf{P}_f^{1/2}\mathbf{w}\right)\right]. \quad (10.46)$$

The generalized first and second derivatives (e.g., Županski et al., 2008) are

$$\frac{\partial f(\mathbf{w})}{\partial \mathbf{w}} = \mathbf{w} - \mathbf{Z}^T\mathbf{R}^{-1/2}\left[\mathbf{y}^o - H(\mathbf{x})\right] - \mathbf{Q}^T\mathbf{U}^{-1/2}\left[\mathbf{x} - q(\mathbf{x})\right] \quad (10.47)$$

and

$$\frac{\partial^2 f(\mathbf{w})}{\partial \mathbf{w}^2} = \mathbf{I} + \mathbf{Z}^T\mathbf{Z} + \mathbf{Q}^T\mathbf{Q}, \quad (10.48)$$

respectively, where

$$\mathbf{Z}_i = \mathbf{R}^{-1/2}\left[H(\mathbf{x}+\mathbf{p}_i) - H(\mathbf{x})\right] \quad (10.49)$$

and

$$\mathbf{Q}_i = \mathbf{U}^{-1/2}\left[\mathbf{p}_i - (q(\mathbf{x}+\mathbf{p}_i) - q(\mathbf{x}))\right]. \quad (10.50)$$

The vector $\mathbf{p}_i$ is the $i$-th column of the square root ensemble forecast error covariance $\mathbf{P}_f^{1/2}$. Therefore, the above-described ensemble data assimilation algorithm would be altered from the original algorithm through the inclusion of additional gradient and Hessian terms. This is a straightforward procedure, as the additional gradient and Hessian terms are both defined in ensemble space.

The calculation of derivatives (10.47) and (10.48) now involves calculating the quantities defined by (10.50), which implies an application of the balancing operator q. If this calculation includes time filtering (as in digital filter), it would add a nonnegligible computational cost. However, there may be ways to reduce this additional computational cost and produce an efficient algorithm that improves dynamical balances in the analysis. Such a system would have an implicit control of dynamical imbalances in the analysis and ameliorate the adverse impact of covariance localization.

## References


Bocquet M (2016) Localization and the iterative ensemble Kalman smoother. *Quart J Roy Meteor Soc* 142:1075–1089.

Bocquet M, Gurumoorthy KS, Apte A, et al. (2017) Degenerate Kalman filter error covariances and their convergence onto the unstable subspace. *SIAM-ASA J Uncertain Quantif* 5:304–333.

Caron JF, Fillion L (2010) An examination of the incremental balance in a global ensemble-based 3d-var data assimilation system. *Mon Wea Rev* 138:3946–3966.

Carrassi A, Bocquet M, Bertino L, Evensen G (2018) Data assimilation in the geosciences: An overview of methods, issues, and perspectives. *WIREs Clim Change* 9:e535, doi: 10.1002/wcc.535

Gurumoorthy KS, Grudzien C, Apte A, Carrassi A, Jones CKRT (2017) Rank deficiency of Kalman error covariance matrices in linear time-varying system with deterministic evolution. *SIAM J Control Optim* 55:741–759.

Huang X-Y, Lynch P (1993) Diabatic digital-filtering initialization: Application to the HIRLAM model. *Mon Wea Rev* 121:589–603.

Jazwinski AH (1970) *Stochastic Processes and Filtering Theory*. Academic Press, New York, 376 pp.

Kepert JD (2009) Covariance localisation and balance in an ensemble Kalman filter. *Quart J Roy Meteor Soc* 135:1157–1176.

Kondo K, Miyoshi T (2016) Impact of removing covariance localization in an ensemble Kalman filter: Experiments with 10 240 members using an intermediate agcm. *Mon Wea Rev* 144:4849–4865.

Lynch P, Huang X-Y (1992) Initialization of the HIRLAM model using a digital filter. *Mon Wea Rev* 120:1019–1034.

Mitchell HL, Houtekamer PL, Pellerin G (2002) Ensemble size, balance, and model-error representation in an ensemble Kalman filter. *Mon Wea Rev* 130:2791–2808.

Nicolis G (1995) *Introduction to Nonlinear Science*. Cambridge University Press, New York, 254 pp.

Palatella L, Carrassi A, Trevisan A (2013) Lyapunov vectors and assimilation in the unstable subspace: Theory and applications. *J Phys A: Math Theor* 46:254020.

Pikovsky A, Politi A (1998) Dynamic localization of lyapunov vectors in spacetime chaos. *Nonlinearity* 11:1049.

Polavarapu S, Ren S, Clayton AM, Sankey D, Rochon Y (2004) On the relationship between incremental analysis updating and incremental digital filtering. *Mon Wea Rev* 132:2495–2502.

Thépaut J-N, Courtier P (1991) Four-dimensional variational data assimilation using the adjoint of a multilevel primitive-equation model. *Quart J Roy Meteor Soc* 117: 1225–1254.

Trevisan A, Palatella L (2011) Chaos and weather forecasting: The role of the unstable subspace in predictability and state estimation problems. *Int J Bifurc Chaos* 21: 3389–3415.

Zou X, Navon IM, Le Dimet F-X (1992) Incomplete observations and control of gravity waves in variational data assimilation. *Tellus A* 44:273–296.
Županski D, Hou AY, Zhang SQ, et al. (2007) Applications of information theory in ensemble data assimilation. *Quart J Roy Meteor Soc* 133:1533–1545.
Županski M (2005) Maximum likelihood ensemble filter: Theoretical aspects. *Mon Wea Rev* 133:1710–1726.
Županski M, Navon IM, Županski D (2008) The Maximum Likelihood Ensemble Filter as a non-differentiable minimization algorithm. *Quart J Roy Meteor Soc* 134:1039–1050.
Županski M, Fletcher SJ, Navon IM, et al. (2006) Initiation of ensemble data assimilation. *Tellus A* 58:159–170.

# Part IV

## Applications

# 11

# Sensitivity Analysis and Adaptive Observation

## 11.1 Introduction

*Sensitivity*, in its simplest meaning, denotes quantitative changes in the projected output (say, $\mathbf{Y} = \mathbf{F}(\mathbf{X})$) due to given changes (perturbations) in the input (say, $\mathbf{X}$), which can be formulated in terms of NLP (i.e., $\Delta\mathbf{Y}/\Delta\mathbf{X}$): It can be approximated by the first-order derivatives, for a small $\Delta\mathbf{X}$, as

$$\frac{\Delta\mathbf{Y}}{\Delta\mathbf{X}} \approx \frac{\partial\mathbf{Y}}{\partial\mathbf{X}} = \left.\frac{\partial\mathbf{F}}{\partial\mathbf{X}}\right|_{\mathbf{X}_0}, \tag{11.1}$$

where $\Delta\mathbf{Y} = \mathbf{F}(\mathbf{X}_0 + \Delta\mathbf{X}) - \mathbf{F}(\mathbf{X}_0)$. In meteorology, this concept can be used to assess the change in a weather element at a downstream location due to a change in another (or the same) weather element at an upstream location and at an earlier time. For example, how much would the temperature ($T$) change in Chicago at $t = 12$ h due to a current change of humidity ($q$) in Denver (i.e., at $t = 0$ h)? This can be calculated through

$$\frac{\Delta T_{\text{Chicago}}^{12\text{ h}}}{\Delta q_{\text{Denver}}^{0\text{ h}}} \approx \frac{\partial T_{\text{Chicago}}^{12\text{ h}}}{\partial q_{\text{Denver}}^{0\text{ h}}}, \tag{11.2}$$

which describes sensitivity of temperature in Chicago 12 hours later with respect to the current humidity in Denver.

As numerical weather prediction is a time marching problem in a 3D grid space, whose practical utility depends on a number of factors with various uncertainty sources, the spatial and temporal evolution of uncertainties in input parameters can be examined through *sensitivity analysis* (SA). That is, SA quantifies uncertainties in the output caused by uncertainties in the corresponding input, which is useful for the following tasks (see Razavi and Gupta, 2015; Devak and Dhanya, 2017) but not limited to

- constructing a response surface that represents the model responses due to the given input variations, depicting the areas of high sensitivity in the entire parameter space (e.g., Fronzek et al., 2018);
- identifying and ranking the influential parameters, i.e., parameter screening (e.g., Ciric et al., 2012; Ji et al., 2018; Xing et al., 2021);
- assessing interactions among various parameters (e.g., Fenwick et al., 2014; Razavi and Gupta, 2015);

- diagnosing the behavior and reliability of the given model (e.g., Bosnić and Kononenko, 2008; Shin and Choi, 2018); and
- identifying insensitive (noninfluential) parameters to reduce the model/analysis complexity (e.g., Degenring et al., 2004).

In terms of scope, SA can be divided into *local* and *global* procedures (Cacuci et al., 2005):

- local SA examines the changes of the model response by changing a parameter around a given point or trajectory or a specific nominal value (i.e., local) in the parameter/state space (e.g., Heynen et al., 2013; Rakovec et al., 2014; Che et al., 2019); and
- global SA explores all possible changes in the model response due to interactions among the parameters as well as changes in individual parameters by spanning a wide spectrum of the entire parameter/state space (i.e., global) through subsequent application of local SA (e.g., Nossent et al., 2011; Ji et al., 2018; Xing et al., 2021).

The derivative representation of sensitivity, as in (11.1), is based on the Taylor series expansion (see Section 4.2), which is valid for a small perturbation around a nominal value and hence is essentially a local approximation. Therefore, local SA typically involves differentiation of the model equations or outputs with respect to the input parameters/states – in the form of either finite differences or partial derivatives. Global SA is mainly based on the sampling concept because a quantitative variation of the model output is evaluated from the variations in a sample set of input parameters/states within a specified parameter/state space.

In terms of methodology, SA is based on either *deterministic* or *statistical* approaches – the former mostly for local SA while the latter for both local and global SA (Cacuci et al., 2005):

- Deterministic approaches include the following (Ionescu-Bujor and Cacuci, 2004):

  1. the *brute-force* method that computes the output differences between simulations from the base and perturbed inputs through a finite difference scheme (e.g., Chen et al., 2019);
  2. the *decoupled direct* method that solves the sensitivity equations, derived from the original model equations (e.g., Dunker et al., 2002; Wang et al., 2011);
  3. *Green's function* method that replaces the differential sensitivity equations with a set of integrals of time-dependent sensitivities (e.g., Hwang et al., 1978; Brix et al., 2015);
  4. the *forward* or *tangent linear SA* method using the tangent linear model (TLM) (e.g., Tromble et al., 2016; Yang et al., 2021); and
  5. the *adjoint SA* method using the ADJM (e.g., Chu and Tan, 2010; Doyle et al., 2012, 2019).

- Statistical approaches are classified into the following (Cacuci and Ionescu-Bujor, 2004):

  1. the *sampling-based* methods, including random sampling, importance sampling, and Latin hypercube sampling (e.g., Helton et al., 2006; He et al., 2018);

2. the *reliability* methods of both first- and second-order algorithms (e.g., Karamchandani and Cornell, 1992);
3. the *variance-based* methods such as correlation ratio-based methods (e.g., Manache and Melching, 2008), the Fourier amplitude sensitivity test (FAST) (Collins and Avissar, 1994), extended FAST (e.g., Saltelli et al., 1999; Zhao and Tiede, 2011), and Sobol's method (e.g., Sobol', 1993; Saltelli et al., 2010; Marzban, 2013; Houle et al., 2017; He et al., 2018); and
4. the *screening design* methods, including the local and global one-at-a-time (OAT) experiments (e.g., Morris, 1991; Campolongo et al., 2007).

In a general approach, based on the variational and optimal control theories (see Colloquy 7.3), sensitivity of the optimization system can be obtained with respect to the control parameters, e.g., initial conditions, model error, empirical parameters, observations, etc. (e.g., Cacuci, 1981; Županski, 1995; Le Dimet et al., 1997; Shutyaev et al., 2018). An ensemble-based SA has also been suggested in order to assess forecast sensitivity to observations (e.g., Liu and Kalnay, 2008; Li et al., 2010; Kalnay et al., 2012).

The derivative-based SA is largely considered a local procedure; however, the derivatives can be also used as a global sensitivity measure (e.g., Sobol' and Kucherenko, 2009; Cleaves et al., 2019). In this chapter, we focus on the derivative-based sensitivity analyses, which are categorized as a deterministic approach and a local procedure, especially using the TLM (i.e., forward SA) and the ADJM (i.e., adjoint SA). More details on other SA methods are referred to in some textbooks (e.g., Cacuci et al., 2005; Saltelli et al., 2008) or review articles (e.g., Cacuci and Ionescu-Bujor, 2004; Ionescu-Bujor and Cacuci, 2004; Razavi and Gupta, 2015; Pianosi et al., 2016; Devak and Dhanya, 2017).

In numerical weather prediction, whose solutions are flow dependent, large forecast errors in a region of interest may stem from an upstream region with analysis errors. Sensitivity information, especially from the adjoint SA, is useful in identifying such an upstream region in the context of adaptive observations (Langland, 2005a; Majumdar, 2016), which will be also discussed in this chapter.

## 11.2 Deterministic Sensitivity Analysis

### *11.2.1 Forward Sensitivity Coefficient*

From (11.1) and (11.2), we note that the humidity at Denver $t = 0$ h affects the weather elements in Chicago at $t = 12$ h, not only temperature but also humidity, horizontal/vertical winds, cloud, precipitation, etc. To examine how much the Chicago weather elements are affected by the humidity in Denver 12 hours ago, one may want to evaluate the following derivatives:

$$\frac{\partial T^{12\text{ h}}_{\text{Chicago}}}{\partial q^{0\text{ h}}_{\text{Denver}}}, \quad \frac{\partial q^{12\text{ h}}_{\text{Chicago}}}{\partial q^{0\text{ h}}_{\text{Denver}}}, \quad \frac{\partial \mathbf{v}^{12\text{ h}}_{\text{Chicago}}}{\partial q^{0\text{ h}}_{\text{Denver}}}, \quad \frac{\partial q_c{}^{12\text{ h}}_{\text{Chicago}}}{\partial q^{0\text{ h}}_{\text{Denver}}}, \quad \frac{\partial q_r{}^{12\text{ h}}_{\text{Chicago}}}{\partial q^{0\text{ h}}_{\text{Denver}}}, \ldots, \tag{11.3}$$

where $T$ is temperature (in °F), $q$ is humidity (say, mixing ratio of water vapor; in g kg$^{-1}$), $\mathbf{v}$ is a 3D wind vector, and $q_c$ and $q_r$ are mixing ratios of cloud water and rainwater (in g kg$^{-1}$), respectively.

The *forward SA* focuses on how the dependent variables change in any future time (i.e., $\mathbf{Y}(t_n)$) given the change in a specific input variable (i.e., $X_j(t_0)$ for fixed $j$). We can perform the forward SA in a linear framework using the TLM (Eq. (4.18)). Let us set a component of $\delta\mathbf{X}_0$ to unity, with all other components 0, that is,

$$\delta X^0_{k=j} = 1 \text{ and } \delta X^0_{k\neq j} = 0 \tag{11.4}$$

for $k = 1, \ldots, j, \ldots, m$. Then, we have

$$\delta X^n_k = \frac{\partial F^{n-1}_k}{\partial X^0_j} = \frac{\partial X^n_k}{\partial X^0_j} \tag{11.5}$$

for fixed $j$ and $k = 1, \ldots, m$. This indicates that the TLM solutions at any intermediate time ($t_n$) represent the sensitivities of all dependent variables at that time with respect to one specific independent variable at the initial time ($t_0$).

In general, we can define a dependent variable vector $\mathbf{J}$ whose components are scalar functions of model states (forecast error, total precipitable water, etc.) at a given time (i.e., $X^n_i$) such that their gradients $\partial J^n_k / \partial X^n_i$ exist. We can represent it, in vector form, as

$$\mathbf{J}^n = \mathbf{P}^n\left(\mathbf{X}^n\right), \tag{11.6}$$

with its TLM as

$$\begin{aligned}\delta\mathbf{J}^n &= \mathbf{Q}^n\delta\mathbf{X}^n \\ &= \mathbf{Q}^n\mathbf{L}^n\delta\mathbf{X}^0,\end{aligned} \tag{11.7}$$

where the elements of $\mathbf{Q}^n$ are expressed as $q^n_{ij} = \partial P^n_i / \partial X^n_j$ and those of $\mathbf{L}^n$, the tangent linear propagator, as $l^n_{ij} = \partial X^n_i / \partial X^0_j$. Using the same initialization as in Eq. (11.4), we have

$$\delta J^n_k = \frac{\partial P^n_k}{\partial X^0_j} = \frac{\partial J^n_k}{\partial X^0_j}. \tag{11.8}$$

Noting that $\mathbf{J}$ includes the model state $\mathbf{X}$ itself, we can generally define the *forward sensitivity coefficient* (FSC) ($S_F$), from Eqs. (11.5) and (11.8), as

$$S_F = \frac{\partial J^n_k}{\partial X^0_j}, \quad (j \text{ fixed; } k = 1, \ldots, j, \ldots, m;\ n = 0, \ldots, N), \tag{11.9}$$

where $n = N$ represents the model verification time, $t_N$. Figure 11.1 depicts the concept of FSCs: Changes in dependent variables $\mathbf{X}$ or $\mathbf{J}$ at $t_n$ are obtained by a change in a specific independent variable, say $X_{j=5}$, at $t_0$ by running the TLM ($\mathbf{L}$) from $t_0$ to $t_n$. In other words, FSCs are obtained by one forward run of the TLM for a specific independent variable $X_j$: It also requires a simultaneous run of the nonlinear model (NLM) because the TLM solutions evolve along the NLM solution trajectories.

Equation (11.9) indicates that obtaining sensitivities with respect to all components of the independent variable vector requires us to run the TLM as many times as the number of independent variables; thus, the forward SA is efficient only when the number of independent variables is much less than the number of dependent variables. Nevertheless,

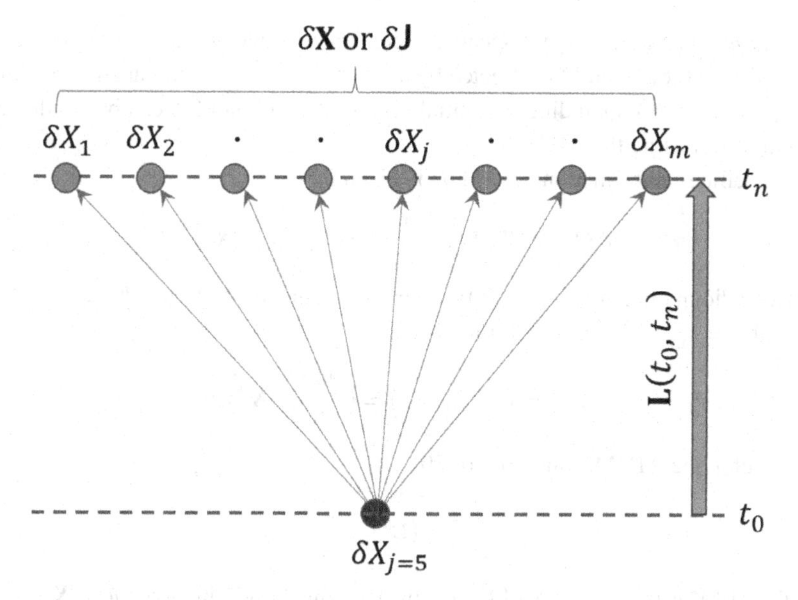


Figure 11.1 Concept of FSCs: $\frac{\partial X_k^n}{\partial X_5^0}$ or $\frac{\partial J_k^n}{\partial X_5^0}$ $(k = 1, \ldots, j = 5, \ldots, m)$ are computed by one forward run of the NLM and the TLM, respectively, from $t_0$ to $t_n$.

it is an effective tool to check the responses of all model states as a function of changes in a specific control parameter, e.g., changes in future weather/climate elements due to a change of $CO_2$ in various scenarios, the impact of missing observations in a single variable on various model states, etc.

The FSCs have been used in the forward SA (e.g., Bischof et al., 1996; Park and Droegemeier, 1999; Schwinger et al., 2010), parameter screening (e.g., Backman et al., 2017), overparameterization reduction (e.g., Martinec et al., 2015), uncertainty analysis (e.g., Zhao and Mousseau, 2012; Yang et al., 2021), data assimilation (e.g., Lakshmivarahan and Lewis, 2010; Tromble et al., 2016), etc.

### *11.2.2 Adjoint Sensitivity Coefficient*

Note that, from (11.1) and (11.2), the humidity at Denver $t = 0$ h is not the only factor that affects the temperature in Chicago at $t = 12$ h: The latter can also be changed by other weather elements, currently observed in Denver, e.g., temperature itself, horizontal/vertical winds, cloud, precipitation, etc. To examine how much the temperature in Chicago is affected by the Denver weather elements 12 hours ago, one may want to evaluate the following derivatives, with the same notation in (11.3):

$$\frac{\partial T_{\text{Chicago}}^{12\text{ h}}}{\partial q_{\text{Denver}}^{0\text{ h}}},\ \frac{\partial T_{\text{Chicago}}^{12\text{ h}}}{\partial T_{\text{Denver}}^{0\text{ h}}},\ \frac{\partial T_{\text{Chicago}}^{12\text{ h}}}{\partial \mathbf{v}_{\text{Denver}}^{0\text{ h}}},\ \frac{\partial T_{\text{Chicago}}^{12\text{ h}}}{\partial q_c{}_{\text{Denver}}^{0\text{ h}}},\ \frac{\partial T_{\text{Chicago}}^{12\text{ h}}}{\partial q_r{}_{\text{Denver}}^{0\text{ h}}},\ \ldots \tag{11.10}$$

The *adjoint SA* examines how a specific dependent variable (e.g., temperature) at current time (i.e., $Y_k(t_N)$ for a fixed $k$) is affected by the independent variables at any previous time (i.e., $\mathbf{X}(t_n)$ for $n < N$), including the initial time $t_0$. The adjoint SA can be conducted in a linear framework using the ADJM.

For a specific scalar function $J_k^n$, (11.7) is given by

$$\delta J_k^n = \left\langle \mathbf{Q}_k^n, \mathbf{L}^n \delta \mathbf{X}^0 \right\rangle = \left\langle \left(\mathbf{L}^n\right)^* \mathbf{Q}_k^n, \delta \mathbf{X}^0 \right\rangle$$

through the adjoint operator (see (4.28)), where the elements of $\mathbf{Q}_k^n$ are $q_{kj}^n = \partial P_k^n / \partial X_j^n$ for $k$ fixed and $j = 1, \ldots, k, \ldots, m$. Noting that

$$\delta J_k^n = \left\langle \nabla J_k^n, \delta \mathbf{X}^0 \right\rangle = \left\langle \frac{\partial J_k^n}{\partial \mathbf{X}^0}, \delta \mathbf{X}^0 \right\rangle,$$

we can develop the ADJM, following (4.30):

$$\frac{\partial J_k^n}{\partial \mathbf{X}^0} = \left(\mathbf{L}^n\right)^* \frac{\partial J_k^n}{\partial \mathbf{X}^n}. \tag{11.11}$$

Noting that $(\mathbf{L}^n)^*$ is the transpose of $\mathbf{L}^n$ and by denoting the adjoint variable as $\hat{\mathbf{X}} = \partial J_k / \partial \mathbf{X}$ for $k$ fixed, we can determine $\hat{\mathbf{X}}^0 = \partial J_k^n / \partial \mathbf{X}^0$ through a recursive system, following (4.36) and (4.37),

$$\hat{\mathbf{X}}^{n-1} = \left(\mathbf{G}^{n-1}\right)^T \hat{\mathbf{X}}^n, \tag{11.12}$$

where $\mathbf{G}$ is given in (4.15). This can be solved backward starting from the verification time, $t = t_N$, with the terminal condition

$$\hat{\mathbf{X}}^N = \frac{\partial J_k^N}{\partial \mathbf{X}^N}. \tag{11.13}$$

For any intermediate time $t_n < t_N$, the solutions of the ADJM (11.12) are given by

$$\hat{\mathbf{X}}^n = \frac{\partial J_k^N}{\partial \mathbf{X}^n}, \tag{11.14}$$

which are the sensitivities of a chosen dependent variable $J_k$ at the verification time with respect to all independent variables at any earlier (intermediate or initial) time. We define the *adjoint sensitivity coefficient* (ASC) ($S_A$), based on (11.14), as

$$S_A = \frac{\partial J_k^N}{\partial X_j^n}, \quad (k \text{ fixed};\ j = 1, \ldots, k, \ldots, m;\ n = N, \ldots, 0). \tag{11.15}$$

Figure 11.2 depicts the concept of ASCs: A change at $t_N$ in a state (say, $X_{k=5}$) or a diagnotic function (say, $J$) is obtained by changes in all independent variables $\mathbf{X}$ at $t_n$ by running the ADJM ($\mathbf{L}^*$) backward from $t_N$ to $t_n$. In other words, ASCs are obtained by one backward run of the ADJM for a specific dependent variable: It also requires a forward run of the NLM to save the basic states (see Section 4.8).

To obtain sensitivities with respect to all components of the dependent variable vector, based on (11.15), one needs as many ADJM runs as the number of dependent variables; thus,

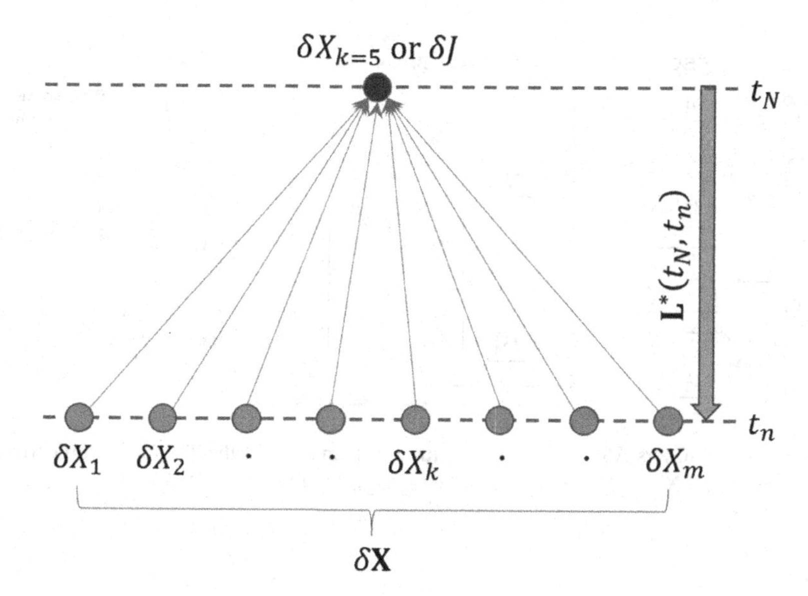


Figure 11.2 Concept of ASCs: $\frac{\partial X_5^N}{\partial X_j^n}$ or $\frac{\partial J_5^N}{\partial X_j^n}$ $(j = 1, \ldots, k = 5, \ldots, m)$ are computed by one forward run of the NLM and one backward run of the ADJM, respectively, from $t_N$ to $t_n$.

the adjoint SA is efficient only when the number of independent variables greatly exceeds that of the dependent variables.

The ASCs have been used for the adjoint SA (e.g., Errico and Vukićević, 1992; Park and Droegemeier, 1999, 2000; Losch and Heimbach, 2007; Heimbach and Bugnion, 2009; Chu and Tan, 2010; Doyle et al., 2012, 2019), variational data assimilation (VAR) (e.g., Navon et al., 1992; Zou et al., 1993; Park and Županski, 2003; Zheng et al., 2018), parameter estimation (e.g., Nguyen et al., 2016), observation impacts on data assimilation (e.g., Jung et al., 2013; Lorenc and Marriott, 2014; Zhang et al., 2015), adaptive observations (e.g., Langland et al., 1999a; Zhong et al., 2007), etc.

### 11.2.3 Sensitivity Coefficients and Jacobian

Sensitivity coefficients are generally provided by a Jacobian matrix. For example, from (11.1), a mapping $\mathbf{F}: \mathcal{R}^n \to \mathcal{R}^m$ gives the Jacobian matrix $\nabla\mathbf{F}(\mathbf{X}) \in \mathcal{R}^{m\times n}$:

$$\frac{\partial \mathbf{F}}{\partial \mathbf{X}} = \nabla\mathbf{F}(\mathbf{X}) = \begin{pmatrix} \frac{\partial F_1}{\partial X_1} & \cdots & \frac{\partial F_1}{\partial X_n} \\ \vdots & \ddots & \vdots \\ \frac{\partial F_m}{\partial X_1} & \cdots & \frac{\partial F_m}{\partial X_n} \end{pmatrix}_{m\times n} . \tag{11.16}$$

Consider an independent variable vector $\mathbf{X} = (X_1, X_2, X_3)$ and a dependent variable vector $\mathbf{J} = (J_1, J_2)$. Here, the component of $\mathbf{J}$ can be $\mathbf{X}$ itself, or any scalar function of the components of $\mathbf{X}$ such that its gradient $\partial J_k/\partial X_j$ exists. Assume that our goal is to obtain the Jacobian

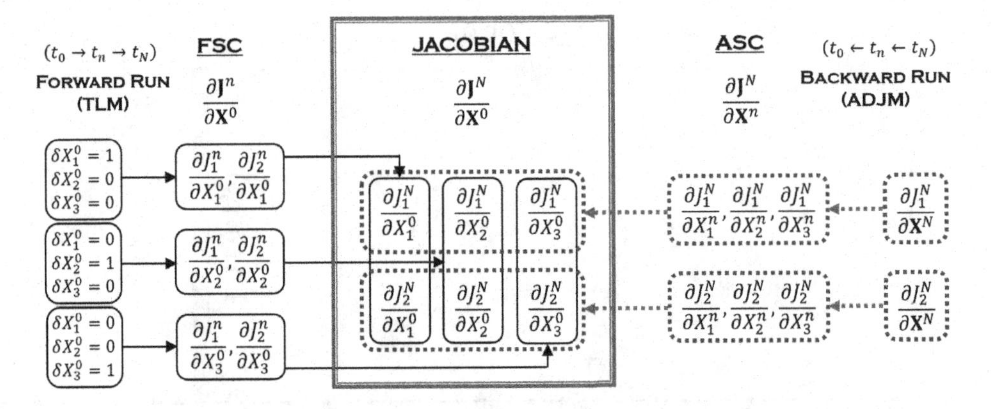


Figure 11.3 FSCs vs ASCs in computing the Jacobian through the TLM and ADJM runs, respectively, for $\mathbf{X} = (X_1, X_2, X_3)$ and $\mathbf{J} = (J_1, J_2)$. Modified from Park (1996). ©Seon Ki Park, 1996.

$$\nabla \mathbf{J} = \frac{\partial \mathbf{J}(t_N)}{\partial \mathbf{X}(t_0)} = \frac{\partial \mathbf{J}^N}{\partial \mathbf{X}^0} = \begin{pmatrix} \frac{\partial J_1^N}{\partial X_1^0} & \frac{\partial J_1^N}{\partial X_2^0} & \frac{\partial J_1^N}{\partial X_3^0} \\ \frac{\partial J_2^N}{\partial X_1^0} & \frac{\partial J_2^N}{\partial X_2^0} & \frac{\partial J_2^N}{\partial X_3^0} \end{pmatrix}, \tag{11.17}$$

where $t_0$ is the initial time and $t_N$ is the final (verification) time. In this example, the Jacobian has six components: One can evaluate the full Jacobian in the context of FSC and ASC – through three forward runs of the TLM or two backward runs of the ADJM, respectively, each with different initial conditions (see Figure 11.3). Both the TLM and ADJM require one forward run of the NLM to save the basic states. It is evident that the ADJM is efficient for obtaining $\nabla \mathbf{J}$ when the number of independent variables is much larger than that of the dependent variables, and vice versa for the TLM.

### 11.2.4 Relative Sensitivity Coefficients

One of the goals in SA is to identify and rank the influential parameters, which essentially requires us to assess the relative importance of input parameters on the sensitivity fields. As both the FSC and ASC are dimensional quantities, they are not feasible in addressing this problem. Thus, we introduce *relative sensitivity coefficients* (RSCs) ($S_R$), normalized by appropriate quantities. The following RSC represents a sensitivity, $\frac{\partial J_k}{\partial X_j}$, normalized by its NLM counterparts (i.e., $J_k$ and $X_j$), and hence nondimensional (Park, 1996):

$$S_{R1} = \frac{\partial J_k}{\partial X_j} \bigg/ \frac{J_k}{X_j} = \frac{\partial \ln J_k}{\partial \ln X_j}, \tag{11.18}$$

where $j$ and $k$ are the indices for independent and dependent variables, respectively. It represents a percentage change in $J_k$ for a 1% change in $X_j$: the FSC and ASC depict the change in $J_k$ with respect to unit increase in $X_j$. This definition, however, may lead to misleading or uninformative information in many cases (Park, 1996). For instance, $J_k$

(say, wind speed) may be 0, making $S_{R1}$ singular (i.e., no information). When $X_k$ (say, cloud water mixing ratio) is 0, $S_{R1}$ becomes 0 though the nonnormalized sensitivity $\frac{\partial J_k}{\partial X_j}$ is nonzero. Therefore, the sensitivity can be normalized by a statistical property, such as the mean, variance, or standard deviation of $J_k$ and $X_j$.

When plotted on a time-space diagram, small sensitivities at any given time are often ignored though they may have important information on local characteristics. Park (1996) defined another RSC, called the *local sensitivity coefficient* (LSC) ($S_{R2}$), where the sensitivity is normalized by its domain maximum absolute value at a given time:

$$S_{R2} = \frac{\partial J_k}{\partial X_j} \Big/ \max_D \left| \frac{\partial J_k}{\partial X_j} \right|, \tag{11.19}$$

where $D$ indicates the spatial domain of interest.

Computing FSCs is expensive for 3D models, which have a large number of independent variables. In practice, one may be interested in the responses of model outputs for given perturbations inserted in the input variables at specific areas rather than the grid variables themselves. By introducing an artificial perturbation parameter, $e$, as an independent variable to the original NLM, Park and Droegemeier (2000) suggested the calculation of FSCs with respect to the imposed perturbations directly. Suppose that we perturb an input variable (say, the water vapor mixing ratio $q$) by a factor $e$,

$$q(x,y,z,t;e) = (1+e)q(x,y,z,t). \tag{11.20}$$

Note that any quantity $J$ affected by $q$ implicitly depends on $e$. Then, the following derivative

$$\left. \frac{\partial J(x,y,z,t;e)}{\partial e} \right|_{e=0} \tag{11.21}$$

can be interpreted as the sensitivity of $J$ to a *uniform perturbation* in $q$, evaluated at the reference state, $J(e=0)$. Park and Droegemeier (2000) also defined the following RSC:

$$S_{R3} = \frac{\partial J}{\partial e} \Big/ \frac{J}{e}, \tag{11.22}$$

representing a percentage change in $J$ due to a 1% change in $e$. Although the formulation in (11.20) may not cover the whole spectrum of possible errors and their nonlinear interactions, it can handle both bias and random errors.

**Colloquy 11.1**

**Semi-dimensional RSCs**

The RSC can also be defined as a dimensional quantity. An et al. (2016) defined the RSC as

$$S_{R4} = \frac{\partial J_k}{\partial X_j} X_j, \tag{11.23}$$

making the unit of $S_{R4}$ the same as that of $J_k$; thus, the sensitivity magnitudes can be compared in the relative sense. Similarly, Schwinger et al. (2010) defined the RSC as

$$S_{R5} = \frac{\partial J_k}{\partial X_j}\frac{X_j}{100}, \quad (11.24)$$

which can be interpreted as the changes in $J_k$ resulting from a 1% change in the input $X_j$.

### 11.2.5 Examples of Sensitivity Analysis

Figure 11.4 shows the SA results in a cloud model from Park and Droegemeier (2000) – ASCs and RSCs of accumulated rainfall (ARF) at $t = 70$ min to all model variables at $t = 30$ min, soon after the occurrence of maximum updraft. The ASCs ($S_A$ in (11.15)) in Figure 11.4a depict that the ARF is largely affected by the moisture variables in low levels below 1.75 km: It changes by the largest amount due to a unit increase in $Q_r$ at 0.5 km, followed by that in $Q_c$ at 0.5 km and in $Q_v$ at 0.75 km. The RSCs ($S_{R1}$ in (11.18)) in Figure 11.4b, however, reveal that the comparison among ASCs is potentially misleading: The low-level $T$, especially at 0.75 km (near cloud base), exerts the largest impact on ARF – the latter increases by ~67% due to a 1% increase in the former. The second largest RSC, with a much smaller magnitude, is due to $Q_v$ at the same level. This implies that accurate temperature measurements or retrievals, especially near cloud base and in the lower part of the cloud, are essentially required for storm-scale numerical prediction (Park and Droegemeier, 2000).

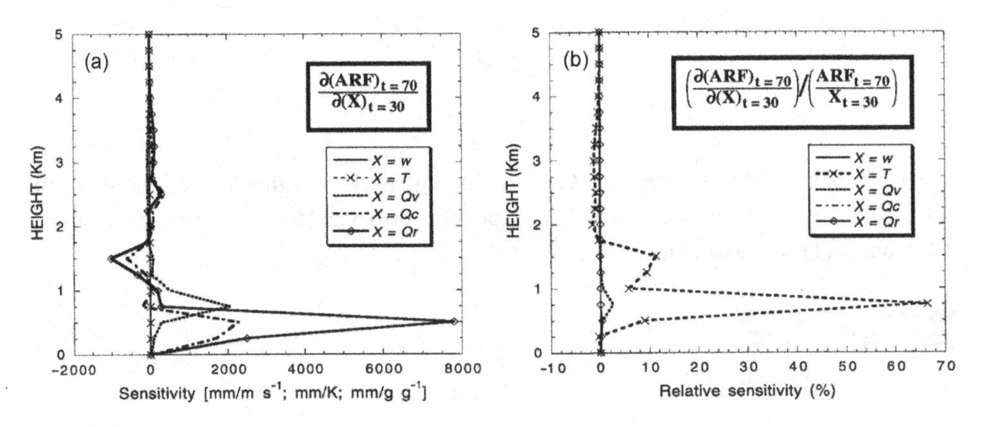


Figure 11.4 Sensitivities in a cloud model: (a) ASCs and (b) RSCs of accumulated rainfall (ARF) at $t = 70$ min with respect to vertical velocity ($w$), cloud temperature ($T$), and mixing ratios of water vapor ($Q_v$), cloud water ($Q_c$), and rainwater ($Q_r$) at $t = 30$ min. Units of ASCs in (a) are mm (m s$^{-1}$)$^{-1}$ for $w$, mm K$^{-1}$ for $T$, and mm (g g$^{-1}$)$^{-1}$ for $Q_v$, $Q_c$, and $Q_r$. From Park and Droegemeier (1999). ©American Meteorological Society. Used with permission.

Figure 11.5 shows the time evolution of the FSCs, based on (11.21), and RSCs ($S_{R3}$ in (11.22)) of the cost function ($J$), typically used in 4DVAR, to perturbations in potential temperature ($e_\theta$) and pressure ($e_p$) inside a convective storm (see Park and Droegemeier, 2000). Among the FSCs, only the largest two are represented: $\frac{\partial J}{\partial e_\theta}$ and $\frac{\partial J}{\partial e_p}$. At the time of error insertion ($t = 80$ min), $J$ is affected by $p$ much larger than by $\theta$; however, after about 10 min, $\theta$ influences $J$ more than $p$ does. To compare the relative importance for different times, the FSCs are normalized following (11.22). Note that the FSCs of $J$ to $e_p$ increase very slowly with time; however, the RSCs of $J$ to $e_p$ show a maximum at the error insertion time and then rapidly decrease toward the verification time ($t = 110$ min). In contrast, the RSCs of $J$ with respect to the $e_\theta$ increase logarithmically with time, exceeding that to $e_p$ after 10 min of error insertion. In the context of data assimilation, this implies that the errors in $p$ at the earlier stage of assimilation result in larger changes in the cost function than that at the later stage of assimilation, and vice versa for $\theta$. As the cost function also represents the forecast error, the SA results also indicate that errors in temperature have larger influence than those in pressure on the forecast error over a longer forecast period.

Errico and Vukićević (1992) developed a mesoscale moist ADJM and demonstrated its applications to various adjoint SA. As an example, they defined $J$ as the forecast error of surface pressure ($p_s$):

$$J = p_s(\text{forecast}) - p_s(\text{verification}) \tag{11.25}$$

at $t = 0$ at the point $P$ in Figure 11.6, where the minimum sea level pressure is located in the NLM forecast. The final conditions at $t = 0$ for the adjoint variables are $\hat{p}_s = 1$ and all others are 0. Figure 11.6 shows the ASCs ($S_A$ in (11.15)) of $J(t = 0)$ with respect to hydrostatic geopotential $z$ ($t = -36$) at $\sigma = 0.4$. For $J$ in (11.25), the ASC is given by the ADJM solution, $\hat{z}$, as

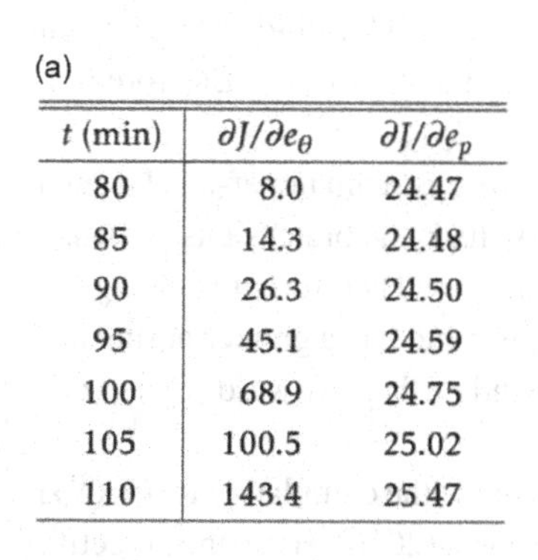

(a)

| $t$ (min) | $\partial J/\partial e_\theta$ | $\partial J/\partial e_p$ |
|---|---|---|
| 80 | 8.0 | 24.47 |
| 85 | 14.3 | 24.48 |
| 90 | 26.3 | 24.50 |
| 95 | 45.1 | 24.59 |
| 100 | 68.9 | 24.75 |
| 105 | 100.5 | 25.02 |
| 110 | 143.4 | 25.47 |

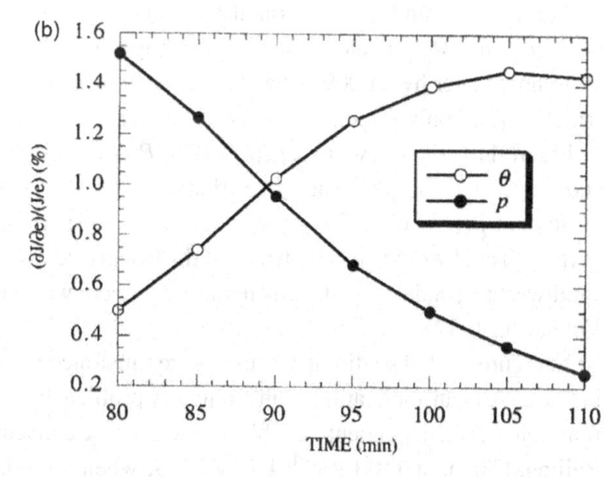


Figure 11.5 Temporal variation of (a) FSCs from (11.21) and (b) RSCs ($S_{R3}$ in (11.22)) of the cost function ($J$) with respect to perturbations of potential temperature ($e_\theta$) and pressure ($e_p$), inserted at $t = 80$ min inside the cloud. Modified from Park and Droegemeier (2000). ©American Meteorological Society. Used with permission.

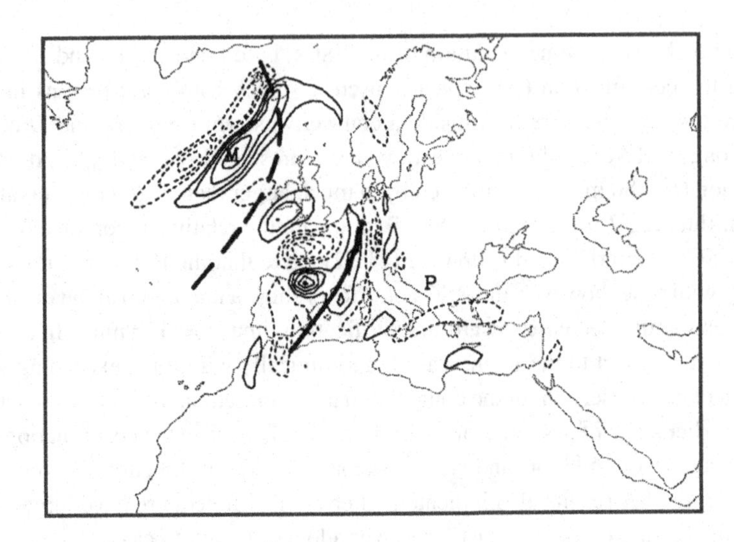


Figure 11.6 The ASCs of $J$ at the verification time ($t = 0$) with respect to perturbations of $z$ at $\sigma = 0.4$ (level $k = {}^{7}/_{2}$) introduced 36 hours earlier with a contour interval of 0.0002 hPa m$^{-1}$. Negative contours are dashed, and the 0 contour is omitted. Point $P$ denotes the location where $J$ is evaluated. Heavy solid and dashed lines represent the upstream 500 hPa trough and ridge, respectively. From Errico and Vukićević (1992). ©American Meteorological Society. Used with permission.

$$\hat{z}_{i,j,k}(t) = \frac{\partial p_s(t = 0;\ \text{point } P)}{\partial z_{i,j,k}(t)}, \tag{11.26}$$

where $i, j,$ and $k$ are indices for the $x, y,$ and $z$ directions; thus, the ASCs in Figure 11.6 quantifies the change in $p_s$ at point $P$ at the verification time ($t = 0$) for the unit changes in $z$ at $\sigma = 0.4$ and any horizontal location at $t = -36$. For instance, $\hat{z} = 0.00097$ hPa m$^{-1}$ at point $M$ indicates that a unit change (i.e., 1 m) in $z$ at $M$ would change $p_s$ at $P$ 36 hours later by 0.00097 hPa. This change in $p_s$ is exact in terms of the TLM forecast, which is valid only for small perturbations in $z$.

Figure 11.6 also shows that $p_s(t = 0)$ at $P$ is sensitive to the upstream upper-level trough and ridge at $t = -36$. When perturbations in $z\,(t = -36)$ had algebraic signs opposite those of $\hat{z}$, $p_s(t = 0)$ at $P$ would decrease (i.e., negative change). Therefore, the change in $p_s(t = 0)$ at $P$ would be negative: 1) if the trough were sharper and had a greater northeast–southwest tilt; and/or 2) if the upstream ridge were weaker and farther eastward (Errico and Vukićević, 1992).

In Figure 11.7, the adjoint SA results are illustrated in terms of the cumulative RSCs ($S_{R4}$ in (11.23)), in an application to an air quality problem by An et al. (2016). Here, the objective function $J$ (i.e., dependent variable) is the average concentration of black carbon (BC) over Beijing, China, at 1100 LST[1] 4 July 2008, when the highest BC concentration occurred: the independent variable is the hourly gridded offline emissions intensity ($q$). The effect of

[1] Local standard time.

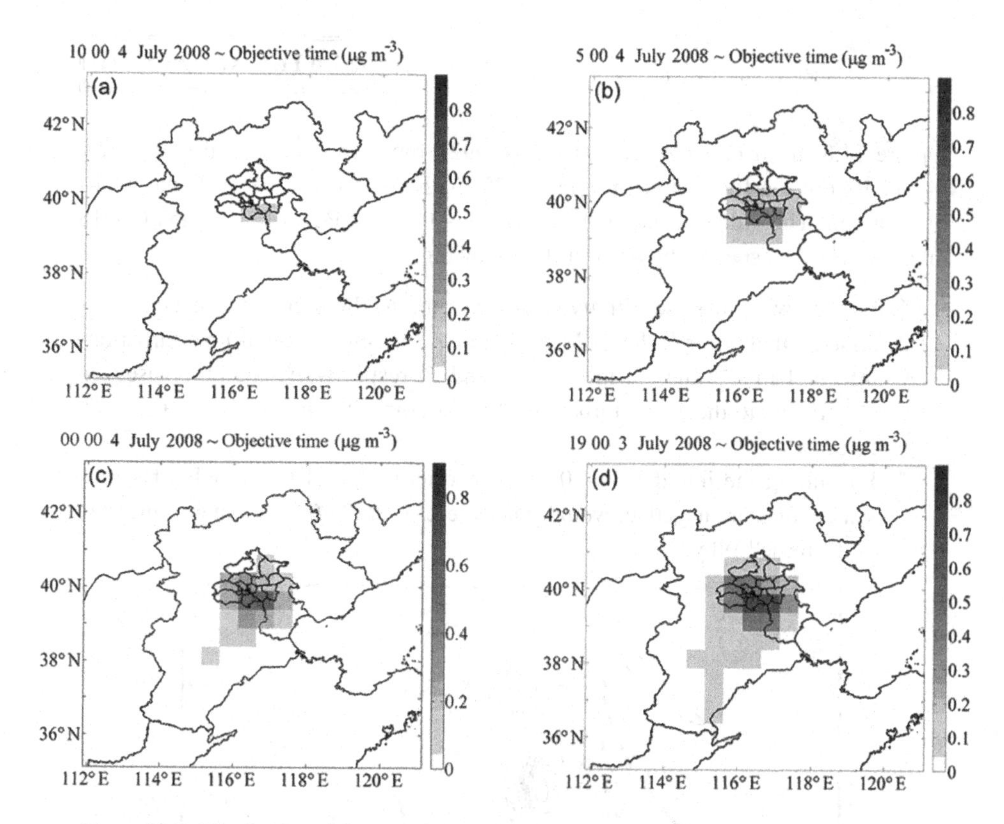


Figure 11.7 Distribution of the cumulative RSC ($S_{R4}$ in (11.23)) of $J$, i.e., the concentration of BC at Beijing, China, at 1100 LST 4 July 2008 (highest BC concentration occurred) with respect to $q$, i.e., the hourly gridded offline emissions intensity: $\frac{\partial J}{\partial q} q$ (in μg $m^{-3}$). Panels (a–d) are 1, 6, 11, and 16 h cumulative RSCs. From An et al. (2016). ©Xing Qin An et al. 2016. Distributed under CC BY 3.0 License.

BC emission is accumulated along an inverse time series, and hence is represented as the cumulative (adjoint) RSCs in Figure 11.7. The $S_{R4}$ values, accumulated from the previous hour (i.e., $t = -1$ h at 1000 LST 4 July), are 0.05–0.1 μg $m^{-3}$ over a small area in Beijing (Figure 11.7a). When the sensitivities are accumulated back to the previous 6 h (i.e., up to 0500 LST 4 July), the cumulative $S_{R4}$ values have increased with a maximum of 0.3–0.4 μg $m^{-3}$ over a significantly enlarged area (Figure 11.7b). As the backward cumulative time increases, the area of influence further expands and the BC impact gets intensified, as shown in Figures 11.7c and 11.7d, eventually leading to the maximum $S_{R4}$ of ~0.7 μg $m^{-3}$ at $t = -16$ h (i.e., 1900 LST 3 July; Figure 11.7d). In the context of air quality control, this implies that reducing BC emission at a ratio of $N\%$ from 1900 LST 3 July to the objective time (1100 LST 4 July) over the sensitivity areas (grid cells) could result in an average $\sim (N\%) \times 0.7$ g $m^{-3}$ decrease of the BC concentration over the objective region (i.e., Beijing) at 1100 LST 4 July 2008 (see An et al., 2016).

**Practice 11.1 Adjoint SA (I)**

Figure 11.8, from Errico and Vukićević (1992), represents $\frac{\partial J}{\partial v}$, i.e., the ASC ($S_A$ in (11.15)) of the forecast error $J$ in (11.25) with respect to the meridional winds $v$ 36 h ago. The values at point $A$ and $B$ are $-0.0023$ hPa (m s$^{-1}$)$^{-1}$ and 0.0018 hPa (m s$^{-1}$)$^{-1}$, respectively. Answer the following:

1. Explain what these sensitivity values (i.e., the ADJM solution $\hat{v}$) mean.
2. Assume that you run the TLM from $t = -36$ with the initial $v$ perturbations ($\delta v$) of 1 m s$^{-1}$ and $-2$ m s$^{-1}$ at $A$ and $B$, respectively, and otherwise 0 m s$^{-1}$. Estimate the $p_s$ perturbation ($\delta p_s$) at point $P$ valid at $t = 0$ (i.e., 36-h forecast).
3. By putting the initial $\delta v > 0$ at the northern edge of the trough where the sensitivities are negative, would you expect $p_s$ at $P$ 36 hours later to increase or decrease? Why?

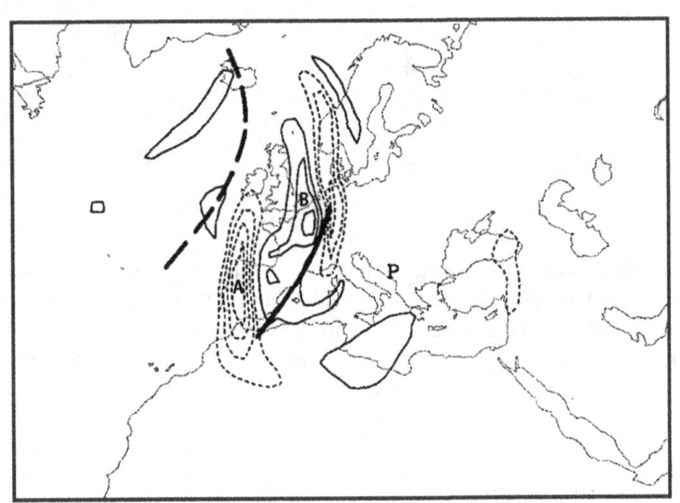


Figure 11.8: Same as in Figure 11.6 but with respect to $v$ at $\sigma = 0.35$ with a contour interval of 0.0005 hPa (m s$^{-1}$)$^{-1}$. Modified from Errico and Vukićević (1992). ©American Meteorological Society. Used with permission.

**Practice 11.2 Adjoint SA (II)**

Doyle et al. (2014) showed Figure 11.9 that depicted the ASCs ($S_A$ in (11.15)) of the intensity of an extratropical cyclone, represented in terms of 36-h kinetic energy (KE) (wind strength), with respect to the initial sea surface temperature (SST). Answer the following:

1. Assume that the $S_A$ value off the west coast of Portugal is 0.29 m$^2$ s$^{-2}$ K$^{-1}$. Explain how the storm intensity would change.

2. What are the implications of the ASCs for the intensification and propagation of the storm?
3. Discuss the difference in the SST sensitivities and the ground temperature sensitivities, and explain its implication.

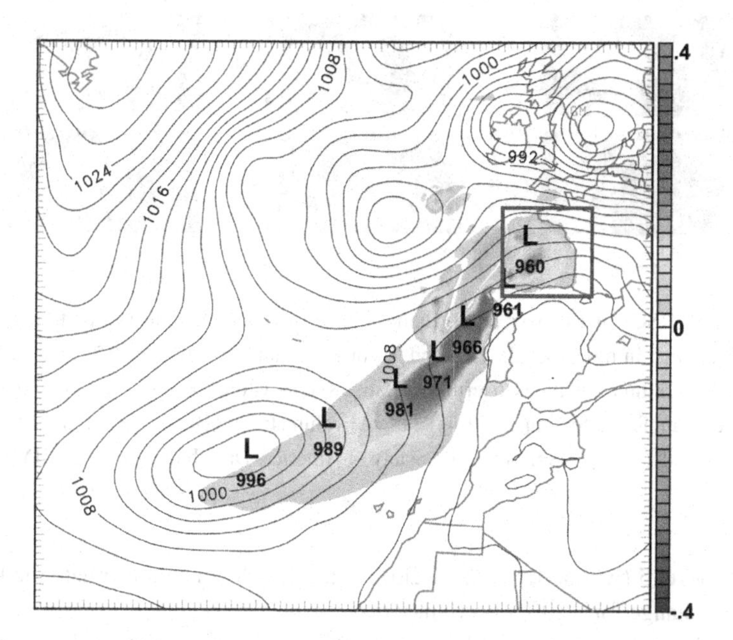


Figure 11.9 The ASCs of KE with respect to SST (color shaded with interval every 0.02 $m^2$ $s^{-2}$ $K^{-1}$), and sea level pressure (SLP) (contoured with interval 2 hPa) from the analysis, valid at the initial time and displayed for a subdomain on the coarse mesh ($\Delta x = 45$ km). The box indicates a fine mesh ($\Delta x = 15$ km) subdomain (600 × 600 $km^2$; 40 × 40 grid cells) where KE is calculated in the lowest 860 m. Over land regions, the sensitivity to the ground surface temperature is displayed. The positions of the storm center are represented with "L" at every 6 h with the minimum central SLP values. Modified from Doyle et al. (2014). ©American Meteorological Society. Used with permission.

**Practice 11.3 Forward SA (RSC)**

Schwinger et al. (2010) calculated the forward RSCs ($S_{R5}$ in (11.24)) of latent heat fluxes (LE) over bare ground with respect to perturbations in initial soil moisture ($\theta_0$) applied in different model levels and soil types in terms of relative soil moisture contents ($\theta_R = \theta / \theta_s$), and showed their temporal variations in Figure 11.3. Soil layers are considered to be dry for $\theta_R \leq 5$ and wet for $\theta_R \geq 0.7$. Answer the following:

1. Describe the characteristic differences of the sensitivities in terms of soil types, depths, and time (days).

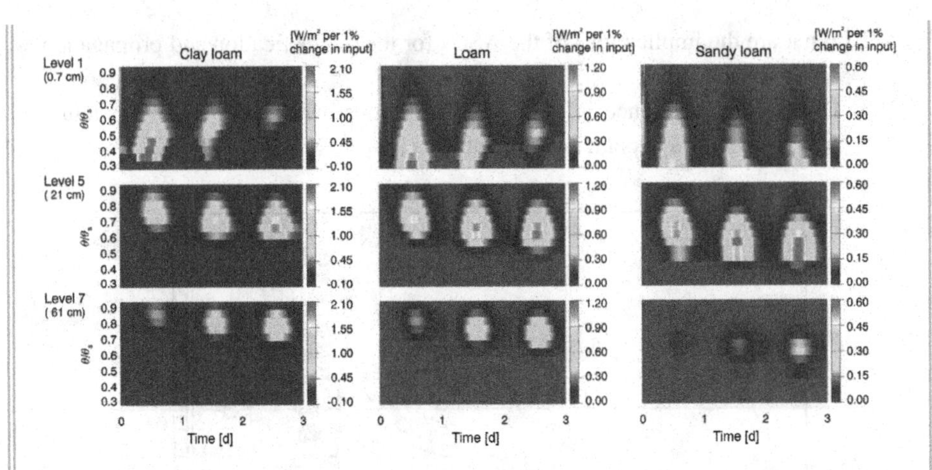


Figure 11.10 RSCs of LE over bare ground with respect to soil moisture perturbations (in W m$^{-2}$) applied in model levels 1, 5, and 7, corresponding to depths of 0.7, 21, and 61 cm, respectively, for three soil types in terms of relative soil moisture contents ($\theta_R = \theta/\theta_s$) where $\theta$ is volumetric moisture content and $\theta_s$ is saturated volumetric moisture content. Modified from Schwinger et al. (2010). ©Soil Science Society of America. Distributed under CC BY-NC-ND 4.0 License.

2. At level 5 (with a depth of 21 cm) for the clay loam soil, estimate the largest LE change for a 20% uncertainty in $\theta_0$.
3. At level 1 (with the shallowest depth of 0.7 cm), estimate the maximum possible changes in LE for a 20% uncertainty in $\theta_0$ for the dry clay loam and dry loam soils, respectively.

## 11.3 Adaptive Observations

The *Glossary of Meteorology* (AMS, 2021) defines "adaptive observations" as follows:

> (Also called *targeted observations.*) Observational data obtained specifically to improve model initial conditions for a numerical forecast of a selected weather feature, or to optimize a measure of forecast outcome (e.g., error).

Analysis errors in any upstream region can bring about large forecast errors in the area of interest in numerical weather prediction. The forecast errors can be reduced by enhancing observations in such an upstream region with significant analysis errors. *Targeted observing* is a process to improve analyses in selected (target) regions of the atmosphere by assimilating additional atmospheric observations, thus diminishing the uncertainty in forecasting high-impact weather events: This is also referred to as *adaptive observing* because the observing network components can be utilized adaptively to manage the time or location of additional observations, including both *in situ* and remotely sensed measurements (Langland, 2005a).

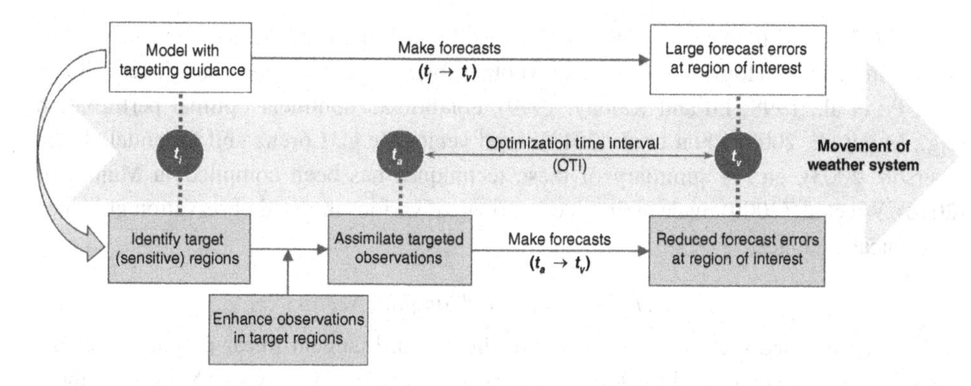


Figure 11.11 Schematic diagram of adaptive observations.

Adaptive observation is based on an assumption that, for a given weather event, a perturbation with a specific structure grows faster than that with a random structure, due to characteristic dynamic and physical processes that interact with the local environment as the event and the perturbation evolve. One should first identify the upstream target regions in which analysis errors have a high probability of growing into large forecast errors in the region of interest (i.e., verification area). Additional specific observations are then decided (i.e., targeted) and assimilated to reduce analysis errors in the target regions. Furthermore, one can also target observations in regions of large initial analysis errors that may develop into large forecast errors (Majumdar, 2016). Therefore, adaptive observation is strongly linked to SA because its main task involves targeting the areas and observations that exert high impact (sensitivity) to the forecast errors or aspects in the verification area.

### *11.3.1 Strategies and Applications*

Figure 11.11 illustrates the process of adaptive observations. The target regions are identified at $t_i$ using any targeting guidance and making forecasts up to verification time $t_v$ when forecast errors are large in the region of interest. After enhancing (deploying) observations in the target regions, we assimilate the targeted observations at the analysis time $t_a$ and make forecasts from $t_a$ up to $t_v$ to verify the forecast aspects and errors at the region of interest. The period $(t_a, t_v)$ is called the *optimization time interval* (OTI). We generally expect that the forecast errors are reduced by applying this adaptive observation technique when a high-impact weather system moves from the (upstream) target regions to the (downstream) verification area.

To identify the target regions where the special observations should be made, several approaches have been adopted as follows: 1) *adjoint-based* methods, including the adjoint SA (e.g., Langland et al., 1999a; Dong et al., 2009; Doyle et al., 2012, 2014), the adjoint-derived sensitivity steering vector (e.g., Wu et al., 2007), and singular vectors (e.g., Buizza and Montani, 1999; Kim et al., 2011); 2) *ensemble-based* methods, including the ensemble Kalman filter (e.g., Jung et al., 2012), the ensemble transformation (e.g., Bishop and Toth, 1999), the ensemble transform Kalman filter (e.g., Bishop et al., 2001; Majumdar et al., 2002b), and the ensemble SA (e.g., Ancell and Hakim, 2007; Garcies and Homar, 2009; Limpert and Houston, 2018); and 3) other methods such as the quasi-inverse TLM (e.g., Pu et al., 1998; Pu and Kalnay, 1999), conditional nonlinear optimal perturbation (e.g., Mu et al., 2009; Zhou et al., 2013), bred vectors (e.g., Lorenz and Emanuel, 1998; Aberson, 2003), etc. A summary of these techniques has been compiled in Majumdar (2016). Wu et al. (2009) made an intercomparison of various targeted observation guidance for tropical cyclones.

#### *11.3.1.1 Adjoint SA and Singular Vectors*

In the adjoint-based methods, the adjoint of the forward tangent linear propagator of the numerical model is required for targeting sensitive areas, using ASCs or SVs; thus, these methods describe the evolution of *linear* perturbations. We introduce the methods based on ASCs and SVs, which have been commonly used for adaptive observations.

Using the ASCs and the adjoint optimal perturbations, Doyle et al. (2019) found that the intensity and precipitation of a damaging extratropical cyclone over the United Kingdom were sensitive to the lower- and mid-tropospheric water vapor fields at the initial time ($t_0 = -36$ h), especially along the sloped frontal region at the poleward edge of the atmospheric river. The optimal perturbations at $t_0$ were oriented along the sloping front and tilted upshear, and subsequently became vertically oriented as they grew from the mean state energy, resulting in increases over the verification area by up to 14 m s$^{-1}$ and 15 mm in the 900-hPa wind speeds and the accumulated precipitation, respectively, in 36 h.

Singular vectors (see Appendix A and Colloquys 11.2 and 11.3) have been widely used to diagnose the sensitive regions to deploy supplementary observations (e.g., Buizza and Montani, 1999; Thorpe and Petersen, 2006). Using both the TLM and ADJM, i.e., in a linear sense, the leading SV represents the most rapidly growing perturbation over a given forecast period (say, from the analysis time $t_a$ to the verification time $t_v$), thus often called the optimal perturbation (Regan and Mahesh, 2019).

**COLLOQUY 11.2**

**Singular vectors for adaptive observations (I)**

Based on the TLM formulation in (4.19)–(4.21), from which the linear perturbation $\delta\mathbf{X}$ is replaced by $\mathbf{x}$ for convenience, the evolution of $\mathbf{x}$ from the analysis time ($t_a$) to the verification time ($t_v$) is given by in terms of the tangent linear propagator $\mathbf{L} = \mathbf{L}(t_a, t_v)$ as

$$\mathbf{x}^v = \mathbf{L}\mathbf{x}^a, \tag{11.27}$$

where $\mathbf{x}^v = \mathbf{x}(t_v)$ and $\mathbf{x}^a = \mathbf{x}(t_a)$. Note that, through the singular value decomposition (SVD) and the eigenvalue decomposition (EVD) (see (A.29)–(A.35) in Appendix A), the SVs of $\mathbf{L}$ are related to the eigenvectors of $\mathbf{L}^*\mathbf{L}$, where $\mathbf{L}^* = \mathbf{L}^T$ is the adjoint (or conjugate transpose) of $\mathbf{L}$.

By applying SVD to $\mathbf{L}$, and following (A.29), we have

$$\mathbf{L} = \mathbf{U}\boldsymbol{\Sigma}\mathbf{V}^T, \tag{11.28}$$

where $\mathbf{U} = (\mathbf{u}_1, \ldots, \mathbf{u}_N)$ and $\mathbf{V} = (\mathbf{v}_1, \ldots, \mathbf{v}_N)$ are orthonormal matrices with column vectors $\{\mathbf{u}_i, \mathbf{v}_i : i = 1, N\}$, and $\boldsymbol{\Sigma}$ is a diagonal matrix with elements $\{\sigma_i : i = 1, N\}$. Here, $\mathbf{u}_i$ and $\mathbf{v}_i$ are the left (final) and the right (initial) *singular vectors*, respectively, and $\sigma_i$ are *singular values* with the order $\sigma_1 \geq \sigma_2 \geq \cdots \geq \sigma_N$. Then, the initial SVs propagate into the final SVs via

$$\mathbf{L}\mathbf{v}_i = \sigma_i \mathbf{u}_i. \tag{11.29}$$

We now apply EVD to $\mathbf{L}^*\mathbf{L}$, a symmetric and positive definite operator having mutually orthogonal eigenvectors, following (A.33). It turns out that the right (left) SVs of $\mathbf{L}$ correspond to the eigenvectors of $\mathbf{L}^*\mathbf{L}$ ($\mathbf{L}\mathbf{L}^*$) and the squared singular values of $\mathbf{L}$ are the eigenvalues of $\mathbf{L}^*\mathbf{L}$ and $\mathbf{L}\mathbf{L}^*$:

$$\mathbf{L}^*\mathbf{L}\mathbf{v}_i = \sigma_i^2 \mathbf{v}_i \text{ and } \mathbf{L}\mathbf{L}^*\mathbf{u}_i = \sigma_i^2 \mathbf{u}_i. \tag{11.30}$$

**COLLOQUY 11.3**

**Singular vectors for adaptive observations (II)**

For targeting the areas where the perturbations amplify for a specific time interval, we define a norm to describe the magnitude of perturbations ($\mathbf{x}$) in terms of inner products as

$$\|\mathbf{x}\|_E^2 \equiv \langle \mathbf{x}, \mathbf{E}\mathbf{x} \rangle, \tag{11.31}$$

where $\mathbf{E}$ is a specific norm used for adaptive observations and can be defined separately for different times (e.g., $\mathbf{E}_a$ for $t_a$ and $\mathbf{E}_v$ for $t_v$). The growth rate of perturbations from $t_a$ to $t_v$ is measured, using (11.27), as the following (see also Diaconescu and Laprise, 2012):

$$\sigma^2 = \frac{\|\mathbf{x}^v\|_{E_v}^2}{\|\mathbf{x}^a\|_{E_a}^2} = \frac{\langle \mathbf{L}\mathbf{x}^a, \mathbf{E}_v \mathbf{L}\mathbf{x}^a \rangle}{\langle \mathbf{x}^a, \mathbf{E}_a \mathbf{x}^a \rangle} = \frac{\langle \mathbf{L}^*\mathbf{E}_v \mathbf{L}\mathbf{x}^a, \mathbf{x}^a \rangle}{\langle \mathbf{x}^a, \mathbf{E}_a \mathbf{x}^a \rangle}. \tag{11.32}$$

The SVs that maximize (11.32), i.e., the most rapidly growing perturbations, are obtained as the solution of the eigenvalue problem, for $\mathbf{v}_i^a = \mathbf{v}_I(t_a)$:

$$\mathbf{L}^*\mathbf{E}_v \mathbf{L}\mathbf{v}_i^a = \sigma^2 \mathbf{E}_a \mathbf{v}_i^a, \tag{11.33}$$

and by introducing a *local projection operator* $\mathbf{P}$ (Buizza, 1994) that sets the perturbation vector to 0 outside the region of interest,

$$\mathbf{L}^*\mathbf{P}^*\mathbf{E}_v \mathbf{P}\mathbf{L}\mathbf{v}_i^a = \sigma^2 \mathbf{E}_a \mathbf{v}_i^a. \tag{11.34}$$

Equations (11.33) and (11.34) can be solved by one forward integration of the TLM, followed by one backward integration of the ADJM. Equation (11.34) is relevant especially for the limited-area models and in the particular case of adaptive observations (Diaconescu and Laprise, 2012); thus, its SVs are often called *targeted SVs* (Buizza and Montani, 1999). The leading SV (say, $\mathbf{v}_1$) maximizes (11.32) over the interval $(t_a, t_v)$.

For the purpose of adaptive observations, various norms with a basic form in (11.31), over the time interval $(t_a, t_v)$, have been used. The norm can be defined in terms of error covariances, $\mathbf{E}(t_a) = (\mathbf{P}^a)^{-1}(t_a)$ and $\mathbf{E}(t_v) = (\mathbf{P}^f)^{-1}(t_v)$, where $\mathbf{P}^a$ is the analysis error covariance (AEC) and $\mathbf{P}^f$ is the forecast error covariance. The SVs based on this norm are denoted the targeted AEC optimals (Majumdar et al., 2006): This inverse AEC is considered to be the most appropriate norm at $t_a$ because the initial conditions are constrained by the observing network and the assimilation process (e.g., Ehrendorfer and Tribbia, 1997; Palmer et al., 1998; Diaconescu and Laprise, 2012). The total energy (TE) can be defined as a norm through (11.31), for $\mathbf{E}(t_a) = \mathbf{E}(t_v) = \mathbf{E}$, by summing up the grid point values of the kinetic and potential energy; then, the eigenvalues in (11.30) measure the growth of TE related to the corresponding SVs, i.e., total-energy SVs (TESVs) (e.g., Buizza and Montani, 1999; Kim et al., 2011). Leutbecher et al. (2002) used a norm given by the Hessian (i.e., the second derivative) of the cost function in the VAR, as an estimate of the inverse AEC, yielding the Hessian singular vectors (HSVs). As an alternative to the full AEC (i.e., HSV) technique, Gelaro et al. (2002) defined the norm based on the inverse of the analysis error variance estimate from the 3DVAR, producing the variance SVs (VARSVs). Several studies have compared various SV-based approaches within the context of adaptive observations (Palmer et al., 1998; Gelaro et al., 2002; Leutbecher et al., 2002, etc.).

Figure 11.12 illustrates the comparison between the TESVs and HSVs by Leutbecher et al. (2002). The sensitive (target) regions were identified from a weighted average of the first five SVs for both TESVs (Figure 11.12) and HSVs (Figure 11.12). As the distribution of the initial SVs depends on the initial norm, the target regions based on the two norms show distinct differences: The TESV target region covers the eastern part of North America with the peak sensitivity over the US east coast and the adjacent sea while the HSV target region appears mostly over the sea with the peak at the central Atlantic. This is mainly because HSVs take the observation network into consideration, including the inhomogeneous distribution of observations, whereas the TESVs are based on a spatially uniform metric (Leutbecher et al., 2002; Thorpe and Petersen, 2006). Note that initial perturbations are constrained over an area with dense observations such as the east coast of North America.

To assess the reliability of the linear estimates via SVs, the optimal zone for observations (OZO) (black dashed box) is selected by integrating the full nonlinear model (NLM) over 14 test regions with 40 additional soundings. The most sensitive HSV areas matched well with the OZO, evidencing the high reliability of the HSV-based targeting method. The evolution of the forecast errors, shown in Figure 11.12 in terms of the energy metric, also demonstrates the superiority of the HSVs to the TESVs in targeting the sensitive regions.

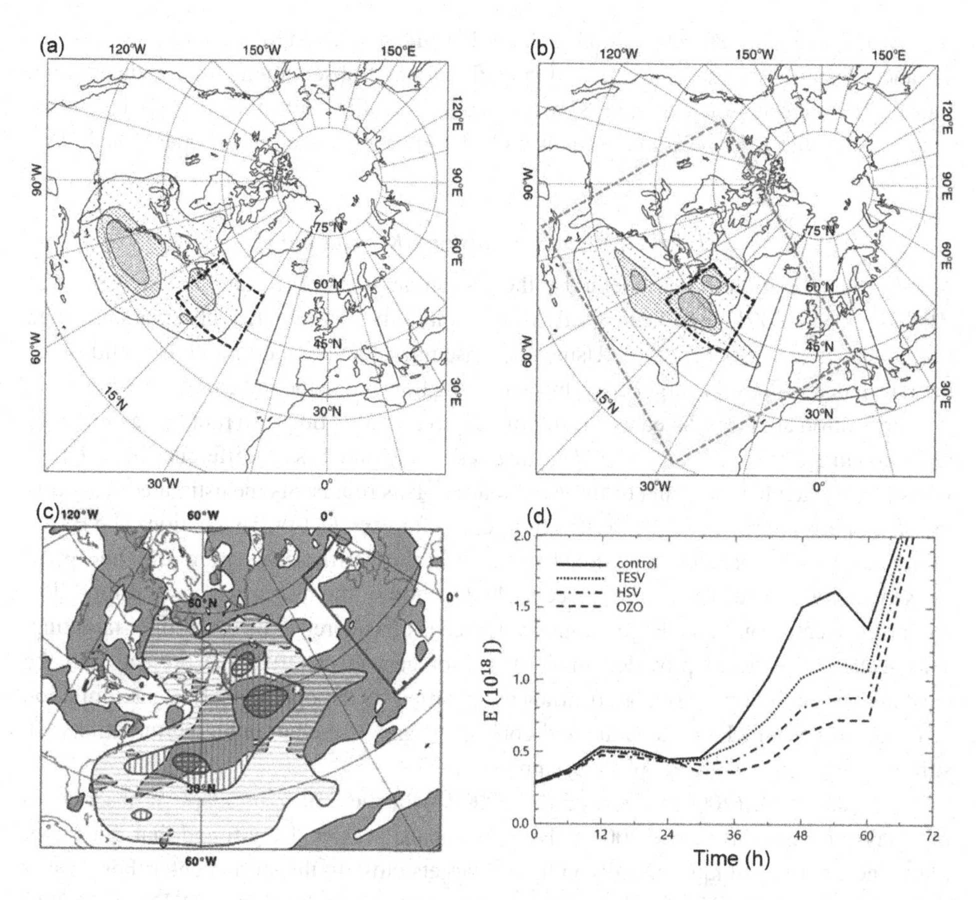


Figure 11.12 Target (sensitive) regions (stippled) based on (a) total-energy singular vectors (TESVs) and (b) Hessian singular vectors (HSVs), (c) target regions (striped and hatched) enlarged from (b) (large dashed box) with cloud layers (shaded), and (d) evolution of the forecast error, represented as the total energy (E in $10^8$ J) of the error fields, which is integrated from the surface to 500 hPa in the verification region, for experiments with the control run, the TESV target, the HSV target, and the optimal zone for observing (OZO) (small dashed box in (a) and (b)) target. In (a)–(c), the sizes of the small, medium, and large target regions are $(0.9, 3, 12) \times 10^6$ km$^2$, respectively, and the solid boxes indicate the verification region (20°W–20°E, 35–60°N). (a), (b), and (d) are modified from Leutbecher et al. (2002) ©Royal Meteorological Society 2002 and (c) is modified from Thorpe and Petersen (2006) ©Cambridge University Press 2006. Used with permission.

Thorpe and Petersen (2006) indicated that the HSV sensitive regions largely overlapped with cloud layers (Figure 11.12c), and addressed that, given the poor resolution of satellite measurements in the cloud layers, alternative adaptive observations, e.g., with *in situ* measurements, should be considered.

Some methods are regarded as extensions of the adjoint-based techniques that account for certain properties of the observations and data assimilation procedures: the Hessian reduced-rank estimate (e.g., Leutbecher, 2003) that predicts the potential reduction of forecast error variance, resulting from deployment of specific targeted observations, by a reduced-rank state estimation using the leading HSVs; and the Kalman filter sensitivity (KFS) (e.g., Oger et al., 2012) that uses AECs consistent with the error estimates of an operational VAR scheme.

#### *11.3.1.2 Ensemble Transform Kalman Filter*

Among the ensemble-based methods, the ensemble transform Kalman filter (ETKF) (Bishop et al., 2001) has been used most commonly for targeted observations (e.g., Majumdar et al., 2002b, 2006). Using the ensemble forecast perturbations valid at $t_v$, the ETKF determines the target areas by estimating how much the forecast error variance in the verification area is reduced as a result of adaptive observations. To rapidly calculate the reduction in error variance, the *signal covariance* is calculated as the difference between the sums of the diagonal elements of the AEC matrix: This represents the estimated reduction in forecast error variance in the verification region for each test-probe location. The ETKF "summary map" represents this signal variance in the verification region at $t_v$ and gives the sensitivity of forecast error variance to the location of the test-probe observations: The test-probe location with the largest signal variance is considered to be optimal for targeting. This technique is valid provided that: 1) the forecast error structures project onto the ensemble perturbations; 2) error covariances specified by the operational data assimilation scheme and the ETKF are accurate and consistent; and 3) the ensemble perturbations are sufficiently small to validate linear dynamics.

Majumdar et al. (2006) compared the ETKF with the TESV in adaptive observation of 2-day forecasts during the 2004 Atlantic hurricane season. They found that, for major hurricanes, both techniques usually indicated targets close to the storm center: For weaker tropical cyclones, the TESV selected similar targets to those from the ETKF in only 30% of the cases. As noted by Buizza (1994), the ETKF accounts for initial condition errors, implying that observations to reduce initial uncertainty of the storm are apt to be more useful than other observations; in contrast, TESVs do not consider initial condition errors, and hence the storm-related uncertainty is not accounted for. The advantages of the ETKF include low cost compared to the adjoint-based techniques, use of nonlinear ensemble forecasts, and ability to find target locations serially given an ensemble data assimilation scheme (Aberson, 2003).

#### *11.3.1.3 Conditional Nonlinear Optimal Perturbations*

Mu et al. (2009) applied the conditional nonlinear optimal perturbation (CNOP) technique (see Colloquy 11.4) – a natural extension of SV into the nonlinear regime – to adaptive observations for tropical cyclones. Note that SVs can approximate the evolution of initial perturbations, in a linear sense, whose behavior is valid only for sufficiently small perturbations and short-time forecasts. Initial perturbations, which show the largest growth in a nonlinear regime, are represented by CNOPs. They compared the first SVs (FSVs) and CNOPs and found that the difference in their structures depends on the constraint, metrics, and the basic state. The CNOP-based perturbations had a larger impact on the forecasts in the verification area as well as the tropical cyclones than the FSV-based perturbations; thus, a reduction of initial errors by CNOPs provided more benefits than that by FSVs.

COLLOQUY 11.4

**CNOP for adaptive observations**

The model state **X** evolves from the analysis time $t_a$ to the verification time $t_v$ through a nonlinear propagator **M**, following

$$\mathbf{X}^v = \mathbf{M}\mathbf{X}^a. \tag{11.35}$$

An initial perturbation $\mathbf{x}_C^a$ of $\mathbf{X}^a$ is called CNOP if and only if the cost function $J$ satisfies

$$J\left(\mathbf{x}_C^a\right) = \max_{\|\mathbf{x}\|_{E_a}^2 \le C} J\left(\mathbf{x}^a\right), \tag{11.36}$$

where

$$J\left(\mathbf{x}^a\right) = \left(\mathbf{PM}\left(\mathbf{X}^a + \mathbf{x}^a\right) - \mathbf{PM}\left(\mathbf{X}^a\right)\right)^T \mathbf{E}_v \left(\mathbf{PM}\left(\mathbf{X}^a + \mathbf{x}^a\right) - \mathbf{PM}\left(\mathbf{X}^a\right)\right) \tag{11.37}$$

and $\|\mathbf{x}\|_{E_a}^2 = \langle \mathbf{x}, \mathbf{E}_a \mathbf{x} \rangle \le C$ is a constraint condition of initial perturbations with the presumed positive constant $C$ representing the magnitude of the initial uncertainty. The operator **P** represents the local projection to take 0 values outside the target region. The norms $\mathbf{E}_a$ and $\mathbf{E}_v$ at $t_a$ and $t_v$, respectively, may be the same depending on the physical problem.

For the chosen norm, the CNOP is the global maximum of $J$, representing the initial perturbation whose nonlinear evolution attains the maximum of $J$ at $t_v$. The *local* CNOP is defined as an initial perturbation that reaches a local maximum of $J$ around a particular point in phase space. The first guess of $\mathbf{x}^a$ is usually taken as the difference between the model solutions at $t_1$ and $t_2$, i.e., $\mathbf{x}^a = \mathbf{X}(t_1) - \mathbf{X}(t_2)$, within the interval $(t_a, t_v)$, which should be adjusted to satisfy the constraint condition. Various methods to obtains CNOPs, including the adjoint-based and adjoint-free methods, the intelligent optimization method, and the unconstrained optimization method, are compiled in Wang et al. (2020).

Figure 11.13 compares CNOP and FSV, in terms of the vertically integrated energy (VIE) and wind components, in adaptive observations for Typhoon Nida (2004) (see Chen, 2011). A total of 15 dropwindsondes (■ + ☆) were deployed every 150–200 km in a circular pattern with its center close to the typhoon's central position: they measured wind speed, wind direction, height, temperature, dewpoint temperature, and relative humidity below 196 hPa. For the targeted observation experiments, both CNOP and FSV used just four dropwindsonde data (marked as ☆), near the areas of maximum VIE, with the sondes #2–5 for CNOP and #10–13 for FSV, respectively. Note that, around the data points, the sensitive areas are quite different: the high magnitudes of VIE and wind of the CNOP are found

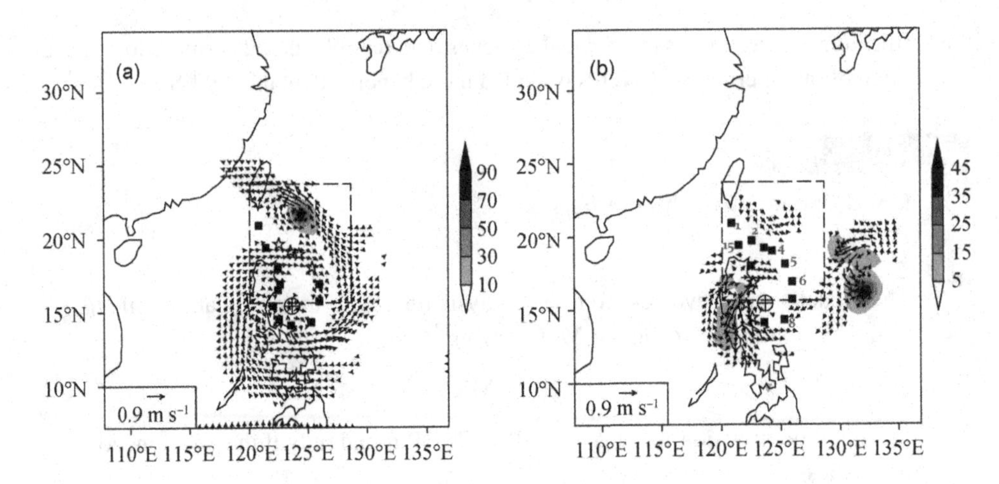


Figure 11.13 VIE (shaded; J kg$^{-1}$) and wind (vector; m s$^{-1}$) component at the level $\sigma = 0.7$ of (a) CNOP and (b) FSV. The verification area is represented in the dashed box. Among the 15 dropwindsonde observations, numbered clockwise starting from the northernmost point (shown in (b)), only 4 observations (pentagrams) are used for adaptive observations (i.e., the sondes 2–5 for CNOP and 10–13 for FSV). The symbol ⊕ indicates the initial position of the center. Modified from Chen (2011). © Institute of Atmospheric Physics, Chinese Academy of Sciences.

mostly at the north while those of the FSV are observed at the west of the initial typhoon center (marked as ⊕), thus making different choices for the targeted data points.

Assimilation of the targeted dropsondes in both the CNOP and FSV sensitive areas improved the track forecasts: For the 24-h and 36-h forecasts, the CNOP (FSV) observations reduced the track error by 63.6% (8.1%) and 35.0% (7.5%), respectively. Adding all the dropsondes also improved the track forecasts by 39.4% (24-h forecast) and 28.5% (36-h forecast) but less significantly compared with the CNOP-based forecasts (Chen, 2011). Various applications of the CNOP method to the adaptive observations, especially for tropical cyclone prediction, are referred to in Zhou et al. (2013).

Overall, the targeting methods have generally demonstrated the usefulness to provide reliable guidance in adaptive observations, though each of them has its own limitations. They have been practically implemented in real time during many field campaigns (see Section 11.3.3) and been continually improved to provide better estimates of analysis and forecast error variances. Various crucial issues regarding adaptive observations are referred to in Langland (2005a) and Majumdar (2016).

### *11.3.2 Forecast Sensitivity to Observations*

Another important aspect of adaptive observations is to understand what kind of observation give the largest impact on forecast error reduction. *Observation-space targeting* produces guidance in observation space by yielding an *observation sensitivity vector* for the targeted and regular observations at a future time to estimate the potential impact of the observing configuration on an estimate of forecast error variance (Langland, 2005a).

Using the adjoint approach, Baker and Daley (2000) derived the observation sensitivity vector (see Colloquy 11.5) to compute the sensitivities of a forecast or analysis aspect with respect to any or all of the observing systems by types, locations, or channels. Furthermore, utilizing the observation sensitivity framework, Langland and Baker (2004) developed a procedure to evaluate the *observation impact* (see Colloquy 11.6), which is a product of the innovation vector and the adjoint observation sensitivity vector obtained from the data assimilation process. Equation (11.c24) gives a total estimate for all observations and can be partitioned into particular subsets of interest.

**COLLOQUY 11.5**

**Observation sensitivity**

Baker and Daley (2000) defined the observation and background ASCs by rewriting the linear analysis equation (1.50) as

$$\mathbf{x}^a = (\mathbf{I} - \mathbf{KH})\,\mathbf{x}^b + \mathbf{K}\mathbf{y}^o, \tag{11.38}$$

with the typical notations as in Section 1.4.4. Then, the sensitivities of the analysis to the observations and the background are derived as

$$\frac{\partial \mathbf{x}^a}{\partial \mathbf{y}^o} = \mathbf{K}^T \ \text{and} \ \frac{\partial \mathbf{x}^a}{\partial \mathbf{x}^b} = \mathbf{I} - \mathbf{H}^T\mathbf{K}^T, \tag{11.39}$$

respectively. As the Kalman gain $\mathbf{K}$ is given by (1.57), its transpose $\mathbf{K}^T$ is written as

$$\mathbf{K}^T = \left(\mathbf{HP}^b\mathbf{H}^T + \mathbf{R}\right)^{-1} \mathbf{HP}^b.$$

To associate (11.36) with the adaptive observation problem, a cost function $J$ is defined, which is a scalar measure of the forecast aspects over the verification region. Then, the gradient of $J$ with respect to the initial conditions (i.e., analysis), $\frac{\partial J}{\partial \mathbf{x}^a}$, is defined as the *analysis sensitivity vector*. Similarly, the *observation sensitivity vector*, $\frac{\partial J}{\partial \mathbf{y}^o}$, defines the sensitivity of the forecast aspect to the observations and is formulated as

$$\frac{\partial J}{\partial \mathbf{y}^o} = \frac{\partial \mathbf{x}^a}{\partial \mathbf{y}^o}\frac{\partial J}{\partial \mathbf{x}^a} = \mathbf{K}^T\frac{\partial J}{\partial \mathbf{x}^a} = \left(\mathbf{HP}^b\mathbf{H}^T + \mathbf{R}\right)^{-1} \mathbf{HP}^b\frac{\partial J}{\partial \mathbf{x}^a}, \tag{11.40}$$

whereas the background sensitivity vector, $\frac{\partial J}{\partial \mathbf{x}^b}$, is the sensitivity of the forecast aspect to the background field, formulated as

$$\frac{\partial J}{\partial \mathbf{x}^b} = \frac{\partial \mathbf{x}^a}{\partial \mathbf{x}^b}\frac{\partial J}{\partial \mathbf{x}^a} = \left(\mathbf{I} - \mathbf{H}^T\left(\mathbf{HP}^b\mathbf{H}^T + \mathbf{R}\right)^{-1} \mathbf{HP}^b\right)\frac{\partial J}{\partial \mathbf{x}^a}. \tag{11.41}$$

Finally, the difference between the analysis sensitivity vector and the background sensitivity vector is given by

$$\frac{\partial J}{\partial \mathbf{x}^a} - \frac{\partial J}{\partial \mathbf{x}^b} = \mathbf{H}^T \frac{\partial J}{\partial \mathbf{y}^o}, \tag{11.42}$$

which is denoted the *analysis space projection of the observation sensitivity vector*.

**Colloquy 11.6**

## Observation impact

Langland and Baker (2004) derived a procedure to quantify the observation impact as follows. The errors of two forecasts, $\mathbf{x}^f$ and $\mathbf{x}^g$, can be measured against an analysis $\mathbf{x}^v$ available at verification time $t_v$ using the following energy norms

$$e_f = \left\langle \left(\mathbf{x}^f - \mathbf{x}^v\right), \mathbf{E}\left(\mathbf{x}^f - \mathbf{x}^v\right) \right\rangle \text{ and } e_g = \left\langle \left(\mathbf{x}^g - \mathbf{x}^v\right), \mathbf{E}\left(\mathbf{x}^g - \mathbf{x}^v\right) \right\rangle, \tag{11.43}$$

where $\mathbf{E}$ is a matrix of energy weighting coefficients for dry TE.

One may consider that a short-term forecast (say, 6 h) of $\mathbf{x}^g$ serves as the background field $\mathbf{x}^b$ to produce the analysis field $\mathbf{x}^a$ for the $\mathbf{x}^f$ forecast. The difference between the errors of $\mathbf{x}^f$ and $\mathbf{x}^g$ is given by $\Delta e_f^g = e_f - e_g$. Then, we can define the cost functions

$$J_f = \frac{1}{2} e_f \text{ and } J_g = \frac{1}{2} e_g \tag{11.44}$$

and the corresponding first-order derivatives

$$\frac{\partial J_f}{\partial \mathbf{x}^f} = \mathbf{E}\left(\mathbf{x}^f - \mathbf{x}^v\right) \text{ and } \frac{\partial J_g}{\partial \mathbf{x}^g} = \mathbf{E}\left(\mathbf{x}^g - \mathbf{x}^v\right). \tag{11.45}$$

Then, using (11.43) and (11.45),

$$\Delta e_f^g = \left\langle \left(\mathbf{x}^f - \mathbf{x}^g\right), \left(\frac{\partial J_f}{\partial \mathbf{x}^f} + \frac{\partial J_g}{\partial \mathbf{x}^g}\right) \right\rangle. \tag{11.46}$$

The difference between forecast trajectories $f$ and $g$ at the analysis time is $\left(\mathbf{x}^a - \mathbf{x}^b\right)$. The adjoint model maps $\frac{\partial J_f}{\partial \mathbf{x}^f}$ to $\frac{\partial J_f}{\partial \mathbf{x}^a}$ and $\frac{\partial J_g}{\partial \mathbf{x}^g}$ to $\frac{\partial J_g}{\partial \mathbf{x}^b}$ along trajectories $f$ and $g$, respectively. Then, an estimate of $\Delta e_f^g$ in analysis space can be written as

$$\delta e_f^g = \left\langle \left(\mathbf{x}^a - \mathbf{x}^b\right), \left(\frac{\partial J_f}{\partial \mathbf{x}^a} + \frac{\partial J_g}{\partial \mathbf{x}^b}\right) \right\rangle. \tag{11.47}$$

Using the linear analysis equation (1.50) and the adjoint operator property, the *observation impact* in observation space is given by

$$\delta e_f^g = \left\langle \left(\mathbf{y}^o - \mathbf{H}\mathbf{x}^b\right), \mathbf{K}^\mathrm{T}\left(\frac{\partial J_f}{\partial \mathbf{x}^a} + \frac{\partial J_g}{\partial \mathbf{x}^b}\right) \right\rangle = \left\langle \left(\mathbf{y}^o - \mathbf{H}\mathbf{x}^b\right), \frac{\partial J_f^g}{\partial \mathbf{y}^o} \right\rangle. \tag{11.48}$$

**Practice 11.4 Observation impact**

Derive (11.c24) from (11.c23) using the relation (Langland and Baker, 2004)

$$\frac{\partial J_f^g}{\partial \mathbf{x}^a} = \frac{\partial J_f}{\partial \mathbf{x}^a} + \frac{\partial J_g}{\partial \mathbf{x}^b}$$

and the adjoint operator property in (4.28).

Some examples of applying the observation impact have been introduced by (Baker and Langland, 2009): monitoring systematic nonbeneficial impact by observing systems to help fix potential problems in processing data; identifying beneficial and nonbeneficial channels for satellite data assimilation; checking potential problems in a data assimilation system by finding unusual nonbeneficial impact due to specific channels of satellite instruments; assessing the observation impact for all observations, from the global observation network, assimilated by a prediction system; and the observation impact per observation for a data assimilation system as a function of the observation time in the assimilation window.

Figure 11.14, from Baker and Langland (2009), depicts an example of channel selection for the radiance observations from the AIRS[2] and AMSU[3] onboard the NASA/Aqua

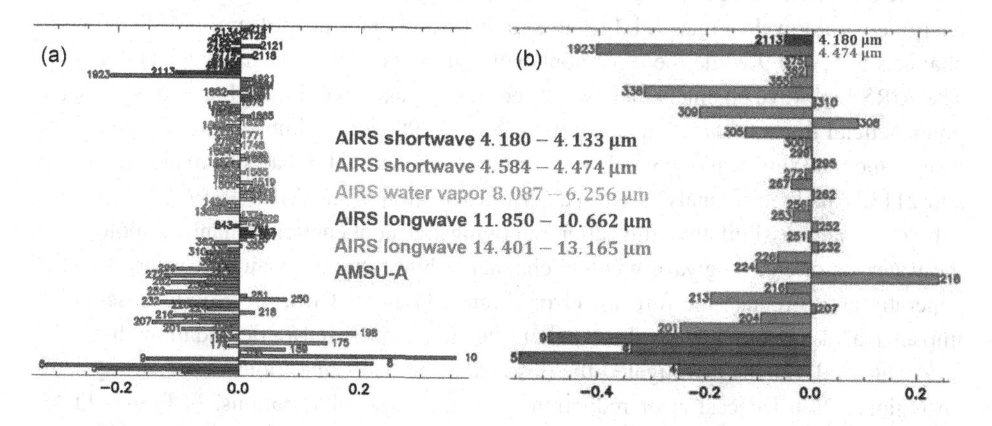


Figure 11.14 Observation impact, in terms of moist static energy-weighted total error norm (in J kg$^{-1}$), of the AIRS and AMSU channels for the NAVDAS/NOGAPS system (a) by assimilating all the channels during August 19–25, 2006, and (b) by assimilating only beneficial channels during August 15–26, 2006. Negative (positive) values indicate a reduction (an increase) in the 24-h global forecast error. Channel numbers are listed beside the corresponding error bars for the AMSU microwave channels (AMSU-A) and the AIRS channels of shortwave $CO_2$ (4.180–4.133 μm and 4.584–4.474 μm), water vapor (8.087–6.256 μm), longwave window (11.850–10.662 μm), and longwave $CO_2$ (14.401–13.165 μm). Modified from Baker and Langland (2009). ©Springer-Verlag Berlin Heidelberg 2009. Used with permission.

[2] Advanced Infrared Sounder.

[3] Advanced Microwave Sounding Unit.*****

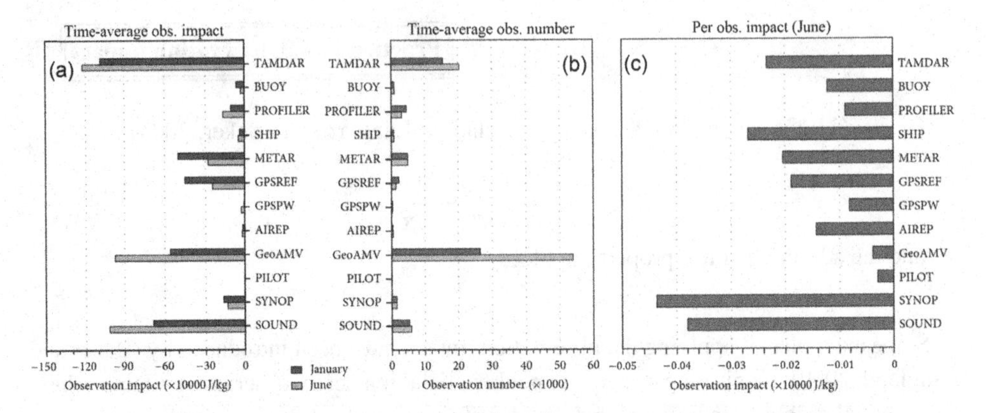


Figure 11.15 (a) Total observation impact (in J $kg^{-1}$) and (b) the corresponding assimilated observation number of the types of observing systems in June (gray) and January (black) 2010, and (c) per-observation impact (in J $kg^{-1}$) for June 2010, obtained by dividing (a) by (b). All the fields in (a)–(c) are time-averaged for a period of two weeks. Experiments are done using the Weather Research and Forecasting data assimilation (WRFDA) system over the continental United States. Modified from Zhang et al. (2015). 

satellite for the NAVDAS[4]/NOGAPS[5] system for an assimilation run using the moist static energy error norm. In Figure 11.14a, the largest beneficial impact from AMSU is due to channels 5, 6, and 9 while the most nonbeneficial impact comes from channels 8 and 10. The AIRS longwave channels mostly exerted strong beneficial impact but with significant nonbeneficial impact from channels 169, 175, and 198. Varied impacts are shown for the water vapor and shortwave channels with prominent beneficial impacts from channels 1923 and 2113. This kind of analysis is used for modification of the AIRS and AMSU channel subsets. A new assimilation run, after removing the nonbeneficial channels along with the water vapor and longwave window channels, shows that the majority of the channels generally act to reduce the forecast error (Figure 11.14b). Thus, the adjoint observation impact analyses is valuable for the channel selection and the sensor degradation check.

Zhang et al. (2015) investigated the observation impacts of various observing systems on regional 24-h forecast error reduction and their seasonal variations. In Figure 11.15, negative impacts (i.e., reductions in forecast errors) are generally obtained by all the observing systems, for both January and June, with the largest impacts by TAMDAR,[6] followed by SOUND (rawinsonde), GeoAMV,[7] METAR,[8] GPSREF,[9] PROFILER (wind profilers), SYNOP (surface synoptic observations from land stations), etc. (Figure 11.15a). Significant decreases (increases) in the observation impacts from summer to winter are recognized by TAMDAR, SOUND, and GeoAMV (METAR and GPSREF). Relatively

[4] Naval Research Laboratory (NRL) Atmospheric Variational Data Assimilation System.
[5] Navy Operational Global Atmospheric Prediction System.
[6] Tropospheric Airborne Meteorological Data Reporting.
[7] Geostationary satellite atmospheric motion vector (AMV).
[8] METeorological Aerodrome Report.
[9] Global Positioning System (GPS) refractivity.

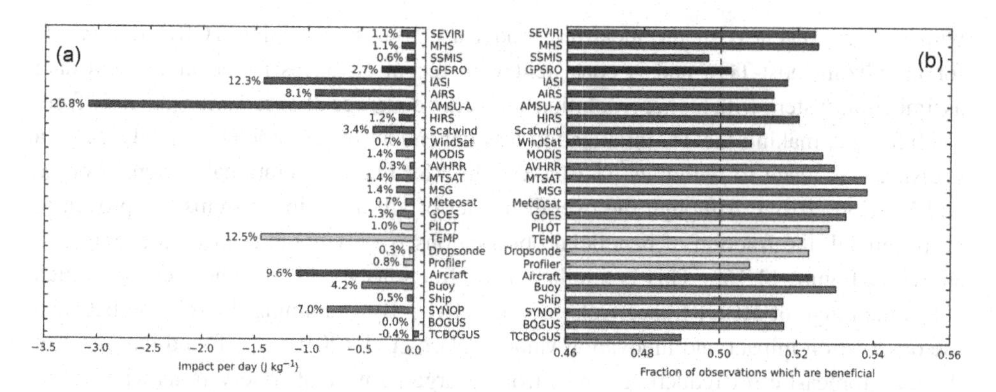


Figure 11.16 (a) The 24-h observation impact per day by observation type for the period from 1800 UTC 22 August to 1200 UTC 29 September 2010 and (b) the fraction of beneficial 24-h observation impact by observation type for the full trial period. Shaded bars represent various observing systems: satellite radiances (SEVIRI, MHS, SSMIS, IASI, AIRS, AMSU-A, and HIRS) and GPS Radio Occultations (GPSRO); scatterometer winds (Scatwind and Wind-Sat); AMVs (MODIS, AVHRR, MTSAT, MSG, Meteosat, and GOES); sondes and wind prolers (PILOT, TEMP, Dropsonde, and Proler); aircraft; surface observations (Buoy, Ship, and SYNOP) and synthetic observations (BOGUS and TCBOGUS). Percentage values in (a) give the fraction of the total impact. Modified from Lorenc and Marriott (2014). ©Crown Copyright 2013. Used with permission.

small reductions in total forecast errors by AIREP,[10] GPSPW,[11] BUOY (surface synoptic observations from buoy), and SHIP (surface synoptic observations from ship) are attributed to their sparse coverage with limited data amount (Figure 11.15b). The observation impacts normalized by the corresponding observation numbers, shown in Figure 11.15c, indicate that SYNOP surface observations have the greatest (normalized) observation impact, followed by SOUND, TAMDAR, GeoAMV, METAR, and GPSREF. This implies that every single SYNOP observation contains more important information in order to reduce the total forecast error than any single observation of other observing system.

Lorenc and Marriott (2014) also compared the observation impacts among various observing systems in the global 4DVAR system of the UK Met Office, using a moist energy norm. Figure 11.16a shows the average observation impact per day per observing system. The observing systems contributing most to the 24-h forecast error reduction include AMSU-A, TEMP (radiosondes), IASI,[12] Aircraft, AIRS,[13] and SYNOP. The exceptionally large impact by AMSU-A, compared with a similar study in Gelaro et al. (2010), is attributed to the use of a moist energy norm and assimilation of additional radiances from MetOp-A[14] and channels 1, 2, and 12–14. The detrimental impact by TCBOGUS[15] may be associated

[10] Aircraft Report.
[11] GPS Precipitable Water.
[12] Infrared Atmospheric Sounding Interferometer.
[13] Atmospheric Infrared Sounder.
[14] Meteorological Operational satellite-A.
[15] Tropical cyclone (TC) BOGUS: Wind observations synthesized automatically from TC warning center advisory reports (Lorenc and Marriott, 2014).

with inappropriateness of the idealized model used for synthesizing TCBOGUS winds for initializing post-TCs, and is consistent with recent advances in the model and data assimilation system with higher resolution and better coverage of observations, mainly from satellite data, making the TC intialization less dependent on TCBOGUS: In July 2012, a decision was made to withdraw TCBOGUS the Met Office operational system (Lorenc and Marriott, 2014). Although the overall impact of all observing systems has proven to be beneficial, the fraction of beneficial observations is only about 51.5% on average, as shown in Figure 11.16b: This is similar to the fractions found for other centers, which are in the range of 50–54% (see Gelaro et al., 2010). By examining the relation between the observation impact and innovation value, Gelaro et al. (2010) also found that most of the total forecast error reduction comes from a large number of observations with small-to-moderate-sized innovations and forecast impacts, and not from outliers with very large innovations.

Daescu and Langland (2013) extended the adjoint-based observation sensitivity and forecast impact assessment to diagnosis and tuning of parameters in the error covariances of both observation and background. Through this technique, one can identify the error covariance parameters with a potentially large reduction in the forecast error, which can then be used to make appropriate tuning on those parameters. For example, positive (negative) sensitivities to the observation error covariance coefficient for a specific instrument indicate that reducing (inflating) the values of the observation error variances of that instrument exerts a beneficial impact on the forecasts.

Overall, the observation sensitivity and impact analyses, based on the adjoint of a given data assimilation system, serve as useful guidance for observation-space targeting to diagnose the relative value of various observing systems, to identify potential problems with the observation network, and to compare characteristic performance of different data assimilation systems. Furthermore, an effective tuning process of the error covariance parameters can be implemented to reduce the forecast errors through the guidance from the forecast sensitivities to the observation and background error covariances.

### *11.3.3 Targeting Field Programs*

The forecast skill of numerical weather prediction (NWP) increases significantly through global/regional field experiments that utilize all available data from various observing systems, including remotely sensed and *in situ* observations, and employ the advanced model and data assimilation system. A global field experiment was conducted in 1979 under the Global Atmospheric Research Programme (GARP), which was named the *First GARP Global Experiment* (FGGE) (Gosset, 1979): It provided the momentum for a leap in NWP skill (Uppala et al., 2004). About 25 years after FGGE, a decade-long global field and research program of the WMO/WWRP,[16] named *THe Observing system Research and Predictability EXperiment* (THORPEX), was established in 2003 and formally commenced in 2005, by incorporating contemporary advances in numerical modeling, data assimilation,

[16] World Meteorological Organization/World Weather Research Program.

ensemble prediction, and adaptive observations, with the goal of accelerating improvements in the forecasting of high-impact weather events (see Shapiro and Thorpe, 2004a, 2004b; Parsons et al., 2017).

The various adaptive observation strategies explained above (see Section 11.3.1) have been applied to a number of regional field campaigns:

- the *Fronts and Atlantic Storm-Track Experiments* (FASTEX) (Joly et al., 1999) took place from 5 January to 27 February 1997 over the North Atlantic with ~400 dropsondes deployed though 19 intensive observation periods (IOPs) and 6 lesser observation periods (LOPs), focusing on the life cycle of mid-latitude cyclones, practical feasibility of adaptive observations, and air–sea turbulent exchange, through which adaptive observations were practiced using the adjoint SA, TESV, ensemble transform, and quasi-inverse TLM (e.g., Bergot, 1999; Bishop and Toth, 1999; Buizza and Montani, 1999; Langland et al., 1999a; Pu and Kalnay, 1999; Fourrié et al., 2002);
- the *North Pacific Experiment* (NORPEX-98) (Langland et al., 1999b) was conducted during 14 January and 27 February 1998 in the northeast Pacific with deployment of ~700 dropsondes, dedicated entirely to targeted observation, especially on the issue of observational sparsity over the North Pacific basin that contributes to forecast failures for winter-season storms over North America – mainly through TESV and ETKF (see Majumdar et al., 2002a);
- the *Winter Storm Reconnaissance* (WSR) is an annual program operated by NOAA,[17] collecting dropsonde data in January–March from 1999 (operational since 2001) over data sparse (sensitive) oceanic regions to improve forecasts of potentially high-impact weather events over the United States, by employing the ETKF as the targeting guidance (Hamill et al., 2013);
- the *Dropwindsonde Observations for Typhoon Surveillance near the Taiwan Region* (DOTSTAR) (Wu et al., 2005) is an annual tropical targeting field program since 2003 (first mission on 1 September 2003), which has completed 85 missions for 69 typhoons with 1303 dropsondes deployed up to 2020, mainly using the adjoint-derived sensitivity steering vector (ADSSV) (Wu et al., 2007) as the operational targeting guidance.

In connection with THORPEX, several regional campaigns under international cooperation were performed:

- the *Atlantic-THORPEX Regional Campaign* (A-TReC) occurred from 13 October to 12 December 2003, aiming to test the feasibility of a real-time quasi-operational targeting of multiple observing systems over the North Atlantic, through which some targeting strategies were tested, including SVs with the analysis error variance norms (Gelaro et al., 2002; Leutbecher et al., 2002) and ETKF (Majumdar et al., 2002b), along with observation sensitivity (Cardinali and Buizza, 2004) and observation impact (Langland, 2005b; Fourrié et al., 2006);

[17] National Oceanic and Atmospheric Administration.

- the *THORPEX Pacific Asian Regional Campaign* (T-PARC) (Parsons et al., 2008) was performed from 1 August to 3 October in 2008 for the summer phase and from mid-January to the end of February in 2009 for the winter phase: the summer phase cooperated with the *Tropical Cyclone Structure* (TCS08) ( Elsberry and Harr, 2008) field experiment, addressing tropical cyclones' formation, intensification, structure change, and extratropical transition, and its downstream impacts, with more than 1,500 dropsonde deployment and adaptive observations using SVs, ensemble KF, ETKF, adjoint SA, ADSSV, and ensemble variance (e.g., Kim et al., 2010; Chen et al., 2011; Kim et al., 2011; Doyle et al., 2012; Jung et al., 2012), whereas the winter phase aimed at improving 3–6 day forecasts for high-impact weather events for North America targeting with ETKF (e.g., Majumdar et al., 2010).
- the *MEDiterranean EXperiment* (MEDEX) (Jansa et al., 2014) was carried out in two phases (2000–2005 and 2006–2010), aiming for better forecasting and understanding of cyclones that bring high-impact weather in the Mediterranean, with sensitive zones calculated via ensemble sensitivity and KFS (e.g., Garcies and Homar, 2009; Oger et al., 2012): This resulted in a subprogram in the second phase, from 30 September to 20 December 2009, specific to "data targeting" (i.e., adaptive observations) – the Data Targeting System-MEDEX (DTS-MEDEX-2009) – which used the DTS from ECMWF, based on SVs, and also constructed sensitivity climatologies by calculating the ensemble sensitivity (Jansa et al., 2011);
- the *HYdrological cycle in the Mediterranean EXperiment* (HyMeX) (Drobinski et al., 2014) is an extension of MEDEX as a 10-year program from 2012 focusing on the water cycle in the Mediterranean with special interest in high-impact weather events: the first special observation period (SOP1) (5 September–6 November 2012) was dedicated to heavy precipitation and flash-flooding while SOP2 (February–March 2013) aimed to study intense air–sea exchanges and dense water formation, using the sensitive area predictions provided by the DTS through SVs (from ECMWF) and ETKF (from the Met Office and Météo-France); and
- the *Convective and Orographically-induced Precipitation Study* (COPS) was conducted in cooperation with the *European THORPEX Regional Campaign* (E-TReC07), with a field phase from 01 June to 31 August 2007 in an area of moderate terrain over southwestern Germany and northeastern France, aiming to improve probabilistic quantitative precipitation forecasting of orographically-induced convective precipitation: It performed 18 IOPs over 37 operation days and 8 additional SOPs and was provided with daily sensitive location predictions through E-TReC07 using several versions of adjoint sensitivities, moist SVs, and ETKF (see Wulfmeyer et al., 2011).

A tabular summary of the field campaigns that adopted the adaptive observation strategies is referred to Majumdar (2016).

As T-PARC required a measurement strategy for the extratropical transition and downstream impacts of tropical cyclones, it employed the *driftsonde* system – a balloon-borne *in situ* observation system with a gondola loading 20–50 dropsondes (38 for T-PARC) – to deploy dropsondes that measure atmospheric temperature, humidity, pressure, and winds from the lower stratosphere to the surface (Figure 11.17a): This is ideal for collecting data

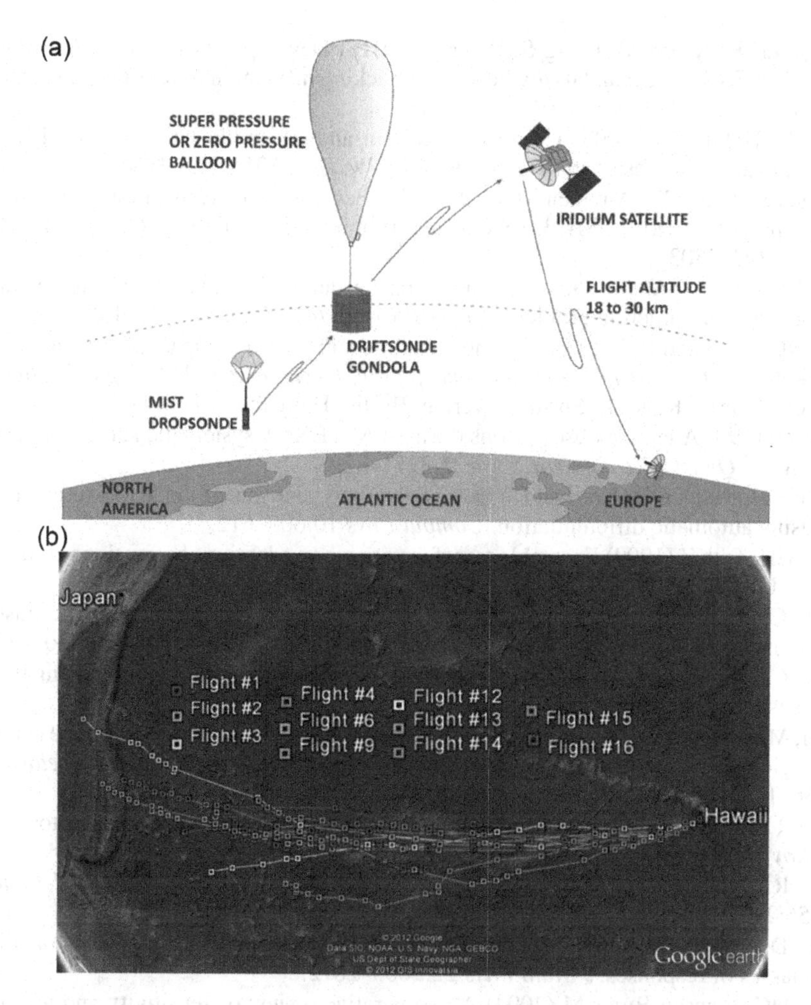


Figure 11.17 (a) The driftsonde system concept and (b) locations of dropsondes (squares) deployed from the driftsondes that were launched from Hawaii during T-PARC. Modified from Cohn et al. (2013). ©American Meteorological Society. Used with permission.

over oceans and remote data-void regions to fill critical gaps in conventional data coverage (Cohn et al., 2013). During T-PARC, out of 16 driftsondes launched from Hawaii (the Big Island) between 15 August and 30 September 2008, 13 driftsondes made successful flights to the primary T-PARC region (Figure 11.17b), yielding 253 good dropsonde soundings.

## References

Aberson SD (2003) Targeted observations to improve operational tropical cyclone track forecast guidance. *Mon Wea Rev* 131:1613–1628.

AMS (2021) Glossary of Meteorology: Adaptive observations. American Meteorological Society, https://glossary.ametsoc.org/wiki/Adaptive_observations

An XQ, Zhai SX, Jin M, Gong S, Wang Y (2016) Development of an adjoint model of GRAPES-CUACE and its application in tracking influential haze source areas in north China. *Geosci Model Dev* 9:2153–2165.

Ancell B, Hakim GJ (2007) Comparing adjoint-and ensemble-sensitivity analysis with applications to observation targeting. *Mon Wea Rev* 135:4117–4134.

Backman J, Wood CR, Auvinen M, et al. (2017) Sensitivity analysis of the meteorological preprocessor MPP-FMI 3.0 using algorithmic differentiation. *Geosci Model Dev* 10:3793–3803.

Baker NL, Daley R (2000) Observation and background adjoint sensitivity in the adaptive observation-targeting problem. *Quart J Roy Meteor Soc* 126:1431–1454.

Baker NL, Langland RH (2009) Diagnostics for evaluating the impact of satellite observations. In *Data Assimilation for Atmospheric, Oceanic and Hydrologic Applications*, (eds.) Park SK, Xu L, Springer-Verlag, Berlin, Heidelberg, 177–196.

Bergot T (1999) Adaptive observations during FASTEX: A systematic survey of upstream flights. *Quart J Roy Meteor Soc* 125:3271–3298.

Bischof CH, Pusch GD, Knoesel R (1996) Sensitivity analysis of the MM5 weather model using automatic differentiation. *Comput Phys* 10:605–612.

Bishop CH, Toth Z (1999) Ensemble transformation and adaptive observations. *J Atmos Sci* 56:1748–1765.

Bishop CH, Etherton BJ, Majumdar SJ (2001) Adaptive sampling with the ensemble transform Kalman filter. Part I: Theoretical aspects. *Mon Wea Rev* 129:420–436.

Bosnić Z, Kononenko I (2008) Estimation of individual prediction reliability using the local sensitivity analysis. *Appl Intell* 29:187–203.

Brix H, Menemenlis D, Hill C, et al. (2015) Using Green's Functions to initialize and adjust a global, eddying ocean biogeochemistry general circulation model. *Ocean Modell* 95:1–14.

Buizza R (1994) Localization of optimal perturbations using a projection operator. *Quart J Roy Meteor Soc* 120:1647–1681.

Buizza R, Montani A (1999) Targeting observations using singular vectors. *J Atmos Sci* 56:2965–2985.

Cacuci DG (1981) Sensitivity theory for nonlinear systems. II. Extensions to additional classes of responses. *J Math Phys* 22:2803–2812.

Cacuci DG, Ionescu-Bujor M (2004) A comparative review of sensitivity and uncertainty analysis of large-scale systems—II: Statistical methods. *Nucl Sci Eng* 147:204–217.

Cacuci DG, Ionescu-Bujor M, Navon IM (2005) *Sensitivity and Uncertainty Analysis, (vol. II): Applications to Large-Scale Systems*. CRC Press, Boca Raton, FL, 368 pp.

Campolongo F, Cariboni J, Saltelli A (2007) An effective screening design for sensitivity analysis of large models. *Environ Modell Software* 22:1509–1518.

Cardinali C, Buizza R (2004) Observation sensitivity to the analysis and the forecast: A case study during ATreC targeting campaign. In *Proceedings of the First THORPEX International Science Symposium*, 6–10 December 2004, Montreal, WMO/TD-No. 1237, WWRP/THORPEX No. 06.

Che Y, Zhang M, Li Z, et al. (2019) Energy balance model of mass balance and its sensitivity to meteorological variability on Urumqi River Glacier No. 1 in the Chinese Tien Shan. *Sci Rep* 9:1–13.

Chen B-Y (2011) Observation system experiments for Typhoon Nida (2004) using the CNOP method and DOTSTAR data. *Atmos Oceanic Sci Lett* 4:118–123.

Chen S-G, Wu C-C, Chen J-H, Chou K-H (2011) Validation and interpretation of adjoint-derived sensitivity steering vector as targeted observation guidance. *Mon Wea Rev* 139:1608–1625.

Chen TF, Chang KH, Lee CH (2019) Simulation and analysis of causes of a haze episode by combining CMAQ-IPR and brute force source sensitivity method. *Atmos Environ* 218:117006, doi:10.1016/j.atmosenv.2019.117006

Chu K-K, Tan Z-M (2010) Mesoscale moist adjoint sensitivity study of a Mei-yu heavy rainfall event. *Adv Atmos Sci* 27:1415–1424.

Ciric C, Ciffroy P, Charles S (2012) Use of sensitivity analysis to identify influential and non-influential parameters within an aquatic ecosystem model. *Ecol Modell* 246: 119–130.

Cleaves HL, Alexanderian A, Guy H, Smith RC, Yu M (2019) Derivative-based global sensitivity analysis for models with high-dimensional inputs and functional outputs. *SIAM J Sci Comput* 41:A3524–A3551.

Cohn SA, Hock T, Cocquerez P, et al. (2013) Driftsondes: Providing in situ long-duration dropsonde observations over remote regions. *Bull Amer Meteor Soc* 94:1661–1674.

Collins DC, Avissar R (1994) An evaluation with the Fourier amplitude sensitivity test (FAST) of which land-surface parameters are of greatest importance in atmospheric modeling. *J Clim* 7:681–703.

Daescu DN, Langland RH (2013) The adjoint sensitivity guidance to diagnosis and tuning of error covariance parameters. In *Data Assimilation for Atmospheric, Oceanic and Hydrologic Applications* (vol. II), (eds.) Park SK, Xu L, Springer-Verlag, Berlin, Heidelberg, 205–232.

Degenring D, Froemel C, Dikta G, Takors R (2004) Sensitivity analysis for the reduction of complex metabolism models. *J Process Control* 14:729–745.

Devak M, Dhanya CT (2017) Sensitivity analysis of hydrological models: Review and way forward. *J Water Clim Change* 8:557–575.

Diaconescu EP, Laprise R (2012) Singular vectors in atmospheric sciences: A review. *Earth Sci Rev* 113:161–175.

Dong P, Zhong K, Zhao S (2009) Study on adjoint-based targeted observation of mesoscale low on Meiyu front. In *Data Assimilation for Atmospheric, Oceanic and Hydrologic Applications*, (eds.) Park SK, Xu L, Springer-Verlag, Berlin, Heidelberg, 253–268.

Doyle JD, Reynolds CA, Amerault C (2019) Adjoint sensitivity analysis of high-impact extratropical cyclones. *Mon Wea Rev* 147:4511–4532.

Doyle JD, Amerault C, Reynolds CA, Reinecke PA (2014) Initial condition sensitivity and predictability of a severe extratropical cyclone using a moist adjoint. *Mon Wea Rev* 142:320–342.

Doyle JD, Reynolds CA, Amerault C, Moskaitis J (2012) Adjoint sensitivity and predictability of tropical cyclogenesis. *J Atmos Sci* 69:3535–3557.

Drobinski P, Ducrocq V, Alpert P, et al. (2014) HyMeX: A 10-year multidisciplinary program on the Mediterranean water cycle. *Bull Amer Meteor Soc* 95:1063–1082.

Dunker AM, Yarwood G, Ortmann JP, Wilson GM (2002) The decoupled direct method for sensitivity analysis in a three-dimensional air quality model – Implementation, accuracy, and efficiency. *Environ Sci Tech* 36:2965–2976, doi:10.1021/es0112691

Ehrendorfer M, Tribbia JJ (1997) Optimal prediction of forecast error covariances through singular vectors. *J Atmos Sci* 54:286–313.

Elsberry RL, Harr PA (2008) Tropical Cyclone Structure (TCS08) field experiment science basis, observational platforms, and strategy. *Asia-Pac J Atmos Sci* 44:209–231.

Errico RM, Vukićević T (1992) Sensitivity analysis using an adjoint of the PSU-NCAR mesoseale model. *Mon Wea Rev* 120:1644–1660.

Fenwick D, Scheidt C, Caers J (2014) Quantifying asymmetric parameter interactions in sensitivity analysis: Application to reservoir modeling. *Math Geosci* 46:493–511.

Fourrié N, Doerenbecher A, Bergot T, Joly A (2002) Adjoint sensitivity of the forecast to TOVS observations. *Quart J Roy Meteor Soc* 128:2759–2777.

Fourrié N, Marchal D, Rabier F, Chapnik B, Desroziers G (2006) Impact study of the 2003 North Atlantic THORPEX Regional Campaign. *Quart J Roy Meteor Soc* 132: 275–295.

Fronzek S, Pirttioja N, Carter TR, et al. (2018) Classifying multi-model wheat yield impact response surfaces showing sensitivity to temperature and precipitation change. *Agric Syst* 159:209–224.

Garcies L, Homar V (2009) Ensemble sensitivities of the real atmosphere: Application to mediterranean intense cyclones. *Tellus A* 61:394–406.

Gelaro R, Rosmond T, Daley R (2002) Singular vector calculations with an analysis error variance metric. *Mon Wea Rev* 130:1166–1186.

Gelaro R, Langland RH, Pellerin S, Todling R (2010) The THORPEX observation impact intercomparison experiment. *Monthly Weather Review* 138:4009–4025.

Gosset B (1979) First GARP Global Experiment. *WMO Bull* 28:5–17.

Hamill TM, Yang F, Cardinali C, Majumdar SJ (2013) Impact of targeted Winter Storm Reconnaissance dropwindsonde data on midlatitude numerical weather predictions. *Mon Wea Rev* 141:2058–2065.

He F, Posselt DJ, Narisetty NN, Zarzycki CM, Nair VN (2018) Application of multivariate sensitivity analysis techniques to AGCM-simulated tropical cyclones. *Mon Wea Rev* 146:2065–2088.

Heimbach P, Bugnion V (2009) Greenland ice-sheet volume sensitivity to basal, surface and initial conditions derived from an adjoint model. *Ann Glaciol* 50:67–80.

Helton JC, Johnson JD, Sallaberry CJ, Storlie CB (2006) Survey of sampling-based methods for uncertainty and sensitivity analysis. *Reliab Eng Syst Saf* 91:1175–1209.

Heynen M, Pellicciotti F, Carenzo M (2013) Parameter sensitivity of a distributed enhanced temperature-index melt model. *Ann Glaciol* 54:311–321.

Houle ES, Livneh B, Kasprzyk JR (2017) Exploring snow model parameter sensitivity using Sobol' variance decomposition. *Environ Modell Software* 89:144–158.

Hwang J-T, Dougherty EP, Rabitz S, Rabitz H (1978) The Green's function method of sensitivity analysis in chemical kinetics. *J Chem Phys* 69:5180–5191.

Ionescu-Bujor M, Cacuci DG (2004) A comparative review of sensitivity and uncertainty analysis of large-scale systems—I: Deterministic methods. *Nucl Sci Eng* 147: 189–203.

Jansa A, Alpert P, Arbogast P, et al. (2014) MEDEX: A general overview. *Nat Hazards Earth Syst Sci* 14:1965–1984.

Jansa A, Arbogast P, Doerenbecher A, et al. (2011) A new approach to sensitivity climatologies: The DTS-MEDEX-2009 campaign. *Nat Hazards Earth Syst Sci* 11:2381–2390.

Ji D, Dong W, Hong T, et al. (2018) Assessing parameter importance of the weather research and forecasting model based on global sensitivity analysis methods. *J Geophys Res Atmos* 123:4443–4460.

Joly A, Browning KA, Bessemoulin P, et al. (1999) Overview of the field phase of the Fronts and Atlantic Storm-Track EXperiment (FASTEX) project. *Quart J Roy Meteor Soc* 125:3131–3163.

Jung B-J, Kim HM, Zhang F, Wu CC (2012) Effect of targeted dropsonde observations and best track data on the track forecasts of Typhoon Sinlaku (2008) using an ensemble Kalman filter. *Tellus A* 64:14984, doi:doi.org/10.3402/tellusa.v64i0.14984

Jung B-J, Kim HM, Auligné T, et al. (2013) Adjoint-derived observation impact using WRF in the western North Pacific. *Mon Wea Rev* 141:4080–4097.

Kalnay E, Ota Y, Miyoshi T, Liu J (2012) A simpler formulation of forecast sensitivity to observations: Application to ensemble Kalman filters. *Tellus A* 64:18462, doi:10.3402/tellusa.v64i0.18462

Karamchandani A, Cornell CA (1992) Sensitivity estimation within first and second order reliability methods. *Struct Saf* 11:95–107.

Kim HM, Kim SM, Jung BJ (2011) Real-time adaptive observation guidance using singular vectors for Typhoon Jangmi (200815) in T-PARC 2008. *Wea Forecasting* 26:634–649.

Kim Y-H, Jeon E-H, Chang D-E, Lee H-S, Park J-I (2010) The impact of T-PARC 2008 dropsonde observations on typhoon track forecasting. *Asia-Pac J Atmos Sci* 46: 287–303.

Lakshmivarahan S, Lewis JM (2010) Forward sensitivity approach to dynamic data assimilation. *Adv Meteorol* 2010:375615, doi:10.1155/2010/375615

Langland RH (2005a) Issues in targeted observing. *Quart J Roy Meteor Soc* 131:3409–3425.

Langland RH (2005b) Observation impact during the North Atlantic TReC-2003. *Mon Wea Rev* 133:2297–2309.

Langland RH, Baker NL (2004) Estimation of observation impact using the NRL atmospheric variational data assimilation adjoint system. *Tellus A* 56:189–201.

Langland RH, Gelaro R, Rohaly GD, Shapiro MA (1999a) Targeted observations in FASTEX: Adjoint-based targeting procedures and data impact experiments in IOP17 and IOP18. *Quart J Roy Meteor Soc* 125:3241–3270.

Langland RH, Toth Z, Gelaro R, et al. (1999b) The North Pacific Experiment (NORPEX-98): Targeted observations for improved North American weather forecasts. *Bull Amer Meteor Soc* 80:1363–1384.

Le Dimet Fn, Ngodock Hn, Luong B, Verron J (1997) Sensitivity analysis in variational data assimilation. *J Meteor Soc Japan* 75:245–255.

Leutbecher M (2003) A reduced rank estimate of forecast error variance changes due to intermittent modifications of the observing network. *J Atmos Sci* 60:729–742.

Leutbecher M, Barkmeijer J, Palmer TN, Thorpe AJ (2002) Potential improvement to forecasts of two severe storms using targeted observations. *Quart J Roy Meteor Soc* 128:1641–1670.

Li H, Liu J, Kalnay E (2010) Correction of "Estimating observation impact without adjoint model in an ensemble Kalman filter." *Quart J Roy Meteor Soc* 136:1652–1654.

Limpert GL, Houston AL (2018) Ensemble sensitivity analysis for targeted observations of supercell thunderstorms. *Mon Wea Rev* 146:1705–1721.

Liu J, Kalnay E (2008) Estimating observation impact without adjoint model in an ensemble Kalman filter. *Quart J Roy Meteor Soc* 134:1327–1335.

Lorenc AC, Marriott RT (2014) Forecast sensitivity to observations in the Met Office global numerical weather prediction system. *Quart J Roy Meteor Soc* 140:209–224.

Lorenz EN, Emanuel KA (1998) Optimal sites for supplementary weather observations: Simulation with a small model. *J Atmos Sci* 55:399–414.

Losch M, Heimbach P (2007) Adjoint sensitivity of an ocean general circulation model to bottom topography. *J Phys Oceanogr* 37:377–393.

Majumdar SJ (2016) A review of targeted observations. *Bull Amer Meteor Soc* 97: 2287–2303.

Majumdar SJ, Bishop CH, Buizza R, Gelaro R (2002a) A comparison of ensemble-transform Kalman-filter targeting guidance with ECMWF and NRL total-energy singular-vector guidance. *Quart J Roy Meteor Soc* 128:2527–2549.

Majumdar SJ, Bishop CH, Etherton B, Toth Z (2002b) Adaptive sampling with the ensemble transform Kalman filter. Part II: Field program implementation. *Mon Wea Rev* 130:1356–1369.

Majumdar SJ, Sellwood KJ, Hodyss D, Toth Z, Song Y (2010) Characteristics of target areas selected by the ensemble transform Kalman filter for medium-range forecasts of high-impact winter weather. *Mon Wea Rev* 138:2803–2824.

Majumdar SJ, Aberson SD, Bishop CH, et al. (2006) A comparison of adaptive observing guidance for Atlantic tropical cyclones. *Mon Wea Rev* 134:2354–2372.

Manache G, Melching CS (2008) Identification of reliable regression- and correlation-based sensitivity measures for importance ranking of water-quality model parameters. *Environ Modell Software* 23:549–562.

Martinec Z, Sasgen I, Velímský J (2015) The forward sensitivity and adjoint-state methods of glacial isostatic adjustment. *Geophys J Int* 200:77–105.

Marzban C (2013) Variance-based sensitivity analysis: An illustration on the Lorenz'63 model. *Mon Wea Rev* 141:4069–4079.

Morris MD (1991) Factorial sampling plans for preliminary computational experiments. *Technometrics* 33:161–174.

Mu M, Zhou F, Wang H (2009) A method for identifying the sensitive areas in targeted observations for tropical cyclone prediction: Conditional nonlinear optimal perturbation. *Mon Wea Rev* 137:1623–1639.

Navon IM, Zou X, Derber J, Sela J (1992) Variational data assimilation with an adiabatic version of the NMC spectral model. *Mon Wea Rev* 120:1433–1446.

Nguyen VT, Georges D, Besançon G (2016) State and parameter estimation in 1-D hyperbolic PDEs based on an adjoint method. *Automatica* 67:185–191.

Nossent J, Elsen P, Bauwens W (2011) Sobol' sensitivity analysis of a complex environmental model. *Environ Modell Software* 26:1515–1525.

Oger N, Pannekoucke O, Doerenbecher A, Arbogast P (2012) Assessing the influence of the model trajectory in the adaptive observation Kalman Filter Sensitivity method. *Quart J Roy Meteor Soc* 138:813–825, doi:10.1002/qj.950

Palmer TN, Gelaro R, Barkmeijer J, Buizza R (1998) Singular vectors, metrics, and adaptive observations. *J Atmos Sci* 55:633–653.

Park SK (1996) *Sensitivity Analysis of Deep Convective Storms*. PhD thesis, University of Oklahoma, Norman, OK, 245 pp.

Park SK, Droegemeier KK (1999) Sensitivity analysis of a moist 1D Eulerian cloud model using automatic differentiation. *Mon Wea Rev* 127:2180–2196.

Park SK, Droegemeier KK (2000) Sensitivity analysis of a 3D convective storm: Implications for variational data assimilation and forecast error. *Mon Wea Rev* 128:140–159.

Park SK, Županski D (2003) Four-dimensional variational data assimilation for mesoscale and storm-scale applications. *Meteor Atmos Phys* 82:173–208.

Parsons DB, Harr P, Nakazawa T, Jones S, Weissmann M (2008) An overview of the THORPEX-Pacific Asian Regional Campaign (T-PARC) during August–September 2008. In *Extended Abstracts, 28th Conference on Hurricanes and Tropical Meteorology*, American Meteorological Society, Orlando, FL, 1–6.

Parsons DB, Beland M, Burridge D, et al. (2017) THORPEX research and the science of prediction. *Bull Amer Meteor Soc* 98:807–830.

Pianosi F, Beven K, Freer J, et al. (2016) Sensitivity analysis of environmental models: A systematic review with practical workflow. *Environ Modell Software* 79:214–232.

Pu Z-X, Kalnay E (1999) Targeting observations with the quasi-inverse linear and adjoint NCEP global models: Performance during FASTEX. *Quart J Roy Meteor Soc* 125:3329–3337.

Pu Z-X, Lord SJ, Kalnay E (1998) Forecast sensitivity with dropwindsonde data and targeted observations. *Tellus A* 50:391–410.

Rakovec O, Hill MC, Clark MP, et al. (2014) Distributed Evaluation of Local Sensitivity Analysis (DELSA), with application to hydrologic models. *Water Resour Res* 50: 409–426.

Razavi S, Gupta HV (2015) What do we mean by sensitivity analysis? The need for comprehensive characterization of "global" sensitivity in Earth and Environmental systems models. *Water Resour Res* 51:3070–3092.

Regan MA, Mahesh K (2019) Adjoint sensitivity and optimal perturbations of the low-speed jet in cross-flow. *J Fluid Mech* 877:330–372.

Saltelli A, Tarantola S, Chan KP-S (1999) A quantitative model-independent method for global sensitivity analysis of model output. *Technometrics* 41:39–56.

Saltelli A, Annoni P, Azzini I, et al. (2010) Variance based sensitivity analysis of model output. Design and estimator for the total sensitivity index. *Comput Phys Commun* 181:259–270.

Saltelli A, Ratto M, Andres T, et al. (2008) *Global Sensitivity Analysis: The Primer*. John Wiley & Sons, Chichester, West Sussex, 304 pp.

Schwinger J, Kollet SJ, Hoppe CM, Elbern H (2010) Sensitivity of latent heat fluxes to initial values and parameters of a land-surface model. *Vadose Zone J* 9:984–1001.

Shapiro MA, Thorpe AJ (2004a) THORPEX: A global atmospheric research programme for the beginning of the 21st century. *WMO Bull* 53:222–226.

Shapiro MA, Thorpe AJ (2004b) *THORPEX International Science Plan. Version III*. Tech. Rep. WMO/TD-No. 1246, WWRP/THORPEX No. 02, WMO, Geneva, 55 pp.

Shin M-J, Choi YS (2018) Sensitivity analysis to investigate the reliability of the grid-based rainfall-runoff model. *Water* 10:1839, doi:10.3390/w10121839

Shutyaev V, Le Dimet F-X, Parmuzin E (2018) Sensitivity analysis with respect to observations in variational data assimilation for parameter estimation. *Nonlinear Processes Geophys* 25:429–439.

Sobol' IM (1993) Sensitivity analysis for non-linear mathematical models. *Math Modell Comput Exp* 1:407–414.

Sobol' IM, Kucherenko S (2009) Derivative based global sensitivity measures and their link with global sensitivity indices. *Math Comput Simul* 79:3009–3017.

Thorpe AJ, Petersen GN (2006) Predictability and targeted observations. In *Predictability of Weather and Climate*, (eds.) Palmer T, Hagedorn R, Cambridge University Press, Cambridge, 561–583.

Tromble EM, Lakshmivarahan S, Kolar RL, Dresback KM (2016) Application of the forward sensitivity method to a GWCE-based shallow water model. *J Mar Sci Eng* 4:73, doi:10.3390/jmse4040073

Uppala S, Simmons AJ, Kallberg P (2004) Global numerical weather prediction – an outcome of FGGE and a quantum leap for meteorology. *WMO Bull* 53:207–212.

Wang Q, Mu M, Sun G (2020) A useful approach to sensitivity and predictability studies in geophysical fluid dynamics: Conditional non-linear optimal perturbation. *Natl Sci Rev* 7:214–223.

Wang X, Zhang Y, Hu Y, et al. (2011) Decoupled direct sensitivity analysis of regional ozone pollution over the Pearl River Delta during the PRIDE-PRD2004 campaign. *Atmos Environ* 45:4941–4949, doi:10.1016/j.atmosenv.2011.06.006

Wu C-C, Chen J-H, Lin P-H, Chou K-H (2007) Targeted observations of tropical cyclone movement based on the adjoint-derived sensitivity steering vector. *J Atmos Sci* 64:2611–2626.

Wu C-C, Lin P-H, Aberson S, et al. (2005) Dropwindsonde observations for typhoon surveillance near the taiwan region (dotstar): An overview. *Bull Amer Meteor Soc* 86:787–790.

Wu C-C, Chen J-H, Majumdar SJ, et al. (2009) Intercomparison of targeted observation guidance for tropical cyclones in the northwestern Pacific. *Mon Wea Rev* 137: 2471–2492.

Wulfmeyer V, Behrendt A, Kottmeier C, et al. (2011) The Convective and Orographically-induced Precipitation Study (COPS): The scientific strategy, the field phase, and research highlights. *Quart J Roy Meteor Soc* 137:3–30.

Xing Y, Shao D, Ma X, Zhang S, Jiang G (2021) Investigation of the importance of different factors of flood inundation modeling applied in urbanized area with variance-based global sensitivity analysis. *Sci Total Environ* 772:145327, doi:10.1016/j.scitotenv.2021.145327

Yang J, Kim J-H, Jiménez PA, et al. (2021) An efficient method to identify uncertainties of WRF-Solar variables in forecasting solar irradiance using a tangent linear sensitivity analysis. *Sol Energy* 220:509–522.

Zhang X, Wang H, Huang X-Y, Gao F, Jacobs NA (2015) Using adjoint-based forecast sensitivity method to evaluate TAMDAR data impacts on regional forecasts. *Adv Meteorol* 2015:427616, doi:10.1155/2015/427616

Zhao H, Mousseau VA (2012) Use of forward sensitivity analysis method to improve code scaling, applicability, and uncertainty (CSAU) methodology. *Nucl Eng Des* 249: 188–196.

Zhao J, Tiede C (2011) Using a variance-based sensitivity analysis for analyzing the relation between measurements and unknown parameters of a physical model. *Nonlinear Processes Geophys* 18:269–276.

Zheng T, French NH, Baxter M (2018) Development of the WRF-CO2 4D-Var assimilation system v1.0. *Geosci Model Dev* 11:1725–1752.

Zhong K, Dong P, Zhao S, Cai Q, Lan W (2007) Adjoint-based sensitivity analysis of a mesoscale low on the Mei-yu front and its implications for adaptive observation. *Adv Atmos Sci* 24:435–448.

Zhou F, Qin X, Chen B, Mu M (2013) The advances in targeted observations for tropical cyclone prediction based on conditional nonlinear optimal perturbation (CNOP) method. In *Data Assimilation for Atmospheric, Oceanic and Hydrologic Applications* (vol. II) (eds.) Park SK, Xu L, Springer-Verlag, Berlin, Heidelberg, 577–607.

Zou X, Navon IM, Sela JG (1993) Variational data assimilation with moist threshold processes using the NMC spectral model. *Tellus A* 45:370–387.

Županski M (1995) An iterative approximation to the sensitivity in calculus of variations. *Mon Wea Rev* 123:3590–3604.

# 12
# Satellite Data Assimilation

Satellite data are the major source of observations for meteorological applications, and therefore deserve a special attention in the context of data assimilation. Their main appeal is that satellites cover all parts of the globe and provide information in areas not accessible by any other observation type, such as polar areas and oceans. Satellite data that can be assimilated include radiances (often defined in terms of brightness temperature) and products (often referred to as retrievals). Products refer to partially processed satellite observations, such as aerosol optical depth (AOD) and atmospheric motion vectors (AMV). It has been generally assumed that radiances have smaller observation errors than satellite products, because of lessened processing, which potentially implies their higher information content. However, the computational cost and complexity of satellite radiance observation operators can be considerable, making their use in data assimilation somewhat less attractive. On the other hand, given that the quality of retrieval algorithms has been steadily improving and the relatively low cost of their observation operators, the use of satellite products in data assimilation is becoming more competitive with satellite radiances. There are also other satellite observations that are not classified as radiances or products, such as Global Positioning System (GPS) radio occultation (RO). We refer to products and GPS RO as nonradiance satellite observations.

Our discussion is presented in the context of variational, ensemble, and hybrid ensemble-variational data assimilation, since they represent the most relevant methodologies used today. We begin by presenting the current status of satellite radiance assimilation (Section 12.1), and then discuss several major challenges of all-sky radiance assimilation in Section 12.2. Satellite product assimilation is described in Section 12.1.2, and in Section 12.3 we summarize the issues and look at the future of satellite data assimilation. Our presentation underlines a special interest for using satellite data assimilation for high-impact weather areas, such as tropical cyclones and severe weather outbreaks, which are mostly covered by clouds.

## 12.1 Current Status

We now give an overview of data assimilation methodologies currently used in research and operations. Assimilation of satellite radiances and GPS RO is the backbone of today's operational data assimilation, however with increasing relevance of satellite products.

### 12.1.1 Assimilation of Satellite Radiances

One can distinguish between clear-sky and all-sky satellite radiance assimilation. The "clear-sky" refers to assimilation of cloud-free regions only, while "all-sky" implies assimilation of all regions, with or without clouds. It is also important to know if there are precipitation-affected areas that are observed, since precipitation causes scattering, which introduces higher nonlinearity and complexity of the radiance observation operators. Although actual satellite radiance observations from a satellite instrument are by definition all-sky radiances, the satellite radiance observation operator can be considerably simplified in the case of clear-sky radiances, somewhat less simplified in the case of cloudy radiances, and reaching its highest complexity for precipitation-affected radiances.

In cloudy and precipitation scenes, the radiance observation operator is effectively a forward radiative transfer model, which generally includes absorption, emission, and scattering. In cloud-free and precipitation-free scenes, on the other hand, there is no scattering and the radiance observation operator reduces to optical transmittance. The relative simplicity of the clear-sky radiance observation operator makes it computationally more efficient and therefore more applicable to operational numerical weather prediction (NWP).

Since a radiance observation is all-sky, an additional procedure is required in clear-sky data assimilation in order to avoid cloudy and precipitation-affected radiance observations. This procedure, referred to as *cloud clearing* (e.g., McNally and Watts, 2003), also increases the uncertainty of the clear-sky radiance observation operator. Since all-sky radiance assimilation does not require cloud clearing the all-sky observation operator may have some advantages. Unfortunately, given additional complexities of all-sky radiance assimilation it also has an increased uncertainty.

Data assimilation of clear-sky satellite radiances started in the mid-1990s (e.g., Andersson et al., 1994; Derber and Wu, 1998; McNally et al., 2000; Harris and Kelly, 2001), with considerable expansion therafter (e.g., Okamoto et al., 2005; McNally et al., 2006; Okamoto and Derber, 2006; Weng, 2007; Kazumori et al., 2008; Bouchard et al., 2010; Kazumori, 2014). This technique blends important information about temperature and moisture in the atmosphere and has a great impact on data assimilation in the southern hemisphere where observations over the ocean are sparse. Introducing the assimilation of all-sky satellite radiances brought one of the major advantages of the variational data assimilation (VAR) method, compared to previously used optimal interpolation.

In principle, all-sky radiances can bring important new information about clouds, prompting research efforts to include and improve all-sky radiance assimilation. However, even with a wealth of information that all-sky radiances could bring to the prediction system, only limited research effort was initially invested in assimilation of these observations. Most efforts include the use of variational methods (e.g., Vukicevic et al., 2004; Bauer et al., 2006, 2010; Geer and Bauer, 2010; Geer et al., 2010; Polkinghorne and Vukicevic, 2011) with some applications within ensemble and hybrid variational-ensemble methods (e.g., Županski et al., 2011a, 2011b).

The relevance of all-sky radiance assimilation is widely recognized and discussed (e.g., Errico et al., 2007a, 2007b; Auligné et al., 2011; Bauer et al., 2011). The nonlinearity and high resolution of cloud microphysical processes have direct implications on all-sky radiance assimilation, further burdened by the increased computational cost of the all-sky radiance observation operator.

Particularly relevant is the pioneering work by the satellite research group at the European Centre for Medium Range Weather Forecast (ECMWF) (e.g., Bauer et al., 2010; Geer et al., 2010) leading to the first operational assimilation of all-sky radiances since 2009. This research paved the way for more intense efforts to include all-sky satellite radiance assimilation in operational NWP (e.g., Geer et al., 2012, 2017; Kazumori et al., 2016; Kazumori, 2019; Zhu et al., 2016).

### *12.1.2 Assimilation of Nonradiance Satellite Observations and Satellite Products*

A useful and widely assimilated nonradiance satellite observation includes GPS RO and bending angles (Eyre, 1994; Zou et al., 1995; Matsumura et al., 1999; Palmer et al., 2000). It was noticed only by serendipity that GPS signals are impacted by atmosphere and that atmospheric impact can be transformed into information about temperature and moisture. In addition, it was found that GPS measurements have almost negligible bias and therefore can produce very accurate assimilation. A side effect of this low-bias property is that RO assimilation helps in reducing the bias of radiance observations as well. At present, GPS RO is an intrinsic part of operational NWP data assimilation (Zou et al., 2000; Cucurull, 2010; Poli et al., 2010). Bending angles have been assimilated in the past, but their observation operator includes additional processing compared to RO and is consequently less used.

Another, more recent nonradiance satellite observation is lightning (Goodman et al., 2013). The lightning mapper satellite instrument records lightning strikes, which can be normalized by time and space units to produce a lightning flash rate. The lightning assimilation is using a relationship between lightning flash rates and atmospheric variables, mostly temperature and cloud ice, as the observation operator (Ştefănescu et al., 2013; Apodaca et al., 2014).

One can also assimilate satellite products derived from satellite measurements. Although in principle they may result in increased uncertainty due to the additional retrieval procedure, such observations can still be very valuable. The most commonly used retrieval is the AMV wind retrieval that produces winds from satellite measurements (e.g., Holmlund, 1998; Rohn et al., 2001; Bormann and Thépaut, 2004; Wu et al., 2014). The AMV retrievals are commonly assimilated in operational NWP. Other examples of retrievals that have been assimilated include water content or hydrometeor profile retrievals (e.g., Wu et al., 2016), or soil moisture retrievals (e.g., Liu et al., 2011; Maggioni et al., 2012). The observation operator for assimilation of soil moisture retrievals is simply an interpolation since soil moisture is a standard variable of land surface models. This is also true for assimilation of hydrometeor retrievals, but for assimilation of integrated water content a transformation from temperature, moisture, and hydrometeors is needed. With the advancements of retrieval techniques, the quality of retrieved satellite products has been steadily increasing.

## 12.2 All-sky Radiance Assimilation Challenges

The challenges of all-sky satellite radiance assimilation in principle originate due to their relation to clouds and precipitation. Observing and simulating clouds is challenging in it own right. Cloud microphysical processes are highly nonlinear, discontinuous, and characterized by very small spatial scales of the order of hundreds of meters to a kilometer, requiring high resolution and complex modeling. This is magnified in data assimilation, being a method that combines information from observations and from prediction models. Consequently, data assimilation of clouds and precipitation is also nonlinear and high-dimensional. High spatiotemporal resolution of cloud processes has a direct consequence on computations, posing an additional challenge in practical applications. Precipitating clouds introduce even more difficulty due to complex scattering that cannot be neglected. There is also the additional computational cost of radiance observation operator associated with scattering.

Problems related to accurate and efficient assimilation of all-sky radiances are fundamentally related to each other, but one could distinguish the challenges related to data assimilation, simulation and prediction, and computation. Simulation and prediction of clouds is related to the ability of prediction models to represent clouds, and the complexity of the employed microphysics. Although this clearly impacts data assimilation, it does so mostly in the context of smoothers, such as 4DVAR, since they include a forecast model in assimilation. The computational challenges of all-sky radiance assimilation we refer to are caused by a high spatiotemporal resolution of cloud microphysical processes, as well as by a necessity to include cloud scattering processes in the forward radiative transfer model used as the observation operator. Computational restrictions impact the choices one could have regarding methodology and algorithms used in data assimilation. Here we focus on some of these fundamental issues related to assimilation of all-sky satellite radiances.

The challenges of all-sky radiance data assimilation are defined as the components of data assimilation that are particularly exposed by assimilation of all-sky radiances and related cloud-resolving scales. They can be all traced back to clouds and precipitation and range from methodological to computational: (i) forecast error covariance; (ii) correlated observation errors; (iii) nonlinearity and nondifferentiability; and (iv) non-Gaussian errors.

### *12.2.1 Forecast Error Covariance*

The fundamental role of forecast error covariance (see Section 8.2) extends its relevance to all-sky satellite data assimilation as well. It represents one of the main differences between variational- and ensemble-based data assimilation methodologies: it is time independent (i.e., static) in variational methods, while it is flow dependent in ensemble methods. Time dependence is generally an advantage for applications at cloud scales due to typically fast-changing dynamical processes. Examples of such processes include cloud microphysical processes, in a hurricane or in severe storm outbreaks. However, ensemble data assimilation at such high resolution also implies a severely reduced-rank approximation to the forecast error covariance. This deficiency can be considerably improved by error covariance localization techniques (e.g., Hamill et al., 2001; Houtekamer and Mitchell, 2001), but it is still a limitation.

It is important to understand how static, but full-rank covariance has a different impact from flow-dependent, but reduced-rank error covariance in all-sky radiance assimilation. The structure of forecast error covariance can be used to assess the quality of data assimilation, given the fundamental role of forecast error covariance in data assimilation. In geoscience applications, such as weather, climate, and hydrology, the structure of true forecast error covariance can be very complex. In all-sky radiance assimilation, the structure of forecast error covariance with respect to cloud microphysical variables is especially important since cloud variables are inputted into the radiative transfer model. There are numerous processes that involve interactions between cloud variables, mathematically represented by cross-correlations in forecast error covariance. There are also interactions between cloud and standard dynamical variables, such as temperature, pressure, and wind, which also imply cross-correlations between these variables. Such complex cross-correlations between numerous variables are most efficiently described using a block matrix form of the forecast error covariance

$$\mathbf{P}_f = \begin{bmatrix} \mathbf{P}_{dd} & \mathbf{P}_{cd} \\ \mathbf{P}_{cd} & \mathbf{P}_{cc} \end{bmatrix}, \tag{12.1}$$

where the index $d$ refers to dynamical variables, and the index $c$ to cloud variables (e.g., cloud ice, snow, rain, etc.). $\mathbf{P}_f$ is a symmetric matrix, thus, $\mathbf{P}_{cd} = \mathbf{P}_{cd}^T = \mathbf{P}_{dc}$. In this example we assume that dynamical variables include temperature, pressure, and wind, and that cloud variables include cloud ice, snow, and rain. The block covariance matrices in (12.1) are

$$\mathbf{P}_{dd} = \begin{bmatrix} \mathbf{P}_{T,T} & \mathbf{P}_{T,p} & \mathbf{P}_{T,v} \\ \mathbf{P}_{T,p} & \mathbf{P}_{p,p} & \mathbf{P}_{p,v} \\ \mathbf{P}_{T,v} & \mathbf{P}_{p,v} & \mathbf{P}_{v,v} \end{bmatrix}, \tag{12.2}$$

$$\mathbf{P}_{cc} = \begin{bmatrix} \mathbf{P}_{ice,ice} & \mathbf{P}_{ice,snow} & \mathbf{P}_{ice,rain} \\ \mathbf{P}_{ice,snow} & \mathbf{P}_{snow,snow} & \mathbf{P}_{snow,rain} \\ \mathbf{P}_{ice,rain} & \mathbf{P}_{snow,rain} & \mathbf{P}_{rain,rain} \end{bmatrix}, \text{ and} \tag{12.3}$$

$$\mathbf{P}_{cd} = \begin{bmatrix} \mathbf{P}_{ice,T} & \mathbf{P}_{ice,p} & \mathbf{P}_{ice,v} \\ \mathbf{P}_{snow,T} & \mathbf{P}_{snow,p} & \mathbf{P}_{snow,v} \\ \mathbf{P}_{ice,T} & \mathbf{P}_{snow,p} & \mathbf{P}_{rain,v} \end{bmatrix}. \tag{12.4}$$

The main diagonal blocks (12.2) and (12.3) are symmetric matrices, while the off-diagonal block matrix (12.4) is not symmetric.

The elements of block matrices (12.2), (12.3) and (12.4) are themselves 3D matrices, so one can appreciate the immense complexity represented by a deceivingly simple form of the matrix (12.1). Since these matrices also represent the uncertainty of model variables, one can quickly realize that elements of these matrices are fundamentally time dependent. The natural formation and decay of clouds will have a profound impact on the elements of matrices (12.3) and (12.4). These matrices have all elements essentially equal to 0 for clear skies as a consequence of nonexistent cloud interactions. When clouds begin to form, however, the block matrices begin to fill in with nonzero elements and eventually challenge data assimilation to adequately represent such complex processes and interactions in error covariance. Because of that there is still a limited capability of existing data assimilation methodologies to accurately address the structures in (12.1)–(12.4).

In the context of VAR, it is extremely difficult to represent cross-correlations between cloud variables, as well as between cloud and dynamical variables. One reason is that the forecast error covariance is modeled and typically does not represent time-dependent information about interactions. An exception is the 4DVAR method which includes some time dependence through tangent linear and adjoint model integration. However, the forecast error covariance defined at the initial time of assimilation is still modeled as in 3DVAR. For dynamical variables one can identify simplified relations such as hydrostatic, geostrophic, and similar balance constraints that are commonly used in modeling cross-variable interactions (e.g., Parrish and Derber, 1992). Unfortunately, balance constraints are poorly known or unknown at cloud scales, apparently creating a serious difficulty for VAR to represent cloud-variable cross-correlations in (12.3) and dynamical-cloud correlations in (12.4). One should acknowledge, however, that at least in principle it may be possible to successfully model all these cross-variable correlations. A practical solution is to assume a regular (i.e., isotropic and homogeneous) correlation for the diagonal blocks in (12.3), i.e., to predefine correlation functions for $\mathbf{P}_{ice,ice}$, $\mathbf{P}_{snow,snow}$, and $\mathbf{P}_{rain,rain}$, and thus avoid modeling more complex cross-correlations. In this case the matrices $\mathbf{P}_{cd}$ and $\mathbf{P}_{cc}$ become

$$\mathbf{P}_{cc} = \begin{bmatrix} \mathbf{P}_{ice,ice} & 0 & 0 \\ 0 & \mathbf{P}_{snow,snow} & 0 \\ 0 & 0 & \mathbf{P}_{rain,rain} \end{bmatrix} \quad \text{and} \tag{12.5}$$

$$\mathbf{P}_{cd} = \begin{bmatrix} 0 & 0 & 0 \\ 0 & 0 & 0 \\ 0 & 0 & 0 \end{bmatrix}. \tag{12.6}$$

One should be aware that even the block diagonal matrix (12.5) is difficult to model correctly given a relatively limited knowledge about cloud variables' autocorrelations.

Ensemble-based forecast error covariance has a better chance of capturing these complex inter-variable correlations since ensemble forecasting includes all variables and is time dependent. The inherent problem for ensemble error covariance is the number of degrees of freedom (DOF) captured by ensembles. Even with error covariance localization the low-rank approximation of ensemble error covariance may not allow accurate representation of all cross-correlations. Fortunately, $O(10)$ ensembles appear to be sufficient to represent at least some inter-variable correlations. Typically, a correlation between variables at nearby points is well represented even by a limited size ensemble. This property, coupled with adequate covariance localization, essentially allows ensemble forecast error covariance to include all terms defined by (12.1)–(12.4), although with reduced accuracy. The practical problem is how to distinguish between "good" and "bad" correlations, and the answer is not yet clear. More aggressive localization may have the appearance of better controlling cross-variable correlations, but it may adversely impact the dynamical balance of the analysis and could prevent some important correlations in the vertical for well-developed cloud systems. A possible remedy could be to introduce a set of localizing function parameters that depend on the synoptic situation, the variable, and the vertical level. The adverse consequence of such an approach is that the number of unknown parameters in a data assimilation system would effectively increase and introduce additional DOF. Estimated either online or precalculated, these localization parameters would still pose a challenge for data assimilation. Their estimation is likely to be one of the main concerns in the future development of error covariance localization techniques.

Although the algebraic description of the all-sky forecast error covariance presented above is helpful, it also helps to visually examine the covariance structure. As seen in Chapter 9, the structure of error covariance can be inspected by plotting columns of the forecast error covariance, also related to "single-observation" data assimilation experiments. Lets define a vector with all zero elements except for the $i$th element with the value one

$$\mathbf{z}_i = \begin{bmatrix} 0_1 & \cdots & 0_{i-1} & 1_i & 0_{i+1} & \cdots & 0_{N_S} \end{bmatrix}^T, \tag{12.7}$$

where the index refers to a grid point and a variable (e.g., index of the state vector) and $N_S$ is the dimension of state vector. After multiplying vector $\mathbf{z}_i$ by matrix $\mathbf{P}_f$, one obtains the $i$th column of the forecast error covariance matrix

$$\mathbf{c}_i = \mathbf{P}_f \mathbf{z}_i = \begin{bmatrix} \mathbf{f}_1^i & \cdots & \mathbf{f}_{i-1}^i & \mathbf{f}_i^i & \mathbf{f}_{i+1}^i & \cdots & \mathbf{f}_{N_S}^i \end{bmatrix}^T \tag{12.8}$$

with $\mathbf{f}_j^i$ representing the $i$th column value at location $j$. Note that location refers to a grid point and variable. Following Thépaut et al. (1996) and Huang et al. (2009), one can derive the analysis increment for single observation at $i$th point

$$\mathbf{x}^a - \mathbf{x}^f = P_f[\mathbf{y}^o - H(\mathbf{x})]_i \tag{12.9}$$

Applying the matrix-vector product in (12.9), and using (12.8)

$$\mathbf{x}^a - \mathbf{x}^f = [\mathbf{y}^o - H(\mathbf{x})]_i \mathbf{c}_i, \tag{12.10}$$

i.e., the analysis increment is simply the $i$th column of the forecast error covariance scaled by the observation increment. Therefore, a column of the forecast error covariance can be interpreted as analysis response, and thus give a physical meaning to the structure of forecast error covariance.

An example of examining the structure of ensemble error covariance for cloud snow is shown in Figures 12.1 and 12.2. The plotted field is the analysis response to a single observation of cloud snow at 650 hPa. This corresponds to a hypothetical observation of high-frequency microwave all-sky satellite radiance that is sensitive to cloud snow and ice. The results are obtained using the Weather Research and Forecasting (WRF) model

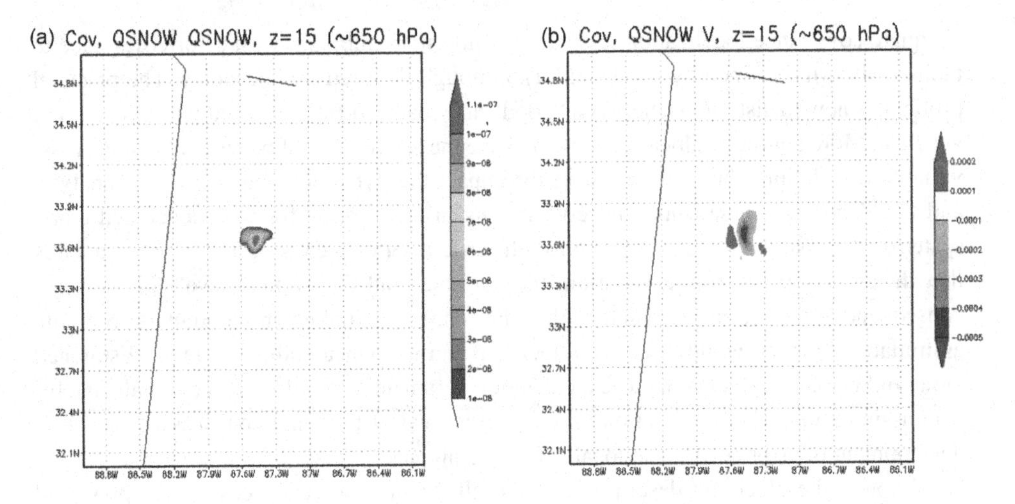


Figure 12.1 Horizontal analysis response to a single observation of cloud snow at 650 hPa on 09 September 2012 at 1800 UTC: (a) cloud snow analysis increment and (b) north–south wind analysis increment. The results are shown for the inner nest at 3 km horizontal resolution. From Županski (2013). ©Springer-Verlag Berlin Heidelberg 2013. Used with permission.

(e.g., Skamarock et al., 2005) at 9 km/3 km resolution and the maximum likelihood ensemble filter (MLEF) data assimilation algorithm (e.g., Županski, 2005; Županski et al., 2008). The control variables in data assimilation include dynamical variables (e.g., perturbation pressure, perturbation height, perturbation potential temperature, and winds) and cloud variables (e.g., cloud ice, cloud snow, cloud water, graupel, and water vapor). A horizontal map of the analysis increment for cloud snow and for the north–south wind component at the level of the cloud snow observation are shown in Figures 12.1a and 12.1b. The snow analysis has a strong positive response to snow observation (Figure 12.1a), as expected. It is also interesting to note that cloud snow observation impacts wind (Figure 12.1b), a dynamical variable, corresponding to $\mathbf{P}_{snow,v}$ component of the forecast error covariance from (12.4). This indicates that the structure of ensemble forecast error covariance enables all-sky radiance observations to impact dynamical variables in realistic systems. A relatively symmetric response appears to be similar to the modeled error covariance structure (e.g., Parrish and Derber, 1992; Wu et al., 2002), possibly suggesting that such covariance components can be successfully modeled in variational methods.

Cloud snow and rain analysis responses in the vertical are shown in Figure 12.2, corresponding to the components $P_{snow,snow}$ and $P_{snow,rain}$. A well-defined cloud snow response centered at the observation location (Figure 12.2a) can be seen. The response is confined to a few levels above and below the observation, again suggesting that modeling such autocovariance components may be possible. However, the response of cloud rain (Figure 12.2b) exposes a potential difficulty for modeling cross-variable correlations, such as snow-rain in this example. One problem is to create a noncentered response of rain to cloud snow observation. Although this is mathematically possible (e.g., Gaspari and Cohn, 1999), there are several difficulties in applying it in practice. One problem is that there is

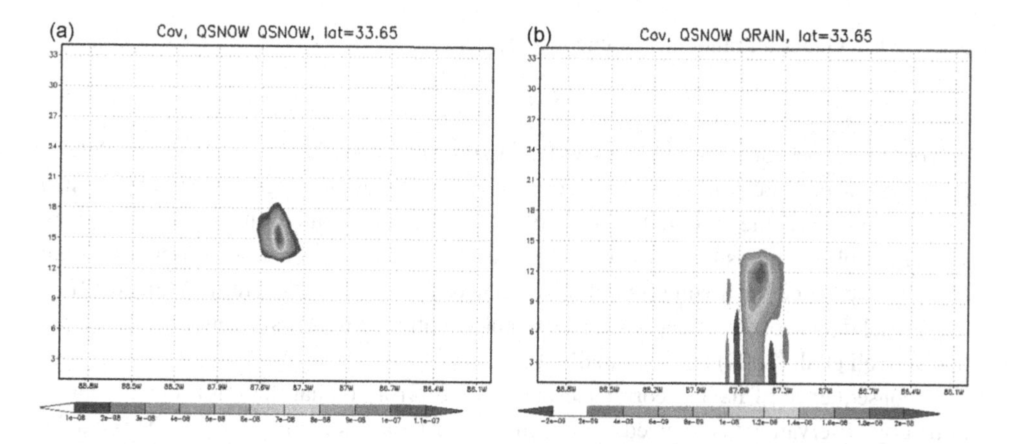


Figure 12.2 Vertical analysis response to a single observation of snow at 650 hPa on 09 September 2012 at 1800 UTC for: (a) cloud snow analysis increment and (b) rain analysis increment. The results are shown for the inner nest at 3 km horizontal resolution. From Županski (2013). ©Springer-Verlag Berlin Heidelberg 2013. Used with permission.

very limited knowledge of cloud-variable correlation statistics. The most difficult problem, however, may be related to the flow dependence of these correlations. It is clear that the existence of cloud rain and snow depends on the current temperature conditions that change with time and thus require additional flow-dependent parameters to be introduced to the modeling function and eventually estimated. In other words, the melting level is space and time dependent and therefore difficult to be represented by a preset function. The forecast error covariance can have a very different structure depending on the methodology used to represent the error covariance, and therefore the choice of methodology will impact satellite radiance data assimilation. Even within the same methodology one can choose different localization parameters of a preset correlation function in variational methods or of the localization function in ensemble methods, effectively implying a large number of possible choices for forecast error covariance. This apparent multitude of possible choices for the forecast error covariance creates a problem since it implies a nonunique analysis solution. The "optimal" choice of forecast error covariance may be good for overall data assimilation performance but may not adequately address all-sky radiance observations since they are generally confined to a smaller local area of intense dynamical development and thus their global impact is relatively small. A possible way to address the problem of nonuniqueness of forecast error covariance is to reduce the number of additional parameters, or to search for a new methodology that would not have this problem.

### *12.2.2 Correlated Observation Errors*

The inclusion of cloudy and precipitation-affected radiances in data assimilation can significantly increase the number of observations and their density. This opens several new data assimilation issues such as computational overhead and observation error correlations, respectively. A commonly used observation equation is

$$\mathbf{y}^o = H(\mathbf{x}) + \varepsilon, \tag{12.11}$$

where $\varepsilon$ is a Gaussian random variable $N(0, \mathbf{R})$ and

$$\mathbf{R} = \left\langle \varepsilon\varepsilon^T \right\rangle. \tag{12.12}$$

The observation error covariance matrix $\mathbf{R}$ is assumed diagonal in typical data assimilation, implying uncorrelated observation increments. This assumption greatly simplifies data assimilation since the required inverse of $\mathbf{R}$ is easily computed and is also relatively accurate if observations are not very close to each other. However, when observations are densely distributed the assumption of uncorrelated observation errors may not be justified. Additional details of the impact of correlated observation in data assimilation can be found in Cardinali et al. (2004) and Cardinali (2013).

A consequence of having correlated observation errors is that the information content of nearby observations is reduced compared to their independent information. This can be further formalized by using the mathematical framework of Shannon information theory (Shannon and Weaver, 1949, see also Section 2.7). Let $Y_1$ and $Y_2$ represent random observation errors for two nearby observations. Using a general relationship between entropy $H$ and mutual information $I$ (e.g., Cover and Thomas, 2006)

$$I_E\left(Y_1; Y_2\right) = H_E\left(Y_1\right) + H_E\left(Y_2\right) - H_E\left(Y_1, Y_2\right) \tag{12.13}$$

as well as

$$I_E\left(Y_1; Y_2\right) = I_E\left(Y_1; Y_1\right) + I_E\left(Y_2; Y_2\right) - H_E\left(Y_1, Y_2\right). \tag{12.14}$$

Since by definition $H\left(Y_1, Y_2\right) \geq 0$ for arbitrary random variables $Y_1$ and $Y_2$, we have

$$I_E\left(Y_1; Y_2\right) \leq I_E\left(Y_1; Y_1\right) + I_E\left(Y_2; Y_2\right). \tag{12.15}$$

The last expression (12.15) implies that the mutual information of dependent variables is smaller than the mutual information of independent variables. Since in the Gaussian framework the notion of dependence is directly related to correlations, one can say that correlated observations bring less information than uncorrelated observations. This implies a need to account for correlated observation in all-sky satellite radiance assimilation.

**Example 12.1 Entropy of a 2D Gaussian random innovation vector**

Let a 2D innovation vector $\mathbf{d}_o = \mathbf{y}^o - H(\mathbf{x})$ have a Gaussian PDF $N(0, \mathbf{R})$. From (2.56) one can obtain the bivariate PDF

$$\varphi(\mathbf{d}_o) = \frac{1}{\sqrt{(2\pi)^2 \det(\mathbf{R})}} e^{-\frac{1}{2}(\mathbf{d}_o)^T \mathbf{R}^{-1}(\mathbf{d}_o)}.$$

The entropy can be derived by applying (2.47) with the above bivariate PDF to obtain

$$H_E(\mathbf{d}_o) = \frac{1}{2} \ln\left[(2\pi)^2 \det(\mathbf{R})\right] + 1. \tag{12.16}$$

**Example 12.2 Observation error correlations and innovation entropy**

Consider a $2 \times 2$ observation error covariance matrix

$$\mathbf{R} = \sigma_o^2 \begin{pmatrix} 1 & \rho \\ \rho & 1 \end{pmatrix}, \tag{12.17}$$

where $\rho$ denotes the correlation coefficient and $\sigma_o$ is the standard deviation. The entropy of innovation vector can be obtained from (12.16)

$$H_E(\mathbf{d}_o) = \frac{1}{2} \ln \left[ (2\pi)^2 \sigma_o^4 (1 - \rho^2) \right] + 1. \tag{12.18}$$

If it is incorrectly assumed that observation errors are not correlated, the observation error covariance is diagonal

$$\mathbf{R}^{approx} = \begin{pmatrix} \sigma_o^2 & 0 \\ 0 & \sigma_o^2 \end{pmatrix}. \tag{12.19}$$

The corresponding approximate entropy is then

$$H_E^{approx}(\mathbf{d}_o) = \frac{1}{2} \ln \left[ (2\pi)^2 \sigma_o^4 \right] + 1. \tag{12.20}$$

The change of information due to incorrectly assuming diagonal observation error covariance is obtained by taking the difference between (12.20) and (12.19)

$$H_E^{approx}(\mathbf{d}_o) - H_E(\mathbf{d}_o) = -\frac{1}{2} \ln \left[ 1 - \rho^2 \right]. \tag{12.21}$$

Note that mutual information of the covariance (12.17), obtained using (9.9) is

$$I_E = -\frac{1}{2} \ln \left[ 1 - \rho^2 \right]. \tag{12.22}$$

Therefore, the change of information due to incorrectly using diagonal matrix $\mathbf{R}^{approx}$ is equal to mutual information of the correlated matrix $\mathbf{R}$

$$H_E^{approx}(\mathbf{d}_o) - H_E(\mathbf{d}_o) = I_E. \tag{12.23}$$

**Practice 12.1 Quantifying information of correlated observations**

Consider a $2 \times 2$ observation error covariance matrix (12.17) with standard deviation $\sigma_o = 1$ and two options for correlation coefficients (a) $\rho = 0.9$ (strongly correlated) and (b) $\rho = 0.1$ (weakly correlated). Quantify information in terms of entropy for each correlation coefficient option.

**Practice 12.2 Quantifying information discrepancy**

As in Practice 12.1, consider a $2 \times 2$ observation error covariance matrix (12.17) with standard deviation $\sigma_o = 1$ and two options for correlation coefficients: (a) $\rho = 0.9$ and (b) $\rho = 0.1$. Quantify information discrepancy for each correlation coefficient option.

Observation error correlations in data assimilation can be addressed in several ways:

1. *Increase observation errors*: If observation density is high, reduce the impact of dense observations by increasing the observation error.
2. *Observation thinning*: If observation density is high, thin observations by selecting every $n$th observation. The reduced number of observations could keep the original error or have some error adjustment.
3. *Cut-off based selection* (Fertig et al., 2007): Based on an empirical estimate of observation correlations one can design an algorithm to select which radiance observations to assimilate and which to reject.
4. *Eigenvalue decomposition* (Parrish and Cohn, 1985; Anderson, 2003): For a nondiagonal $R$ work in eigenvector space where the errors are diagonal, i.e., $\mathbf{R} = \mathbf{S}\boldsymbol{\Lambda}\mathbf{S}^T$. Introduce the change of variable $\mathbf{S}^T[\mathbf{y}^o - H(\mathbf{x})] = \mathbf{S}^T\varepsilon$ to obtain transformed error as $\mathbf{R}_S = \left\langle \left(\mathbf{S}^T\varepsilon\right)\left(\mathbf{S}^T\varepsilon\right)^T \right\rangle = \mathbf{S}^T\left\langle \varepsilon\varepsilon^T \right\rangle \mathbf{S} = \mathbf{S}^T\mathbf{R}\mathbf{S} = \boldsymbol{\Lambda}$. Since the transformed observation error is diagonal, a standard data assimilation algorithm with diagonal observation covariance can still be applied.
5. *Direct application of the inverse*: This can be accomplished by directly calculating the matrix inverse or indirectly by using a matrix-vector product. In both cases one needs to assume the correlation properties. Given a large number of observations the former approach may be computationally prohibitive. The latter approach may be computationally feasible and is described as follows. Assume that $\mathbf{R} = \mathbf{DCD}$, where $\mathbf{D}$ is the diagonal matrix of observation error standard deviation and $\mathbf{C}$ is the correlation matrix. $\mathbf{C} = \mathbf{E}\mathbf{E}^T$ can be decomposed using the unique symmetric square root $\mathbf{E}$. The inverse square root of the correlated observation error covariance is $\mathbf{R}^{-1/2} = (\mathbf{DE})^{-1} = \mathbf{E}^{-1}\mathbf{D}^{-1}$. Since the inverse of a symmetric positive definite matrix is symmetric and positive definite, $\mathbf{E}^{-1}$ can be modeled using a simple correlation matrix such as Toeplitz (e.g., Golub and Van Loan, 2013) and therefore avoid the calculation of the inverse $\mathbf{E}^{-1}$. In practice the method would be applied using a matrix-vector product such as $\mathbf{R}^{-1/2}[\mathbf{y}^o - H(\mathbf{x})]$, which makes the approach feasible even for a large number of observations.

Approaches (1) and (2) still assume a diagonal matrix $\mathbf{R}$. They only address correlations by adjusting the observation errors (1) or the density (and number) of observations (2) to match the desired impact of observations. If the observations errors are indeed correlated, approach (1) is implicitly using a top-hat function instead of a true correlation function. Approach (2) implies that nearby observations have similar information content (i.e., homogeneity) which may not be true for observations of clouds and precipitation given that the quality of radiance observation depends on the scan angle, for example. Approach (3) implicitly assumes a nondiagonal **R** but it employs this information only to select radiance observations to be assimilated, still using the original diagonal-based assimilation framework.

Approaches (4) and (5) both assume nondiagonal (e.g., correlated) observation errors. Approach (4) is mathematically more general than (5), since it can be applied to an arbitrary **R**, while approach (5) requires simplified **R** correlations in order to be practical. However, approach (4) also needs an assumption about the eigenvalue threshold. For example, the inverse square root of $\mathbf{R} = \mathbf{S}\boldsymbol{\Lambda}\mathbf{S}^T$ is $\mathbf{R}^{-1/2} = \mathbf{S}\boldsymbol{\Lambda}^{-1/2}\mathbf{S}^T$. Practical calculation of $\boldsymbol{\Lambda}^{-1/2}$ requires a threshold value in order to avoid the division by 0. Unfortunately, the smallest values of $\boldsymbol{\Lambda}$ are exactly those that are most important for the inverse, making the decision about the threshold difficult.

There are additional computational issues that arise due to the use of a radiative transfer operator for all-sky radiances. The inclusion of cloud and precipitation scattering processes required for all-sky radiance calculations adds considerably to the computational cost of data assimilation (e.g., Stephens, 1994). Coupled with a significant increase of the number of radiance observations, the cost of all-sky radiance calculations can be much larger than the cost of clear-sky radiances. This directly impacts the cost of data assimilation and needs to be taken into account. Although not discussed here and still in the domain of research, artificial intelligence (AI) methods may offer the means to reduce the computational cost of all-sky radiance observation operators.

### 12.2.3 Nonlinearity and Nondifferentiability

Cloud microphysical processes and the radiative transfer observation operator for all-sky radiances are known to be nonlinear (e.g., Errico et al., 2007a). One could address nonlinearity by choosing a fundamentally different methodology, such as particle filters (PFs) (e.g., Gordon et al., 1993; Xiong et al., 2006; van Leeuwen, 2009) or by improving the conditioning of the Hessian in minimization (e.g., Axelsson and Barker, 1984; Županski et al., 2008). There are also the so-called linear channels (e.g., frequencies) that do not have strong nonlinearity and thus can be treated using linear or weakly nonlinear methods. Variational methods are generally equipped to address nonlinearity using iterative minimization of the cost function. Standard ensemble Kalman filtering methods do not address the observation nonlinearity specifically, which prompted a development of hybrid variational-ensemble methods (e.g., Županski, 2005; Wang et al., 2007), ensemble iterative Kalman filters (e.g., Gu et al., 2007), and a refinement of the ensemble Kalman filter (e.g., Evensen, 2003).

We discuss here the improved conditioning of minimization in more detail since the majority of practical data assimilation algorithms today use numerical optimization/minimization to solve nonlinear problems. The most commonly used minimization algorithms in data assimilation are unconstrained algorithms, such as the nonlinear conjugate gradient algorithm or the quasi-Newton algorithms (e.g., Luenberger and Ye, 2016). Hessian preconditioning (e.g., Axelsson and Barker, 1984; Županski, 1996) can be efficiently used

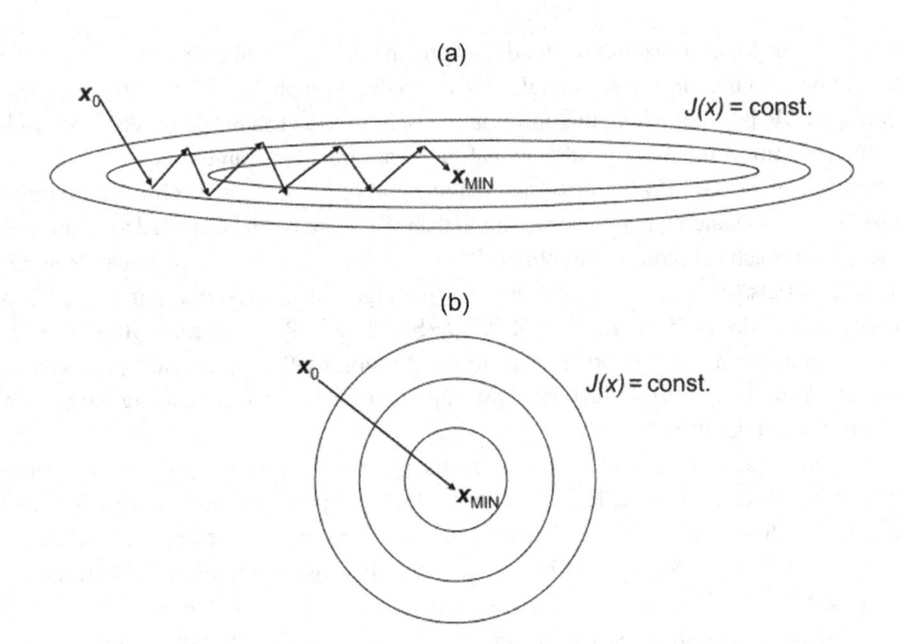


Figure 12.3 Impact of Hessian preconditioning on minimization: (a) physical space and (b) preconditioned space. In this example of a quadratic cost function, it is shown how an ideal preconditioning can change the cost function so that the minimum is reached in a single minimization iteration. From Županski (2013). ©Springer-Verlag Berlin Heidelberg 2013. Used with permission.

in such minimization algorithms and is therefore relevant for all-sky satellite radiance assimilation. Its general role is in speeding up minimization and effectively reducing the condition number of the Hessian matrix, achieved by introducing a change of variable. The ideal impact of preconditioning is illustrated in Figure 12.3, which shows the change of a quadratic cost function contours from an elongated ellipse to a circle. Starting minimization from an arbitrary point will lead to numerous minimization iterations for the original ellipsoidal cost function (Figure 12.3a), while a single iteration will be sufficient for a preconditioned minimization problem (Figure 12.3b).

An additional role of preconditioning, of special importance in practical applications but often disregarded, is to provide a "balanced" reduction of the cost function. This means that minimization produces a change of control variables that is in agreement with the actual weather situation. From an algorithmic viewpoint, one would like to achieve a comparable percentage of adjustment for each physical variable, with the idea that even when minimization does not have sufficient time to reach the mathematical convergence it would still produce an acceptable physical solution. For example, consider temperature and horizontal wind. Assume that the initial guess is dynamically balanced, which is generally true given that it is produced by a forecast. In the case where the temperature component of the cost function is perfectly preconditioned, but the wind component is not preconditioned at all, the analysis after the first minimization iteration would have a fully adjusted temperature but practically unchanged winds. Since wind and temperature were in dynamical balance before minimization, the produced analysis after the first iteration would be unbalanced and eventually create noise in the ensuing forecast. This situation can be visualized from Figure 12.3 with the temperature converging according to Figure 12.3b and the wind slowly converging as shown in Figure 12.3a. After the first minimization iteration the temperature will reach its optimal value, but the wind would be still far from the optimum. The important implication is that such a problem can be efficiently resolved by adequate Hessian preconditioning.

Consider now the impact of Hessian preconditioning on all-sky satellite radiance assimilation. Assimilation of all-sky satellite radiance would be most beneficial if cloud variables were defined as control variables since they have the most direct impact on the radiative transfer operator. In the case where the cloud-variable component of the cost function is not adequately preconditioned, however, all-sky radiances will not be utilized well, eventually producing an unbalanced analysis. Assuming that other dynamical variables were well preconditioned, it is likely that the ensuing forecast will treat the initial adjustments of clouds and precipitation created by the analysis as noise and effectively filter out the adjustments in the forecast, only because of inadequate preconditioning. This problem is real, and it can have serious consequences for data assimilation.

In practical high-dimensional variational methods the forecast error covariance is used to precondition minimization instead of a full Hessian, due to the high computational cost. Given that the preset static covariance is not able to correctly represent cross-variable correlations for cloud variables, while the preset covariance of dynamical variables has a relatively well-defined cross-variable structure, the overall preconditioning will be off balance. This also impacts hybrid ensemble-variational methods since in practice the variational component is used for calculating the analysis. The above examples illustrate the important role of Hessian preconditioning in assimilation of all-sky radiances and indirectly suggest that the preconditioning method should include a knowledge of dynamics.

While the nonlinearity of the all-sky radiance operator has been often acknowledged, the nondifferentiability of this operator is typically not discussed and thus requires attention. The nondifferentiability of the observation operator $H$ becomes an issue in the case of all-sky radiances because of the on–off switch in the radiative transfer operator used to decide which branch to take: the cloudy branch that normally includes scattering effects, or noncloudy branch without scattering. Given that this decision depends on atmospheric parameters such as cloud mixing ratios and temperature, the forward radiative transfer operator has a discontinuity in the function value and derivative, implying a discontinuity of the cost function. One can write the radiative transfer operator for all-sky radiances as

$$H(\mathbf{x}) = \begin{cases} c(\mathbf{x}) & \mathbf{x} \in C \\ s(\mathbf{x}) & \mathbf{x} \notin C, \end{cases} \tag{12.24}$$

where $C$ represents the state subspace corresponding to clear-sky conditions, $c$ is the clear-sky component, and $s$ is the cloudy component of the observation operator. The point where the state can cross between clear and cloudy conditions is the discontinuity point, and thus

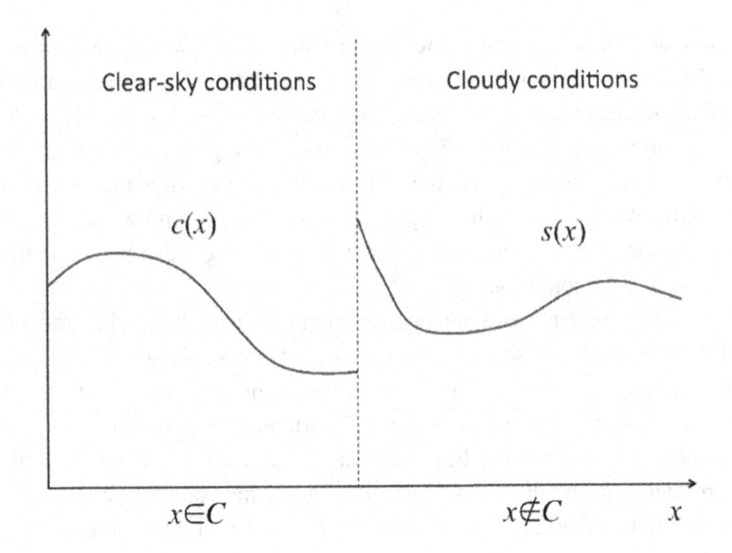


Figure 12.4 Discontinuous all-sky radiative transfer operator defined by Eq. (12.24). The left area of the figure represents clear-sky conditions and the right area corresponds to cloudy conditions. In principle, the function value and its derivatives all can experience a discontinuity. From Županski (2013). ©Springer-Verlag Berlin Heidelberg 2013. Used with permission.

the operator $H$ has two branches. This is visualized in Figure 12.4, indicating that the function value, and all its derivatives, can have a discontinuity.

The discontinuity of the observation operator creates an apparent problem for variational methods since they are commonly using gradient-based minimization (e.g., Navon, 1986). A nonexistent gradient at the point of transition from clear-sky to cloudy conditions prevents correct performance of minimization, eventually resulting in an incorrect minimizing solution. Since the Kalman filter related methods can be also described in terms of gradient-based minimization, (e.g., Jazwinski, 1970), with or without iterative minimization, they consequently have similar difficulties. This implies that the nondifferentiability of the all-sky observation operator is a problem of data assimilation in general.

There are several ways that could be used to address discontinuity, most obvious being: (i) neglect it; (ii) apply smoothing; and (iii) use nondifferentiable minimization algorithms. Although the first option may not be mathematically correct, it does not require any additional effort. Since the discontinuity point is in the area of transition from clear-sky to cloudy conditions, the discontinuity problem may be confined to only those geographical areas, allowing minimization to perform well in the rest of the domain. However, since discontinuity also impacts the line-search algorithm (i.e., finding the optimal step-size), it may have additional adverse impact on minimization. Consequently, neglecting the discontinuity of the all-sky radiance observation operator may be acceptable in some situations, but not in general and definitely not in operational practice. Option (ii) has been successfully applied within 4DVAR methods in the cases of parameterization schemes (e.g., Županski and Mesinger, 1995). The approach consists of changing the original operator by introducing a differentiable and continuous function in place of an on–off switch, thus preventing code branching. In choosing the adequate smoothing function and parameters it is important to maintain the comparable skill and accuracy of the original operator. This could be difficult, and it requires an extensive preparation of the code and extensive testing. The third option (iii) is the most correct approach, since it does not alter the original observation operator and it addresses the true problem, which is the minimization algorithm performance. There are numerous minimization algorithms that can address nondifferentiability, some of those developed as an extension of gradient-based algorithms. Encouraging results obtained using this approach in data assimilation have been reported by Steward et al. (2012).

However, it is likely that the choice of approach to address nondifferentiability will depend on the actual assimilation problem and the amount of work required to implement the changes. Important message from this section is that nondifferentiability of all-sky observation operator should not be overlooked. Once the problem is identified, one can proceed to solutions (i)–(iii), or take an alternative approach.

### *12.2.4 Satellite Radiance Bias and Non-Gaussian Errors*

Commonly used data assimilation algorithms are based on the assumption of Gaussian errors with zero mean (i.e., no bias). Some observation types, such as satellite radiances, generally show a bias (e.g., Derber and Wu, 1998; McNally, 2009; Zhu et al., 2016) Therefore, the assimilation of satellite radiances requires a way of accounting for the bias. Radiance biases can be related to inadequate or inaccurate instrument calibration, prediction model biases in representing atmospheric and surface variables, and possibly to the radiative transfer model used to simulate radiances (Saunders et al., 2013). A common approach to address the existence of radiance bias is to apply a method to correct this bias. The most widely used method for satellite radiance bias correction in data assimilation is the so-called variational bias correction (VBC) (Harris and Kelly, 2001; Dee and Uppala, 2009). The method consists of adding a new term to the cost function. The satellite bias term is typically represented as a linear regression of predictors that include satellite geometry (e.g., viewing angle) and atmospheric precursors (e.g., thickness, skin temperature, surface wind speed):

$$\mathbf{b}(\varphi, \mathbf{x}) = \sum_i \beta_i \mathbf{p}_i(\mathbf{x}) = \mathbf{p}^T \boldsymbol{\beta}, \tag{12.25}$$

where $\mathbf{p}$ is the predictor vector and $\boldsymbol{\beta}$ is the vector of regression coefficient. The coefficients of linear combination of this regression are added as control variables to minimization, thus creating an augmented control variable. Although this method has been used with great success, there are channels that are not well controlled using this technique. One reason may be the nonlinearity of some channels with respect to the chosen predictors, implying that the selection of predictors is critical for successful application of this method. Also, current operational practice still includes mostly clear-sky, not all-sky radiance assimilation, and sometimes the same predictors are used for both clear- and all-sky radiances. However, the atmospheric predictors used for a clear sky may not be the optimal choice for cloudy conditions. Even if adequate atmospheric predictors can be found, this approach still requires a lot of experimentation and fundamental knowledge of interactions between clouds and satellites. The augmented control variable in minimization may be technically difficult to implement depending on the existing minimization setup. It was also suggested in several papers (e.g., Errico et al., 2007a) that, if the actual observation error has a skewed PDF that resembles lognormal distribution, then one could pose the problem in terms of a logarithm of the original variable making it a Gaussian-like variable to satisfy the Gaussian error assumption of data assimilation. Although this is a feasible solution, it was shown to be nonunique (e.g., Fletcher and Županski, 2007).

Satellite radiance biases are almost exclusively examined within Gaussian data assimilation since most of the data assimilation algorithms are still based on Gaussian error assumptions. However, satellite radiance observation error statistics indicate a skewness for some instruments and channels that may be attributed to non-Gaussian PDFs (e.g., Okamoto and Derber, 2006; Errico et al., 2007a; Bauer et al., 2010). It has been generally acknowledged that there are observations and/or control variables with non-Gaussian errors and that a non-Gaussian data assimilation framework may be required (e.g., Abramov and Majda, 2004; Fletcher and Županski, 2006a, 2006b, Bocquet et al., 2010).

However, developing and eventually implementing a non-Gaussian data assimilation system is challenging and not sufficiently investigated, especially in the context of operational NWP. Therefore, one could distinguish several possible approaches to deal with non-Gaussian errors of all-sky radiances: (i) neglect the problem; (ii) apply Gaussian assumption but introduce bias correction; (iii) apply a change of variable to convert from a non-Gaussian to a Gaussian framework; and (iv) use a non-Gaussian data assimilation framework.

Option (i) is the simplest, and thus the easiest. If one chooses to assimilate only channels with approximately Gaussian observation errors, it may be still possible to use the original Gaussian data assimilation framework. However, this approach may leave out important observation information. It also requires a good knowledge of the observation error statistics by channels, which could take time and effort to accumulate.

Option (ii) is commonly used in operations (e.g., Harris and Kelly, 2001; Dee and Uppala, 2009). In the case where some radiance channels are nonlinear with respect to the chosen predictors, one could try to introduce quadratic or higher nonlinear regression for those channels and estimate its coefficients by augmenting the control variable.

We already mentioned that option (iii) introduces nonuniqueness, but that does not necessarily prevent its use in practical problems. Option (iv) may be the most complete since it addresses the true problem of having non-Gaussian errors in data assimilation. It was shown that ensemble data assimilation could be defined in terms of non-Gaussian errors within hybrid variational-ensemble methodology (e.g., Fletcher and Županski, 2006a), or within PFs (e.g., van Leeuwen, 2009). Implementing new methodology is a slow process, however, and the ultimate decision about the approach for handling non-Gaussian errors will depend on a desired application and the time required for development. Addressing non-Gaussian all-sky radiance observation errors inherently implies a need for better handling of nonlinearity. Therefore, if the available data assimilation algorithm is not very good for handling nonlinear operators, it is probably better to avoid introducing non-Gaussian errors.

## 12.3 Summary and Future

All-sky satellite radiance assimilation is a difficult problem that puts data assimilation methodology and algorithmic solutions that are used today to the test. Given the great value of satellite radiances, and in particular all-sky radiances, in data assimilation, using a "short-cut" should not be acceptable. We discussed several critical aspects of successful all-sky radiance data assimilation, with emphasis on forecast error covariance, nonlinearity, nondifferentiability, non-Gaussianity, and correlated observation errors. The focus of this chapter was on how variational and ensemble data assimilation can handle these challenging problems. In conclusion, both methods have their advantages and disadvantages for all-sky radiance assimilation, and the widely accepted compromise is to use hybrid variational-ensemble methods. One can also adopt other methodologies, such as PFs, that are better equipped to address the difficulties noted in variational and ensemble methods.

We did not describe in detail all issues related to all-sky radiances, such as verification, or algorithmic details related to a specific methodology, but they also have to be taken into account. It is also likely that there may be research issues that we are not aware of at present but may become important as all-sky radiance assimilation becomes more widespread.

An important implication of the presented challenges is that the development of any new data assimilation methodology that is better suited for all-sky satellite radiance assimilation has to be comprehensive. For example, solving nonlinear issues cannot be properly done without addressing non-Gaussian errors or without utilizing the full power of Hessian preconditioning. An improved definition of forecast error covariance will not be fully beneficial unless it is combined with superior Hessian preconditioning that can maximize the benefit of nonlinear minimization and also improve the estimation of analysis uncertainty. All said, it is important to balance between developing the all-inclusive data assimilation algorithms and keeping the low complexity of data assimilation at the same time. Often, the increased complexity is achieved by adding new parameters to be estimated, without sufficient knowledge of their statistical properties, however, so this should be avoided if possible. There are applications that may accept simple solutions to some of the issues, but it is important not to dismiss an issue before its impact is well understood. For example, before developing and applying a widespread treatment of uncorrelated observation errors, it may be beneficial to first evaluate the potential impact of correlations or to investigate statistics of observations errors.

COLLOQUY 12.1

**Nonlinear closed form solution for variational satellite bias correction**

The use of variational satellite bias correction methodology in ensemble Kalman filter was introduced by Miyoshi et al. (2010). As with other variational correction schemes, this method represents radiance bias using (12.25) and introduces the augmented control vector $\mathbf{z} = (\mathbf{x}, \boldsymbol{\beta})$, where $\mathbf{x}$ is the standard state vector. The considered augmented cost function is quadratic, i.e., with linear observation operator $\mathbf{H}$, and the incremental approach was applied to address the nonlinearities. They suggest an iterative solution that alternates between solving for $\delta\mathbf{x}$ and $\delta\beta$, although in practice only one iteration was performed.

Since a general data-assimilation problem is nonlinear, it is of interest to include the nonlinear observation operator $H$ in radiance bias correction as well. In this colloquy we discuss the nonlinear variational radiance bias correction formulation in ensemble data assimilation by extending the work of Miyoshi et al. (2010). However, instead of using an incremental approach, we rely on the MLEF algorithm (Županski et al., 2008) and directly apply an iterative minimization to minimize the nonlinear augmented cost function

$$f(\mathbf{x}, \beta) = \frac{1}{2}\left[\mathbf{x} - \mathbf{x}^f\right]^T \mathbf{P}_f^{-1}\left[\mathbf{x} - \mathbf{x}^f\right] + \frac{1}{2}\left[\beta - \beta^f\right]^T \mathbf{B}_f^{-1}\left[\beta - \beta^f\right]$$
$$+ \frac{1}{2}\left[\mathbf{y}^o - H(\mathbf{x}) - \mathbf{p}^T\beta\right]^T \mathbf{R}^{-1}\left[\mathbf{y}^o - H(\mathbf{x}) - \mathbf{p}^T\beta\right],$$

where $\mathbf{B}$ is the error covariance of bias correction coefficients. Following the derivative-free approach of MLEF that includes all nonlinear terms of the cost function, one considers the first variation of the above cost function in terms of $\delta\mathbf{x}$ and $\delta\beta$

$$\delta\mathbf{x} : f(\mathbf{x} + \delta\mathbf{x}, \beta) - f(\mathbf{x}, \beta)$$
$$\delta\beta : f(\mathbf{x}, \beta + \delta\beta) - f(\mathbf{x}, \beta).$$

The top equation above is essentially identical to the standard MLEF analysis solution, except with the additional bias correction. The bias correction is formally included by substituting $H(\mathbf{x})$ by $H(\mathbf{x}) + \mathbf{p}^T\beta$ in the MLEF analysis equation (see Section 8.4.2). The remaining question is how to estimate the optimal $\boldsymbol{\beta}$ that could be used in the analysis equation. The first variation of the cost function for the $\delta\beta$ component results in

$$f(\mathbf{x}, \beta + \delta\beta) - f(\mathbf{x}, \beta) = (\delta\beta)^T\left[\mathbf{B}^{-1}(\beta - \beta^f) - \mathbf{p}\mathbf{R}^{-1}\mathbf{d}\right]$$
$$+ \frac{1}{2}(\delta\beta)^T\left[\mathbf{B}^{-1} + \mathbf{p}\mathbf{R}^{-1}\mathbf{p}^T\right](\delta\beta),$$

where $\mathbf{d} = \mathbf{y}^o - H(\mathbf{x}) - \mathbf{p}^T\beta$. The first term is related to the gradient and the second term to the Hessian. Note that since radiance bias correction is a linear function, the true derivative and the generalized derivative defined in Županski et al. (2008) are the same. To solve the above equation, we set the gradient to 0

$$\mathbf{B}^{-1}(\beta - \beta^f) - \mathbf{p}\mathbf{R}^{-1}(\mathbf{y}^o - H(\mathbf{x}) - \mathbf{p}^T\beta) = 0. \tag{12.26}$$

After some processing the solution is

$$\beta - \beta^f = \left(\mathbf{B}^{-1} + \mathbf{p}\mathbf{R}^{-1}\mathbf{p}^T\right)^{-1}\mathbf{p}\mathbf{R}^{-1}\left[\mathbf{y}^o - H(\mathbf{x}) - \mathbf{p}^T\beta^f\right]. \tag{12.27}$$

Since the matrix $\mathbf{B}^{-1} + \mathbf{p}\mathbf{R}^{-1}\mathbf{p}^T$ is low-dimensional (number of predictors is typically 10–20), the inversion is straightforward. A numerically more stable solution can be found if the inversion $\mathbf{B}^{-1}$ is avoided. This can be accomplished by transforming the above matrix and at the same time maintaining the symmetry of the matrix that needs to be inverted

$$\beta - \beta^f = \mathbf{B}^{1/2}\left(\mathbf{I} + \mathbf{C}^T\mathbf{C}\right)^{-1}\mathbf{C}^T\mathbf{R}^{-1/2}\left[\mathbf{y}^o - H(\mathbf{x}) - \mathbf{p}^T\beta^f\right], \tag{12.28}$$

where

$$\mathbf{C} = \mathbf{R}^{-1/2}\mathbf{p}^T\mathbf{B}^{1/2}. \tag{12.29}$$

The equations (12.28) and (12.29) represent the solution for coefficients $\beta$. Although iterations may still be required, substituting $\mathbf{x}$ by $\mathbf{x}^f$ in the above solution should produce a very good estimate of optimal $\beta$

$$\beta = \beta^f + \mathbf{B}^{1/2}\left(\mathbf{I} + \mathbf{C}^T\mathbf{C}\right)^{-1}\mathbf{C}^T\mathbf{R}^{-1/2}\left[\mathbf{y}^o - H(\mathbf{x}^f) - \mathbf{p}^T\beta^f\right]. \tag{12.30}$$

The above value of $\beta$ could then be used in the standard analysis solution to correct radiance bias in the current data-assimilation cycle.

## References


Abramov RV, Majda AJ (2004) Quantifying uncertainty for non-Gaussian ensembles in complex systems. *SIAM J Sci Comput* 26:411–447.

Anderson JL (2003) A local least squares framework for ensemble filtering. *Mon Wea Rev* 131:634–642.

Andersson E, Pailleux J, Thépaut JN, et al. (1994) Use of cloud-cleared radiances in three/four-dimensional variational data assimilation. *Quart J Roy Meteor Soc* 120:627–653.

Apodaca K, Županski M, DeMaria M, Knaff JA, Grasso LD (2014) Development of a hybrid variational-ensemble data assimilation technique for observed lightning tested in a mesoscale model. *Nonlin Processes Geophys* 21:1027–1041.

Auligné T, Lorenc A, Michel Y, et al. (2011) Toward a new cloud analysis and prediction system. *Bull Amer Meteor Soc* 92:207–210.

Axelsson O, Barker VA (1984) *Finite Element Solution of Boundary Value Problems. Theory and Computation.* Academic Press, Orlando, FL, 432 pp.

Bauer P, Geer AJ, Lopez P, Salmond D (2010) Direct 4d-var assimilation of all-sky radiances. Part I: Implementation. *Quart J Roy Meteor Soc* 136:1868–1885.

Bauer P, Ohring G, Kummerow C, Auligne T (2011) Assimilating satellite observations of clouds and precipitation into nwp models. *Bull Amer Meteor Soc* 92:ES25–ES28.

Bauer P, Lopez P, Salmond D, et al. (2006) Implementation of 1D+ 4D-Var assimilation of precipitation-affected microwave radiances at ECMWF. II: 4D-Var. *Quart J Roy Meteor Soc* 132:2307–2332.

Bocquet M, Pires CA, Wu L (2010) Beyond Gaussian statistical modeling in geophysical data assimilation. *Mon Wea Rev* 138:2997–3023.

Bormann N, Thépaut JN (2004) Impact of MODIS polar winds in ECMWF's 4DVAR data assimilation system. *Mon Wea Rev* 132:929–940.

Bouchard A, Rabier F, Guidard V, Karbou F (2010) Enhancements of satellite data assimilation over antarctica. *Mon Wea Rev* 138:2149–2173.

Cardinali C (2013) Observation influence diagnostic of a data assimilation system. In *Data Assimilation for Atmospheric, Oceanic and Hydrologic Applications* (vol. II), (eds.) Park SK, Xu L, Springer-Verlag, Berlin, Heidelberg, 89–110.

Cardinali C, Pezzulli S, Andersson E (2004) Influence-matrix diagnostic of a data assimilation system. *Quart J Roy Meteor Soc* 130:2767–2786.

Cover TM, Thomas JA (2006) *Elements of Information Theory*. 2nd ed., Wiley-Interscience, Hoboken, NJ, 748 pp.

Cucurull L (2010) Improvement in the use of an operational constellation of GPS radio occultation receivers in weather forecasting. *Wea Forecasting* 25:749–767.

Dee DP, Uppala S (2009) Variational bias correction of satellite radiance data in the era-interim reanalysis. *Quart J Roy Meteor Soc* 135:1830–1841.

Derber JC, Wu WS (1998) The use of TOVS cloud-cleared radiances in the NCEP SSI analysis system. *Mon Wea Rev* 126:2287–2299.

Errico RM, Bauer P, Mahfouf JF (2007a) Issues regarding the assimilation of cloud and precipitation data. *J Atmos Sci* 64:3785–3798.

Errico RM, Ohring G, Weng F, et al. (2007b) Assimilation of satellite cloud and precipitation observations in numerical weather prediction models: Introduction to the jas special collection. *J Atmos Sci* 64:3737–3741.

Evensen G (2003) The ensemble Kalman filter: Theoretical formulation and practical implementation. *Ocean Dyn* 53:343–367.

Eyre JR (1994) *Assimilation of radio occultation measurements into a numerical weather prediction system*. Tech. Memo. 199, ECMWF, Reading, 35 pp.

Fertig EJ, Hunt BR, Ott E, Szunyogh I (2007) Assimilating non-local observations with a local ensemble Kalman filter. *Tellus A* 59:719–730.

Fletcher SJ, Županski M (2006a) A data assimilation method for log-normally distributed observational errors. *Quart J Roy Meteor Soc* 132:2505–2519.

Fletcher SJ, Županski M (2006b) A hybrid multivariate normal and lognormal distribution for data assimilation. *Atmos Sci Lett* 7:43–46.

Fletcher SJ, Županski M (2007) Implications and impacts of transforming lognormal variables into normal variables in var. *Meteorol Z* 16:755–765.

Gaspari G, Cohn SE (1999) Construction of correlation functions in two and three dimensions. *Quart J Roy Meteor Soc* 125:723–757.

Geer AJ, Bauer P (2010) *Enhanced Use of All-sky Microwave Observations Sensitive to Water Vapour, Cloud and Precipitation*. Tech. Memo. 620, ECMWF, Reading, 41 pp.

Geer AJ, Bauer P, English S (2012) *Assimilating AMSU-A temperature sounding channels in the presence of cloud and precipitation*. Tech. Memo. 670, ECMWF, Reading, 41 pp.

Geer AJ, Bauer P, Lopez P (2010) Direct 4D-Var assimilation of all-sky radiances. Part II: Assessment. *Quart J Roy Meteor Soc* 136:1886–1905.

Geer AJ, Baordo F, Bormann N, et al. (2017) The growing impact of satellite observations sensitive to humidity, cloud and precipitation. *Quart J Roy Meteor Soc* 143: 3189–3206.

Golub GH, Van Loan CF (2013) *Matrix Computations*. 4th ed., Johns Hopkins University Press, Baltimore, MD, 756 pp.

Goodman SJ, Blakeslee RJ, Koshak WJ, et al. (2013) The goes-r geostationary lightning mapper (glm). *Atmos Res* 125:34–49.

Gordon NJ, Salmond DJ, Smith AFM (1993) Novel approach to nonlinear/non-Gaussian Bayesian state estimation. In *IEE Proceedings F (Radar and Signal Processing)*, vol. 140, IET, 107–113.

Gu Y, Oliver DS (2007) An iterative ensemble Kalman filter for multiphase fluid flow data assimilation. *SPE J* 12:438–446.

Hamill TM, Whitaker JS, Snyder C (2001) Distance-dependent filtering of background error covariance estimates in an ensemble Kalman filter. *Mon Wea Rev* 129:2776–2790.

Harris BA, Kelly G (2001) A satellite radiance-bias correction scheme for data assimilation. *Quart J Roy Meteor Soc* 127:1453–1468.

Holmlund K (1998) The utilization of statistical properties of satellite-derived atmospheric motion vectors to derive quality indicators. *Wea Forecasting* 13:1093–1104.

Houtekamer PL, Mitchell HL (2001) A sequential ensemble Kalman filter for atmospheric data assimilation. *Mon Wea Rev* 129:123–137.

Huang X-Y, Xiao Q, Barker DM, et al. (2009) Four-dimensional variational data assimilation for WRF: Formulation and preliminary results. *Mon Wea Rev* 137:299–314.

Jazwinski AH (1970) *Stochastic Processes and Filtering Theory*. Academic Press, New York, 376 pp.

Kazumori M (2014) Satellite radiance assimilation in the JMA operational mesoscale 4DVAR system. *Mon Wea Rev* 142:1361–1381.

Kazumori M (2019) Assimilation experiments of microwave and infrared radiance data in JMA Global Numerical Weather Prediction System. In *IGARSS 2019–2019 IEEE International Geoscience and Remote Sensing Symposium*, 4738–4740, doi:10.1109/IGARSS.2019.8898781

Kazumori M, Geer AJ, English SJ (2016) Effects of all-sky assimilation of GCOM-W/AMSR2 radiances in the ECMWF numerical weather prediction system. *Quart J Roy Meteor Soc* 142:721–737.

Kazumori M, Liu Q, Treadon R, Derber JC (2008) Impact study of AMSR-E radiances in the NCEP global data assimilation system. *Mon Wea Rev* 136:541–559.

Liu Q, Reichle RH, Bindlish R, et al. (2011) The contributions of precipitation and soil moisture observations to the skill of soil moisture estimates in a land data assimilation system. *J Hydrometeor* 12:750–765.

Luenberger DG, Ye Y (2016) *Linear and Nonlinear Programming*. 4th ed., Springer, Cham, 546 pp.

Maggioni V, Reichle RH, Anagnostou EN (2012) The impact of rainfall error characterization on the estimation of soil moisture fields in a land data assimilation system. *J Hydrometeor* 13:1107–1118.

Matsumura T, Derber JC, Yoe JG, Vandenberghe F, Zou X (1999) *The inclusion of GPS limb sounding data into NCEP's Global Data Assimilation System*. Office Note 426, NOAA/NCEP, Silver Spring, MD, 76 pp.

McNally AP (2009) The direct assimilation of cloud-affected satellite infrared radiances in the ECMWF 4d-var. *Quart J Roy Meteor Soc* 135:1214–1229.

McNally AP, Watts PD (2003) A cloud detection algorithm for high-spectral-resolution infrared sounders. *Quart J Roy Meteor Soc* 129:3411–3423.

McNally AP, Derber JC, Wu W, Katz BB (2000) The use of TOVS level-1B radiances in the NCEP SSI analysis system. *Quart J Roy Meteor Soc* 126:689–724.

McNally AP, Watts PD, Smith JA, et al. (2006) The assimilation of AIRS radiance data at ECMWF. *Quart J Roy Meteor Soc* 132:935–957.

Miyoshi T, Sato Y, Kadowaki T (2010) Ensemble Kalman filter and 4D-Var intercomparison with the Japanese operational global analysis and prediction system. *Mon Wea Rev* 138:2846–2866.

Navon IM (1986) A review of variational and optimization methods in meteorology. In *Variational Methods in Geosciences,* (ed.) Sasaki YK, Elsevier, Amsterdam, 29–34.

Okamoto K, Derber JC (2006) Assimilation of ssm/i radiances in the ncep global data assimilation system. *Mon Wea Rev* 134:2612–2631.

Okamoto K, Kazumori M, Owada H (2005) The assimilation of atovs radiances in the jma global analysis system. *J Meteor Soc Japan* 83:201–217.

Palmer PI, Barnett JJ, Eyre JR, Healy SB (2000) A nonlinear optimal estimation inverse method for radio occultation measurements of temperature, humidity, and surface pressure. *J Geophys Res Atmos* 105:17513–17526.

Parrish DF, Cohn SE (1985) *A Kalman Filter for a Two-Dimensional Shallow-Water Model: Formulation and Preliminary Experiments.* Office Note 304, NOAA/NWS/NMC, Silver Spring, MD, 64 pp.

Parrish DF, Derber JC (1992) The National Meteorological Center's spectral statistical-interpolation analysis system. *Mon Wea Rev* 120:1747–1763.

Poli P, Healy S, Dee D (2010) Assimilation of Global Positioning System radio occultation data in the ECMWF ERA–Interim reanalysis. *Quart J Roy Meteor Soc* 136: 1972–1990.

Polkinghorne R, Vukicevic T (2011) Data assimilation of cloud-affected radiances in a cloud-resolving model. *Mon Wea Rev* 139:755–773.

Rohn M, Kelly G, Saunders RW (2001) Impact of a new cloud motion wind product from meteosat on nwp analyses and forecasts. *Mon Wea Rev* 129:2392–2403.

Saunders RW, Blackmore TA, Candy B, Francis PN, Hewison TJ (2013) Monitoring satellite radiance biases using NWP models. *IEEE Trans Geosci Remote Sens* 51:1124–1138.

Shannon CE, Weaver W (1949) *The Mathematical Theory of Communication.* University of Illinois Press, Chicago, IL, 117 pp.

Skamarock WC, Klemp JB, Dudhia J, et al. (2005) *A Description of the Advanced Research WRF version 2.* NCAR Tech. Note NCAR/TN-468+STR, NCAR, Boulder, CO, 88 pp.

Ştefănescu R, Navon IM, Fuelberg H, Marchand M (2013) 1D+4D-VAR data assimilation of lightning with WRFDA system using nonlinear observation operators. *arXiv:13061884 [mathOC].*

Stephens GL (1994) *Remote Sensing of the Lower Atmosphere: An Introduction.* Oxford University Press, New York, 544 pp.

Steward JL, Navon IM, Županski M, Karmitsa N (2012) Impact of non-smooth observation operators on variational and sequential data assimilation for a limited-area shallow-water equation model. *Quart J Roy Meteor Soc* 138:323–339.

Thépaut J-N, Courtier P, Belaud G, Lemaître G (1996) Dynamical structure functions in a four-dimensional variational assimilation: A case study. *Quart J Roy Meteor Soc* 122:535–561.

van Leeuwen PJ (2009) Particle filtering in geophysical systems. *Mon Wea Rev* 137: 4089–4114.

Vukicevic T, Greenwald T, Županski M, et al. (2004) Mesoscale cloud state estimation from visible and infrared satellite radiances. *Mon Wea Rev* 132:3066–3077.

Wang X, Hamill TM, Whitaker JS, Bishop CH (2007) A comparison of hybrid ensemble transform Kalman filterâŁ"OI and ensemble square-root filter analysis schemes. *Mon Wea Rev* 136:5116–5131.
Weng F (2007) Advances in radiative transfer modeling in support of satellite data assimilation. *J Atmos Sci* 64:3799–3807.
Wu T-C, Liu H, Majumdar SJ, Velden CS, Anderson JL (2014) Influence of assimilating satellite-derived atmospheric motion vector observations on numerical analyses and forecasts of tropical cyclone track and intensity. *Mon Wea Rev* 142:49–71.
Wu T-C, Županski M, Grasso LD, et al. (2016) The GSI capability to assimilate TRMM and GPM hydrometeor retrievals in HWRF. *Quart J Roy Meteor Soc* 142:2768–2787.
Wu W-S, Purser RJ, Parrish DF (2002) Three-dimensional variational analysis with spatially inhomogeneous covariances. *Mon Wea Rev* 130:2905–2916.
Xiong X, Navon IM, Uzunoglu B (2006) A note on the particle filter with posterior Gaussian resampling. *Tellus A* 58:456–460.
Zhu Y, Liu E, Mahajan R, et al. (2016) All-sky microwave radiance assimilation in NCEP's GSI analysis system. *Mon Wea Rev* 144:4709–4735.
Zou X, Kuo Y-H, Guo Y-R (1995) Assimilation of atmospheric radio refractivity using a nonhydrostatic adjoint model. *Mon Wea Rev* 123:2229–2250.
Zou X, Wang B, Liu H, et al. (2000) Use of GPS/MET refraction angles in three-dimensional variational analysis. *Quart J Roy Meteor Soc* 126:3013–3040.
Županski D, Mesinger F (1995) Four-dimensional variational assimilation of precipitation data. *Mon Wea Rev* 123:1112–1127.
Županski D, Zhang SQ, Županski M, Hou AY, Cheung SH (2011a) A prototype WRF-based ensemble data assimilation system for dynamically downscaling satellite precipitation observations. *J Hydrometeor* 12:118–134.
Županski D, Županski M, Grasso LD, et al. (2011b) Assimilating synthetic GOES-R radiances in cloudy conditions using an ensemble-based method. *Int J Remote Sens* 32:9637–9659.
Županski M (1996) A preconditioning algorithm for four-dimensional variational data assimilation. *Mon Wea Rev* 124:2562–2573.
Županski M (2005) Maximum likelihood ensemble filter: Theoretical aspects. *Mon Wea Rev* 133:1710–1726.
Županski M (2013) All-sky satellite radiance data assimilation: Methodology and challenges. In *Data Assimilation for Atmospheric, Oceanic and Hydrologic Applications* (vol. II), (eds.) Park SK, Xu L, Springer-Verlag, Berlin, Heidelberg, 465–488.
Županski M, Navon IM, Županski D (2008) The Maximum Likelihood Ensemble Filter as a non-differentiable minimization algorithm. *Quart J Roy Meteor Soc* 134:1039–1050.

# Part V

Appendices

# Appendix A

## Linear Algebra and Functional Analysis

### A.1 Introduction

The mathematical foundation of data assimilation includes linear algebra and functional analysis, probability and statistics, as well as some additional topics in calculus. In this chapter we will present a basic overview of the mathematical foundation of data assimilation, with emphasis on practical relevance for data assimilation. We assume that the reader is familiar with elementary algebra, as well as with vectors and matrices. We also implicitly assume that objects, such as vectors and matrices, are real.

We begin by outlining basic mathematical ingredients of data assimilation.

### A.2 Vector Space and Functional Analysis

Vector space, also referred to as linear space, is a field, defined using a set of vectors, which satisfies the addition and scalar multiplication rules. Given a set of vectors $U = (\mathbf{u}, \mathbf{v}, \mathbf{w})$ defined over a field $F$, we assume that the following rules are satisfied:

1. Addition is commutative

$$\mathbf{u} + \mathbf{v} = \mathbf{v} + \mathbf{u}. \tag{A.1}$$

2. Addition is associative

$$\mathbf{u} + (\mathbf{v} + \mathbf{w}) = (\mathbf{u} + \mathbf{v}) + \mathbf{w}. \tag{A.2}$$

3. Addition with zero vector

$$\mathbf{u} + \mathbf{0} = \mathbf{u}. \tag{A.3}$$

4. Addition of negative number

$$\mathbf{u} + (-\mathbf{u}) = \mathbf{0}. \tag{A.4}$$

5. Multiplication by a real scalar $\alpha$

$$\alpha(\mathbf{u} + \mathbf{v}) = \alpha\mathbf{u} + \alpha\mathbf{v}. \tag{A.5}$$

The vector set $U$ defined over the field $F$ with rules (A.1)–(A.5) is called a vector field. It is a main subject of linear algebra, which also deals with linear mappings between vector spaces. A linear mapping, or linear transformation, is a mapping between two vector spaces $L: U \to V$ that preserves the structure of the vector spaces. Given vectors $\mathbf{u}, \mathbf{v} \in U$ and scalars $\alpha, \beta \in F$ the following set of rules defines a linear transformation $L$:

$$L(\mathbf{u}+\mathbf{v}) = L(\mathbf{u}) + L(\mathbf{v}), \tag{A.6}$$

$$L(\alpha\mathbf{u}) = \alpha L(\mathbf{u}), \text{ and} \tag{A.7}$$

$$L(\alpha\mathbf{u}+\beta\mathbf{v}) = L(\alpha\mathbf{u}) + L(\beta\mathbf{v}) = \alpha L(\mathbf{u}) + \beta L(\mathbf{v}). \tag{A.8}$$

Any physical field, such as temperature, pressure, aerosol, chemical constituents, satellite radiance, is mathematically represented as a vector. A set of these vectors forms a vector space. [Example 1.1 gives further clarification.]

Another important mathematical component of data assimilation is functional analysis. Functional analysis studies vector spaces with additional properties related to norms and geometry, mathematically referred to as topology of vector spaces.

Among the most important special properties of vector spaces in relation to data assimilation are norm, distance, and orthogonality. They are intrinsically connected with the notion of the inner product. Given a real vector space $U$, the inner product is a function $z: U \times U \to F$ satisfying

$$\begin{aligned} &1.\ z(\alpha\mathbf{u}+\beta\mathbf{v}, \mathbf{w}) = \alpha z(\mathbf{u}, \mathbf{w}) + \beta z(\mathbf{v}, \mathbf{w}) \\ &2.\ z(\mathbf{u}, \mathbf{u}) \geq 0 \\ &3.\ \text{If } z(\mathbf{u}, \mathbf{u}) = 0 \text{ then } \mathbf{u} = \mathbf{0}. \end{aligned} \tag{A.9}$$

The function name is typically not used and the inner product between vectors $\mathbf{u}$ and $\mathbf{v}$ is denoted $\langle \mathbf{u}, \mathbf{v} \rangle$. Implicitly, the inner product defines a component-wise product between two vectors. Given two vectors $\mathbf{u} = (u_1, \ldots, u_M)$ and $\mathbf{v} = (v_1, \ldots, v_M)$ their inner product is

$$\langle \mathbf{u}, \mathbf{v} \rangle = \mathbf{u}^T v = u_1 v_1 + \cdots + u_M v_M = \sum_{i=1}^{M} u_i v_i, \tag{A.10}$$

where the superscript $T$ denotes the transpose and $M$ is the dimension of vectors $\mathbf{u}$ and $\mathbf{v}$. From (A.10) it is clear that inner product is a scalar.

The inner product can be related to a norm of a vector, defined as

$$\|\mathbf{u}\| = \langle \mathbf{u}, \mathbf{u} \rangle^{1/2}. \tag{A.11}$$

A norm most commonly used in data assimilation is a so-called $L^p$ norm

$$\|\mathbf{u}\| = \left( \sum_{i=1}^{M} |u_i|^p \right)^{1/p}, \tag{A.12}$$

where line brackets define absolute value and $p \geq 1$. Among the norms in this family, three norms are most often calculated

$$\begin{aligned} &1.\ \text{Absolute value norm: } \|\mathbf{u}\|_{p=1} = |u_1| + |u_2| + \cdots + |u_M|, \\ &2.\ \text{Quadratic norm: } \|\mathbf{u}\|_{p=2} = \sqrt{u_1^2 + u_2^2 + \cdots + u_M^2}, \text{ and} \\ &3.\ \text{Maximum (infinity) norm: } \|\mathbf{u}\|_{p=\infty} = \max\left(|u_1|, |u_2|, \ldots, |u_M|\right). \end{aligned} \tag{A.13}$$

From (A.9b) it follows that the norm cannot be negative. Therefore, one can use the norm to define a distance between vectors $u$ and $v$

$$d(\mathbf{u}, \mathbf{v}) = \|\mathbf{u} - \mathbf{v}\|. \tag{A.14}$$

With definition (A.14) one can now quantify a distance between two vectors. [See Example 1.3.]

The inner product can be also used to define orthogonality. Two vectors are orthogonal if their inner product is 0

$$\langle \mathbf{u}, \mathbf{v} \rangle = 0 \quad \Leftrightarrow \quad \mathbf{u} \text{ is orthogonal to } \mathbf{v}. \tag{A.15}$$

[Example A.4 gives further clarification of orthogonality.]

The inner product, norm, and distance introduce a metric of a vector space, and thus allow comparison between vectors and characterization of vector length or magnitude. They also allow geometric interpretation of a relationship between vectors. For example, one can define the cosine of an angle between two vectors

$$\cos(\phi) = \frac{\mathbf{u}^T \mathbf{v}}{\|\mathbf{u}\| \cdot \|\mathbf{v}\|}. \tag{A.16}$$

Vector spaces with a defined norm metric are called normed vector spaces and are generally the mathematical spaces used in data assimilation. Most common spaces are finite-dimensional Euclidian spaces, and their generalization to infinite dimensions, the Hilbert space. These spaces have their norm defined through the inner product, as described above. Another related mathematical space, the Banach Space, is also a normed space but the norm is not defined through the inner product.

In data assimilation, and in this book, we assume that all vector spaces are real, with the norm defined through the inner product. Therefore, the considered vector space is typically Euclidian, with implied extensions to Hilbert space.

Let $\chi$ be a real space of dimension $N$ and let vectors $\{\mathbf{c}_i \in \chi : i = 1, N\}$ belong to that space. Vectors $\mathbf{c}_i$ are *linearly independent* if for any scalar $\{\alpha_i \neq 0 : i = 1, N\}$

$$\alpha_1 \mathbf{c}_1 + \cdots + \alpha_N \mathbf{c}_N \neq 0. \tag{A.17}$$

If the sum (A.17) is equal to 0, vectors $\mathbf{c}_i$ are *linearly dependent*. A minimal set of linearly independent vectors that spans a space is called the *basis*. Space $\chi \in \mathrm{R}^N$ has an $N$-dimensional set of basis vectors.

Since basis vectors span a space, any vector from that space can be represented as a linear combination of basis vectors. Therefore, given basis vectors $(\mathbf{e}_1, \ldots, \mathbf{e}_N) \in \chi$ and vector $\mathbf{b} \in \chi$, there are scalars $\{\gamma_i : i = 1, N\}$ such that

$$\mathbf{b} = \gamma_1 \mathbf{e}_1 + \cdots + \gamma_N \mathbf{e}_N. \tag{A.18}$$

Consequently, if a vector can be represented by a linear combination of basis vectors of space $\chi$, then this vector belongs to that space.

In principle, basis vectors can have any angle between them. When the angle between basis vectors is always a right angle, the basis vectors are mutually orthogonal. Such a basis is called an *orthogonal* basis. An orthogonal basis with unit norm basis vectors, i.e.,

$\{\|\mathbf{e}_i\| = 1\colon\ i = 1, N\}$, is called the *orthonormal* basis. If vectors $(\mathbf{e}_1, \ldots, \mathbf{e}_N) \in \chi$ form an orthonormal basis, then

$$\langle \mathbf{e}_i, \mathbf{e}_j \rangle = \mathbf{e}_i^T \mathbf{e}_j = \delta_{ij} = \begin{cases} 1 & \text{for} \quad i = j \\ 0 & \text{for} \quad i \neq j, \end{cases} \tag{A.19}$$

where $\delta$ is a Dirac delta function. The most known orthonormal basis is the standard basis, consisting of unit vectors along the directions of a Cartesian coordinate system. Another well-known basis can be formed by trigonometric functions, used in various forms for Fourier expansion or for JPEG compression, for example. Important to understand is that different basis sets can be used to represent the same space. Note that it is possible to have multiple sets of basis vectors that describe the same space. In fact, there are an infinite number of basis vector sets that span the same space. Examples of different basis sets are included in Example 1.5.

We now return to specification of vector components. Vector notation by components was used earlier without explaining what the components mean. The basis definition can provide a mechanism for such explanation. First, recall that an outer product of vectors creates a matrix, i.e., a linear transformation. Given basis vectors $(\mathbf{e}_1, \ldots, \mathbf{e}_N) \in \chi$ one can form a matrix $\mathbf{E} = \sum_{i=1}^{N} \mathbf{e}_i \mathbf{e}_i^T$. Note that matrix $\mathbf{E}$ is symmetric ($\mathbf{E} = \mathbf{E}^T$) and idempotent ($\mathbf{E}^2 = \mathbf{E}$). These two conditions imply that $\mathbf{E}$ is an orthogonal projection operator. Assuming $\mathbf{E} \neq \mathbf{0}$, the norm of this matrix is $\|\mathbf{E}\| = 1$ (see Example 1.6).

Further assuming that this basis is orthonormal, for an arbitrary $N$-dimensional vector $b$ we calculate

$$\mathbf{a} = \mathbf{E}\mathbf{b} = \sum_{i=1}^{N} \mathbf{e}_i \mathbf{e}_i^T \mathbf{b} = \sum_{i=1}^{N} \gamma_i \mathbf{e}_i = \gamma_1 \mathbf{e}_1 + \cdots + \gamma_N \mathbf{e}_N, \tag{A.20}$$

where $\gamma_i = \mathbf{e}_i^T \mathbf{b}$. Since vector $\mathbf{a}$ is represented as a linear combination of the basis vectors of $\chi$, then $\mathbf{a} \in \chi$. As we have already seen, both the standard basis and the trigonometric basis can describe the same 2D space. Let $(\mathbf{g}_1, \ldots, \mathbf{g}_N) \in \chi$ be another orthonormal basis of space $\chi$. One can define matrix $\mathbf{G} = \sum_{i=1}^{N} \mathbf{g}_i \mathbf{g}_i^T$ and calculate

$$\mathbf{c} = \mathbf{G}\mathbf{b} = \sum_{i=1}^{N} \mathbf{g}_i \mathbf{g}_i^T \mathbf{b} = \sum_{i=1}^{N} \mu_i \mathbf{g}_i = \mu_1 \mathbf{g}_1 + \cdots + \mu_N \mathbf{g}_N, \tag{A.21}$$

where $\mu_i = \mathbf{g}_i^T \mathbf{b}$. Following the same argument as for vector $\mathbf{a}$, since vector $\mathbf{c}$ is represented as a linear combination of the basis vectors of $\chi$, then $\mathbf{c} \in \chi$. Therefore, vector $\mathbf{b} \in \chi$ can have two representations, $\mathbf{a}$ or $\mathbf{c}$, depending on the basis set.

We have already introduced matrices $\mathbf{E}$ and $\mathbf{G}$. In relation to the linear mapping defined by (A.6)–(A.8), matrix is a linear mapping relative to a specific basis set. There are several groups of matrices and matrix operations that are commonly used in data assimilation: symmetric, diagonal, and identity matrices with operations such as transpose and inversion. A transpose of matrix $\mathbf{A}$ is a matrix denoted $\mathbf{A}^T$ that reflects its elements over the main diagonal, i.e., $[\mathbf{A}^T]_{ij} = [\mathbf{A}]_{ji}$. Given a matrix

$$\mathbf{A} = \begin{pmatrix} a_{11} & a_{12} & \cdots & a_{1N} \\ a_{21} & a_{22} & & \\ \vdots & & \ddots & \vdots \\ a_{N1} & & \cdots & a_{NN} \end{pmatrix}, \tag{A.22}$$

its *transpose* is

$$\mathbf{A}^T = \begin{pmatrix} a_{11} & a_{21} & \cdots & a_{N1} \\ a_{12} & a_{22} & & \\ \vdots & & \ddots & \vdots \\ a_{1N} & & \cdots & a_{NN} \end{pmatrix}. \tag{A.23}$$

An important property of a transpose in Euclidian and Hilbert spaces is that the *transpose* of a transpose is equal to the original matrix, i.e., $\left(\mathbf{A}^T\right)^T = \mathbf{A}$. If matrix $\mathbf{A}$ is invertible, a similar property is also valid, i.e., $\left(\mathbf{A}^{-1}\right)^{-1} = \mathbf{A}$.

A *symmetric* matrix is a matrix that is equal to its transpose ($\mathbf{A} = \mathbf{A}^T$), which implies $a_{ij} = a_{ji}$. Therefore, a symmetric matrix $\mathbf{S}$, derived from $\mathbf{A}$, is

$$\mathbf{S} = \begin{pmatrix} a_{11} & a_{12} & \cdots & a_{1N} \\ a_{12} & a_{22} & & \\ \vdots & & \ddots & \vdots \\ a_{1N} & & \cdots & a_{NN} \end{pmatrix}. \tag{A.24}$$

A matrix is *orthogonal* if all its column vectors are mutually orthogonal. If additionally the norm of column vectors is 1, then the matrix is *orthonormal*. If matrix $\mathbf{Q}$ with columns $\{\mathbf{q}_i : i = 1, N\}$, written in *column notation* as $\mathbf{Q} = (\mathbf{q}_1, \ldots, \mathbf{q}_N)$, is orthonormal, then it satisfies the property $\mathbf{q}_i^T \mathbf{q}_j = \delta_{ij}$. An orthonormal matrix also has its transpose equal to its inverse, i.e., $\mathbf{Q}^T = \mathbf{Q}^{-1}$ (see Example A.5).

A *diagonal* matrix is defined as matrix with elements defined on its main diagonal only, and all other elements equal to 0. A diagonal matrix $\mathbf{D}$, derived from $\mathbf{A}$, is

$$\mathbf{D} = \begin{pmatrix} a_{11} & 0 & \cdots & 0 \\ 0 & a_{22} & & \\ \vdots & & \ddots & \vdots \\ 0 & & \cdots & a_{NN} \end{pmatrix}. \tag{A.25}$$

An *identity* matrix $\mathbf{I}$ is a diagonal matrix with all diagonal elements equal to 1, i.e.,

$$\mathbf{I} = \begin{pmatrix} 1 & 0 & \cdots & 0 \\ 0 & 1 & & \\ \vdots & & \ddots & \vdots \\ 0 & & \cdots & 1 \end{pmatrix}. \tag{A.26}$$

Given matrices $\mathbf{A}$, $\mathbf{B}$, and $\mathbf{C}$, the *transpose of their product* is

$$(\mathbf{ABC})^T = \mathbf{C}^T \mathbf{B}^T \mathbf{A}^T, \tag{A.27}$$

i.e., the order of matrices in the product is reversed. Assuming these matrices are invertible, the *inverse of the product* is

$$(\mathbf{ABC})^{-1} = \mathbf{C}^{-1}\mathbf{B}^{-1}\mathbf{A}^{-1}. \tag{A.28}$$

As for the transpose, the order of matrices in the product is reversed.

Another important matrix used in data assimilation is a *symmetric positive (semi)definite* matrix. Given an arbitrary vector $\mathbf{x}$, a symmetric matrix $\mathbf{P}$ is positive semidefinite if $\mathbf{x}^T\mathbf{P}\mathbf{x} \geq 0$ and strictly positive definite if $\mathbf{x}^T\mathbf{P}\mathbf{x} > 0$. The error covariance matrix is one important example of a symmetric positive semidefinite matrix.

A symmetric matrix can be represented as an *inner product* of matrices ($\mathbf{P} = \mathbf{X}^T\mathbf{X}$), or as an *outer product* of matrices ($\mathbf{P} = \mathbf{Y}\mathbf{Y}^T$), where $\mathbf{X}, \mathbf{Y}$ are matrices. This can be easily demonstrated using (A.27). Note that $\left(\mathbf{X}^T\mathbf{X}\right)^T = (\mathbf{X})^T\left(\mathbf{X}^T\right)^T = \mathbf{X}^T\mathbf{X}$, which implies $\mathbf{P}^T = \mathbf{P}$. Similarly $\left(\mathbf{Y}\mathbf{Y}^T\right)^T = \left(\mathbf{Y}^T\right)^T(\mathbf{Y})^T = \mathbf{Y}\mathbf{Y}^T$.

When a symmetric matrix $\mathbf{P}$ is represented as an outer product of matrices, $\mathbf{P} = \mathbf{Y}\mathbf{Y}^T$, then matrix $\mathbf{Y}$ is a *square root* of $\mathbf{P}$. Note that the matrix square root is not unique since there is an infinite number of square root matrices. For example, given an orthonormal matrix $\mathbf{Q}$, matrix $\mathbf{Y}\mathbf{Q}^T$ is also a square root of $\mathbf{P}$ since $(\mathbf{Y}\mathbf{Q}^T)(\mathbf{Y}\mathbf{Q}^T)^T = \mathbf{Y}\mathbf{Q}^T\left(\mathbf{Q}^T\right)^T\mathbf{Y}^T = \mathbf{Y}\mathbf{Q}^T\mathbf{Q}\mathbf{Y}^T = \mathbf{Y}\mathbf{Y}^T = \mathbf{P}$. This could be extended to an arbitrary number of matrices $\mathbf{Q}_i$. However, there is a unique *symmetric* square root matrix. The square root in the form of a lower (upper) *triangular matrix* that is calculated from a Cholesky factorization of $\mathbf{P}$ is also unique.

Among all possible basis sets, there is a unique basis set in which some matrices can be expressed as diagonal matrices. This process is called *diagonalization* and has an important role in data assimilation. Singular value decomposition (SVD) represents matrix $A$ as a product of three matrices, $\mathbf{U}$, $\mathbf{V}$, and $\mathbf{\Sigma}$:

$$\mathbf{A} = \mathbf{U}\mathbf{\Sigma}\mathbf{V}^T, \tag{A.29}$$

where $\mathbf{U} = (\mathbf{u}_1, \ldots, \mathbf{u}_N)$ and $\mathbf{V} = (\mathbf{v}_1, \ldots, \mathbf{v}_N)$ are orthonormal matrices with column vectors $\{\mathbf{u}_i, \mathbf{v}_i : i = 1, N\}$, and $\mathbf{\Sigma}$ is a diagonal matrix with elements $\{\sigma_i : i = 1, N\}$. The column vectors are referred to as the *singular vectors* and the elements of the matrix $\Sigma$ as *singular values*. The matrices $\mathbf{U}$ and $\mathbf{V}$ are also referred to as the *left* and *right* singular vector matrices, respectively. The SVD of matrix $\mathbf{A}$ implies a procedure for finding a specific basis formed by singular vectors that can transform the matrix $\mathbf{A}$ into a diagonal matrix, denoted $\Sigma$. Using the orthogonality of $\mathbf{U}$ and $\mathbf{V}$, the matrix $\mathbf{A}$ can also be written as

$$\mathbf{A} = \sum_{i=1}^{N} \sigma_i \mathbf{u}_i \mathbf{v}_i^T. \tag{A.30}$$

If the singular values are ordered from largest to smallest ($\sigma_1 \geq \sigma_2 \geq \cdots \geq \sigma_N$), then one can create a *truncated SVD* by keeping only the leading (i.e., the largest) singular values and vectors. For $M < N$ we can write

$$\mathbf{A} \approx \mathbf{C} = \sum_{i=1}^{M} \sigma_i u_i v_i^T, \tag{A.31}$$

where **C** is a truncated SVD approximation to **A**. The error of that approximation in $L^2$ norm can be easily quantified

$$\|\mathbf{A} - \mathbf{C}\| = \sigma_{M+1}, \tag{A.32}$$

i.e., it is equal to the largest singular value not included in the truncation. Closely related to SVD is eigenvalue decomposition (EVD) of a real symmetric matrix **B**:

$$\mathbf{B} = \mathbf{W}\mathbf{\Lambda}\mathbf{W}^T, \tag{A.33}$$

where $\mathbf{W} = (\mathbf{w}_1, \ldots, \mathbf{w}_N)$ is an orthonormal matrix, with column vectors $\{\mathbf{w}_i : i = 1, N\}$ referred to as *eigenvectors* and $\mathbf{\Lambda}$ is a diagonal matrix with elements $\{\lambda_i : i = 1, N\}$ referred to as *eigenvalues*. The EVD is also called spectral decomposition. Both SVD and EVD have numerous applications and important theoretical consequences. For example, let $\mathbf{B} = \mathbf{A}\mathbf{A}^T$. Using the SVD of **A**, we can write **B** as

$$\mathbf{B} = (\mathbf{U}\mathbf{\Sigma}\mathbf{V}^T)(\mathbf{U}\mathbf{\Sigma}\mathbf{V}^T)^T = \mathbf{U}\mathbf{\Sigma}\mathbf{V}^T\mathbf{V}\mathbf{\Sigma}\mathbf{U}^T = \mathbf{U}\mathbf{\Sigma}^2\mathbf{U}^T = \sum_{i=1}^{N} \sigma_i^2 \mathbf{u}_i \mathbf{u}_i^T. \tag{A.34}$$

Using the orthogonality of **W**, the matrix **B** can also be written as

$$\mathbf{B} = \sum_{i=1}^{N} \lambda_i \mathbf{w}_i \mathbf{w}_i^T. \tag{A.35}$$

If eigenvalues are ordered from the largest to the smallest ($\lambda_1 \geq \lambda_2 \geq \cdots \geq \lambda_N$), then one can create a *truncated EVD*, as was done for SVD, by keeping only the leading (i.e., the largest) eigenvalues and vectors. For $M < N$ we can approximate **B** by

$$\mathbf{B} \approx \sum_{i=1}^{M} \lambda_i \mathbf{w}_i \mathbf{w}_i^T. \tag{A.36}$$

By comparing (A.34) with (A.35), we see that the SVD and EVD are identical for $\mathbf{u}_i = \mathbf{w}_i$ and $\lambda_i = \sigma_i^2$. Therefore, eigenvectors of a real symmetric matrix are equal to left singular vectors of its square root and eigenvalues are equal to squared singular values. One of the consequences of this relationship is that real symmetric matrices are nonnegative, as for an arbitrary vector **b**

$$\mathbf{b}^T\mathbf{B}\mathbf{b} = \mathbf{b}^T\left(\sum_{i=1}^{N} \sigma_i^2 \mathbf{u}_i \mathbf{u}_i^T\right) b = \sum_{i=1}^{N} \sigma_i^2 (\mathbf{u}_i^T \mathbf{b})^2 \geq 0. \tag{A.37}$$

Let **B** denote a covariance matrix and **A** its square root. For an arbitrary vector **b**,

$$\mathbf{a} = \mathbf{B}\mathbf{b} = \sum_{i=1}^{N} \lambda_i \mathbf{w}_i \mathbf{w}_i^T \mathbf{b} = \sum_{i=1}^{N} \sigma_i^2 \mathbf{u}_i \mathbf{u}_i^T \mathbf{b} = \sum_{i=1}^{N} \theta_i \mathbf{u}_i$$

$$\mathbf{c} = \mathbf{A}\mathbf{b} = \sum_{i=1}^{N} \sigma_i \mathbf{u}_i \mathbf{v}_i^T \mathbf{b} = \sum_{i=1}^{N} \tau_i \mathbf{u}_i,$$

where $\theta_i = \sigma_i^2 \mathbf{u}_i^T \mathbf{b}$ and $\tau_i = \sigma_i \mathbf{v}_i^T \mathbf{b}$. Since both vectors are a linear combination of left singular vectors/eigenvectors of **A/B**, they belong to the space of the covariance matrix. Therefore, using a square root covariance matrix or a full covariance matrix product with a vector will produce vectors in the same space.

There are several matrix equalities that have a wide array of applications in data assimilation. Let **P** and **R** be invertible real symmetric positive definite matrices with dimensions $N \times N$ and $M \times M$, respectively. Let **H** be an $M \times N$ real matrix. Then the following equalities hold:

$$\begin{aligned}
&1.\ (\mathbf{P}^{-1} + \mathbf{H}^T\mathbf{R}^{-1}\mathbf{H})^{-1} = \mathbf{P} - \mathbf{P}\mathbf{H}^T(\mathbf{H}\mathbf{P}\mathbf{H}^T + \mathbf{R})^{-1}\mathbf{H}\mathbf{P}\\
&2.\ (\mathbf{P}^{-1} + \mathbf{H}^T\mathbf{R}^{-1}\mathbf{H})^{-1} = \mathbf{P}^{1/2}(\mathbf{I} + \mathbf{P}^{T/2}\mathbf{H}^T\mathbf{R}^{-1}\mathbf{H}\mathbf{P}^{1/2})^{-1}\mathbf{P}^{T/2}\\
&3.\ (\mathbf{P}^{-1} + \mathbf{H}^T\mathbf{R}^{-1}\mathbf{H})^{-1} = (\mathbf{I} + \mathbf{P}\mathbf{H}^T\mathbf{R}^{-1}\mathbf{H})^{-1}\mathbf{P}\\
&4.\ (\mathbf{P}^{-1} + \mathbf{H}^T\mathbf{R}^{-1}\mathbf{H})^{-1}\mathbf{H}^T\mathbf{R}^{-1} = \mathbf{P}\mathbf{H}^T(\mathbf{H}\mathbf{P}\mathbf{H}^T + \mathbf{R})^{-1}\\
&5.\ (\mathbf{I} + \mathbf{P}\mathbf{H}^T\mathbf{R}^{-1}\mathbf{H})^{-1}\mathbf{P}\mathbf{H}^T\mathbf{R}^{-1} = \mathbf{P}\mathbf{H}^T(\mathbf{H}\mathbf{P}\mathbf{H}^T + \mathbf{R})^{-1}.
\end{aligned} \tag{A.38}$$

## A.3 Covariance and Correlation Matrices

We have already introduced the covariance matrix as a symmetric, positive semi-definite real matrix. Forecast error covariance has a fundamental role in data assimilation, as discussed in other chapters. Here we will formally derive the relationship between covariance and correlation matrices that is useful in data assimilation and elsewhere.

Correlation is a nondimensional matrix, while covariance is a matrix with physical dimensions. In general, correlation represents dependence between variables, and can be written for $N$-dimensional state as

$$\mathbf{C} = \begin{pmatrix} 1 & \rho_{12} & \cdots & \rho_{1N} \\ \rho_{12} & 1 & & \rho_{2N} \\ \vdots & & \ddots & \vdots \\ \rho_{1N} & \rho_{2N} & \cdots & 1 \end{pmatrix}, \tag{A.39}$$

where $\rho_{ij}$ is a correlation coefficient between variables at points $i$ and $j$, and the correlation of a variable with itself at the same point ($i = j$) is equal to 1. The covariance matrix, on the other hand, can be written as

$$\mathbf{P} = \begin{pmatrix} p_{11} & p_{12} & \cdots & p_{1N} \\ p_{12} & p_{22} & & p_{2N} \\ \vdots & & \ddots & \vdots \\ p_{1N} & p_{2N} & \cdots & p_{NN} \end{pmatrix}, \tag{A.40}$$

where the elements $p_{ij}$ include physical dimensions.

For clarity, consider a 2D problem defined at a single grid point with two variables. Let the first variable be surface pressure, denoted by the index 1, and the second variable is specific humidity, denoted by the index 2. A covariance between two variables is

$$\mathbf{P} = \begin{pmatrix} \sigma_1^2 & \rho_{12}\sigma_1\sigma_2 \\ \rho_{12}\sigma_1\sigma_2 & \sigma_1^2 \end{pmatrix}, \tag{A.41}$$

where $\sigma_1$ and $\sigma_2$ are standard deviations of variables 1 and 2, respectively. Related to (A.40) one has $p_{ii} = \sigma_i^2$. Important to note is that standard deviations include physical dimensions, while correlation is nondimensional. For example, let the standard deviation of surface pressure error be 1 hPa = 100 Pa ($\sigma_1 = 100$), while the standard deviation of specific humidity be 0.001 $\text{kg kg}^{-1}$ ($\sigma_2 = 10^{-3}$). Also, let the correlation between pressure and specific humidity be 0.5 ($\rho_{12} = 0.5$). Substituting this into (A.41) gives covariance with elements

$$\mathbf{P} = \begin{pmatrix} 10000 & 5 \\ 5 & 0.000001 \end{pmatrix}. \tag{A.42}$$

This matrix has 10 orders of magnitude of difference of its diagonal elements and is difficult to interpret in terms of dependence between variables. This is why it is of interest to transform the covariance matrix using the correlation matrix.

Therefore, let us decompose matrix (A.41) into two components: with physical units (standard deviation matrix) and without physical units

$$\mathbf{P} = \begin{pmatrix} \sigma_1 & 0 \\ 0 & \sigma_2 \end{pmatrix} \begin{pmatrix} 1 & \rho_{12} \\ \rho_{12} & 1 \end{pmatrix} \begin{pmatrix} \sigma_1 & 0 \\ 0 & \sigma_2 \end{pmatrix}. \tag{A.43}$$

Now the correlation matrix in the middle is nondimensional, scaled, and reveals the structure of covariance. The outside diagonal matrices are with physical dimensions, and thus can widely vary in magnitude. The relationship (A.43) can be generalized to an $N$-dimensional problem

$$\mathbf{P} = \mathbf{SCS}, \tag{A.44}$$

where $\mathbf{C}$ is the correlation matrix given by (A.39) and

$$\mathbf{S} = \begin{pmatrix} \sigma_1 & \cdots & & 0 \\ \vdots & \sigma_2 & & \\ & & \ddots & \vdots \\ 0 & & \cdots & \sigma_N \end{pmatrix} \tag{A.45}$$

is a diagonal matrix with standard deviation as elements. Formula (A.44) has a wide use in data assimilation.

## A.4 State Vector, Control Vector, and Uncertainty

One of the main goals of data assimilation is to find an optimal analysis that is often understood as the initial conditions of the forecast model. The analysis is then defined as a vector that includes all model variables that require initial conditions. Such a vector is generally referred to as a state vector or state variable. Consider a 3D atmospheric grid-point forecast model with $N_x$, $N_y$, $N_z$ grid points in the $x$, $y$, and $z$ directions, respectively.

In general, the initial conditions of temperature, pressure, winds, humidity, and cloud hydrometeor species are required. Let us assume there are $N_{ic}$ such variables. Then, the state (analysis) vector will include $N_{ic}$ variables at $N_x \cdot N_y \cdot N_z$ grid points, i.e., a total of $N_{state} = N_x \cdot N_y \cdot N_z \cdot N_{ic}$ elements. In practice, however, one may want to use data assimilation to adjust only a few variables. For example, one can neglect the impact of cloud hydrometeor initial conditions. The number of variables becomes $N_{var}$, where $N_{var} \leq N_{ic}$. The dimension of such state vector is now $N_S = N_x \cdot N_y \cdot N_z \cdot N_{var}$.

Although the ultimate goal of the analysis is to produce the initial conditions at model grid points, for practical reasons the analysis space can be defined in a different grid and for different variables that are related to state variables. Such variables that are possibly defined at a grid that is different from the model grid and/or that are functionally related to state vectors, are referred to as the control variables. We will have a much better understanding of control variables throughout the book, but for now consider east–west and north–south components of a horizontal wind. Since wind is a model variable, it is part of a state vector. However, for practical reasons one may want to define stream-function $\psi$ and velocity potential $\vartheta$ that are related to horizontal wind as

$$u = -\frac{\partial \psi}{\partial y} + \frac{\partial \vartheta}{\partial x} \qquad v = \frac{\partial \psi}{\partial x} + \frac{\partial \vartheta}{\partial y}$$

and adjust those variables in data assimilation. In that case, $(u, v)$ defines the components of a state vector and $(\psi, \vartheta)$ defines the components of a control vector. In some instances, however, state and control vectors can be the same.

It is also important to realize that state and control vectors are not necessarily restricted to initial conditions. One may also want to adjust the empirical parameters of the forecast model, or lateral boundary conditions, or error of model equations. The initial conditions of state and control vectors can therefore be augmented to include some or all of the mentioned additional variables.

In addition to estimating optimal values, it is of interest to estimate their uncertainty, and ultimately the complete probability density function (PDF). Uncertainty can be understood as an error bar of a variable at each model grid point. The source of uncertainty is an error, or insufficient knowledge about a variable. The uncertainty at a certain point can change due to the errors at that point, but also due to errors at other points and other variables. Therefore, uncertainties can develop correlations if there is a dependency between points. In general, such defined uncertainty is a matrix. For example, if a state vector is 2D, i.e., the model has only two points, then state vector is $\mathbf{T} = (T_1, T_2)$. The uncertainty of $\mathbf{T}$, however, is a matrix

$$\mathbf{U}_T = \begin{pmatrix} U_{11} & U_{12} \\ U_{21} & U_{22} \end{pmatrix},$$

where $U_{11}$ denotes the uncertainty at point $T_1$ due to changes of $T_1$, $U_{12}$ represents the uncertainty at point $T_1$ due to changes of $T_2$, etc. Uncertainty is often related to error covariance, but also to entropy, all which are discussed in the later chapters.

## A.5 Overview and Examples

We have seen that data assimilation requires understanding several mathematical topics:

1. Normed spaces and orthogonality.
2. Basis of a vector space.
3. Linear transformations and matrices.
4. Linear algebra.
5. Functional analysis.
6. State vector.
7. Uncertainty.

These topics are of fundamental importance to data assimilation because the goal and tools of data assimilation cannot be defined without understanding the meaning of the underlying principles and the implied assumptions. Overall, the presented topics are used to build basic algebraic tools to address data assimilation.

**Example A.1 Vector representation and addition**

Consider, for example, a grid-point representation of temperature.

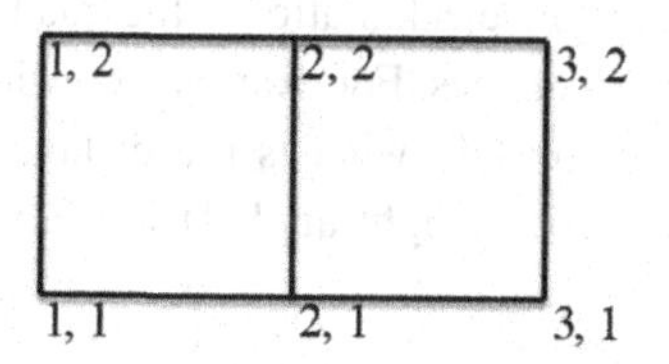


With the first index defining the east–west component and the second index defining the north–south component. The temperature on this grid, denoted **T**, can be written as a vector with components

$$\mathbf{T} = (T_{11}, T_{21}, T_{31}, T_{12}, T_{22}, T_{32})^T. \tag{A.46}$$

Although temperature is often referred to as *scalar* in meteorology, note that mathematically it represents a *vector*. If there is a perturbation of temperature, for example due to warming up during the day, one can also represent it on the same grid

$$\Delta\mathbf{T} = (\Delta T_{11}, \Delta T_{21}, \Delta T_{31}, \Delta T_{12}, \Delta T_{22}, \Delta T_{32})^T. \tag{A.47}$$

The resulting temperature after warming up can be obtained by applying the vector addition rule to obtain

$$\mathbf{T}+\Delta\mathbf{T} = (T_{11}+\Delta T_{11}, T_{21}+\Delta T_{21}, T_{31}+\Delta T_{31}, T_{12}+\Delta T_{12}, T_{22}+\Delta T_{22}, T_{32}+\Delta T_{32})^T. \tag{A.48}$$

Although adding (A.46) and (A.47) appears trivial from a practitioner's point of view, it does reflect the mathematical rules associated with vector addition.

**Example A.2 Distance**

Consider two vectors given by their coordinates $\mathbf{u} = (0,5)^T$ and $\mathbf{v} = (3,1)^T$.

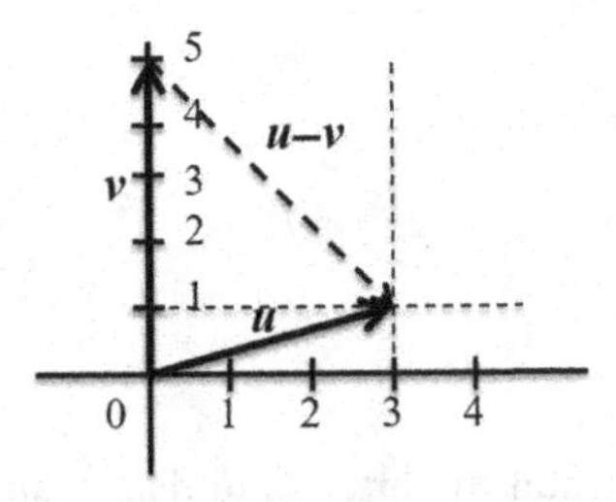


The distance between these two vectors is $d(\mathbf{u}, \mathbf{v}) = \|\mathbf{u} - \mathbf{v}\|$. One can calculate the distance for each of the three norms

1. Absolute value norm : $d(\mathbf{u}, \mathbf{v}) = |0-3| + |5-1| = 7$.
2. Quadratic norm : $d(\mathbf{u}, \mathbf{v}) = \sqrt{(0-3)^2 + (5-1)^2} = 5$.
3. Maximum (infinity) norm : $d(\mathbf{u}, \mathbf{v}) = \max(|0-3|, |5-1|) = 4$.

It is important to note that different norms give different distances. The most commonly used norm for a distance calculation is the quadratic norm and is generally a basis for our "intuition" of distances. For example, consider the following problem: given $\mathbf{u} = (3,0)^T$ and $\mathbf{v} = (0,4)^T$, what is the distance $d(\mathbf{u}, \mathbf{v})$? The coordinate representation of the problem gives a right-angled triangle with sides $u = 3$ and $v = 4$.

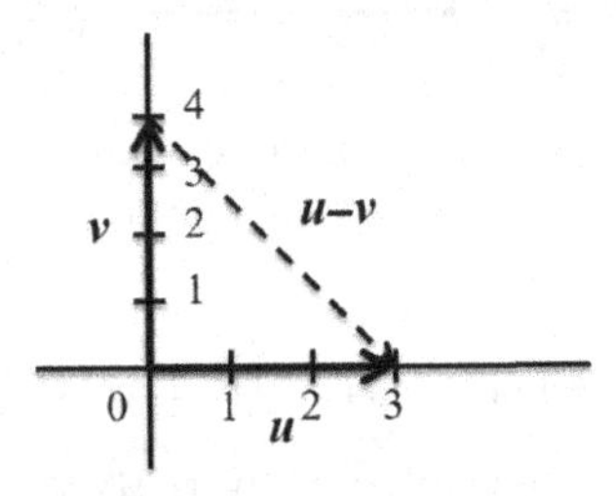


Therefore, we can also use Pythagoras' theorem, which states $u^2 + v^2 = z^2$. The unknown hypotenuse is $z = \sqrt{u^2 + v^2} = \sqrt{3^2 + 4^2} = 5$. Using the quadratic norm distance calculation we obtain $d(\mathbf{u}, \mathbf{v}) = \sqrt{(3-0)^2 + (0-4)^2} = 5$, i.e., the same result as from Pythagoras' theorem. This suggests that our common notion of distance is based on a specific (quadratic) norm, and thus is limited. This is important to know when interpreting the results of data assimilation.

**Example A.3 Basis sets**

In a 2D Cartesian system, for example, the standard basis form vectors $\mathbf{e}_1 = (1,0)^T$ and $\mathbf{e}_2 = (0,1)^T$. It is clear that the standard basis is orthonormal, since

$$\langle \mathbf{e}_1, \mathbf{e}_2 \rangle = \mathbf{e}_1^T \mathbf{e}_2 = 1 \cdot 0 + 0 \cdot 1 = 0$$
$$\|\mathbf{e}_1\| = \langle \mathbf{e}_1, \mathbf{e}_1 \rangle^{1/2} = \sqrt{\mathbf{e}_1^T \mathbf{e}_1} = \sqrt{1^2 + 0^2} = 1$$
$$\|\mathbf{e}_2\| = \langle \mathbf{e}_2, \mathbf{e}_2 \rangle^{1/2} = \sqrt{\mathbf{e}_2^T \mathbf{e}_2} = \sqrt{0^2 + 1^2} = 1.$$

Another well-known basis can be formed by trigonometric functions, for example, $\mathbf{e}_1 = (\sin\beta, \cos\beta)^T$ and $\mathbf{e}_2 = (-\cos\beta, \sin\beta)^T$, where $\beta$ is an arbitrary angle. This is also an orthonormal basis

$$\langle \mathbf{e}_1, \mathbf{e}_2 \rangle = \mathbf{e}_1^T \mathbf{e}_2 = -\sin\beta \cdot \cos\beta + \cos\beta \cdot \sin\beta = 0$$
$$\|\mathbf{e}_1\| = \langle \mathbf{e}_1, \mathbf{e}_1 \rangle^{1/2} = \sqrt{\mathbf{e}_1^T \mathbf{e}_1} = \sqrt{(-\sin\beta)^2 + (\cos\beta)^2} = 1$$
$$\|\mathbf{e}_2\| = \langle \mathbf{e}_2, \mathbf{e}_2 \rangle^{1/2} = \sqrt{\mathbf{e}_2^T \mathbf{e}_2} = \sqrt{(\cos\beta)^2 + (\sin\beta)^2} = 1.$$

**Example A.4 Orthogonal projection and its norm**

Let $\mathbf{E} = \sum_{i=1}^N \mathbf{e}_i \mathbf{e}_i^T$ and assume $\mathbf{E} \neq \mathbf{0}$. By taking a transpose

$$\mathbf{E}^T = \sum_{i=1}^N \left(\mathbf{e}_i \mathbf{e}_i^T\right)^T = \sum_{i=1}^N \mathbf{e}_i \mathbf{e}_i^T = \mathbf{E},$$

so the matrix is symmetric. By taking a square

$$\mathbf{E}^2 = \sum_{i=1}^N \left(\mathbf{e}_i \mathbf{e}_i^T\right)^T \sum_{j=1}^N \left(\mathbf{e}_j \mathbf{e}_j^T\right)^T = \sum_{i=1}^N \sum_{j=1}^N \mathbf{e}_i \mathbf{e}_i^T \mathbf{e}_j \mathbf{e}_j^T$$
$$= \sum_{i=1}^N \sum_{j=1}^N \mathbf{e}_i \delta_{ij} \mathbf{e}_j^T = \sum_{i=1}^N \mathbf{e}_i \mathbf{e}_i^T = \mathbf{E},$$

i.e., the matrix is also idempotent. Let us now compute the norm of such a matrix. For an arbitrary vector $\mathbf{b} \neq \mathbf{0}$,

$$\|\mathbf{Eb}\|^2 = \langle \mathbf{Eb}, \mathbf{Eb} \rangle = \left\langle \mathbf{b}, \mathbf{E}^T\mathbf{Eb} \right\rangle = \left\langle \mathbf{b}, \mathbf{E}^2\mathbf{b} \right\rangle = \langle \mathbf{b}, \mathbf{Eb} \rangle \leq \|\mathbf{b}\| \, \|\mathbf{Eb}\|,$$

where we employed $\mathbf{E} = \mathbf{E}^T$, $\mathbf{E}^2 = \mathbf{E}$, and the Cauchy–Bunyakowsky–Schwarz inequality for norms. After dividing by $\|\mathbf{Eb}\|$, we obtain

$$\frac{\|\mathbf{Eb}\|}{\|\mathbf{b}\|} \leq 1, \tag{A.49}$$

which implies that $\|\mathbf{E}\| \leq 1$. At the same time

$$\|\mathbf{Eb}\| = \left\|\mathbf{E}^2\mathbf{b}\right\| = \|\mathbf{E}(\mathbf{Eb})\| \leq \|\mathbf{E}\| \cdot \|\mathbf{Eb}\|.$$

After dividing by $\|\mathbf{Eb}\|$, we obtain

$$\|\mathbf{E}\| \geq 1. \tag{A.50}$$

From (A.49) and (A.50) it follows that $\|\mathbf{E}\| = 1$. Therefore, the norm of an orthogonal projection is equal to 1.

**Example A.5 Orthonormal matrix**

Let $\mathbf{Q} = \begin{pmatrix} \mathbf{q}_1 & \cdots & \mathbf{q}_N \end{pmatrix}$ be an orthonormal matrix, i.e., it satisfies

$$\mathbf{q}_i^T \mathbf{q}_j = \delta_{ij} = \begin{cases} 1 & \text{for} \quad i = j \\ 0 & \text{for} \quad i \neq j \end{cases}. \tag{A.51}$$

Then

$$\mathbf{Q}^T\mathbf{Q} = \begin{pmatrix} \mathbf{q}_1^T \\ \vdots \\ \mathbf{q}_N^T \end{pmatrix} \begin{pmatrix} \mathbf{q}_1 & \cdots & \mathbf{q}_N \end{pmatrix}$$

$$= \begin{pmatrix} \mathbf{q}_1^T\mathbf{q}_1 & \cdots & \mathbf{q}_1^T\mathbf{q}_N \\ \vdots & \ddots & \vdots \\ \mathbf{q}_N^T\mathbf{q}_1 & \cdots & \mathbf{q}_N^T\mathbf{q}_N \end{pmatrix} = \begin{pmatrix} 1 & \cdots & 0 \\ \vdots & \ddots & \vdots \\ 0 & \cdots & 1 \end{pmatrix} = \mathbf{I}, \tag{A.52}$$

where (A.51) was used. Therefore, by multiplying from the right (A.52) by $\mathbf{Q}^{-1}$ we obtain $\mathbf{Q}^T = \mathbf{Q}^{-1}$.

**Practice A.1 Vectors and inner product**

Plot vectors $\mathbf{u} = (3,1)^T$ and $\mathbf{v} = (2,4)^T$ in a Cartesian coordinate system and calculate their inner product.

**Practice A.2 Vectors and orthogonality**

Plot vectors $\mathbf{u} = (3,1)^T$ and $\mathbf{v} = (-1,3)^T$ in a Cartesian coordinate system and show that they are orthogonal.

# Appendix B

## Discretization of Partial Differential Equations

Discretization is a process to convert continuous fluid variables to discrete point values. This section describes some examples of discretizing partial differential equations (PDEs) into finite difference equations (FDEs), based on the Taylor series expansion.

### B.1 Grid-point System

Assume that a variable $u = u(x, y, t)$ is a function of horizontal space, $x$ (zonal) and $y$ (meridional), and time $t$. A grid point $x_i$ is located at a distance of $i\Delta x$ from the origin where $\Delta x$ is the grid size in the $x$-dimension (Figure B.1). Computational grid-point domains are represented in Figure B.2, where $\Delta x$ and $\Delta y$ are grid sizes and $\Delta t$ is a time step. The indices $i$, $j$, and $n$ represent locations of grid points in $x$-, $y$-, and $t$-directions, respectively. Then, a solution $u(x_i, y_j, t^n)$ is generally expressed in the grid-point system as:

$$u(x_i, y_j, t^n) = u\left(i\Delta x, j\Delta y, n\Delta t\right) = u^n_{i,j}.$$

### B.2 Finite Difference Methods

We want to convert a partial differential $\partial u/\partial x$ into a finite difference form in terms of grid-point variables (e.g., $u^n_i$, $u^n_{i-1}$, etc.). By definition, the derivative for $u(x, y, t)$ with respect to $x$ at $\left(x = x_i, y = y_j, t = t^n\right)$ is approximated as

$$\frac{\partial u}{\partial x} = \lim_{\Delta x \to 0} \frac{u(x_i + \Delta x, y_j, t^n) - u(x_i, y_j, t^n)}{\Delta x}$$

for a sufficiently small and finite $\Delta x$. Then, a small variation to $u(x_i, y_j, t^n)$ by $\pm\Delta x$, i.e., $u(x_i \pm \Delta x, y_j, t^n)$, is described by Taylor series expansion as:

$$u(x_i \pm \Delta x, y_j, t^n) = u(x_i, y_j, t^n) \pm \left.\frac{\partial u}{\partial x}\right|^n_{x_i} \Delta x + \underbrace{\frac{1}{2!}\left.\frac{\partial^2 u}{\partial x^2}\right|^n_{x_i} (\Delta x)^2 \pm \frac{1}{3!}\left.\frac{\partial^3 u}{\partial x^3}\right|^n_{x_i} (\Delta x)^3 + \cdots}_{\text{higher-oder terms (HOTs)}} \tag{B.1}$$

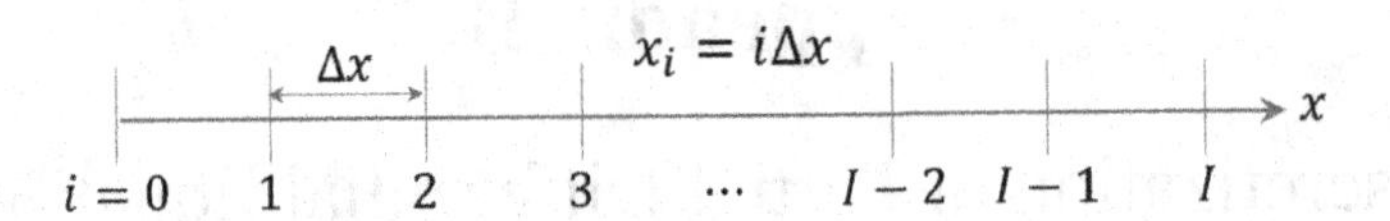


Figure B.1 Grid-point system for the $x$-dimension with the grid index $i = 0, \ldots, I$.

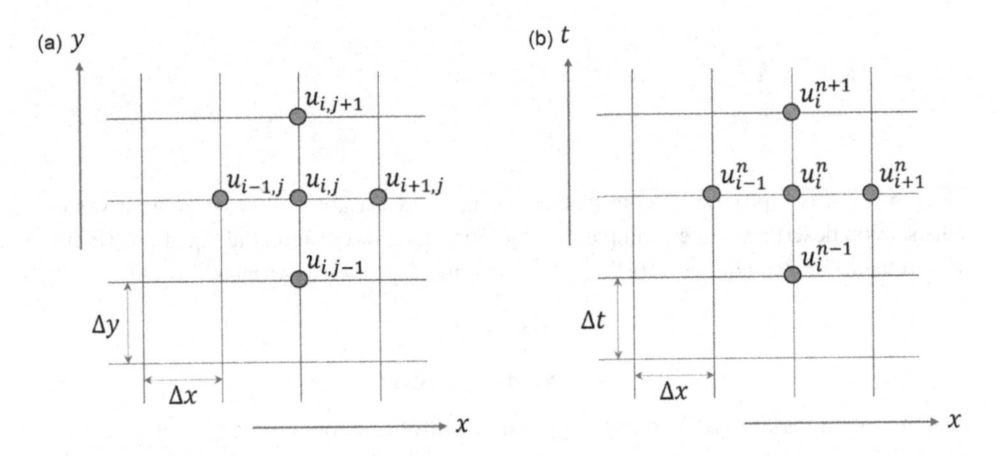


Figure B.2 Grid-point systems for (a) the $x-y$ domain and (b) the $x-t$ domain where the grid indices $i$, $j$, and $n$ are for $x, y$, and $t$, respectively.

### B.2.1 Forward Difference

From (B.1), by taking a positive variation, $\Delta x$, and by neglecting the HOTs:

$$\left.\frac{\partial u}{\partial x}\right|_{x_i}^n = \frac{u(x_i + \Delta x, y_j, t^n) - u(x_i, y_j, t^n)}{\Delta x} - \frac{\Delta x}{2}\left.\frac{\partial^2 u}{\partial x^2}\right|_{x_i}^n - \cdots.$$

Then, in terms of grid-point variables, we have

$$\frac{\partial u}{\partial x} = \frac{u_{i+1,j}^n - u_{i,j}^n}{\Delta x} + O\left(\Delta x\right), \tag{B.2}$$

where $O\left(\Delta x\right)$ is called the truncation error (TE).

### B.2.2 Backward Difference

From (B.1), by taking a negative variation, $-\Delta x$, and by neglecting the HOTs:

$$\left.\frac{\partial u}{\partial x}\right|_{x_i}^n = \frac{u(x_i, y_j, t^n) - u(x_i - \Delta x, y_j, t^n)}{\Delta x} + \frac{\Delta x}{2}\left.\frac{\partial^2 u}{\partial x^2}\right|_{x_i}^n - \cdots.$$

Then, in terms of grid-point variables, we have

$$\frac{\partial u}{\partial x} = \frac{u_{i,j}^n - u_{i-1,j}^n}{\Delta x} + \underbrace{O(\Delta x)}_{TE}. \tag{B.3}$$

### B.2.3 Centered Difference

The centered difference scheme for a first-order derivative $\partial u/\partial x$ is given by

$$\frac{\partial u}{\partial x} = \frac{u_{i+1,j}^n - u_{i-1,j}^n}{2\Delta x} + \underbrace{O\left((\Delta x)^2\right)}_{TE}. \tag{B.4}$$

The centered difference scheme for a second-order derivative $\partial^2 u/\partial x^2$ is described as

$$\frac{\partial^2 u}{\partial x^2} = \frac{u_{i+1,j}^n - 2u_{i,j}^n + u_{i-1,j}^n}{2(\Delta x)^2} + \underbrace{O\left((\Delta x)^2\right)}_{TE} \tag{B.5}$$

while that of a second-order mixed derivative $\partial^2 u/\partial x \partial y$ is given by

$$\frac{\partial^2 u}{\partial x \partial y} = \frac{u_{i+1,j+1}^n - u_{i+1,j-1}^n - u_{i-1,j+1}^n + u_{i-1,j-1}^n}{4\Delta x \Delta y} + \underbrace{O\left((\Delta x)^2, (\Delta y)^2\right)}_{TE}. \tag{B.6}$$

**Practice B.1 Deriving a centered difference scheme**

Derive (B.4), (B.5), and (B.6) using the Taylor series expansion (B.1) for $u(x_i + \Delta x, y_j, t^n)$ and $u(x_i - \Delta x, y_j, t^n)$.

## B.3 Example – 1D Advection Equation

Assume that we have the following equation that describes a flow with nonlinear advection of $Q(x,t)$ by a speed of $u(x,t)$:

$$\frac{\partial Q}{\partial t} + u\frac{\partial Q}{\partial x} = 0. \tag{B.7}$$

We show a few examples of developing FDEs.

### B.3.1 Euler Scheme

By applying the forward difference (B.2) for the time derivative $\partial Q/\partial t$ in (B.6), we have the so-called (forward) Euler scheme given by:

$$\frac{\partial Q}{\partial t} = \frac{Q_i^{n+1} - Q_i^n}{\Delta t} + O\left(\Delta t\right). \tag{B.8}$$

In combination with the backward difference of $\partial Q/\partial x$ (see (B.3)), we develop the forward-in-time backward-in-space (FTBS) scheme as a finite difference approximation to (B.6):

$$\frac{\partial Q}{\partial t} + u\frac{\partial Q}{\partial x} = \frac{Q_i^{n+1} - Q_i^n}{\Delta t} + u_i^n \frac{Q_i^n - Q_{i-1}^n}{\Delta x} + \underbrace{O\left(\Delta x, \Delta t\right)}_{TE}. \tag{B.9}$$

Then, we can obtain the future state $Q_i^{n+1}$ to the first-order accuracy as:

$$\begin{aligned} Q_i^{n+1} &= Q_i^n - \frac{u_i^n \Delta t}{\Delta x}\left(Q_i^n - Q_{i-1}^n\right) \\ &= Q_i^n - \mu\left(Q_i^n - Q_{i-1}^n\right). \end{aligned} \tag{B.10}$$

Here, $\mu \equiv u_i^n \Delta t/\Delta x$ is called the Courant number, which is related to the numerical stability. To ensure stability in numerical solutions, the condition $|\mu|_{max} = \left|(\max_i u_i^n)\Delta t/\Delta x\right| \leq 1$ should be satisfied: this implies that the domain-maximum physical velocity $\max_i u_i^n$ should be smaller than the numerical velocity $\Delta x/\Delta t$.

With the centered difference of $\partial Q/\partial x$ (see (B.4)), we also develop the forward-in-time centered-in-space (FTCS) scheme as:

$$Q_i^{n+1} = Q_i^n - \frac{\mu}{2}\left(Q_{i+1}^n - Q_{i-1}^n\right), \tag{B.11}$$

with TE of $O\left(\Delta t, (\delta x)^2\right)$.

### B.3.2 Leapfrog Scheme

By taking the centered finite differences in both $\partial Q/\partial t$ and $\partial Q/\partial x$, we develop the centered-in-time centered-in-space (CTCS) scheme as:

$$\frac{\partial Q}{\partial t} + u\frac{\partial Q}{\partial x} = \frac{Q_i^{n+1} - Q_i^{n-1}}{2\Delta t} + u_i^n \frac{Q_{i+1}^n - Q_{i-1}^n}{2\Delta x} + \underbrace{O\left((\Delta x)^2, (\Delta t)^2\right)}_{TE},$$

which gives

$$Q_i^{n+1} = Q_i^{n-1} - \mu\left(Q_{i+1}^n - Q_{i-1}^n\right). \tag{B.12}$$

This three-time level scheme obtains $Q_i^{n+1}$ from $Q_i^{n-1}$, thus it is known as the *leapfrog* scheme because of the leaping over the time $n\Delta t$.

For the initial integration, (B.12) requires $Q_i^{-1}$ that does not exist. Therefore, we usually employ the Euler scheme (i.e., FTCS; (B.11)) to obtain $Q_i^1$ based on $Q_i^0$ for the very first integration. Because of this treatment, the leapfrog scheme has a *computational* mode, which is physically not meaningful and often induces oscillations in numerical solutions (see Kalnay, 2003; Pletcher et al., 2013).

### B.3.3 Lax–Friedrich Scheme

From the FTCS scheme (B.11), we take $Q_i^n$ as the average of $Q_{i+1}^n$ and $Q_{i-1}^n$:

$$Q_i^{n+1} = \frac{Q_{i+1}^n + Q_{i-1}^n}{2} - \frac{\mu}{2}\left(Q_{i+1}^n - Q_{i-1}^n\right), \tag{B.13}$$

with TE of $O(\Delta t, \delta x)$. Noting that

$$\frac{Q_{i+1}^n + Q_{i-1}^n}{2} = Q_i^n + \frac{Q_{i+1}^n - 2Q_i^n + Q_{i-1}^n}{2},$$

we can rewrite (B.13) as

$$Q_i^{n+1} = Q_i^n - \frac{\mu}{2}\left(Q_{i+1}^n - Q_{i-1}^n\right) + \frac{1}{2}\left(Q_{i+1}^n - 2Q_i^n + Q_{i-1}^n\right),$$

then finally as

$$\frac{Q_i^{n+1} - Q_i^n}{\Delta t} + u_i^n\left(\frac{Q_{i+1}^n - Q_{i-1}^n}{2\Delta x}\right) = \frac{(\Delta x)^2}{2\Delta t}\left(\frac{Q_{i+1}^n - 2Q_i^n + Q_{i-1}^n}{(\Delta x)^2}\right). \tag{B.14}$$

Equation (B.14) is similar to the FDE of an advection-diffusion equation:

$$\frac{\partial Q}{\partial t} + u\frac{\partial Q}{\partial x} = \nu\frac{\partial Q^2}{\partial x^2},$$

where $\nu = (\Delta x)^2/(2\Delta t)$. Therefore, the Lax–Friedrichs scheme has a property of numerical diffusion which damps the solution magnitude as time integration proceeds.

### B.3.4 Upwind Scheme

For simplicity, assume that the advection velocity is a constant, i.e., $u_i^n = c$, then we define the upwind (or upstream) scheme, depending on the sign of $c$, as:

$$c > 0 \quad \frac{Q_i^{n+1} - Q_i^n}{\Delta t} + c\frac{Q_i^n - Q_{i-1}^n}{\Delta x} \Longrightarrow Q_i^{n+1} = Q_i^n - \mu\left(Q_i^n - Q_{i-1}^n\right),$$

$$c < 0 \quad \frac{Q_i^{n+1} - Q_i^n}{\Delta t} + c\frac{Q_{i+1}^n - Q_i^n}{\Delta x} \Longrightarrow Q_i^{n+1} = Q_i^n - \mu\left(Q_{i+1}^n - Q_i^n\right), \tag{B.15}$$

where $\mu = c\Delta t/\Delta x$. Therefore, in this scheme, the spatial differences are inclined to the "upwind" side, from which the advecting flow originates; thus, the grid-point information is transported from the upwind side through the advection velocity, $c$ (see Figure B.3).

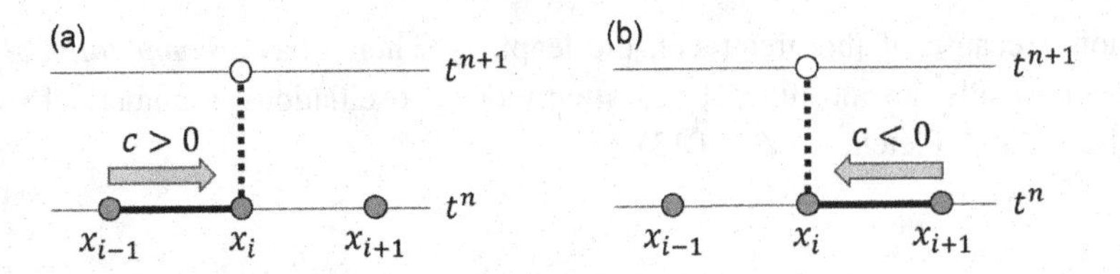


Figure B.3 Schematic diagram of the upwind scheme for (a) $c > 0$ and (b) $c < 0$. Arrows indicate information transports via the advection velocity, $c$.

## References

Kalnay E (2003) *Atmospheric Modeling, Data Assimilation and Predictability*. Cambridge University Press, New York, 341 pp.

Pletcher RH, Tannehill JC, Anderson DA (2013) *Computational Fluid Mechanics and Heat Transfer*. CRC Press, Boca Raton, FL, London, New York, 740 pp.

# Appendix C

## Lab Practice I

In this section, we show practical examples of constructing the derivative codes – the tangent linear and adjoint codes – both by hand and via automatic differentiation (AD), based on the logistic map (e.g., May, 1976; Park, 2003).

### C.1 Logistic Map

Chaotic behavior of solutions can often occur in a simple nonlinear dynamical system such as the *logistic equation*:

$$\frac{dX}{dt} = rX(1 - X). \tag{C.1}$$

The *logistic map*, the discrete counterpart of (C.1), is represented as a recursive quadratic model:

$$X^{n+1} = rX^n(1 - X^n), \tag{C.2}$$

where $n$ is a time index (or an iteration number). We limit our discussion to $r > 1$, where the solutions $X^n$, including the initial condition $X^0$, are within $[0, 1]$: if $X^0 < 0$ or $X^0 > 0$, then $X^n \rightarrow -\infty$ as $n \rightarrow \infty$ (Devaney, 2003). Equation (C.2) has a maximum value $\frac{r}{4}$ at $X = \frac{1}{2}$; thus, $r$ is in the range of $[1, 4]$. Both (C.1) and (C.2) are called the original nonlinear model (NLM).

Note that the solution behavior of (C.2) is controlled by $r$. For a given value of $r$ in the interval $[1, 3]$, the model converges to one stable solution, $1 - \frac{1}{r}$, regardless of the initial condition $X^0$. For $r > 3$, the solutions start to bifurcate, eventually leading to the chaos – period-doubling bifurcation at $3 < r < 3.5699\ldots$ and the onset of chaos at $r \approx 3.5699$ (see more details in Devaney, 2003).

### C.2 Development of the Tangent Linear Model and the Adjoint Model

The tangent linear model (TLM) and the adjoint model (ADJM) of the continuous logistic equation (C.1) are given by

$$\frac{d}{dt}\delta X = r(1 - 2X)\delta X \tag{C.3}$$

and

$$-\frac{d}{dt}\delta X^* = r(1-2X)\delta X^*, \tag{C.4}$$

respectively, where $\delta X$ is the tangent linear variable and $\delta X^*$ is the adjoint variable.

**Practice C.1 Deriving the TLM/ADJM of the logistic equation**

Do the following for the logistic equation (C.1):

1. Derive the TLM (C.3) by taking the first-order term in the Taylor series expansion (see Section 4.3.1).
2. Derive the ADJM (C.4) by employing the adjoint operator approach (see Section 4.4.1.1).

The TLM of the discrete logistic map (C.2) is expressed as

$$\delta X^{n+1} = r\left(1-2X^n\right)\delta X^n, \tag{C.5}$$

where $X^n$ and $\delta X^n$ are the nonlinear trajectory and the tangent linear variable, respectively, at the $n$th iteration. The corresponding ADJM is derived, following Talagrand (1991) and Park (2003) (see also Section 4.8), as:

1. Express the TLM (C.5) in a matrix form, with an input vector $(\delta X^n)^T$ and an output vector $(\delta X^{n+1}, \delta X^n)^T$, as

$$\begin{pmatrix} \delta X^{n+1} \\ \delta X^n \end{pmatrix} = \begin{pmatrix} r(1-2X^n) \\ 1 \end{pmatrix} \left(\delta X^n\right). \tag{C.6}$$

2. The corresponding ADJM is expressed in terms of the transpose of the TLM matrix $(r(1-2X^n)\ \ 1)^T$, with an input vector $\left((\delta X^*)^{n+1}, (\delta X^*)^n\right)^T$ and an output vector $\left((\delta X^*)^n\right)^T$, as

$$\left(\delta X^*\right)^n = \begin{pmatrix} r(1-2X^n) & 1 \end{pmatrix} \begin{pmatrix} (\delta X^*)^{n+1} \\ (\delta X^*)^n \end{pmatrix}. \tag{C.7}$$

3. By expanding (C.7), we obtain the ADJM of (C.5):

$$\left(\delta X^*\right)^n = r\left(1-2X^n\right)\left(\delta X^*\right)^{n+1} + \left(\delta X^*\right)^n. \tag{C.8}$$

### C.3 Tangent Linear Model/Adjoint Model Codes by Hand

In this subsection, the computer codes for the NLM, TLM, and ADJM of the logistic map are provided for tutorial purposes, which are adopted from Park (2003) with slight modification. The subroutine for the NLM is shown in Listing C.1, where "x" is the nonlinear variable (i.e., $X^n$). Listing C.2 shows the code for corresponding TLM with "tlx" representing the tangent linear variable (i.e., $\delta X^n$). Listing C.3 represents the corresponding adjoint code with "adx" depicting the adjoint variable (i.e., $(\delta X^*)^n$).

Listing C.1. Original nonlinear code of the logistic map in Fortran 77.

```
1       subroutine logis (nitr, a, x)
2       integer nitr
3       real x(0:nitr), a
4       do i = 1, nitr
5          x(i) = a*x(i-1)*(1. - x(i-1))
6       enddo
7       return
8       end
```

Listing C.2. Tangent linear code of the logistic map in Fortran 77.

```
1       subroutine logistl (nitr, a, x, tlx)
2       integer nitr
3       real x(0:nitr), tlx(0:nitr), a
4       do i = 1, nitr
5         tlx(i) = a*(1. - 2.*x(i-1))*tlx(i-l)
6       enddo
7       return
8       end
```

Listing C.3. Adjoint code of the logistic map in Fortran 77.

```
1       subroutine logis (nitr, a, x, tlx, adx)
2       integer nitr
3       real x(0:nitr), tlx(0:nitr), adx(0:nitr), a, djdx
4       djdx = 1.
5       do i = nitr, 1, -1
6         adx(i) = 0.
7         adx(i) = adx(i) + djdx
8         adx(i-1) = 0.
9         adx(i-1)= adx(i-1)+a*(1. - 2.*x(i-1))*adx(i)
10        djdx = adx(i-1)
11      enddo
12      return
13      end
```

## C.4 Tangent Linear Model/Adjoint Model Codes Using an Automatic Differentiation Tool

In this subsection, an example is provided to generate the derivative codes, in both forward and reverse modes, of the original nonlinear code of the logistic map by applying an AD tool – named Tapenade (Hascoët and Pascual, 2013). One can download and install the software package from Inria (2021a). Tapenade also provides an online AD engine through which the derivative codes can be generated directly at Inria (2021b). The online tutorial is available at Inria (2021c). In this example, we use the online platform and give a step-by-step introduction on how to use it.

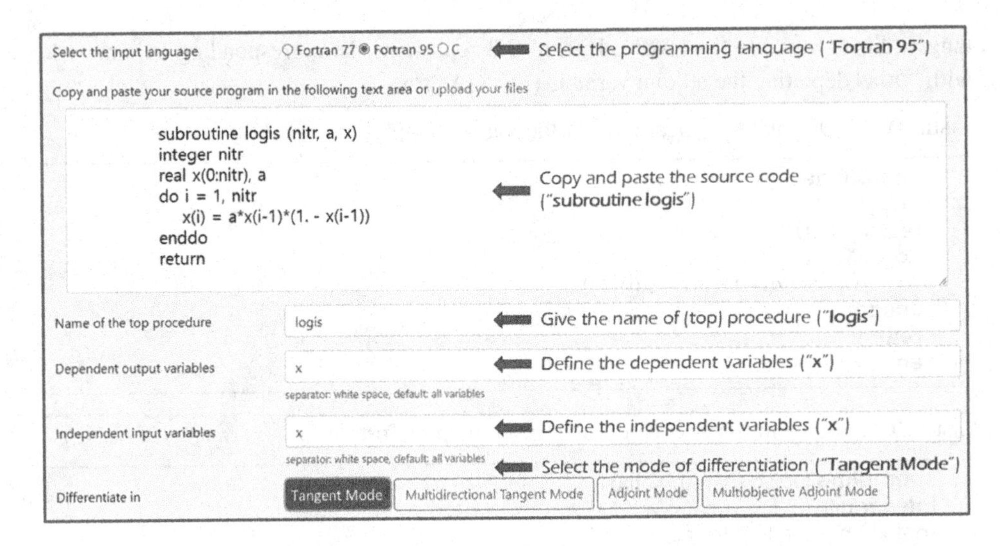


Figure C.1 An example of generating a derivative code of the logistic map through the tangent (forward) mode in the Tapenade AD engine online platform.

Figure C.1 is a screen capture from the Tapenade AD engine where online generation of the derivative codes is available. It illustrates what information one should provide to the AD engine before execution. We also show an example of generating a derivative code out of the original NLM code (see Listing C.1), which are annotated with arrows in Figure C.1. To do this, one can visit the Tapenade AD engine webpage at http://tapenade.inria.fr:8080/tapenade/paste.jsp, with the copy and paste option, to select or fill in the required information as follows:

1. *Select the input language* among three possible options – Fortran 77, Fortran 95, and C. We selected Fortran 95.
2. *Copy and paste the source code* to be differentiated. We put "`subroutine logis`" here, written in Fortran 77. Note that we had chosen Fortran 95 in the previous step. We tried this way to check if the engine produces the output code in Fortran 95 though the input code is written in Fortran 77.
3. *Key in the name of top procedure*. As we put the subroutine itself, we keyed in just the subroutine title (i.e., "`logis`").
4. *Define the dependent and independent variables*, which are "`x`" in our case.
5. Then, *choose the desired differentiation mode* among four available options, and click it. We have chosen the "Tangent Mode."

The output codes are shown in Listings C.4, C.5, and C.6. The Tapenade AD engine also has an option to upload the source code files (see Inria, 2021b).

Listing C.4 is the NLM code: It was originally written in Fortran 77 (see Listing C.1) but is reproduced by Tapenade in Fortran 95. Listing C.5 is the TLM code, generated by Tapenade via the "Tangent Mode" (i.e., forward mode) under the subroutine title

"LOGIS_D." Here, "xd" represents the tangent linear variable. Listing C.6 represents "SUBROUTINE LOGIS_B", the Tapenade-generated ADJM code through the "Adjoint Mode" (i.e., reverse mode). The adjoint variable is specified as "xb."

Listing C.4. *Original nonlinear code of the logistic map, generated by Tapenade in Fortran 95.*

```
1  !      Generated by TAPENADE     (INRIA, Ecuador team)
2  ! Tapenade 3.16 (develop) - 16 Mar 2021 14:40
3  !
4  SUBROUTINE LOGIS(nitr, a, x)
5    IMPLICIT NONE
6    INTEGER :: nitr
7    REAL :: x(0:nitr)
8    INTEGER :: i
9    REAL :: a
10   DO i = 1, nitr
11         x(i) = a*x(i-1)*(1.-x(i-1))
12   END DO
13   RETURN
14 END SUBROUTINE LOGIS
```

Listing C.5. *Tangent linear code of the logistic map, generated by Tapenade in Fortran 95.*

```
1  !      Generated by TAPENADE     (INRIA, Ecuador team)
2  ! Tapenade 3.16 (develop) - 16 Mar 2021 14:40
3  !
4  ! Differentiation of logis in forward (tangent) mode:
5  !   variations of useful results: x
6  !   with respect to varying inputs: x
7  !   RW status of diff variables: x:in-out
8  SUBROUTINE LOGIS_D(nitr, a, x, xd)
9    IMPLICIT NONE
10   INTEGER :: nitr
11   REAL :: x(0:nitr)
12   REAL :: xd(0:nitr)
13   INTEGER :: i
14   REAL :: a
15   DO i = 1, nitr
16         xd(i) = a*(1.-2*x(i-1))*xd(i-1)
17         x(i) = a*x(i-1)*(1.-x(i-1))
18   END DO
19   RETURN
20 END SUBROUTINE LOGIS_D
```

Listing C.6. Adjoint code of the logistic map, generated by Tapenade in Fortran 95.

```
1  !      Generated by TAPENADE     (INRIA, Ecuador team)
2  ! Tapenade 3.16 (develop) - 16 Mar 2021 14:40
3  !
4  ! Differentiation of logis in reverse (adjoint) mode:
5  !   gradient of useful results: x
6  !   with respect to varying inputs: x
7  !   RW status of diff variables: x:in-out
```

```
8  SUBROUTINE LOGIS_B(nitr, a, x, xb)
9    IMPLICIT NONE
10   INTEGER :: nitr
11   REAL :: x(0:nitr)
12   REAL :: xb(0:nitr)
13   INTEGER :: i
14   REAL :: a
15   DO i = 1, nitr
16       CALL PUSHREAL4(x(i))
17       x(i) = a*x(i-1)*(1.-x(i-1))
18   END DO
19   DO i = nitr, 1, -1
20       CALL POPREAL4(x(i))
21       xb(i-1) = xb(i-1) + ((1.-x(i-1))*a-x(i-1)*a)*xb(i)
22       xb(i) = 0.0
23   END DO
24 END SUBROUTINE LOGIS_B
```

Listing C.6 also includes extra routines such as "`PUSHREAL4`" and "`POPREAL4`": these are the by-products in generating the reverse-mode derivative codes using Tapenade. These routines are defined in a separate library that consists of two files – "`adBuffer.f`" (in Fortran) and "`adStack.c`" (in C). The files are included in "ADFirstAidKit.tar," which is downloadable from the differentiation result page after the Tapenade AD engine has been executed, by clicking on the "Download PUSH/POP" button. One needs to compile both `adBuffer.f` and `adStack.c` and link the reverse-differentiated executable to this library. See more details in Inria (2021c).

**Practice C.2**

By visiting the Tapenade AD engine webpage at Inria (2021b) and following the guide in Figure C.1 (see also Inria, 2021c), generate the tangent linear code (i.e., Listing C.5) and the adjoint code (i.e., Listing C.6) of the original (nonlinear) logistic map code (i.e., Listing C.1). Then, download the derivative codes and by-products (i.e., ADFirstAidKit.tar). Write a main program code to include the subroutines and make an executable file by compiling and linking all necessary files.

## C.5 Suggested Experiments

The following experiments are suggested by employing the derivative codes of the logistic map, produced by either hand or the Tapenade AD engine, along with the nonlinear code. One needs to construct a main program to call the necessary subroutines as in Practice C.2.

### *C.5.1 Checking the Correctness of TLM/ADJM (I)*

One can use the TLM/ADJM codes to check their correctness by conducting the following experiments:

***Exp***-1: Using the TLM code, perform the TLM correctness test, following Eq. (4.25) (see Table 4.1) for $r = 1.5$ and $\delta X^0 = 0.2$.

***Exp***-2: Using the ADJM code, perform the gradient accuracy test, following Eq. (4.40) (see Figure 4.1) for $r = 1.5$ and $\delta X^0 = 0.2$.

### *C.5.2 Behavior of TLM Solutions*

The characteristics of TLM solutions (i.e., linear perturbation) can be explored, using the TLM code, in comparison with the nonlinear perturbation (NLP) ($\Delta X^n$), i.e., the difference between two NLM solutions starting from nearby initial conditions (say, $X_1^0$ and $X_2^0$, thus having $\Delta X^n = X_2^n - X_1^n$). Conduct the following experiments for $r = 3.55$ (period-8 bifurcation):

***Exp***-3: 1) Run the NLM code up to $n = 60$ using two nearby initial conditions – $X_1^0 = 0.200$ and $X_2^0 = 0.201$, then take the difference between the solutions of two NLM runs; 2) run the TLM code for $\delta X^0 = 0.001$; and 3) plot $\Delta X^n$ and $\delta X^n$ and discuss the behaviors of TLM and NLP solutions, especially on the similarity between the solutions and the periodicity.

***Exp***-4: Repeat ***Exp***-3 up to $n = 100$ for different sets of perturbation initial conditions: 1) larger perturbations with $\Delta X^0 = \delta X^0 = 0.02$ (i.e., $X_1^0 = 0.20$ and $X_2^0 = 0.22$); and 2) smaller perturbations with $\Delta X^0 = \delta X^0 = 0.0001$ (i.e., $X_1^0 = 0.2000$ and $X_2^0 = 0.2001$). Then, plot $\Delta X^n$ and $\delta X^n$, separately, for $n = [60, 100]$, and discuss the results.

***Exp***-5: Run the TLM code up to $n = 80$ for $\delta X^0 = 0.001$ using different NLM initial conditions: 1) $X^0 = 0.2000$; 2) $X^0 = 0.2005$; and 3) $X^0 = 0.2010$. Plot $\delta X^n$ for the three cases and discuss the behavior of TLM solutions for different NLM initial conditions.

### *C.5.3 Checking the Correctness of TLM/ADJM (II)*

One can also check the accuracy of the TLM/ADJM system, following Park (2003), especially for a simple model. Note that the TLM solution evolves from the initial condition $\delta \mathbf{X}^0$ as in (4.18) as

$$\delta \mathbf{X}^n = \mathbf{G}^{n-1}\mathbf{G}^{n-2}\cdots\mathbf{G}^1\mathbf{G}^0\delta\mathbf{X}^0, \tag{C.9}$$

while the ADJM solution is retrospectively determined starting from the final condition $(\delta \mathbf{X}^*)^N$ as in (4.37) as

$$(\delta \mathbf{X}^*)^n = (\mathbf{G}^n)^T\left(\mathbf{G}^{n+1}\right)^T\cdots\left(\mathbf{G}^{N-2}\right)^T\left(\mathbf{G}^{N-1}\right)^T(\delta\mathbf{X}^*)^N. \tag{C.10}$$

By defining a canonical unit vector, $\mathbf{e} = \delta_i$ (i.e., $e_{k=i} = 1$ and $e_{k\neq i} = 0$ for the $k$th component), $\delta \mathbf{X}^0$ and $(\delta \mathbf{X}^*)^N$ are given by

$$\delta \mathbf{X}^0 = e_j \text{ and } \left(\delta \mathbf{X}^*\right)^N = e_k,$$

respectively. For all $t_0 \leq t_n \leq t_N$,

$$\left((\delta\mathbf{X}^*)^n\right)^T \delta\mathbf{X}^n = \left\{\left((\delta\mathbf{X}^*)^N\right)^T \mathbf{G}^{N-1}\mathbf{G}^{N-2}\dots\mathbf{G}^{n+1}\mathbf{G}^n\right\}\left\{\mathbf{G}^{n-1}\mathbf{G}^{n-2}\dots\mathbf{G}^1\mathbf{G}^0\delta\mathbf{X}^0\right\}$$

$$= (e_k)^T \frac{\partial \mathbf{X}^T}{\partial \mathbf{X}^0} e_j = \frac{\partial X_k(t_N)}{\partial X_j(t_0)}, \tag{C.11}$$

for fixed $k$ and $j$. Thus, by combining the TLM and ADJM, one can evaluate the sensitivity coefficient, i.e., $\partial X_k(t_N)/\partial X_j(t_0)$, which is the same for all $n$. With this background, conduct the following experiment:

***Exp*-6**: Using both TLM and ADJM codes, evaluate $X^n$, $\delta X^n$, $(\delta X^*)^n$, and $\left((\delta X^*)^n\right)^T \delta X^n$ for $r = 1.5$ and $X^0 = 0.2$ up to $n = 25$. Set the TLM initial condition $\delta X^0$ and the ADJM final condition $(\delta X^*)^N$ to 1.0 (see Table 1 in Park, 2003).

## References


Devaney R (2003) *An Introduction to Chaotic Dynamical Systems*. 2nd ed., CRC Press, Boca Raton, FL, 335 pp.

Hascoët L, Pascual V (2013) The Tapenade automatic differentiation tool: Principles, model, and specification. *ACM Trans Math Softw* 39:20, doi:10.1145/2450153.2450158

Inria (2021a) Download Tapenade. `https://tapenade.gitlabpages.inria.fr/userdoc/build/html/download.html`

Inria (2021b) TAPENADE Automatic Differentiation Engine. `www-tapenade.inria.fr:8080/tapenade/`

Inria (2021c) The Tapenade Tutorial. `https://tapenade.gitlabpages.inria.fr/userdoc/build/html/tapenade/tutorial.html`

May RM (1976) Simple mathematical models with very complicated dynamics. *Nature* 261:459–467.

Park SK (2003) Behavior of tangent linear and adjoint solution in a chaotic dynamical system. *Korean J Atmos Sci* 6:47–53.

Talagrand O (1991) The use of adjoint equations in numerical modelling of the atmospheric circulation. In *Automatic Differentiation of Algorithms: Theory, Implementation, and Application*, (eds.) Griewank A, Corliss GF, SIAM, Philadelphia, PA, 169–180.

# Appendix D

## Lab Practice II

In this section we discuss practical aspects of data assimilation (DA) development, with suggested lab exercises. In particular, we present major arguments using the maximum likelihood ensemble filter (MLEF) as an example.

### D.1 Data Assimilation Terminology

The terminology of DA is important, and has the aim of providing common language that should ease communication between researchers. Unfortunately, there is no well-defined set of terms that everyone accepts, as DA is still being developed and some terminology becomes outdated or confusing. Therefore, we will try to define here the DA terminology used in this book.

A DA *cycle* is a time interval between two consecutive analyses. As an example, let a DA cycle be 6 hours, meaning that analysis will be produced every 6 hours, at regular times: 0000, 0600, 1200, and 1800 UTC. In order to produce the analysis at 1200 UTC, for example, it is common to assimilate observations from 0900–1500 UTC, i.e., centered at the analysis time, but shifted by 3 hours from the DA cycle. This defines the DA window. In this situation we would say that the DA window is +/– 3 hours, or 6 hours. Therefore, the assimilation *window* is the maximum time span of observations that can be assimilated in a given cycle.

The assimilation *interval* can be defined as the time between the start and end of a forecast used in a given DA cycle. Therefore, it is determined by the desired span of observation times, (say that it is needed for the first guess ...) and therefore, can be termed "forecast" interval as well. In this example, a forecast from 0600 to 1500 UTC is needed to produce an adequate first guess, implying an assimilation interval of 9 hours. The relationships between cycle, window, and interval are depicted in Figure D.1.

As another example, one can choose to assimilate observations up to the end of a DA cycle, i.e., to not go beyond the end time of the cycle. This can happen if there are observations only at exact times of the analyses or in the operational environment when there is no time to wait for new observations. In that case all mentioned definitions of interval, cycle, and window would become identical, from time $t-1$ to $t$. The cycle is still identical to the one shown in Figure D.1, while the interval is now shorter and identical to the cycle as there is no need to calculate a longer forecast. This automatically implies that the DA window is of the same length as the cycle and the interval.

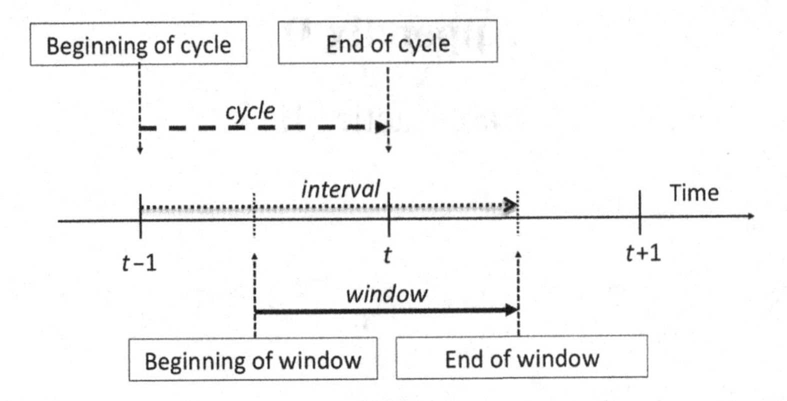


Figure D.1 Specification of DA (dashed), window (solid), and interval (dotted).

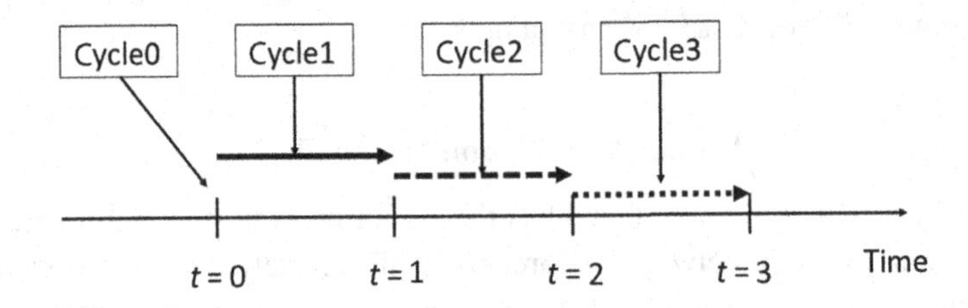


Figure D.2 Ordering of DA cycles.

DA cycles follow each other, with cycle 0 representing the initiation process. A schematic of DA cycling with defined terminology is presented in Figure D.2.

## D.2 Data Assimilation Steps

We define cycling of DA as a recursive application of DA, as explained in Chapter 2. A DA algorithm has three important steps, in the following order: (1) initiation step; (2) forecast step; and (3) analysis step. This could be represented as a simple algorithm:

The complexity of each step depends on the DA methodology, but the steps defined above are common to all DA algorithms. The initiation step is used to define initial conditions before DA begins. This may include only the control initial conditions as in variational methods, or ensemble initial conditions in case of ensemble and hybrid data assimilation (HYB). A characteristic of the *initiation* step is that it is performed only once, at the beginning of the DA cycling. Once the initiation is completed it is followed by a *forecast* step used to define the guess value for the analysis step. In ensemble methods the forecast step also defines forecast uncertainty. In variational methods forecast is only related to control deterministic forecast, while in ensemble DA the forecast implies ensemble forecasts. Finally, the *analysis* step is when the observations are assimilated, eventually

**Algorithm D.1** DA algorithm

```
/*index k denotes cycle number                                                    */
1 begin
2   Initiation
                       /*Choose initial conditions and initial uncertainty, at cycle k = 0*/
3   repeat
4     for k = 1 to kmax do
5       Forecast from cycle k - 1 to k
                       /*Compute forecast guess and forecast uncertainty*/
6       Analysis at cycle k
                       /*Compute analysis and analysis uncertainty*/
7     endfor
8   until end of cycling
9 end
```

producing the optimal state (e.g., analysis) and its uncertainty (ensemble methods). In continuing the cycling of DA, the forecast and analysis steps are interchangeably applied. One of the most important characteristics of such DA cycling is the exchange of information between different steps of the DA algorithm, which is applied via Bayesian inference, i.e., a process of learning from past performances. In the case of variational DA this is performed only in terms of the optimal state, while in the case of ensemble DA this is performed for uncertainties as well.

The initiation step of DA includes choosing initial conditions and initial uncertainty. The goal of this step is twofold: (i) to produce initial conditions that are our best estimate of a true dynamical state; and (ii) to produce initial perturbations that will project on an unstable dynamical space and therefore grow in time. The first goal is desirable since it will assure that the forecast from such an initial state will also be very close to the true dynamical state. The second goal is related to the theory of dynamical systems and DA, as explained in Chapter 10, and it implies that one would like to have the forecast uncertainty defined over the dynamically unstable subspace of the model (e.g., Trevisan and Palatella, 2011), therefore producing uncertainties that will grow with time. Therefore, the initiation process has a similar goal of finding the uncertainty subspace that most likely belongs to the unstable subspace of the prediction model.

The choice of initial conditions would optimally be an interpolation from an existing analysis produced by another model. For example, one can use the analysis of a global modeling system interpolated to the grid of the model used in DA. In some situations, however, the outside global modeling system may not have all the variables required for the desired DA. This is often the case with coupled DA. For example, the global system used for interpolation may not have the chemistry or land surface variables required for

DA. In this case one should examine other ways of producing the initial conditions for all the required variables, such as using another model or starting with predefined values.

The choice of initial uncertainty is clearly relevant to ensemble-based DA systems, but it does not have an impact in some methods, such as variational, for example. However, as most methods of DA include probabilistic forecasting, it is important to specify the initial conditions for such forecasts.

If the initial conditions obtained by interpolating from another model's analysis are denoted $\mathbf{x}^0$ and there are $N$ initial perturbations denoted $\{\mathbf{p}_i^0 : i = 1, N\}$, then one can form $N$ initial conditions for probabilistic forecast $\{\mathbf{x}_i^0 = \mathbf{x}^0 + \mathbf{p}_i^0 : i = 1, N\}$. The superscript *0* denotes the initial time. In typical applications under Gaussian assumption the initial perturbation is equivalent to a column of a square root error covariance

$$\mathbf{P}_0^{1/2} = \begin{pmatrix} \mathbf{p}_1^0 & \cdots & \mathbf{p}_N^0 \end{pmatrix}. \tag{D.1}$$

### D.2.1 Initiation Using Random Perturbations

A natural and simple choice would be to follow the Monte Carlo approach and create $N$ *random* perturbations. The problem with applying random perturbations in a complex dynamical system is that many of those perturbations can have negligible growth over time. In reality, random perturbations may project onto some growing perturbation directions, but these perturbations can be very small in magnitude implying a need for a longer forecast to develop into physically meaningful perturbations. Since the cycle length is generally short, there is likely no time for random perturbations to develop.

This could be improved if the perturbed forecast begins at a much earlier time. Even in this case, however, one cannot be sure that all random perturbations will grow sufficiently, so this approach needs special attention. In any case, such a procedure may be costly for high-resolution models as it requires long integration of ensemble forecasts.

The random approach to initiation is illustrated in Figure D.3. Note that it is a simple approach, but it likely requires long ensemble integration and special attention to make sure that the resulting initial perturbations are dynamically relevant and realistic in magnitude. Given $N$ random perturbations $\omega_i$ at time $t = 0$ one can define the mean and the covariance as sample estimates

$$\overline{\omega} = \frac{1}{N}\sum_{i=1}^{N} \omega_i \tag{D.2}$$

$$\mathbf{P}_0 = \frac{1}{N-1}\sum_{i=1}^{N} (\omega_i - \overline{\omega})(\omega_i - \overline{\omega})^T \tag{D.3}$$

so that

$$\mathbf{p}_i^0 = \frac{1}{\sqrt{N-1}}(\omega_i - \overline{\omega}). \tag{D.4}$$

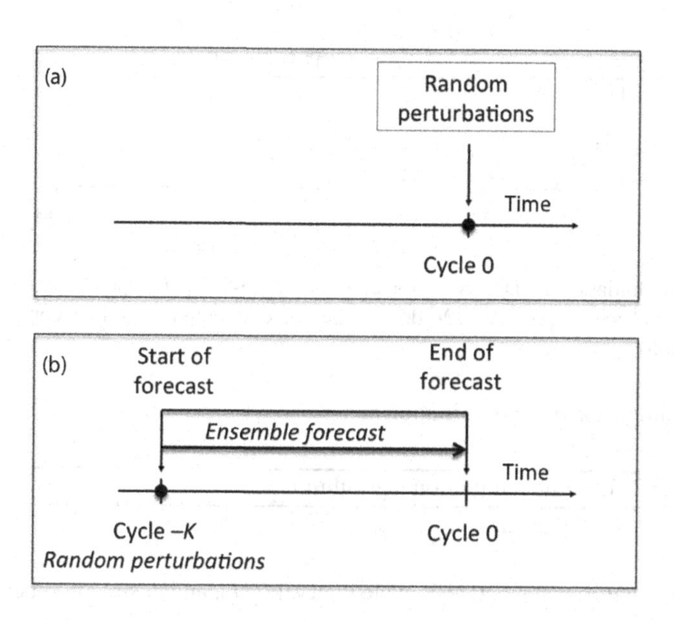


Figure D.3 Initiation of DA using random perturbations: (a) at initiation time and (b) at time before initiation time followed by ensemble forecast to initiation time. The large dot points to the time when random perturbations are created.

These are the columns of the forecast error covariance that are used to create the ensemble initial conditions $\{\mathbf{x}_i^0 = \mathbf{x}^0 + \mathbf{p}_i^0 : i = 1,\ N\}$.

### D.2.2 Initiation Using Forecast Differences

An alternative to random initial perturbations is to integrate a single deterministic model and use the difference between arbitrary forecast outputs and the central forecast time.

Let us define the initiation time as $t = 0$, and time $\tau$ as an integer multiplication of cycle length, i.e., $\tau = m \cdot c_{len}$, where $c_{len}$ is cycle length and $m$ is an integer. Typically, $m = 2$ which implies that $\tau$ is equal to two assimilation cycles. Then, the deterministic forecast is integrated from time $t = -\tau$ to $t = +\tau$. The initial uncertainty is defined using time-shifted forecast differences: Given a forecast integration of length $2\tau$, we define an ensemble perturbation using the difference $\mathbf{x}_i - \mathbf{x}^C$ where $\mathbf{x}$ is a state vector, the forecast output time is $\{t_i : -\tau \leq t_i \leq \tau,\ (i = 1, \ldots, N)\}$, and $N$ is the desired number of ensembles. The index $i$ denotes ensemble member, i.e., $\mathbf{x}_i = \mathbf{x}(t_i)$, and $\mathbf{x}^C = \mathbf{x}(0)$ denotes the central forecast valid at central time at time $t = 0$. The $i$th uncertainty vector is given as $\mathbf{p}_i = \frac{1}{\sqrt{N}}\left(\mathbf{x}_i - \mathbf{x}^C\right)$. If there is an available analysis from another system, denoted $\mathbf{x}^A$, the initial ensemble perturbations can be recentered around that analysis: $\mathbf{x}_i^{IC} = \mathbf{x}^A + \mathbf{p}_i$. Figure D.4 shows a schematic diagram of the procedure used to calculate the initial conditions for ensemble forecasting.

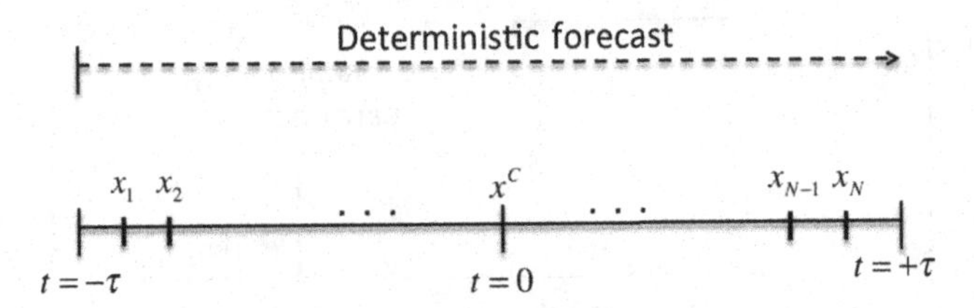


Figure D.4 Initiation of DA using forecast differences. Superscript $C$ denotes the central time and subscripts $1,\ldots,N$ denote the forecast outputs used to compute initial perturbations.

Here is a simple example of initiation code:

**Algorithm D.2** Ensemble initiation algorithm

```
/*index k denotes cycle number                                                  */
1  begin
2      define: date/time at t = 0 (t0), cycle interval (CI), and desired number of
         ensembles (N).
3      define: initiation approach: init = random (using random perturbations and
         ensemble forecast) or init = fcstdiff (use forecast differences in a single
         deterministic forecast).
4      if init = fcstdiff then
5          define: time interval τ for going back from t0 : τ = K · CI
6          Create initial forecast date at t0 − τ
7          Run N forecasts from t0 − τ to t0
8          Save model outputs at t0 as ensemble members
9      else if init = random then
10         define: time interval τ for going back from t0 : τ = K · CI
11         Create initial forecast date at t0 − τ
12         Find equidistant output times required to create N ensembles: Δt ∼ 2·τ/N
13         Run a single deterministic forecast from t0 − τ to t0 + τ
14         Save N model outputs from times t0 − τ, t0 − τ + Δt, t0 − τ + 2Δt, ..., t0 + τ
             as ensemble members
15     end
16 end
```

The forecast step of DA serves to propagate uncertainty from the current analysis time to the future time. In variational DA it consists of integrating a single deterministic forecast, while in ensemble DA it includes integration of several deterministic forecasts. A practical challenge of this step is related to the use of complex and computationally demanding prediction models in DA. First, such models typically have their own executable and require parallel processing. Second, their namelists and input files have to be managed using directives from DA, as they are now a component of DA and cannot be manually submitted.

This requires somewhat more advanced driving through scripts (e.g., shell, Python) or through other computing languages (such as C and Fortran). For example, one can submit a forecast with its own executable from Fortran 90 code using a *SYSTEM* command. For example, a code

```
1  character(len = *) :: command
2  integer :: STATUS
3  integer :: SYSTEM
4
5  command = "./model.exe"
6  STATUS = SYSTEM(command)
```

would imply that a line `./model.exe` is submitted from the shell, where `model.exe` denotes a precompiled model executable. The code will wait until the job has ended before continuing to the next line of the Fortran code.

In the case of multimodel ensembles the algorithmic structure can become even more complex, as one would need to manage several forecast models with their own scripts and executables. These and other challenges have to be resolved when attempting to develop a new DA system, or at least understood when attempting to use an existing DA system. The algorithmic details of the forecast step will likely change with a development of more advanced probabilistic forecasting and/or with an optional development of alternative ways to create the initial guess and its uncertainty, but its purpose will still remain unchanged. Below is a simple forecast step program:

**Algorithm D.3** Forecast step of DA

/*index $k$ denotes cycle number */

1 **begin**
2 **define:** cycle start and end dates/times ($t_{start} = t_{k-1}$ and $t_{end} = t_k$, respectively).
3 **for** $t = t_{start}$ **to** $t_{end}$ **do**
4 Run control deterministic forecast and/or optionally $N$ ensemble forecasts
5 **endfor**
6 **define:** forecast guess from control forecast or from ensembles
7 **define:** forecast uncertainty, static or from ensembles, or combined
8 **end**

The analysis step of DA calculates the analysis using the guess and optionally the uncertainty from the forecast step. In this step one would like to assimilate all observations that can possibly be beneficial to the system. This implies a need to manage the submission of several observation operators and managing their inputs and outputs from the main DA code. Some of the included observation operators can be quite complex and have their own driving script and executable, while other observation operators may be simple enough to require a subroutine only. All this needs to be managed and directed from the main DA code. As suggested for the forecast step this can be done by using a shell script or a code written in common computer language such as Fortran or C.

From the algorithmic point of view, the main challenge of forecast and analysis steps of DA is to direct and manage complex prediction models and observation operators within a single DA system. This means that dates, assimilation intervals, and input namelists, all have to be created on the fly so that DA can proceed.

**Algorithm D.4** Analysis step of DA

/*index $k$ denotes cycle number */

1 **begin**
2   **define:** cycle end dates/times ($t_{end} = t_k$)
    **input:** forecast guess and forecast uncertainty from D.3
    **input:** observations and observation uncertainty
3
4   Produce optimal analysis by direct solution or by iterative minimization
5   Optionally, produce an estimate of analysis uncertainty
6 **end**

## D.3 Other Data Assimilation Considerations

In this section we discuss relevant components of a DA algorithm. When developing a new DA system or trying to understand an available algorithm for DA, it is important to consider several components: (i) DA methodology and its applicability; (ii) relationship between DA cycling and Bayesian inference; (iii) algorithm flexibility; (iv) parallel processing; (v) uncertainty estimation; and (vi) post-processing/verification. The differences between the many DA algorithms are often hidden in the implementation details.

Point (i) includes not only knowing which particular DA methodology the algorithm addresses, but also how this assimilation methodology is implemented.

Point (ii) is about understanding how the cycling of the particular DA algorithm is related to Bayesian inference. As discussed in Chapter 2, Bayesian inference is a by-product of DA, but more importantly the main mechanism that makes DA a "learning" algorithm. By continuously cycling DA over time, the feedback between analysis/forecast and their uncertainties is enhanced. If correctly implemented, more complete feedback implies more powerful DA, as it can better adjust to new conditions and new observations.

Code flexibility (iii) is especially important for more complex and advanced DA systems. It is preferable to have an input, such as a `namelist`, that defines an arbitrary control variable, but without the need to change DA codes. This allows quick adjustment of DA to the new application, or new model, and makes transition from one modeling system to another much easier in terms of DA. This is also advantageous for coupled DA, as additional control variables are easily included. Therefore, when using an already developed DA algorithm, it is important to understand how difficult it would be to add new control variables. For example, it may be concluded that adding cloud hydrometeor variables will improve DA of tropical cyclones. With flexible DA code, one would only need to add these variables to the list before proceeding with the experiments. In less flexible code, however, one may need to spend months trying to add and debug new control variables.

The same is true for new observation operators. In principle, more observations imply better DA, so there is a constant desire to add new observations. Given a predeveloped observation operator, it is beneficial to have the capability of a DA system to include a new observation operator with minimal, if any, change of codes. Such flexible code will be able to include assimilation of an additional observation immediately, and consequently improve DA performance. A less flexible code may require considerable coding and time-consuming effort to achieve the same thing. This is especially relevant for optimizing the insertion of new observation types. After an observation operator for the particular new observation is developed, it can be easily shared with other researchers or operational centers and eventually optimize the performance of DA.

Development of such modular codes is already taking shape within major operational centers (e.g., OOPS at ECMWF and JEDI at NOAA), as well as in various research centers and academia. Exchanging observation operators, minimization algorithms, uncertainty estimation, and other components of a DA algorithm between researchers will eventually allow much faster development and optimization of DA performance.

Parallel processing (iv) is an additional component of a DA algorithm that affects its performance. Most often the codes are parallel and knowing its scaling through processors helps when applying an existing data assimilation code to a problem with much higher dimensions than before, or in general when trying DA in new problem. Increasing the dimension of each horizontal dimensions 3 times will mean increasing the dimension of control variables, and therefore the required calculation almost 10 times. Without a sufficient number of the processors the DA code may not be able to run, so this is important to consider before attempting to change the original dimensions or control variables.

Uncertainty estimation (v) is a less examined component of DA than estimation of the optimal state. This is especially important to know and understand in HYB methods, since a relative weighting between ensemble and static uncertainty is used to define total uncertainty. Giving more weight to static covariance has different consequences on DA compared to giving more weight to ensemble covariance. Within an ensemble system, it is important to know if analysis uncertainty is calculated from a sample of analyses, as in an ensemble Kalman filter (EnKF) and a particle filter (PF), or from an inverse Hessian at the optimum, as in MLEF and implicit PF. If inadequate ensemble spread is noticed, it can be traced back to different causes depending on the uncertainty estimation approach.

Although rarely discussed, post-processing of DA results (vi) is quite relevant for diagnosing DA performance and for disseminating results. In addition to typical performance measures such as root mean squared, maximum absolute errors, and anomaly correlation, it is beneficial to consider entropy, degrees of freedom (DOF) for signal, chi-squared evaluation, and skill versus spread, especially for ensemble and hybrid methods. Entropy is an additional measure of uncertainty that could be evaluated against error standard deviation. DOF is related to change of entropy due to DA so it could also be used to quantify DA performance (e.g., Županski et al., 2007). Chi-squared evaluation has been used for Kalman filters (KFs) (e.g., Ménard et al., 2000) as well as for ensemble DA (Županski, 2005; Županski and Županski, 2006) and can indicate potential problems in the PDF assumption. Skill versus spread (e.g., Whitaker and Loughe, 1998) has been widely used in ensemble DA, but should be relevant for HYB as well. Therefore, when using an existing, or developing a new DA system, it is important to consider verification measures. The versatility of these measures will allow better understanding of DA performance, and potentially lead to components that need improvement.

## D.4 Data Assimilation Code with a 1D Burgers' Equation

This part includes a generic DA algorithm that can be used for an exercise.

### *D.4.1 Overview of the Attached DA Code*

The purpose of including this code is to illustrate various aspects of a DA algorithm using a simple DA code that is easy to understand. The DA code is written in Fortran 90 and it employs shell script (sh) as a driver. The modeling system is a one-dimensional (1D) Burgers' equation that simulates a propagating shock wave (i.e., disturbance). The dimension of this model is 80, as there are 80 grid points. Since there is only one variable, 1D wind, the state dimension is also 80. This implies that forecast error covariance is an $80 \times 80$ matrix.

The observations are defined using an "*identical twin*" experimental setup, in which both DA and nature run are using the same Burgers' model: The nature run has slightly perturbed initial conditions compared to the data assimilation run. Random Gaussian perturbation with observation error $N(0, R)$ is added to the nature run forecast to create observations.

Another option included in the code is the observation coverage. It is possible to assimilate observations at each grid point, or use every 2nd, 3rd, or $N^{\text{th}}$ point. Also, it is possible to define observations in one half of the model domain and no observations in the other half, to simulate missing observation coverage.

This code includes three options for DA: (1) MLEF; (2) KF; and (3) 3D variational (3DVAR) DA. Table D.1 defines basic properties of each option.

Directory structure of the attached code is given in Figure D.5. After copying the `.gz` file, type "`tar -zxvf DA.gz`" to decompress and untar the file. Directories "*DA*" will be

Table D.1. *DA methods included in the attached code. Here, DOF refers to DOF of forecast error covariance: (i) All implies defining all DOF for this system; and (ii) Reduced implies using a small number of ensembles relative to the state dimension.*

| **DA method** | MLEF | KF | 3D-Var |
|---|---|---|---|
| **Covariance** | Flow-dependent | Flow-dependent | Static |
| **DOF** | All, reduced | All | All |

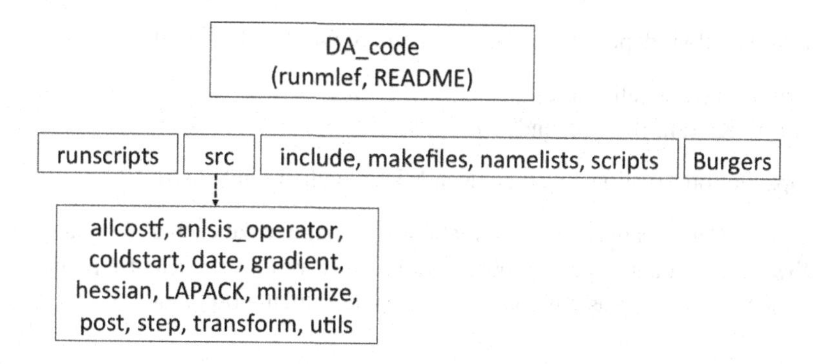


Figure D.5 Directory structure of the DA code.

also formed. After "`cd DA`" there will be several other directories revealed. The *runscripts* is a directory from which the code is submitted, *src/main* is the directory with source codes in Fortran 90, and *scripts* directory contains driving shell scripts. In order to run the DA system, the user needs only the *runscripts* directory. Other directories are only for additional information and code examples, they should not be changed by the user.

The following instructions can also be found inside *README* file in the *DA* directory.

Technical requirements for installing and using the code include: (i) UNIX-based OS with shell-scripting capability; and (ii) Fortran 90–95 compilers (Intel – `ifort` and Portland Group – `pgf90`)

The DA code details and capabilities are as follows: (a) 1D nonlinear Burger's model simulation of a shock wave; (b) nonlinear conjugate-gradient minimization algorithm; (c) choice of the number of ensembles; (d) model simulated observations (identical twin experiment); (e) choice of spatial configuration of observations; (f) choice of the number of analysis cycles; (g) choice of the COLD or WARM start; and (h) interactive capability with guided walk-through compilation, observation setup, and experiment setup.

In this instruction the generic home directory is denoted "*DA*." In actual applications it can be any name the user wants, but the default name and location is "*\$HOME/DA*."

DA directories and explanation: (1) *include* contains files which are used at the compilation time. You need to recompile the DA code each time a file if "*\$HOME/DA/include*" is changed. The file "`first_guess_list.h`" defines the observations' characteristics; (2) *makefiles* contains preset compilation options; (3) *namelists* contains files that are used at the running time; (4) *src* contains a set of source code directories; (5) *scripts* includes all shell scripts used to drive the DA code; (6) *runscripts* directory is used to submit DA codes, including the creation of synthetic observation; and (7) *Burgers* contains all forecast model codes, including the compilation of the model.

Below is a list of suggested experiments with the attached DA algorithm:

1. Impact of ensemble size: Run MLEF and KF options

   ***Exp***-1: MLEF with ensemble size 2 (reduced, flow-dependent)
   ***Exp***-2: MLEF with ensemble size 4 (reduced, flow-dependent)
   ***Exp***-3: MLEF with ensemble size 16 (reduced, flow-dependent)
   ***Exp***-4: KF (all, flow-dependent).

2. Impact of flow-dependent covariance: Run 3DVAR and KF options

   ***Exp***-5: 3DVAR (all, static)
   ***Exp***-6: KF (all, flow-dependent).

3. Impact of observation coverage: Run MLEF with ensemble size 4

   ***Exp***-7: Both subareas with observations (observations at each grid point)
   ***Exp***-8: One subarea without observations (observations at each grid point)
   ***Exp***-9: Both subareas with observations (intermittent observations).

### *D.4.2 Additional Exercises*

Additional exercises written in Fortran 90 include programs "*inner_product_norm*" and "*analysis*."

The program "*inner_product_norm*" computes an inner product, norms, and distances. It is coded in Fortran 90 and presented below.

```
1  !
2      program inner_product_norm
3  !--------
4  ! Program   for computing inner   product  and norms
5  !    (a) Inner    product   between    vectors   x   and  y
6  !    Output:   f
7  !--------
8  !    (b) Norms     L1,  L2,  Linf     for vectors   x    and  y
9  !    Output:   xnorm_L1,  xnorm_L2,  xnorm_Linf
10 !                                            ynorm_L1,  ynorm_L2,  ynorm_Linf
11 !--------
12 !    (c) Distance  between    vectors   x   and  y
13 !    Output:   dist_L1,   dist_L2,   dist_Linf
14 !--------
15
16     implicit    none
17
18     integer, parameter :: izero = 0
19     real, parameter :: zero = 0
20     integer   ::    idim, jdim
21     integer   ::    i, j, k
22     real      ::    f
23     real      ::    xnorm, ynorm
24     real      ::    xnorm_L1, ynorm_L1
25     real      ::    xnorm_L2, ynorm_L2
26     real      ::    xnorm_Linf, ynorm_Linf
27     real      ::    dist_L1
28     real      ::    dist_L2
29     real      ::    dist_Linf
30     real, dimension(:, :), allocatable :: x, y
31
32 !----------------------------------------------------------------------
33     idim = 3
34     jdim = 2
35
36     allocate(x(1:idim, 1:jdim))
```

```
37      allocate(y(1:idim, 1:jdim))
38
39      x(1,1) = 1 ;  y(1, 1) = 2
40      x(2,1) = −1 ;  y(2, 1) = −2
41      x(3,1) = 2 ;  y(3, 1) = 1
42      x(1,2) = −2 ;  y(1, 2) = −1
43      x(2,2) = 4 ;  y(2, 2) = 5
44      x(3,2) = 1 ;  y(3, 2) = −1
45
46      do i = 1, idim
47        do j = 1, jdim
48          write(∗, ∗) "i, j = ", i, j," Input x, y = ", x(i, j), y(i, j)
49        enddo
50      enddo
51   !−−−−−−−−− inner   product   (f)
52   !−−−−−−−−−−−−−−−−−−−−−−−−−−−−−−−−−−−−−−−−−−−−−−−−−−−−−−−
53   !−−−−−−−−−−−−− insert inner product calculation here −−−−−−−−−−−
54   !−−−−−−−−−−−−−−−−−−−−−−−−−−−−−−−−−−−−−−−−−−−−−−−−−−−−−−−
55   !    f = ?
56
57      f = zero
58      do i = 1, idim
59        do j = 1, idim
60          f = f + x(i, j)∗y(i, j)
61        enddo
62      enddo
63   !−−−−−−−−−−−−−−−−−−−−−−−−−−−−−−−−−−−
64      write(∗,∗) "Output: Inner product f = ," f
65   !−−−−−−−−−−−−−−−−−−−−−−−−−−−−−−−−−−−−−−−−−−−−−−−−−−−−−−−
66   !−−−−−−−−−−−−−−−−−−−−− xnorm calculation −−−−−−−−−−−−−−−−−
67   !−−−−−−−−−−−−−−−−−−−−−−−−−−−−−−−−−−−−−−−−−−−−−−−−−−−−−−−
68   !    xnorm_L1 = ?
69   !    xnorm_L2 = ?
70   !    xnorm_Linf = ?
71
72      xnorm_L1 = zero
73      xnorm_L2 = zero
74      xnorm_Linf = zero
75
76      do i = 1, idim
77        do j = 1, jdim
78          xnorm_L1 = xnorm_L1 + abs(x(i, j))
79          xnorm_L2 = xnorm_L2 + x(i, j)∗x(i, j)
80        enddo
81      enddo
82
83      if(xnorm_L2.gt.zero) then
84       xnorm_L2 = sqrt(xnorm_L2)
85      else
86       xnorm_L2 = zero
87      endif
88
89      xnorm_Linf = maxval(x)
90   !−−−−−−−−−−−−−−−−−−−−−−−−−−−−−−−−−−−−−−−−−−−−−−−−−−−−−−−
91      write(∗, ∗) "Output: xnorm_L1 = ," xnorm_L1
```

```
92      write(*, *) "Output: xnorm_L2 = ," xnorm_L2
93      write(*, *) "Output: xnorm_Linf = ," xnorm_Linf
94  !--------------------------------------------------------------
95  !-------------------- ynorm calculation -----------------
96  !--------------------------------------------------------------
97  !    ynorm_L1 = ?
98  !    ynorm_L2 = ?
99  !    ynorm_Linf = ?
100
101     ynorm_L1 = zero
102     ynorm_L2 = zero
103     ynorm_Linf = zero
104
105     do i = 1, idim
106       do j = 1, jdim
107         ynorm_L1 = ynorm_L1 + abs(y(i, j))
108         ynorm_L2 = ynorm_L2 + y(i, j)*y(i, j)
109       enddo
110     enddo
111
112     if(ynorm_L2.gt.zero) then
113      ynorm_L2 = sqrt(ynorm_L2)
114     else
115      ynorm_L2 = zero
116     endif
117
118     ynorm_Linf = maxval(y)
119 !--------------------------------------------------------------
120     write(*, *) "Output: ynorm_L1 = ," ynorm_L1
121     write(*, *) "Output: ynorm_L2 = ," ynorm_L2
122     write(*, *) "Output: ynorm_Linf = ," ynorm_Linf
123 !--------------------------------------------------------------
124 !-------------- insert distance calculation here -------------
125 !--------------------------------------------------------------
126 !    dist_L1 = ?
127 !    dist_L2 = ?
128 !    dist_Linf = ?
129
130     dist_L1 = zero
131     dist_L2 = zero
132     dist_Linf = zero
133
134     do i = 1, idim
135       do j = 1, jdim
136         dist_L1 = dist_L1 + abs(x(i, j) - y(i, j))
137         dist_L2 = dist_L2 + (x(i, j) - y(i, j))*(x(i, j) - y(i, j))
138       enddo
139     enddo
140
141     if(dist_L2.gt.zero) then
142      dist_L2 = sqrt(dist_L2)
143     else
144      dist_L2 = zero
145     endif
146
```

```
147       dist_Linf = maxval(x - y)
148   !-----------------------------------------------------------------
149       write(*, *) "Output: dist_L1 = ," dist_L1
150       write(*, *) "Output: dist_L2 = ," dist_L2
151       write(*, *) "Output: dist_Linf = ," dist_Linf
152   !--------
153       deallocate(x)
154       deallocate(y)
155   !--------
156       end program inner_product_norm
157   !
```

The program "*analysis*" includes the analysis step of a simple 1D problem with state vector dimension of 2. This can be interpreted as a standard DA algorithm at two grid points or a coupled DA algorithm with two variables defined at the same grid point. The analysis solution is using KF equations, and only one of the two dimensions is observed. As above, this can be interpreted as observing one point out of two or observing only one variable. Related to "coupled" interpretation, an important option in program "*analysis*" is the definition of forecast error cross-covariance denoted `sqrtPf(2,1)`. For a 0 value there is no cross-correlation between variables 1 and 2, so the analysis calculations are effectively independent: The observed variable will have updated analysis, while the unobserved variable will remain equal to the initial guess. For a nonzero value there is a coupling between the variables and observation of one variable will impact the analysis of both variables. The code written in Fortran 90 is included below.

```
1    !
2        program analysis
3    !--------
4    ! Program   for  calculating   Kalman   filter   analysis
5    ! (one    point,    two-variables)
6    !
7    !       Input:   guess,    observation    values
8    !                   guess,  observation    uncertainty
9    !   Output:  cost    function  (f,  fb,  fobs)
10   !                 analysis   (xa)
11   !--------
12
13       implicit    none
14
15       integer, parameter :: izero = 0
16       real, parameter :: zero = 0
17       real, parameter :: one = 1
18       integer         ::  dim_state
19       integer         ::  dim_obs
20       real, dimension(:, :), allocatable  ::  sqrtPf
21       real, dimension(:, :), allocatable  ::  Pf
22       real, dimension(:, :), allocatable  ::  Pf_inv
23       real, dimension(:, :), allocatable  ::  sqrtR
24       real, dimension(:, :), allocatable  ::  R
25       real, dimension(:, :), allocatable  ::  R_inv
26       real, dimension(:, :), allocatable  ::  H
27
```

```
28     real,  dimension(:),   allocatable    ::  x
29     real,  dimension(:),   allocatable    ::  xa
30     real,  dimension(:),   allocatable    ::  xf
31     real,  dimension(:),   allocatable    ::  yobs
32     real,  dimension(:),   allocatable    ::  hx
33     real,  dimension(:),   allocatable    ::  xdiff
34     real,  dimension(:),   allocatable    ::  x1
35     real,  dimension(:),   allocatable    ::  ydiff
36     real,  dimension(:),   allocatable    ::  y1
37
38     integer          ::  i, j, k
39     real                 ::  fb, fobs, f
40  !-----------------------------------------------------------------------
41     dim_state = 2
42     dim_obs = 1
43
44     write(*, *) "Input: State vector dimension = ," dim_state
45     write(*, *) "Input: Observation vector dimension = ," dim_obs
46  !-----------------------------------------------------------------------
47  !---------------- state vector and uncertainty ------------
48  !-----------------------------------------------------------------------
49     allocate(xa(1:dim_state))
50     allocate(xf(1:dim_state))
51     allocate(sqrtPf(1:dim_state, 1:dim_state))
52  !-------- forecast     state     vector
53     xf(1) =1000.0
54     xf(2) =     270.0
55  !
56  !-------- square root forecast error covariance
57  !
58  !-------- diagonal
59     sqrtPf(1, 1) = 2.0
60     sqrtPf(2, 2) = 1.0
61
62  !==== BEGIN USER INPUT ====
63  !-------- off-diagonal covariance
64     sqrtPf(2, 1) = zero
65     sqrtPf(2, 1) = 0.5
66  !==== END  USER  INPUT ====
67  !
68     sqrtPf(1, 2) = sqrtPf(2, 1)
69  !
70  !-----------------------------------------------------------------------
71  !------------- observation vector and uncertainty ----------
72  !-----------------------------------------------------------------------
73     allocate(yobs(1:dim_obs))
74     allocate(sqrtR(1:dim_obs, 1:dim_obs))
75
76  !-------- observation vector
77     yobs(1) = 272.0
78     write(*, *) "Input: Observation (temp) = ," yobs
79
80  !-------- observation operator
81     allocate(H(1:dim_obs, 1:dim_state))
82
```

```
83      H(1, 1) = zero
84      H(1, 2) = one
85
86  !-------- square root observation error covariance
87      sqrtR(1, 1)=0.5
88
89      write(*, *) "=== Input: observation error = ," sqrtR
90
91  !------------------------------------------------------------
92  !------------- insert Pf and Pf_inv calculation here -----------
93  !------------------------------------------------------------
94  !    Pf = ?
95  !    Pf_inv = ?
96
97      allocate(Pf(1:dim_state, 1:dim_state))
98
99      Pf = zero
100     do i = 1, dim_state
101       do j = 1, dim_state
102         do k = 1, dim_state
103           Pf(i, j) = Pf(i, j) + sqrtPf(i,k)*sqrtPf(k, j)
104         enddo
105       enddo
106     enddo
107     write(*, *) "     "
108     write(*, *) "==     Pf      =="
109     do i = 1, dim_state
110       do j = 1, dim_state
111         write(*, *) i, j, Pf(i, j)
112       enddo
113     enddo
114     write(*, *) "========"
115     write(*, *) "     "
116 !--------
117     allocate(Pf_inv(1:dim_state, 1:dim_state))
118
119     call matrix_inverse(dim_state, Pf, Pf_inv)
120
121 !------------------------------------------------------------
122 !-------------- insert R and R_inv calculation here -----------
123 !------------------------------------------------------------
124 !    R = ?
125 !    R_inv = ?
126
127     allocate(R(1:dim_obs, 1:dim_obs))
128
129     R = zero
130     do i = 1, dim_obs
131       R(i, i) = sqrtR(i, i)*sqrtR(i, i)
132     enddo
133 !--------
134
135     allocate(R_inv(1:dim_obs, 1:dim_obs))
136
137     R_inv = zero
```

```
138      do i = 1, dim_obs
139        if(R(i, i).gt.zero) then
140          R_inv(i, i) = 1./R(i, i)
141        endif
142      enddo
143
144  !=====================================================
145  !================ cost function calculation ================
146  !=====================================================
147
148      allocate(x(1:dim_state))
149      allocate(hx(1:dim_obs))
150
151  !======================
152  !== at first guess point
153  !======================
154  !      (1) x = xf
155
156      x = xf
157      hx(1) = x(2)
158
159  !--------- background cost function calculation ---------
160
161      call fcost (dim_state, Pf_inv, x, xf, fb)
162
163  !--------- observation cost function calculation ---------
164
165      call fcost (dim_obs, R_inv, yobs, hx, fobs)
166
167  !--------- total cost function calculation ---------
168      f = fb + fobs
169  !---------
170      write(*, 101) fb, fobs, f
171   101 format("====== INITIAL cost function: fb, fobs, f = ," 3F10.5)
172  !=====================
173  !== at analysis point
174  !=====================
175  !      (2) x = xa
176
177      call KF (dim_state, dim_obs, Pf_inv, H, R_inv, xf, yobs, xa)
178
179      x = xa
180      hx(1) = x(2)
181
182  !--------- background cost function calculation ---------
183
184      call fcost (dim_state, Pf_inv, x, xf, fb)
185
186  !--------- observation cost function calculation ---------
187
188      call fcost (dim_obs, R_inv, yobs, hx, fobs)
189
190  !--------- total cost function calculation ---------
191      f=fb + fobs
192
```

```
193      write(*, 102) fb, fobs, f
194   102 format("====== FINAL cost function: fb, fobs, f = ," 3F10.5)
195
196      write(*, *) "      "
197      write(*, *) "=== Input: First guess = ," xf
198      write(*,*) "=== Calculated Analysis = ," xa
199      write(*, *) "      "
200  !=====================
201      deallocate(sqrtPf)
202      deallocate(Pf)
203      deallocate(Pf_inv)
204      deallocate(sqrtR)
205      deallocate(R)
206      deallocate(R_inv)
207      deallocate(xf)
208      deallocate(xa)
209      deallocate(x)
210      deallocate(yobs)
211      deallocate(hx)
212      deallocate(H)
213  !--------
214      end program analysis
215  !
```

```
1   !------------------------------------------------------------
2       subroutine fcost (idim, A_inv, x1, x2, ff)
3   !------------------------------------------------------------
4
5       implicit none
6
7   !---- Input
8       integer          :: idim
9       real,  dimension(1:idim, 1:idim)   ::  A_inv
10      real,  dimension(1:idim)                              ::  x1
11      real,  dimension(1:idim)                              ::  x2
12  !---- Output
13      real                       ::  ff
14  !--------
15      real          ,  parameter       ::  zero=0
16      real,  dimension(:)        ,  allocatable     ::  xdiff
17      real,  dimension(:)        ,  allocatable     ::  x_h
18      integer          ::  i, j
19  !--------
20      allocate(xdiff(1:idim))
21      allocate(x_h(1:idim))
22
23      xdiff = x1 - x2
24      ff = zero
25      do     i = 1, idim
26        x_h(i) = zero
27        do  j = 1, idim
28          x_h(i)=x_h(i)+A_inv(i,j)*xdiff(j)
29        enddo
30        ff = ff + 0.5*xdiff(i)*x_h(i)
31      enddo
```

```
32  !--------
33      end subroutine fcost
34  !
```

```
1   !-------------------------------------------------------------
2       subroutine KF(dim_state, dim_obs, Pf_inv, H, R_inv, xf, yobs, xx)
3   !-------------------------------------------------------------
4   !--------
5   !   Kalman filter analysis
6   !--------
7
8       implicit none
9
10  !---- Input
11      integer          ::  dim_state
12      integer          ::  dim_obs
13      real, dimension(1:dim_state, 1:dim_state) :: Pf_inv
14      real, dimension(1:dim_obs, 1:dim_obs) :: R_inv
15      real, dimension(1:dim_obs, 1:dim_state) :: H
16      real, dimension(1:dim_state) :: xf
17      real, dimension(1:dim_obs) :: yobs :: x2
18  !---- Output
19      real,  dimension(1:dim_state) :: xx
20  !--------
21      real,  parameter        ::  zero = 0
22      real,  dimension(:, :),     allocatable   ::  HT
23      real,  dimension(:, :),     allocatable   ::  HTR_inv
24      real,  dimension(:, :),     allocatable   ::  HTR_invH
25      real,  dimension(:, :),     allocatable   ::  mat_temp
26      real,  dimension(:, :),     allocatable   ::  mat_inv
27      real,  dimension(:)         ,  allocatable   ::  x1_h
28      real,  dimension(:)         ,  allocatable   ::  x2_h
29      real,  dimension(:)         ,  allocatable   ::  x_h
30      integer           ::  i, j, k
31      real                     ::  factor
32  !------------------
33  !-- solution in the form:
34  !-- xx = (Pf_invI + HT*R_inv*H)**(-1)*(Pf_inv*xf + HT*R_inv*yobs)
35  !------------------
36
37  !-- transpose of H
38      allocate(HT(1:dim_state, 1:dim_obs))
39
40      do i = 1, dim_state
41        do j = 1, dim_obs
42          HT(i, j) = H(j, i)
43        enddo
44      enddo
45  !------------------
46  !--    matrices
47
48  !     HT*R_inv
49
50      allocate(HTR_inv(1:dim_state, 1:dim_obs))
51      do i = 1, dim_state
```

```
52        do j = 1, dim_obs
53          HTR_inv(i, j) = HT(i, j)*R_inv(j, j)
54        enddo
55      enddo
56
57  !     HT*R_inv*H
58
59      allocate(HTR_invH(1:dim_state, 1:dim_state))
60      do i = 1, dim_state
61        do j = 1, dim_state
62          HTR_invH(i, j) = zero
63          do k = 1, dim_obs
64            HTR_invH(i, j) = HTR_invH(i, j) + HTR_inv(i, k)*H(k, j)
65          enddo
66        enddo
67      enddo
68
69  !--     (Pf_inv+HTR_invH)**(-1)
70  !--     2 x 2 general inverse
71
72      allocate(mat_temp(1:dim_state, 1:dim_state))
73      allocate(mat_inv(1:dim_state, 1:dim_state))
74
75      mat_temp=Pf_inv + HTR_invH
76
77      call matrix_inverse(dim_state, mat_temp, mat_inv)
78
79  !------------------
80  !--     vectors
81
82      allocate(x1_h(1:dim_state))
83      allocate(x2_h(1:dim_state))
84      allocate(x_h(1:dim_state))
85
86  !--     Pf_inv*xf
87
88      do i = 1, dim_state
89        x1_h(i) = zero
90        do j = 1, dim_state
91          x1_h(i ) = x1_h(i) + Pf_inv(i,j)*xf(j)
92        enddo
93      enddo
94
95  !--     HTR_inv*yobs
96
97      do i = 1, dim_state
98        x2_h(i) = zero
99        do j = 1, dim_obs
100         x2_h(i) = x2_h(i) + HTR_inv(i,j)*yobs(j)
101       enddo
102     enddo
103
104 !--
105     x_h = x1_h + x2_h
106 !--
```

```
107  !-- solution
108  !--
109      do i = 1, dim_state
110        xx(i) = zero
111        do j = 1, dim_state
112          xx(i) = xx(i) + mat_inv(i, j)*x_h(j)
113        enddo
114      enddo
115  !--
116      deallocate(HTR_invH)
117      deallocate(HTR_inv)
118      deallocate(HT)
119      deallocate(mat_temp)
120      deallocate(mat_inv)
121      deallocate(x1_h)
122      deallocate(x2_h)
123      deallocate(x_h)
124  !--------
125      end subroutine KF
126  !
```

```
1   !----------------------------------------------------------------
2       subroutine matrix_inverse(idim, A, A_inv)
3   !----------------------------------------------------------------
4   !------------------
5   !-- 2 x 2 matrix inverse only
6   !------------------
7
8       implicit none
9
10  !---- Input
11      integer                        ::  idim
12      real,  dimension(1:idim, 1:idim)  ::  A
13  !---- Output
14      real,  dimension(1:idim, 1:idim)  ::  A_inv
15  !--------
16      real :: factor
17      real,  parameter        ::  zero = 0
18  !------------------
19
20      factor = A(1, 1)*A(2, 2) - A(1, 2)*A(2, 1)
21
22  !-- check determinant
23      if(factor.eq.zero) then
24        write(*, *) "Need to change matrix (Pf) elements, determinant = 0"
25        stop
26      endif
27
28  !-- inverse
29      A_inv(1, 1) = A(2, 2)/factor
30      A_inv(2, 2) = A(1, 1)/factor
31      A_inv(1, 2) = -A(1, 2)/factor
32      A_inv(2, 1) = -A(2, 1)/factor
33  !--------
34      end subroutine matrix_inverse
35  !
```

## References

Ménard R, Cohn SE, Chang LP, Lyster PM (2000) Assimilation of stratospheric chemical tracer observations using a Kalman filter. Part I: Formulation. *Mon Wea Rev* 128:2654–2671.

Trevisan A, Palatella L (2011) On the Kalman filter error covariance collapse into the unstable subspace. *Nonlinear Processes Geophys* 18:243–250.

Whitaker JS, Loughe AF (1998) The relationship between ensemble spread and ensemble mean skill. *Mon Wea Rev* 126:3292–3302.

Županski D, Županski M (2006) Model error estimation employing an ensemble data assimilation approach. *Mon Wea Rev* 134:1337–1354.

Županski D, Hou AY, Zhang SQ, et al. (2007) Applications of information theory in ensemble data assimilation. *Quart J Roy Meteor Soc* 133:1533–1545.

Županski M (2005) Maximum likelihood ensemble filter: Theoretical aspects. *Mon Wea Rev* 133:1710–1726.

# Index